Techn(
which sets :

■ Compounding of plastic material – we don't do things by halves. WP offers ZSK twin screw compounders from lab size to production lines for throughput rates of up to 38 t/h. Our customers have more than ¼ billion operating hours in the past 30 years with 4.500 WP-compounders. This stands for cost-effective plastics compounding and for high performance and reliability of the ZSK from WP. ■

Werner & Pfleiderer GmbH
D - 70466 Stuttgart / Germany
Telefax: + 0049-711 / 897-3981

₩ WERNER & PFLEIDERER

Rheological Testing Equipment
for Thermoplastics, Thermosets and Elastomers

RHEOGRAPH 2003
a high pressure capillary rheometer to be applied in polymer development.

RHEO-TESTER 1000
the low cost model for rheological materials characterization.

MI-ROBO
a fully automatically operated melt indexer.

ELASTOGRAPH 67.85
a rotorless shear vulcameter with microprocessor.

W E R K S T O F F - P R Ü F M A S C H I N E N   G M B H
D-74722 Buchen/Odenwald · Siemensstraße 2 · Postfach 1261
Telephone (0 62 81) 4 08-0 · Ttx 6281914=Gft · Fax (0 62 81) 4 08 18

Organic Peroxides and Persulphates

for

Polymerisation of Monomers

Curing of unsaturated Polyester Resins

Crosslinking of Polyethylene and Elastomers

Peroxid-Chemie GmbH

LAPORTE
Laporte Organics

Dr.-Gustav-Adolph-Str. 3 · D-82049 Pullach · Germany
Tel.: ++49 (0)89 7 44 22-0
Fax: ++49 (0)89 7 44 22-203

Weingärtner Maschinenbau

OFFERS COMPLETE *WHIRLING*/MILLING MANUFACTURING CELLS ALL OVER THE WORLD

including Individual software for:
* screw design with profile finding
* tool design
* CNC/DNC programming

for the economical production of
SINGLE AND TWIN EXTRUSION SCREWS

by using a flexible modular CNC-machine concept for:
external/internal **WHIRLING** and profiling • peeling • milling • hobbing • turning • grinding • polishing • part inspection

WEINGÄRTNER MASCHINENBAU
Ges.m.b.H.
A-4656 Kirchham 26 · Austria
Tel. 43-7619/2103
Fax 43-7619/2103-41

Testing Polymer Processibility, MFI / MVI, and Purity

The DATA PROCESSING PLASTI-CORDER® PL 2000 is applied worldwide for improved studies of the processibility of thermoplastics, thermosets, elastomers, and other plastic and plastifiable materials with measuring mixers and measuring extruders.

DATA PROCESSING PLASTI-CORDER® PL 2000

It consists of a DYNAMOMETER UNIT with direct electronic torque measurement, an intelligent DATA TRANSFER INTERFACE, and an IBM PC or a compatible computer with efficient pheripheral units. BRABENDER® supplies many comfortable software packages for easy and user-friendly handling of the system as well as for automatic evaluation and correlation analyses of the results. The software is steadily improved and new packages are created, e.g. for evaluation of fusion time, gelation speed, flow-curing behaviour of thermosets in compliance with DIN 53764, vulcanizing behaviour of elastomers, determination of the black incorporation time BIT as well as for tests with measuring extruders for measuring the dynamic viscosity by means of slot or round capillary dies. With flat sheet or film blowing die heads and the opto-electronic FILM QUALITY ANALYZER,

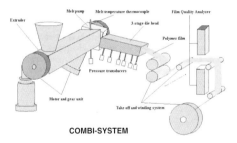

COMBI-SYSTEM

the purity of polymers can be measured. The extruded polymer film is tested for transparency, inhomogeneities, impurities like fisheyes, gels, and agglomerates, and foreign particles in various, absolute size classes that can be defined by the user. The smallest detectable size is 5 µm.

This system can be combined with a special version of the PLASTI-CORDER®, called AUTO-GRADER®, for measuring and printing simultaneously MFI/MVI values, viscosity, and polymer purity. Due to fully automatic and continuous operation, this so-called COMBI-SYSTEM is perfectly suited for online process control in polymer production and polymer processing, but also for laboratory testing.
Samples for trials in our laboratory are welcome.

For further information please contact:

BRABENDER® OHG DUISBURG
Kulturstr. 51-55 · D-47055 Duisburg
Tel. -49-203-7788-0 · Fax -49-203-7788-100 + 101 · Tx. 855603

Agencies in more than 60 countries all over the world

SPECIAL STEELS BROUGHT TO PERFECTION.

FOR STRONG PROFIT YIELDS – BOHLER M 390 ISOMATRIX.

Precision will always cost money. Any lack of precision tends to involve even greater expenses. This is why we recommend the very best steel for your precision work. **BOHLER M 390 ISOMATRIX** is the plastic mould steel that is produced by powder metallurgical methods and features isotropic properties for severely stressed wear parts. Whether you are working in the tool manufacturing or in the plastics processing industries, BOHLER M 390 ISOMATRIX offers you an unequalled quality standard in terms of

- polishability
- dimensional stability
- wear resistance and
- corrosion resistance

For you this means greater productivity and thus higher profits. Be sure to get ahead of competition and consult BOHLER for your next project!

Injection mould for the production of parts for ballpoint pens.
Tool made by Engel, Schwertberg, Austria.

BÖHLER EDELSTAHL GMBH, Mariazeller Strasse 25 , A-8605 Kapfenberg, Austria
Phone.: (3862) 20-7181, Telex: 36612 bok a, Telefax: (3862) 20-7576
BOHLER (U.K.) LTD. EUROPEAN BUSINESS PARK, Taylors Lane, Oldbury, Warley,
West Midlands B69 2DA, Great Britain, Phone (21) 5525681, Telefax (21) 5447623

Saechtling

International

Plastics Handbook

for the Technologist, Engineer and User

Third Edition
by Wilbrand Woebcken

Translated and edited by
John Haim and David Hyatt

Hanser Publishers, Munich Vienna New York

Hanser/Gardner Publications, Inc., Cincinnati

Author:
Prof. Dr.-Ing. Wilbrand Woebcken, Würzburg

Translated and edited by
Dr. John Haim, Bondway Publishing, Turners Hill, West Sussex RH10 4YY, GB
and Dr. David Hyatt, University of North London, London N7 8DB, GB

Distributed in the USA and in Canada by
Hanser/Gardner Publications, Inc.
6600 Clough Pike, Cincinnati, Ohio 45244-4090, USA
Fax: + 1 (513) 527-8950

Distributed in all other countries by
Carl Hanser Verlag
Postfach 86 04 20, 81631 München, Germany
Fax: + 49 (89) 98 12 64

The use of general descriptive names, trademarks etc., in this publication, even if the former are not especially identified, is not to be taken as a sign that such names, as understood by the Trade Marks and Merchandise Marks Act, may accordingly be used freely by anyone.

While the advice and information in this book are believed to be true and accurate at the date of going to press, neither the author nor the editors nor the publisher can accept any legal responsibility for any errors or omissions that may be made. The publisher makes no warranty, express or implied, with respect to the material contained herein.

Die Deutsche Bibliothek - CIP-Einheitsaufnahme

Saechtling, Hansjürgen:
International plastics handbook : for the
technologist, engineer, and user / Saechtling.
Transl. and ed. by John Haim and David Hyatt. –
3. ed. / by Wilbrand Woebcken. – Munich ; Vienna ;
New York : Hanser ; Cincinnati : Hanser/Gardner, 1995
    Einheitssacht.: Kunststoff-Taschenbuch <engl.>
    ISBN 3-446-18172-5
NE: Woebcken, Wilbrand [Bearb.]

    ISBN 1-56990-182-1 (Hanser/Gardner)

All rights reserved. No part of this book may be reproduced or transmitted in any form or by any means, electronic or mechanical, including photocopying, recording or by any information storage and retrieval system, without permission in writing from the publisher.

© 1995 by Carl Hanser Verlag, Munich, Vienna, New York
Printed and bound by R. Oldenbourg, Munich, Germany

Plastic: The Material of Opportunities Requires People Who Can Make Things Happen.

Today, a full service supplier like IBS must be committed in two directions. On the one hand, we understand ourselves as the absolute specialist in processing plastics. Equipped with the best the world market offers as well as the know-how of using it in the best possible way: the latest CAD, CAM and CAE programms, detailed knowledge in tool making and in processing techniques for plastics.

On the other hand, we understand ourselves as an element of an overall concept. For only if you place importance on the final product you can participate in the goal oriented process of development. But being committed to the overall concept means above all service.

Therefore, IBS is able to fulfill all your needs from one source: advice, R&D, design, tool making, injection molding, finishing and assembly, to the complete system.

IBS Brocke GmbH
Lichtenberg
Postfach 1155/65
51589 Morsbach
tel.: 0 22 94 / 697-0
fax: 0 22 94 / 697-155

A 10

Now extended:
Tiebarless machines from 200 - 4000 kN clamping force. The unique HL-clamping system for profitable production.

ENGEL

Daily more than 2000 injection moulding machines from ENGEL-HL tiebarless range prove their advantageous cost - performance ratio in all production areas: ➢ Absolute platen parallelism and rigidity. ➢ Full use of the platen surface. ➢ Free access to the platen area (e.g. for handling, robots, mould change). ➢ 20 - 30% investment advantage through optimum machine productivity. ➢ Beat today´s cost pressure with ENGEL HL!

ISO 9001
CERTIFIED · QUALITY · SYSTEM

From the largest (upper picture) to the smallest (lower picture) tiebarless the perfect cost - performance ratio

Tiebar or tiebarless, is not a question of philosophy, but of practical considerations and profitability

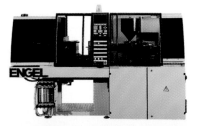

ENGEL VERTRIEBSGESELLSCHAFT m.b.H, A-4311 Schwertb
Tel. 0 72 62 / 620-0, Fax 0 72 62 / 620-308, Telex 6 13 73

Foreword

The Society of Plastics Engineers is pleased to sponsor and endorse the third edition of the successful "International Plastics Handbook" by Hj. Saechtling, which has found wide acceptance among plastics engineers.

SPE, through its Technical Volumes Committee, has long sponsored books on various aspects of plastics. Its involvement has ranged from identification of needed volumes and recruitment of authors to peer review and approval and publication of new books.

Technical competence pervades all SPE activities, not only in the publication of books, but also in other areas such as sponsorship of technical conferences and educational programs. In addition, the Society publishes periodicals including *Plastics Engineering, Polymer Engineering and Science, Polymer Processing and Rheology, Journal of Vinyl Technology and Polymer Composites*, as well as conference proceedings and other publications, all of which are subject to rigorous technical review procedures.

The resource of some 38,000 practicing plastics engineers, scientists, and technologists has made SPE the largest organization of its type worldwide. Further information is available from the Society at 14 Fairfield Drive, Brookfield, Connecticut 06804, USA.

Michael R. Cappelletti
Executive Director
Society of Plastics Engineers

756 Technical Volumes Committee
Claire Bluestein, Chairperson
Captan Associates, Inc.

Preface to the Third Edition

This third English edition of the widely respected "Saechtling" is published almost 60 years after the first edition of its precursor, the "Kunststoff Taschenbuch" by Dr. F. Pabst and almost twenty years after the first international edition. The vigorous development of plastics technology has resulted in a continuous increase in the information suitable for inclusion in a reference work such as this. The fact that the German version has now passed the landmark 25th edition reflects the success of the subsequent editors, Hanjürgen Saechtling and Wilbrand Woebcken, in distilling this information into a comprehensive but manageable form.

This third English edition marks the transfer of the responsibility for compiling and editing the "International Plastics Handbook" from Dr. Hansjürgen Saechtling after 41 years as editor of the German and English editions to Professor Wilbrand Woebcken, Dr. John Haim and Dr. David Hyatt. The text has been substantially revised by the last two and additional material has been contributed by experts too numerous to mention but whose efforts are gratefully acknowledged. Particular thanks are due to Dr. Wolfgang Glenz for compiling the section on specialist literature and the list of tradenames.

The main focus remains the systematic, comprehensive presentation of thermoplastics, thermosets and elastomers and their properties, availability, typical processing technologies and areas of application. Topical subjects include engineering plastics, biodegradable plastics and recycling.

The International Plastics Handbook is primarily a source of information for the sector. It also offers an easy yet comprehensive introduction to the field of plastics technology for students and other tyros.

Commercial names mentioned in the text and in the "Trade Names" index are mostly nationally or internationally registered. For technical reasons ® signs have been omitted and are replaced by this note.

Tabulations of data are believed to be accurate. However the publishers, author and editors cannot be held responsible for any inaccuracies. Consultation of original literature and manufacturers' and processors' documentation before selection of any material by a property-critical value is recommended. As for all such works, the editors would be pleased to receive suggestions for improvements and details of any amendments or corrections readers may think necessary.

W. Woebcken
J. Haim
D. Hyatt

Contents

	Abbreviations.	XI
	A. Country of origin of trade names	XI
	B. Abbreviations for homopolymers, copolymers, intermediates and rubbers.	XI
	C. Abbreviations for rubbers.	XVI
	D. Abbreviations for plasticizers	XVI
	Conversion formulae.	XVI

1. Plastics Technology 1

1.1. General Concepts 1
Thermoplastics, thermosets, reactive resins - Rigid, semirigid, nonrigid plastics

1.2. Polymers for Plastics. 3

1.3. Polymer Modification 5
Inter- and intra-molecular combined polymers - Polymer composites

1.4. Polymer Flow and Elasticity 7

1.5. Molecular Arrangements and Service Temperatures 9
The glassy state of rigid plastics made from amorphous polymers - Partially crystalline thermoplastics - The rubber-like state - Viscous melts

1.6. Properties of Plastics. 13
Range of strength and stiffness - Density - Insulating and construction materials for electrical and electronics engineering - Environmental corrosion resistance - Thermal conductivity and thermal expansion - Combustibility - Design of plastics products - Mass production

1.7. Types of Plastics 15

1.8. Development Trends 18
World production and consumption - Materials developments

2. Literature on Plastics Technology 27
Journals (English and other languages) - Books on plastics technology (Reference books, handbooks, encyclopedias, polymer science textbooks, engineering, processing, material properties, design, polymer science, polymer analysis, dictionaries)

3. Polymer Syntheses; Plastics Processing and Finishing 42

3.1. Polymers and Compounds 42

    3.1.1. Polymerization 42
    Double-bond polymerization - Condensation polymerization - Rearrangement polymerization

    3.1.2. Characteristic Values for Polymers and Plastics. 45

    3.1.3. Auxiliaries and Additives for Plastics 47
    Lubricants or parting agents - Stabilizers - Antistatic agents and conducting additives - Flame retardants - Colorants - Flexibilizers - Coupling agents - Fillers, enhancers and reinforcements - Blowing agents

    3.1.4. Plastics Compounding 54
    Thermoplastic compounds - Thermosetting compounds 55

    3.1.5. Expanded Plastics 55
    Some generalities on cellular plastics

    3.1.6. Foam-plastic Manufacturing. 56
    High and low pressure blowing - Physical and chemical blowing agents - Nucleating agents and pore size regulators - Manufacturing information

IV Contents

- 3.2. No-Pressure and Low-Pressure Molding Processes ... 60
  - 3.2.1. Dipped and Powder-Coated Articles ... 60
  - 3.2.2. Casting and Spraying of Preproducts ... 61
  - 3.2.3. Rotation Molding and Centrifugal Casting ... 63
  - 3.2.4. Low-Pressure Processes for GRP Moldings ... 65
- 3.3. Compression, Injection and Blow Molding ... 71
  - 3.3.1. Introduction ... 71
    High-pressure molding techniques - Designing plastics parts - Shrinkage and tolerances
  - 3.3.2. Injection-Molding Machines ... 76
    Assembly and working cycle - Elements of the injection unit - Elements of the locking unit - The injection-molding process - Ejection of moldings and their aftertreatment - Automation of the injection molding factory
  - 3.3.3. Presses and Ancillary Equipment ... 85
    Types of molding presses - Metering and preheating equipment - Arrangements for finishing moldings
  - 3.3.4. Automated and Large Area Presses ... 87
  - 3.3.5. Molding Tools ... 89
    Materials and production process - Injection molds and gating systems - Compression and transfer molding tools
  - 3.3.6. Molding Techniques for Thermoplastic Materials ... 95
    General injection technique - Injection molding of filled compounds - Multi-component injection molding - Structural foam and controlled internal pressure molding - Transfer-injection stretch molding - Resistance sintering of extremely viscous melts - Blow molding
  - 3.3.7. Molding Techniques for Thermosets and Elastomers ... 114
- 3.4. Production of Semifinished Thermosets ... 119
  - 3.4.1. Casting ... 119
  - 3.4.2. Laminating ... 119
  - 3.4.3. Extrusion and Pultrusion ... 121
- 3.5. Semifinished Thermoplastics: Older Processes ... 121
  - 3.5.1. „Celluloid" Process ... 121
  - 3.5.2. „Astralon" Process ... 121
  - 3.5.3. Pressing Sheets and Blocks ... 122
  - 3.5.4. Casting Films ... 122
- 3.6. Extrusion of Semifinished Thermoplastics ... 122
  - 3.6.1. General ... 122
  - 3.6.2. Construction and Operation of Screw Extruders ... 123
    Types and sizes - Extruder Components - Auxiliary parts between clamp and die - Extruder operating parameters - Processing faults and their causes
  - 3.6.3. Single-Screw Extruders ... 125
    General - Extruders with a forced conveying feed section - Special types of construction
  - 3.6.4. Multi-screw Extruders ... 129
  - 3.6.5. Extrusion Plant for Various Products ... 132
    Pipes and hollow sections - Solid profiles - Foamed profiles - Bristles, filaments and nets - Wire and cable covering - Blown films from annular dies - Sheets, films and coatings with slot dies - Stretch process for flat films - Multi-layer and multi-colored extrusion - Filaments and fibers from films
- 3.7. Webs Made of Thermoplastics ... 141
  - 3.7.1. Films made from Solutions ... 141
  - 3.7.2. Extrusion of Webs ... 141

3.7.3. Calendering of Plastic Film ... 141
3.7.4. Multilayered Webs ... 142
Bonding several layers - Spreading, coating and laminating

3.8. Forming and Joining Semifinished Plastics ... 144
3.8.1. Forming Techniques for Thermoplastics ... 144
Temperature ranges - Embossing, stretching, roll mill pressing, shrinking, annealing
3.8.2. Cold-Forming Thermoplastics ... 146
3.8.3. Hand-Fabricating Thermoforming ... 146
3.8.4. Industrial Stretch Thermoforming ... 148
Principles of thermoforming by machines - Areas of application - Thermoforming machines - Laminating, skin packaging and blister packaging
3.8.5. Forming Materials Other Than Thermoplastics ... 151
3.8.6. Welding of Thermoplastics ... 151
Shop and field welding techniques - Industrial welding techniques
3.8.7. Bonding of Plastics ... 155
3.8.8. Screw, Rivet, Snap-in joints ... 158

3.9. Other Fabrication Processes ... 161
3.9.1. Cutting, Punching and Separating ... 161
3.9.2. Machining ... 161
3.9.3. Surface Treatment ... 162

4. The Individual Plastics ... 165

4.1. Thermoplastics: Raw Materials and Molding Compounds ... 165
4.1.1. Preproducts and Special Products ... 165
4.1.1.1. Monomer reaction resins ... 165
Styrene - Vinyl carbazole - Methacrylates - Caprolactam - Methyl-2-cyano acrylate - Triallyl cyanurate
4.1.1.2. Olefin polymers ... 167
Ethylene (co-)polymers and chlorinated products - Atactic polypropylene (APP) and other α-olefin polymers - Polyisobutylene (PIB) - Ethylene copolymer + Bitumen blends (ECB)
4.1.1.3. Styrene copolymers ... 169
4.1.1.4. Vinyl chloride (co-)polymers ... 170
4.1.1.5. Vinyl acetate (co-)polymers and their derivatives ... 170
PVAC and copolymers - PVAL - Butyral PVB, Formal PVFM
4.1.1.6. Polyvinyl ethers ... 171
4.1.1.7. Polyacrylate resins ... 171
4.1.1.8. Polyamide copolymers and polyamine products ... 172
4.1.1.9. Natural product derivatives ... 172
Rubber conversion products - Cellulose conversion products
4.1.1.10. Dispersions ... 174
4.1.1.11. Water-soluble polymers ... 175
(4.1.2.-4.1.12. Thermoplastics for Molding, Extrusion and Calendering)
4.1.2. Polyolefins ... 176
4.1.2.1. Polyethylene ... 176
Product groups - Processes for synthesis - Commercial types and trade names - Properties - Processing and application
4.1.2.2. Cross-linked polyethylene (PE-X) ... 190
4.1.2.3. PE-C and chlorinated thermoplastic elastomers ... 191
4.1.2.4. Ethylene copolymers ... 191

		E/VA, E/VAC; E/VOH, E/VAL - (E/EA, E/MA, E/AA etc. - Ionomer resins - Copolymers with tetrafluoroethylene	
	4.1.2.5.	Olefinic thermoplastic elastomers (TPE).............	196
	4.1.2.6.	Polypropylene (PP) and copolymers.................	196
	4.1.2.7.	Polybut-1-ene (PB)...............................	201
	4.1.2.8.	Polymethylpentene (PMP)..........................	202
4.1.3.	Styrene-Based Polymers......................................		202
	4.1.3.1.	Homopolymers, copolymers and alloys..............	202
		Structures, properties and polymerization techniques - Products and trade names - General properties - Processing techniques - Fields of application	
	4.1.3.2.	Styrene polymers for foam processes.................	209
4.1.4.	Vinyl Chloride Polymers		211
	(4.1.4.1.-4.1.4.4. VC-polymer raw materials)		
	4.1.4.1.	Syntheses and delivery forms.......................	211
	4.1.4.2.	Molecular weight and field of application	213
	4.1.4.3.	Homopolymers, copolymers and blends	215
	4.1.4.4.	Stabilizer systems and fillers	218
	4.1.4.5.	PVC-U compounds	219
	4.1.4.6.	PVC primary and secondary plasticizers..............	226
	4.1.4.7.	PVC-P compounds................................	226
	4.1.4.8.	Special grades for insulation and coverings............	232
	4.1.4.9.	PVC pastes (plastisols)...........................	232
	4.1.4.10.	Foamable rigid and plasticized PVC	233
4.1.5.	Fluorine-Containing Polymers		234
	PTFE - FEP - PFA, TFA, EPE, TFB - ETFE - PVDF - PVF - E/CTFE - Cross-linkable fluoroelastomers		
4.1.6.	Methylmethacrylate Polymers		240
	4.1.6.1.	MMA polymers	240
	4.1.6.2.	PMMA standard molding compounds................	240
	4.1.6.3.	Copolymers and blends	243
	4.1.6.4.	Polymethacrylimide (PMI) molding compounds........	246
4.1.7.	Polyacetals or Polyoxymethylenes (POM)......................		247
4.1.8.	Polyamides (PA, Nylons) and Thermoplastic Polyurethanes (TPU) ..		251
	4.1.8.1.	Crystallizing PA molding compounds................	252
	4.1.8.2.	Amorphous polyamides	259
	4.1.8.3.	Flexible and thermoplastic elastomeric PA copolymers and PEBA..	260
		PA12 copolymers - Polyethyer-block amides (PEBA)	
	4.1.8.4.	Thermoplastic polyurethane elastomers (TPU).........	262
4.1.9.	Linear (Semi-)Aromatic Polyesters		265
	4.1.9.1.	Polycarbonates (PC)...............................	266
	4.1.9.2.	Polycarbonate + styrene component blends 271	
	4.1.9.3.	Polyalkylene terephthalates (PTP)....................	272
	4.1.9.4.	Thermoplastic poly(ether)ester elastomers	275
	4.1.9.5.	Amorphous polyarylates	277
	4.1.9.6.	Self-reinforced crystalline polyarylates (LCP)	278
	4.1.10.	Linear Polyarylene Ethers (Oxides), Ether ketones, Sulfides and Sulfones.................................	278
		PPE or „PPO" - PEK and PAEK - PPS - PSU, PES	
	4.1.11.	Linear Modified Polyimides........................	284
		PMMI - PAI - PEI - PI-PPS - Malecca	
	4.1.12.	Cellulosics	287
	4.1.12.1.	Cellulose esters (CA, CAB, CP)	287

		4.1.12.2.	Ethyl cellulose	288
	4.1.13.	Thermoplastic Composites		290
		4.1.13.1.	Molding materials	290
		4.1.13.2.	Glass mat reinforced thermoplastic prepregs (GMT)	290
	4.1.14.	Reclaiming, Reprocessing, Recycling		291
		4.1.14.1.	Mechanical separation	292
		4.1.14.2.	Chemical and thermal process	293
		4.1.14.3.	Processing of mixed plastics	294
	4.1.15.	Degradable Plastics		295
		4.1.15.1.	Photodegradable plastics	295
		4.1.15.2.	Biodegradable plastics	295
4.2.	Semi-finished Thermoplastic Material			296
	4.2.1.	Standardized and Hallmarked Piping Material		296
	4.2.2.	Polyolefins		300
		4.2.2.1.	Pipes	300
		4.2.2.2.	Sheets and sheeting	301
		4.2.2.3	Films	301
		4.2.2.4.	Fleeces and flat filaments	303
		4.2.2.5.	Foams	304
	4.2.3.	Styrene Polymers		304
		Standard polystyrene - SAN, SB and ABS sheet - ABS pipe - EPS boards and sheeting		
	4.2.4.	Vinyl Chloride Polymers		306
		4.2.4.1.	Rigid PVC, unplasticized (PVC-U)	306
		Pipes - Chemical engineering - Structural engineering - Consumables, office, drawing, measuring and graphics equipment - Other applications of thermoformable sheet materials - Packaging sheeting and film - Thermally treated PVC-E film - Furniture foils - Blown film		
		4.2.4.2.	Plasticized PVC (PVC-P)	308
		Tubes and profiles - Sheets, webs and band for technical and packaging applications - Flooring, wall and table coverings - Artificial leather and films for welded articles		
		4.2.4.3.	PVC foams	311
		4.2.4.4.	PVC-coated metal sheets	311
	4.2.5.	Fluoropolymers		311
	4.2.6.	Methylmethacrylate polymers, PMI foams		314
		4.2.6.1.	Manufacture of acrylic sheet	314
		4.2.6.2.	Properties of PMMA semi-finished products	315
		4.2.6.3.	Forms of delivery and applications of acrylic sheet	316
		4.2.6.4.	Acrylic foil, manufacture, forms of delivery, applications	316
		4.2.6.5.	Polymethacrylimide (PMI) rigid foam	319
	4.2.7.	Engineering Thermoplastics		321
	4.2.8.	Thermoplastic Polyesters		323
		4.2.8.1.	Polycarbonate (PC)	323
		4.2.8.2.	Polyethylene glycol terephthalate (PETP)	324
		4.2.8.3.	Mixed terephthalate-isophthalate polyesters	325
	4.2.9.	Cellulose Esters		326
		4.2.9.1.	Celluloid and related semi-finished goods	326
		4.2.9.2.	Extruded products from cellulose ester compounds	326
		4.2.9.3.	Cast films for electrotechnical applications	326
4.3.	Electrical Insulation Films and Packaging Materials			329
	4.3.1.	Characteristics and Properties		329
	4.3.2.	Multilayer Packaging Film and Packages		329

		4.3.2.1.	Extruded multilayer composite materials	329
		4.3.2.2.	Coatings, laminates	330
		4.3.2.3.	Special products	330
4.4.	Regenerated Cellulose	331
	4.4.1.	Cellophane	331
	4.4.2.	Foamed Products	331
	4.4.3.	Vulcanized Fiber (VF)	331
4.5.	Artificial Horn Casein Plastics (CS)	333
4.6.	Thermosetting Plastics; Raw Materials, Molding Compounds and Semi-finished Products	333
	4.6.1.	Curable Engineering Resins	333
		4.6.1.1.	Phenol (PF), cresol (CF), xylenol and resorcinol formaldehyde resins	333
		4.6.1.2.	Urea (UF) and melamine (MF) resins	336
		4.6.1.3.	Furane resins	341
		4.6.1.4.	Reactive resins and reactants for casting and low-pressure processing	341
		4.6.1.5.	Unsaturated polyesters (UP)	348
		4.6.1.6.	Epoxy resins (Epoxies, EP)	352
		4.6.1.7.	Special casting resins	356
		4.6.1.8.	Polyurethanes (PU): preproducts and products	356

Chemistry of the polyisocyanates - Processing and application, product-systems - Semi-rigid large RIM moldings - Rigid integral foam RIM-systems - Semi-flexible integral foam RIM systems - PU casting resins - Polymer binder resins - Rigid PU and PU-PIR foams - Flexible molded and block foams

	4.6.2.	Thermosetting Molding Compounds	369
		4.6.2.1.	General characteristics	369
		4.6.2.2.	Phenolic (PF) molding compounds	370
		4.6.2.3.	Aminoplastic molding compounds	374
		4.6.2.4.	Polyester (UP) resin compounds	375
		4.6.2.5.	Epoxy resin compounds	379
		4.6.2.6.	Diallyl phthalate (DAP) molding compounds	380
		4.6.2.7.	Silicone resin molding compounds	381
		4.6.2.8.	Further high-temperature-resistant thermoset molding compounds	381
	4.6.3.	Semifinisheds from Thermosetting Plastics	381
		4.6.3.1.	Cast resin products	381
		4.6.3.2.	Extruded profiles from PF compounds	381
		4.6.3.3.	Drawn fiber reinforced profiles	382
		4.6.3.4.	Wound fiber reinforced profiles	384
		4.6.3.5.	Laminates	384
			Industrial and decorative laminates	
		4.6.3.6.	GRP sheets and sandwiches	393
		4.6.3.7.	Reinforced thermosetting resin pipes (RTRP)	393
4.7.	Foams (Summary)	395
	4.7.1.	Concepts	395
	4.7.2.	Foam Principles	396
	4.7.3.	Blowing Agents	396
	4.7.4.	Properties	397
4.8.	Advanced High-Temperture Plastics	397
	4.8.1.	General Limits to Heat Resistance	397
	4.8.2.	New Routes to High-Temperature Polymers	402

	4.8.3. Linear Polyarylenes	403
	4.8.4. Polyarylene Amides, Aryl Esters and Ethers	403
	4.8.5. Bismaleimide-triazine Resins	404
	4.8.6. Polyimides	404
	4.8.7. Advanced Heterocyclic Aromatic Polymers	406
	4.8.8. Highly Temperature-Resistant and Carbonized Fibers	408
5.	**Related Topics**	**411**
5.1.	Natural and Synthetic Rubber	411
5.2.	Silanes and Silicones (SI)	415
6.	**Standardization**	**418**
6.1.	International and National Organizations for Standardization	418
	6.1.1. The International Standardization Organisation ISO	418
	6.1.2. National Standards Institutes	418
6.2.	Fields of Standardization	420
	6.2.1. Terminology	420
	6.2.2. Standardized Test Methods	422
	6.2.3. The Standardization of Plastics Raw and Molding Materials	422
	6.2.4. National and International Standardization of Semi-finished Products	429
7.	**Tests and Their Significance**	**430**
	Testing areas - Computer aided data bases	
7.1.	Fundamental and Processing Data	433
	7.1.1. Analytical Chemical Methods	433
	7.1.2. Data on Flow Behavior of Thermoplastics That Are Dependent of Degree of Polymerization	434
	7.1.3. Processing Data	435
7.2.	General Physical Properties	437
7.3.	Test Methods for Measuring the Influence of Temperature	439
	7.3.1. The Torsional Pendulum Test	439
	7.3.2. Thermal Analytical Methods (DIN 51005)	443
	7.3.3. Conventional Single-Value Tests	443
	7.3.4. Temperature-Time Limits	443
	7.3.5. Reaction to Fire Tests	444
7.4.	Resistance to Environmental Effects	448
	7.4.1. General Environmental Conditions	448
	7.4.2. Water Absorption	448
	7.4.3. Light and Weathering Resistance	449
	7.4.4. Resistance to Fungi, Bacteria and Animals	449
	7.4.5. Resistance to Chemicals	450
	7.4.6. Resistance to High-Energy Radiation	453
	7.4.7. Resistance to Mechanical Wear, Friction and Sliding	453
7.5.	Migration and Permeation	456
	7.5.1. General Remarks	456
	7.5.2. Gas Transmission	457
	7.5.3. Water Vapor Transmission	458
	7.6. Mechanical Properties	458
	7.6.1. Specimen Preparation	458
	7.6.2. Stress-Strain Properties under Static Load	460
	7.6.2.1. Short-term tests	460

		7.6.2.2. Creep rupture tests	462
	7.6.3.	Long-Term Behavior under Dynamic Stress (Fatigue Tests)	466
	7.6.4.	Impact Resistance Tests	470
	7.6.5.	Hardness Tests	473
7.7.	Some Formulas for the Calculation of Strength		474
7.8.	Computer Aided Design and Manufacturing (CAD/CAM)		476
8.	**Trade Names for Plastics as Raw Materials and Semifinished Goods**		**479**
8.1.	Characteristics of the Register		479
8.2.	Arrangement of the Register		479
	8.2.1.	1st Category: Resins	481
	8.2.2.	2nd Category: Forms of Delivery	484
8.3.	Trade Names		487
Subject Index			**635**

Abbreviations

A. Country of origin of trade names

The abbreviations used for the country of origin of trade names are based on those used in the patent literature.

AT	Austria	IL	Israel
AU	Australia	IN	India
BE	Belgium	IT	Italy
CA	Canada	JP	Japan
CH	Switzerland	NL	Netherlands
CIS	Commonwealth of Independent States	NO	Norway
		PL	Poland
CS	Czechoslovakia*	PT	Portugal
DE	Federal Republic of Germany	RO	Roumania
DK	Denmark	RUS	Russia
ES	Spain	SA	Saudi Arabia
FI	Finland	SE	Sweden
GB	Great Britain	SU	Soviet Union
GR	Greece	TW	Taiwan
HU	Hungary	US	United States of America
IE	Ireland	YU	Yugoslavia*

B. Abbreviations for homopolymers, copolymers, intermediates and rubbers

Comments on the use of abbreviations

The abbreviations are to ISO 1043-1987 (E), ASTM D 1600 and DIN 7728 Part 1 (01.88). The letter "P" for "poly" applies to homopolymers although it can be used for copolymers if its absence would lead to ambiguities. IUPAC rules state that when the "poly" is followed by more than one word, brackets are used. These are often omitted.

In addition to the abbreviation for the basic polymer, up to four important properties can be indicated by letters (without values) as shown in the table below. The letter(s) should be placed directly after the abbreviation separated by a dash.

*no longer used

Abbreviations

Examples of letters designating particular properties

Symbol	Particular properties	Symbol	Particular properties
C	chlorinated (e.g. PVC-C)	M	medium or molecular
D	density	N	normal or novolak (e.g. PF-N)
E	expandable or expanded (e.g. PS-E)	P	plasticized (e.g. PVC-P)
F	flexible or fluid (liquid state)	R	resol
		T	thermoplastic
GP	general purpose	U	ultra or unplasticized
H	high	V	very
I	impact (resistant)	W	weight
L	linear or low (e.g. PE-LLD)	X	cross-linked or cross-linkable

Characterizations relating to composition, polymerization, processing, etc are coded in the "block system (p. 423) as shown in Table 6.3, eg. PVC-S (c.f. Section 4.1.4.1). The special codes for polyamides are summarised on p. 252/253 and Table 4.35.

Fiber reinforced plastics (FRP) follow the same rule, e.g., UP-GF. A polyamide with 35% w/w GF content is designated PA 66, GF 35. GMT stands for glass mat thermoplastics.

The codes for material, form and content of reinforcements, enhancers and fillers are summarized in Table 6.3.

The symbols for monomeric components in copolymers are placed in the order of decreasing weight fraction separated by a "slash" (ISO 1043.1) although this character is sometimes omitted.

The symbols of the basic polymers in blends and alloys are placed in parentheses separated by a " + " although this is still not uniform in the literature.

A/B/A	acrylonitrile/butadiene/acrylate
ABS	acrylonitrile/butadiene/styrene
ACM	acrylic ester rubber
A/E/S	acrylonitrile/ethylene-propylene/styrene
A/MMA	acrylonitrile/methyl methacrylate
APP	atactic polypropylene
ASA	acrylonitrile/styrene/acrylic ester
BMI	bismaleimide
BR	cis-1,4-polybutadiene rubber

CA	cellulose acetate
CAB	cellulose acetobutyrate
CAP	cellulose acetopropionate
CF	cresol-formaldehyde
CMC	carboxymethyl cellulose
CN	cellulose nitrate
CO	epichlorohydrin rubber
CP	cellulose propionate
CR	chloroprene rubber
CSF	casein-formaldehyde
CSM	chlorosulfonated polyethylene
CTA	cellulose triacetate
DAIP	diallylisophthalate
DAP	diallyl phthalate
E/AA	ethylene/acrylic acid copol.
EC	ethyl cellulose
EP	epoxide
ECB	ethylene copol. bitumen
E/CTFE	chlorotrifluoroethylene/ethylene copolymer
E/EA	ethylene ethyl acrylate copol.
E/MA	ethylene/methacrylate copol.
EP	epoxide resin
EPDM	ethylene propylene diene copol.
EPM	ethylene propylene rubber
EPS	expanded polystyrene
E/TFE	ethylene/tetrafluoroethylene copol.
E/VA	ethylene/vinyl acetate copol.
E/VAL	ethylene/vinyl alcohol copol.
FEP	perfluoro(ethylene/propylene)
FF	furan formaldehyde resin
FPM	fluorinated rubber
GR-I	butyl rubber
GR-N	nitrile rubber
GR-S	styrene-butadiene rubber
IIR	butyl rubber
IPDI	isophorone diisocyanate
IR	cis-1,4-polyisoprene rubber
LCP	liquid crystal polymer

MBS	methylmethacrylate/butadiene/styrene
MC	methyl cellulose
MDI	diphenylmethane diisocyanate
MF	melamine-formaldehyde
MMA	methylmethacrylate
MPF	melamine/phenol-formaldehyde
NBR	nitrile rubber
NC	nitrocellulose
NR	natural rubber
PA	polyamide
PAI	polyamide imide
PAN	poly(acrylonitrile)
PAR	polyarylate
PB	polybutene-1
PBA	poly(butylacrylate)
PBI	poly(butyleneterephthalate)
PC	polycarbonate
PCD	polycarbodiimide
PCTFE	poly(chlorotrifluoroethylene)
PDAP	poly(diallylphthalate)
PE	polyethylene
PE-C	chlorinated polyethylene
PE-HD	high density polyethylene
PE-HD-HMW	high density, high molecular weight polyethylene
PE-HD-UHMW	high density, ultra high molecular weight polyethylene
PE-LD	low density polyethylene
PE-LLD	linear low density polyethylene
PE-MD	medium density polyethylene
PEBA	polyether-blockamide copol.
PEEK	poly(etheretherketone)
PEI	poly(etherimide)
PEOX	poly(ethyleneoxide)
PES	poly(ethersulfone)
PET	poly(ethyleneterephthalate)
PF	phenol-formaldehyde
PFA	perfluoro-alkoxyalkane
PFEP	polytetrafluoroethylene propylene
PI	polyimide
PIB	polyisobutylene
PIR	polyisocyanurate

PMI	poly(methacrylimide)
PMMA	poly(methylmethacrylate)
PMP	poly(-4-methylpent-1-ene)
PMS	poly(-α-methylstyrene)
POM	polyoxymethylene; polyformaldehyde; polyacetal
PP	polypropylene
PP-C	chlorinated polypropylene
PPE	poly(phenylenether)
PPMS	poly-p-methylstyrene
PPO	poly(phenyleneoxide) = PPE
PPOX	poly(propyleneoxide)
PPS	poly(phenylenesulfide)
PPSU	poly(phenylenesulfone)
PS	polystyrene
PSU	polysulfone
PTFE	poly(tetrafluoroethylene)
PTP	polyterephthalate
PU	polyurethane
PVAC	poly(vinylacetate)
PVAL	poly(vinylalcohol)
PVB	poly(vinylbutyral)
PVC	poly(vinylchloride)
PVC-C	chlorinated poly(vinylchloride)
PVC-P	plasticized PVC
PVC-U	unplasticized PVC
PVDC	poly(vinylidenechloride)
PVDF	poly(vinylidenefluoride)
PVE	poly(vinylether)
PVF	poly(vinylfluoride)
PVFM	poly(vinylformal), poly(vinylformaldehyde)
PVK	poly(vinylcarbazole)
PVP	poly(vinylpyrrolidone)
SAN	styrene/acrylonitrile copol.
S/B	styrene/butadiene copol.
SBR	styrene butadiene rubber
SI	silicone resin
S/MA	styrene/maleic anhydride
S/MS	styrene/α-methylstyrene
SP	saturated polyester
SR	polysulfide rubber
SP	saturated polyester

xvi *Abbreviations*

TAC	triallylcyanurate
TDI	tolylene diisocyanate
TFA	fluoro-alkoxy terpolymer
TMDI	trimethyhexamethylene diisocyanate
TPU	thermoplastic polyurethane
UF	urea-formaldehyde
UP	unsaturated polyester
VA	vinyl acetate
VC	vinyl chloride
VC/E	vinyl chloride/ethylene copol.
VC/E/MA	vinyl chloride/ethylene/methacrylate copol.
VC/E/VAC	vinyl chloride/ethylene/vinyl acetate copol.
VC/MA	vinyl chloride/methyl acrylate
VC/MMA	vinyl chloride/methyl methacrylate
VC/OA	vinyl chloride/octyl acrylate
VC/VDC	vinyl chloride/vinylidene chloride
VF	vulcanized fiber

C. Abbreviations for rubbers

Such symbols to ASTM 1418/ISO 1629 are summarized in Section 5.1.

D. Abbreviations for plasticizers

Table 4.24 gives examples of symbols for plasticizers to ASTM D 1600 and ISO 1043.3.

Conversion formulae

SI Units				U. S. Units				SI Units
kJ/kg · K	×	0.2388	→	BTU/lb °F	×	4.187	→	kJ/kg · K
W/K · m	×	6.944	→	BTU in/hr ft^2 °F	×	0.144	→	W/K · m
J/m	×	0.01873	→	ft lb/in	×	53.40	→	J/m
kJ/m^2	×	0.4755	→	ft lb/in^2	×	2.103	→	kJ/m^2
m/m K	×	0.5556	→	in/in °F	×	1.8	→	m/m K
MN/m^2 or MPa	×	144.9	→	lb/in^2	×	0.0069	→	MN/m^2 or MPa
Mg/m^3	×	62.50	→	lb/ft^3	×	0.0160	→	Mg/m^3
MV/m	×	25.38	→	V/mil	×	0.0394	→	MV/m

W/K · m = W/mK   MN/m^2 = N/mm^2 MG/m^3 = g/cm^3

Temperature Conversion: °C → °F = (°C × 1.8) + 32   °F → °C = (°F −32) ÷ 1.8
°C → K = °C + 273.15

1. PLASTICS TECHNOLOGY

1.1. General Concepts

Plastics are organic, man-made materials. The components that determine their technological and physical behavior are *polymers*, which are characterized by long chains of repeated "monomeric" units. Polymers used in plastics were originally obtained by the chemical modification of organic natural products (e.g., cellulose, linseed oil and castor oil) but are nowadays mainly produced by the polymerization of relatively simple compounds (*monomers*) derived from the petrochemical industry.

According to ASTM D 883, "(a) plastic is a material that contains as an essential ingredient one or more organic polymeric substances of high molecular weight. It is solid in its finished state and, at some stage in its manufacture or processing into finished articles, can be shaped by flow".

Many plastics are processed in a highly viscous state at high temperatures and pressures. Terms such as *plastics*, *matières plastiques* (French), *Plaste* (German) and *Plastmass* (Russian) have been based on the obsolete definition of that kind of flow as "plastic" flow. Synthetic fibers and textiles, natural and synthetic rubbers and adhesives and paints, which may in some cases accord with the above definition, are not considered plastics. However, there are no sharp boundaries between rubbers and plastic-elastomers, and many bases used for adhesives, paints, fibers, chipboards, etc. are also used for plastics in molded form.

Plastics contain not only polymers but also additives intended to enhance the processing or the physicochemical or mechanical properties. According to ASTM D 883, plastics are divided into the following main groups:

(A) By their chemical and technological behavior.

1. *Thermoplastics*: "A plastic that can be repeatedly softened by heating and hardened by cooling through a temperature range which is characteristic of the plastic and that in the softened state can be shaped by flow into articles by molding or extrusion." Thermoplastic scrap undamaged by thermal degradation can be reprocessed. Thermoplastic parts are weldable and can be thermoformed within a given temperature range. Thermoplastics polymeric molecules are linear or branched and as a rule are soluble in specific organic solvents and can thus be solvent bonded.
2. *Thermosets*: "A plastic that, after having been cured by heat or by other means, is substantially infusible and insoluble." Thermosets cannot be

reprocessed or welded, and only special products are thermoformable to a limited extent. They normally retain their shape from low temperatures to thermal decomposition. In thermosetting, the smaller molecules, generally in the form of a liquid or fusible preproducts, are chemically cross-linked into much larger ones. The molecules of *reactive resins* can be cross-linked with a catalyst at room temperature as well as by elevated temperature. Cross-linked (or cured) polymers can be swelled by organic solvents but cannot be dissolved by them without decomposition.

(B) By their mechanical behavior as tested by specific ASTM methods at room temperature.

1. *Rigid plastics* that have an elastic modulus greater than 700 MPa (100 000 psi);
2. *Semi-rigid plastics* that have an elastic modulus between 70 and 700 MPa (10 000 to 100 000 psi).
3. *Non-rigid plastics* that have an elastic modulus of not more than 70 MPa (10 000 psi).

Some semi-rigid and most non-rigid plastics are regarded as *elastomers*. The ASTM definition: "*Elastomer* – a macromolecular material that returns rapidly at room temperature to approximately its initial dimensions and shape after substantial deformation by a weak stress and subsequent release of the stress" is not generally accepted with respect to its restriction to "room temperature". On the contrary, many people consider that the significantly "rubber-like" state of an elastomer must persist over a service temperature range that is as large as possible, from cryogenic ($<0°C$) to elevated temperatures. That condition is achieved by regulating the cross-linking in elastomeric materials.
Chemically cross-linked elastomers, e.g. vulcanized rubber, remain elastomeric until thermal degradation begins. In that respect they comply with definition A2 above for thermosets.
Thermoplastic elastomers (TPE) possess rather complicated molecular structures containing entanglements of "hard" and "soft" regions. The hard regions consist of linear alignments of the chains below their glass transition temperatures (see p. 4) held together by secondary (van der Waals, dipole, etc.) forces. Above these temperatures, when these forces are not effective, they behave as thermoplastics, as described in definition A1.
Besides thermoplastics and thermosets, which correspond fully with A1 and A2, there are rigid and semi-rigid plastics that never become fluid at elevated temperatures but become rubber-like, e.g., cast acrylics or chemically cross-linked polyethylene (p. 190). The special kind of cross-linking found in ionomers that dissolve at high temperatures is described in Section 4.1.2.2.

1.2. Polymers for Plastics

For practical applications as plastics, polymers must have molecular weights greater than 10^4, distributed (to a greater or lesser extent depending on production conditions), around a mean value. Thus $>10^2$ identical or different monomer units with molecular weights of <100 must be linked together in a single or multi-stage polymerization.

(1) If each of 10^2 to 10^4 monomer units combines with two reactive positions (bifunctional) (Fig. 1.1, a), then long-chain (linear) macromolecules are formed; these are thermoplastics. Materials with extremely long chains ($M_r > 10^6$), e.g., cast acrylics and PE-UHMW, and/or strong secondary valence forces between the molecules (e.g. natural cellulose, PTFE) are not fusible thermoplastics. Linear, half-ladder or ladder polymer chains mainly from aromatic or heterocyclic monomers are boundary cases.
(2) If solitary trifunctional linkages arise on mainly bifunctionally linked main chains, then branched homo- or co-polymer thermoplastics are produced. The number and extent of the side chains determine the behaviour of that group of thermoplastics (Fig. 1.1, b_1 and b_2).
(3) If monomers or intermediate products with predominantly three or more reactive positions (trifunctional) react together, then three-dimensional cross-linked networks are formed of indeterminably high molecular weights.
Hard thermosets or cross-linked reactive resins are densely cross-linked; rubber-like elastomers are only lightly cross-linked (Fig. 1.1, c_1 and c_2).

1.2. Polymers for Plastics

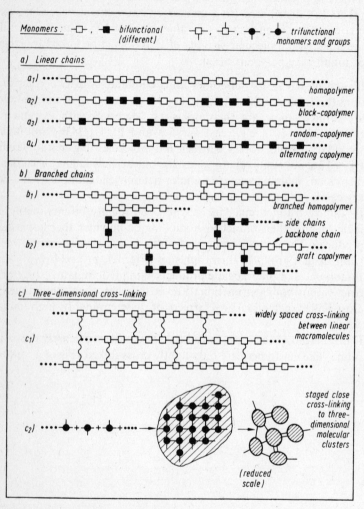

Fig. 1.1. Bi- and tri-functional polymer structures.
a and *b*: thermoplastics and thermoplastic elastomers
c_1: chemically cross-linked elastomers
c_2: thermosetting materials

1.3. Polymer Modification

1.3.1. Inter- and intra-molecular Combined Polymers

Homopolymers composed of monomer units of the same kind (Fig. 1.1, a_1 and b_1) as well as random or immediately alternating copolymers (a_2, a_4) which differ in their chemical structure and/or physical characteristics, generally exhibit limited mutual solubility and are therefore not homogeneously miscible into a single phase. Optimum combinations of polymer properties can be obtained by two processes based on different principles:

1. *Intramolecular*: In *block* polymerization sequences of different monomer units are combined to form linear macromolecules (Fig. 1.1, a_2). In teleblock copolymers very long sequences of one monomer are combined with short end-sequences of another. In multi-block copolymers, different sequences alternate.

In *graft* polymerization (Fig. 1.1, b_2) branched macromolecules are produced by grafting side chains of different monomer units onto the "backbone" chain. Intermediate products can be still-reactive ("living") polymers or oligomers called "macromonomers".

2. *Intermolecular*: Two or more polymers in different proportions are compounded under defined temperature and shear conditions to *(poly) blends or alloys* which are of increasing commercial importance.

Both types of polymer combination have two or more phases on a (sub) microscopic scale. As a rule the polymer component in the minor proportion forms the finely distributed disperse phase, which is embedded in the coherent matrix of the component in excess. Blend technology includes the coupling of the phase boundaries between poorly compatible polymers to prevent demixing during processing in the relatively narrow range that is recommended for them. Too high a processing temperature or shear can cause a phase change which renders the multiphase system ineffective.

For *phase coupling*, special copolymers having groups that promote compatibility with all the main components can be incorporated into the blend as bonding agents, or compatibility promoting groups can be included in the components by copolymerization, i.e., processes 1 and 2 can be combined to some extent.

In the frequent combination of rigid, highly thermostable polymers with flexible elastic polymers to *optimize the impact strength at low temperatures* (blends and/or graft copolymers), the flexible elastic disperse phase, by stabilizing microdeformations in the predominating hard matrix phase, dampens stresses that would cause fracture, in many cases (e.g. PC/ABS, Sec. 4.1.9.2) synergetically down to temperatures below the cold impact strength of both components. Processes such as the freezing of the second phase of such systems are manifested in the torsion pendulum experiment by an inflexion in the shear modulus and an intermediate peak in the loss factors (Fig. 1.4, left). Heat resistant advanced thermoplastic alloys with crystalline and amorphous phases are optimized to lower processing shrinkage and less tendency to deformation than the crystalline component, and better solvent resistance and environmental stress crack resistance than the amorphous component.

In *thermoplastic elastomers* (block copolymers or blends) the predominant flexible component is the coherent matrix phase which determines the properties in the rubber-like state. The amorphous or crystalline rigid domains included as the disperse phase prevent shearing of the molecules against each other up to their glass transition or melting temperatures. Some grades of alloys of an identical group cross the border line between high impact and elastomeric thermoplastic materials depending on their composition. In *ionomers*, ionic bonds between acid groups in macromolecules and metallic ions, which dissolve at higher temperatures, effect a similar temperature-dependent cross-linking. Plastics in which non-miscible components form interpenetrating continuous microphases (*IPN = interpenetrating networks*), are of limited practical importance.

1.3.2. Polymer Composites

Polymer composites are (macro) multiphase plastics materials with a polymer or resin matrix as continuous phase. Compounded plastics can contain as disperse phase specific property-improving additives (e.g. flame retardants, anti-friction additives) or pigments in small quantities, finely distributed gases (Section 3.15), fillers or reinforcing agents of all kinds (Sec. 3.1.3.8) in larger quantities or as the predominant components. Extreme cases of this are polymer-modified concretes (PC, PCC, Sec. 4.6.1.1).

The performance of composites is governed not solely by the properties of the components, but also (largely synergetically) by the development and behavior of the phase boundaries. Optimum bonding between the phases often requires the inclusion of a bonding agent chemically related to both components. The degree of dispersion of property-influencing additives has a considerable influence on their effectiveness. An example is the sudden increase in the electrical conductivity of compounds containing carbon black, known as percolation, at a relatively low concentration of that disperse phase.

Along with traditional thermosetting molding compounds (Chap. 4.6) and laminates (Sec. 4.6.3.5), reinforced thermoplastic compounds (Sec. 4.1.13) are gaining ever greater importance as mechanically and thermally highly durable, low shrinkage engineering plastics composites. For adequate dimensional stability up to their softening point, semicrystalline thermoplastics require combination with stiffening fillers; glassy amorphous thermoplastics with high glass transition temperatures are useable for hard transparent products over a wide temperature range.

Advanced polymer composites (APC) with high strength reinforcing fibers in a computer-calculated structure (Chap. 7.8) with a matrix of high-temperature polymers are lightweight materials for body and structural components in automotive and aerospace applications, heavy duty sports equipment and military purposes.

Graphited high-temperature and corrosion resistant *carbon ceramic* (CFC) is fired from phenolic resin bound carbon fiber preforms. *Geopolymer composites* (Trolit, DE) lie outside the field of organic polymers. With a matrix of reactive $SiO_2 \cdot Al_2O_3$ compounds in an aqueous alkaline medium and fillers, they can be processed into moldings as almost non-shrinking plastics compounds hardening at 100°C, which can be subsequently fired at up to 1000°C for high-temperature applications.

Developments in manufacturing high-strength and rigid reinforcement fibers by crystallizing polymers from solution by spinning followed by multistage, drawing at elevated temperatures are directed, *inter alia*, towards integrally reinforced, impact resistant composites. The first commercial products of this kind are PE-HD fibres with σ_B 1.0–3.5 GPa, E 50–150 GPa (Dyneema, NL, JP). For self-reinforced plastic products made of Liquid Crystal Polymers (LCP) see Sec. 4.1.9.5 and 4.8.4.

1.4. Polymer Flow and Elasticity

No polymer can be volatilized without chemical decomposition. Melts and solutions of polymers with long-chain or branched molecules become more viscous with increasing chain length. Their flow rates through a die (throughput per unit time) increase many times unlike ideal (Newtonian) viscous fluids, where the flow rate is proportional to compressive stress. This behavior is designated pseudoplasticity. The increasing viscosity of solutions $\eta_{rel} = \dfrac{\eta_{solution}}{\eta_{solvent}}$ or their *specific viscosity* ($\eta_{sp.} = \eta_{rel} - 1$) at low shear rates and concentrations *(c)* is also dependent not only on concentration, as for spherically shaped molecules, but to a significant extent on the size and shape of the polymer.

Polymers do not behave as ideal elastic materials in the solid state. Although according to Hooke's law ($\sigma = E\varepsilon$), stress (σ) and strain (ε) should be proportional and the elastic modulus (E) independent of deformation time, the modulus decreases over the long term. This occurs even in the range of small stresses of polymers, of initial stress with strain held constant (relaxation) and

1.4. Polymer Flow and Elasticity

of the corresponding strain with constantly held stress, which does not recover completely after relieving the stress. The extent of relaxation or creep under a given stress is dependent on time and temperature. At higher temperatures polymers relax or creep faster than at lower temperatures. This *viscoelastic* behavior of polymers is clarified by the use of models (Fig. 1.2) that illustrate the numerous ideal elastic elements in the form of springs and the viscous components of the response by cylinders containing a viscous fluid in which movable pistons are immersed (dash pots). Both are linked in parallel as well as working together in tandem. Calculations based on such models and the use of an infinity of elements together provide a complete quantitative understanding of the flow (the rheology) and the viscoelasticity of polymers.

Such treatment forms the basis of designing plastics processing machinery (chapter 3.2 to 3.7) for optimal flow. The time and temperature dependent viscoelastic behavior of polymers determines material properties which are important for the design of plastic components. Practical guidance for the processor is given by the *viscosity number*, derived from dilute polymer solution, which increases with the degree of polymerization. The *melt flow* index (MFI) of molding compounds is another conventional parameter for the flow

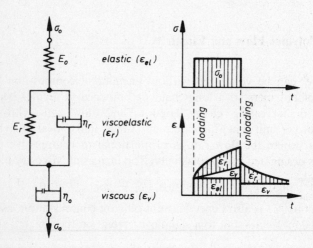

Fig. 1.2. Burger-Kelvin model for viscoelastic behavior E_o, E_r Spring elements for Hooke's elasticity.
η_o, η_r Damping elements for Newtonian flow
σ/t and ε/t cycle:
ε_{el} full elastic, reversible strain
ε_r viscoelastic, time dependent, reversible strain
ε_v viscous, time dependent, irreversible strain

rate of a compound in the processing state; under comparable conditions this relates inversely to the degree of polymerization. For details see paragraph 3.1.2 and 7.1.2.

1.5. Molecular Arrangements and Service Temperatures

At cryogenic temperatures coiled-up chains, molecular packets or networks of all polymers (Fig. 1.3) are tightly frozen in. The polymers are stiff, hard and brittle like glass. As the temperature rises Brownian motion in the polymer molecule affects both the molecule itself and its neighbors and increasing loosening of the system occurs. The polymer becomes less stiff and brittle, until in a limited temperature range, designated as the *glass transition temperature*, T_g, the intermolecular links are lost, insofar as they are not tight chemical cross-links (as in thermosets). The polymers not only become less stiff by some order of magnitude in this range, but other properties such as thermal expansion and specific heat change rapidly. The glass transition in the torsional pendulum test (Sec. 7.3.1) is recognised by the fall of the elastic component of the shear modulus $G(E \sim 3G)$. The approximate midpoint of the temperature range over which the glass transition takes place is T_g, measured as the maximum of the damping d (mechanical loss) during the transition.

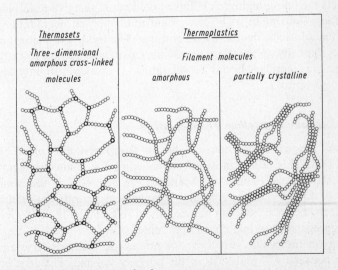

Fig.1.3. Structure of plastics-macromolecular arrangements.
The drawing is enlarged about 10^6 times, opened up and much simplified. Crystallites can also, as a result of folds in the molecular chains, become lamellae oriented tangentially to spherulites.

1.5. Molecular Arrangements and Service Temperatures

The absolute values of T_g are determined by the intramolecular structure of the polymers. These values are of decisive importance in the application of polymers as plastics (Fig. 1.4).

(1) If the T_g of amorphous plastics is high, then they are used below the T_g as rigid plastics.
(2) If the T_g of crystalline thermoplastics is $\leq 0°C$ (32°F), then they are used between the T_g and the melting range of the crystallite, T_m, as rigid or semi-rigid plastics.
(3) Polymers with a T_g (or T_m) $< 0°C$ (32°F) which are considerably less stiff are used as elastomers. Some thermoplastics of Type 1 (e.g., PVC, PMMA) exhibit an elastomeric temperature range before viscous flow. This is used for post-forming of semi-finished articles.

Typical values of T_g, T_m, modulus E and tensile strengths of polymers are given in Fig. 1.4. These measured values cannot be fully used in practice because of thermal degradation and/or the deterioration in the mechanical properties. Only a few unreinforced thermoplastics and thermosets can be continously used at temperatures above 100°C (212°F). Some advanced polymer composites can yield E-values $> 20\,GPa$ ($30 \times 10^5\,lb/in^2$ and strength of $> 200\,MPa$ ($30 \times 10^3\,lb/in^2$)) for extended service at high temperature levels.

1.5.1. The Glassy State of Rigid Plastics Made from Amorphous Polymers

Thermosets including highly cross-linked cold cured reactive-resin plastics are – with elongations at break $\leq 2\%$ – rigid until thermal degradation occurs. Some soften a little below these temperatures. The rigidity of amorphous thermoplastics depends on the more or less irregular cross section of the polymer chains and the secondary bonds between them. It decreases slightly with rising temperatures up to T_g.

1.5.2. Partially Crystalline Thermoplastics

Segments of long-chain polymers combine to form crystallites in sizes between 0.1 and 1 mm. This occurs most effectively when the chains are smooth and regular in structure (see linear and branched PE, Sec. 4.1.2.1, and isotactic PP, Sec. 4.1.2.6) and/or regular molecular arrangement permits the establishment of secondary chemical bonding between them (e.g., PA, Fig. 4.7). Crystallites may be aggregated into bundles or spherulites.

Depending on molecular structure and processing methods, crystallites can constitute between 5% and 95% of the polymer volume. They are linked by disordered entangled chain segments that, with a flexibility arising from a lower T_g, act as intramolecular hinges between the rigid crystallites. These

1.5.2. Partially Crystalline Thermoplastics

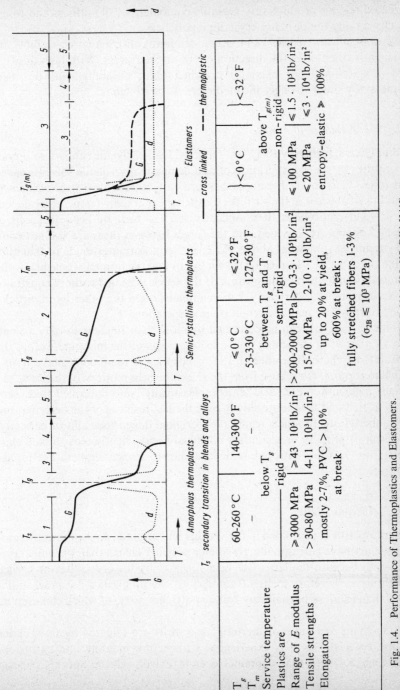

Fig. 1.4. Performance of Thermoplastics and Elastomers.
Dynamic mechanical properties measured by the torsional pendulum method (ASTM D 2236, ISO 537, DIN 53445).
G = elastic component of the shear modulus, d = mechanical loss factor; T_g = glass transition temperature; T_m = crystalline melting range.
Thermal region: (1) glassy elastic, (2) partially crystalline, (3) elastomeric, (4) viscous increased flow until (5) thermal decomposition begins shortly afterward.

disordered regions produce very versatile combinations of stiffness and tough elasticity among the partially crystalline polymers.

Above a significant tensile yield stress, randomly distributed crystallites are permanently oriented in the direction of the applied stress. With high drawing ratios, longitudinally or biaxially stretched films, monofilaments or fibers become many times stronger in the directions of stretch.

1.5.3. The Rubber-like State

Rigid polymers, primarily PVC, are rendered leather-like or rubbery at service temperatures by the incorporation of non or minimally volatile solvents called *plasticizers*. Plasticized thermoplastics are not regarded as true *elastomers* (definition p. 2) because their rubber elasticity is markedly temperature dependent. If rigid amorphous thermoplastics start to melt by heat, the coils of molecular chains are loosened, but the forces between them are still not completely removed somewhat above T_g. Post thermoformings made in that rubber-like state (Sec. 3.8.1) must be frozen by cooling the formed article below the T_g while it is under the forming stress. If the article is heated without constraint near T_g, it resumes its original shape completely. On the other hand, articles molded in fully viscous flow retain their shape above T_g.

Elastomeric materials can be deformed by more than 100% with only a small stress, as a result of stretching the coiled segments of the molecule, before the constraints of microphase (Sec. 1.3.1) or chemical cross-linking make it susceptible to rupture. On release from the stress the heat motions of the stretched molecule segment force it back "entropy-elastically" into the coiled state, very quickly at first and then more slowly over the last residues of the deformation. True rubber-elastic products regain their original dimensions almost perfectly. As a result of primary or secondary chemical bonds, lightly cross-linked elastomers (Fig. 1.1) are entropy-elastic over larger service temperature ranges (Fig. 1.4).

1.5.4. Viscous Melts

All thermoplastics are molded in the viscous molten state. Amorphous thermoplastics must have correspondingly longer polymer chains than partially crystalline ones because the latter are held together by the crystallites at service temperatures. Therefore melts of amorphous thermoplastics are much more viscous than those of partially crystalline polymers, some of which can even be poured.

The long-chain molecules of thermoplastic melts are oriented by flow under pressure, especially as it is experienced in injection molding and extrusion. Molded articles may thus contain frozen-in entropy-elastic and viscoelastic stresses, which can result in warping and distortion. Such stresses can be reduced by tempering the mold or by annealing the moldings. Articles with large residual stresses can lose their shape completely if they are heated freely at the molding temperature.

1.6. Properties of Plastics

1.6.1. Plastics encompass a wide *range of strength and stiffness* (Fig. 1.5); their elastic moduli E, except those of high-strength fiber reinforced composites, are low in comparison with those of metals. Some unreinforced plastics are sensitive to stress-cracking. These characteristics should be taken into account in the design of plastics parts, as should temperature-dependent creep under static or dynamic long-term stress.

1.6.2. The *density* of plastics is low (0.8–2.2 g/cm^3). Mechanical properties of plastics as a function of density are generally better than those of comparable materials. This is also valid for foamed plastics with their extremely low density (<0.2) and for foamed plastics sandwich products and other composites.

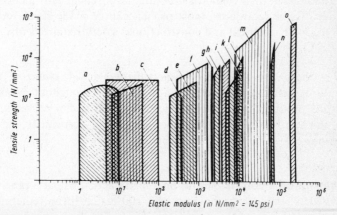

Fig.1.5. Tensile strength and elastic modulus ranges of various materials.
Rubbery elastic materials: (a) Rubber; (b) plasticized PVC; (c) polyurethane elastomers.
Partially crystalline thermoplastics: (d) PTFE; (e) PE; (f) PA.
Glassy thermoplastics: (g) PC; (h) PS, PVC-U, PMMA.
Regenerated cellulose: Vulcanized fiber.
Classical thermosets: (k) Moldings; (l) industrial laminates.
GRP; (m) Mats, fabrics, unidirectional rovings, UP or EP resin matrix.
Metals: (n) Aluminium; (o) steel.

1.6. Properties of Plastics

1.6.3. Plastics are *insulating and construction materials for electrical and electronics engineering*. They meet appropriate international standards for extreme electrical, thermal and mechanical loading. However their resistivity can be modified according to specific demands (Sec. 3.1.3).

1.6.4. The *environmental corrosion resistance* of most plastics is so high that they sometimes require no surface protection in use. Strongly cross-linked thermosets are insoluble in organic solvents. Elastomers swell in chemically related solvents. Thermoplastics are soluble, but some are highly corrosion-resistant materials that are used in the construction of chemical equipment and in piping. The resistance of plastics to weathering is variable. Many plastics are physiologically inert and permitted for contact with food and in medical applications.

1.6.5. The thermal *conductivity* of plastics is very low (extremely low for foamed plastics). *Thermal expansion* is higher than for metals. Only highly filled reinforced plastics have values which (for carbon fibers) can approximate to zero. Service temperature ranges are discussed in Sec. 4.8.1.

1.6.6. All organic materials are combustible. Some plastics are highly flammable on account of their chemical structure. Using chemically incorporated or compounded flame retardants, most plastics can be made flame retardant to meet the requisite electrical and constructional safety requirements.

1.6.7. For the *design of plastics products* there is a wide choice of materials, shapable and glass-clear or permanently colored throughout. Semi-finished and finished products can have glossy, mat or three-dimensionally textured surfaces, imparted by the molding tool. These are permanent and need no after-treatment.

1.6.8. Mass production enables rational manufacture of parts ranging in size from $\ll$ 1g (0.035 oz) to large moldings of $\geq$ 100 kg (220 lb) and (thermoformed) up to 16 m^2 (172 ft^2) in surface area, as well as hollow bodies with capacities of up to 20 000 l (5290 gal). Pipes are extruded with diameters of up to 1.5 m (50 in), films of up to 8 m (315 in) width and sheets of up to 3.5 m (138 in) width. The length of these products is limited only by transport considerations.

Plastics therefore offer a wide scope for application in their main fields of use: precision engineering, electrics and electronics, packaging, domestic appliances, sporting goods, transport and considerations.

1.7. Types of Plastics

Cellulose is a natural polymer of general interest for plastics applications (Table 1.1). Natural resins (colophony, copal) and fatty oils are forerunners of artificial resin lacquers, as is linseed oil for linoleum. Castor oil is a versatile raw material (see Table 1.4, Secs. 21, 22). Cheap plant carbohydrates, protein from maize or soybean and furfural from grain wastes are of basic interest in the synthesis of plastics.

The oldest raw materials for resin synthesis are coal products from bituminous coal tar (benzene, phenol, etc., see Tables 1.2, 1.4). Later, products from the gasification of coal were used (e.g., the Haber-Bosch process yielded formaldehyde and urea, see Tables 1.2, 1.3, 1.4) as well as those made with calcium carbide (melamine, Table 1.2, acetylene, Table 1.3). Today most of the raw materials are derived from oil or petrochemicals (e.g., ethylene, Tables 1.3, 1.4, p. 16/17). Petrochemical products are made commercially as an integral part of the refining of crude oil or by cracking (pyrolysis) at high temperatures.

Plastics can be divided into four groups according to derivation and synthesis. These also reflect their historical development.

1. *Plastics from natural products (Table 1.1)*: Vulcanized fiber (1859), celluloid (about 1870), and artificial horn (1897) are the oldest plastics. Regenerated cellulose was first produced in 1910. Regenerated cellulose and thermoplastics derived from cellulose are in competition with other packaging films and thermoplastics.
2. *"Classical" condensation resins (Table 1.2)*: The processing technology of these thermoseting engineering resins in combination with fillers as high-pressure molding powders and laminating materials stems from the discoveries of Leo H. Baekelands (Bakelite) in about 1910.
3. *Staudinger-polymerized plastics (Table 1.3)*: The work of Hermann Staudinger from 1922 laid the foundations for the systematic synthesis of thermoplastics by the polymerization of unsaturated monomers into long-chain macromolecules. Today polystyrene (1930), polyvinyl chloride (1931) and polyethylene (high-pressure PE-LD, 1939; Ziegler PE-HD, 1953; low-pressure PE-LLD, 1960) with their many homo- and co-polymers and alloys together constitute more than 60% of plastics production in most countries.
4. *Plastics from multifunctional aliphatic and aromatic intermediate products (Table 1.4)*: The synthesis of these plastics, which has occurred within the last four decades, involve new types of reactive intermediates. Specially tailored engineering plastics have been developed by a variety of combinations of specific polyaddition and polycondensation. Bifunctional polymerized products provide new engineering thermoplastics or, alternatively, they act as intermediate products for cross-linking to thermosets or elastomers by new methods, with or without heat and pressure.

Table 1.1. Plastics from Natural Products

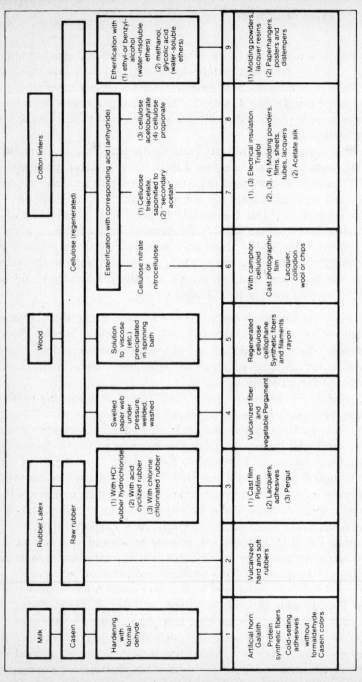

	1	2	3	4	5	6	7	8	9
Source	Milk → Casein	Rubber Latex → Raw rubber			Wood	Cellulose (regenerated)		Cotton linters	
Process	Hardening with formaldehyde		(1) With HCl: rubber hydrochloride (2) With acid: cyclized rubber (3) With chlorine: chlorinated rubber	Swelled paper web under pressure, welded, washed	Solution to viscose (etc.) precipitated in spinning bath	Cellulose nitrate or nitrocellulose	Esterification with corresponding acid (anhydride) (1) Cellulose triacetate, saponified to (2) secondary acetate	(3) cellulose acetobutyrate (4) cellulose propionate	Etherification with (1) ethyl- or benzyl-alcohol (water-insoluble ethers) (2) methanol, glycolic acid (water-soluble ethers)
Products	Artificial horn: Galalith Protein synthetic fibers Cold-setting adhesives without formaldehyde Casein colors	Vulcanized hard and soft rubbers	(1) Cast film Pliofilm (2) Lacquers, adhesives (3) Pergut	Vulcanized fiber and vegetable Pergament	Regenerated cellulose cellophane Synthetic fibers and filaments rayon	With camphor: celluloid Cast photographic film Lacquer, collodion wool or chips	(1), (3) Electrical insulation Triafol (2), (3), (4) Molding powders, films, sheets, tubes, lacquers (2) Acetate silk		(1) Molding powders, lacquer resins (2) Paperhangers, posters and distempers

Table 1.2. Classical Condensation Resins – Thermosets

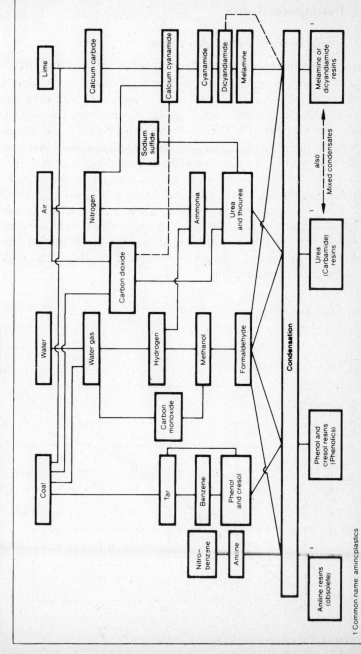

1.8. Development Trends

1.8.1. World Production and Consumption

Against a background of a world production of around 110 000 t/a of "classical" thermosets and derivatives of natural materials, the development of plastics into bulk scale materials started around 1930 with the commercialization of standard thermoplastics based on styrene, vinyl chloride, ethylene, and later also propylene. A transition from coal-based to petroleum-based chemicals and the introduction of new condensation and addition polymers since the 1950s resulted in a doubling of production every five years. In 1979, at the peak of this growth to close to 60 million t/a, the industrial countries of Western Europe, the USA and Japan had about a 90% share. After the subsequent slump in the market in the years of recession, the world volume of plastics production has practically doubled again with a total of approx. 100 million t/a in 1990; in 1994 a worldwide production of 113 million t has been reported. (Fig. 1.6). Table 1.5 gives an overview of the worldwide production, total and per capita consumption.

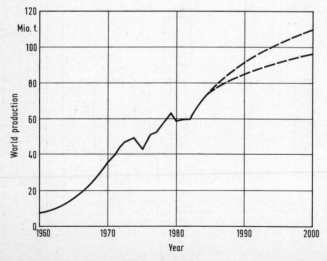

Fig. 1.6. Development of plastics production worldwide.

Table 1.5. Production, Consumption and Consumption per Head of Plastics Materials in 1992 (*Source*: IPAD-International Status Report on Plastics and Japan Plastics)

Country	Production (k tonnes)	Consumption (k tonnes)	Consumption per head kg
Australia	703	919	47.1
Austria	947	754	115.6
Belgium	3332	1517	152.0
Canada	2574	2089	75.7
Denmark	0	543	99.4
Finland	430	346	69.0
France	4746	3998	70.6
Germany	9908	9457	118.2
Hungary	692	306	29.7
India	734	1146	1.3
Israel	261	381	73.3
Italy	3100	4515	79.0
Japan	12580	10800	86.8
Mexico	1409	1267	15.4
Netherlands	3915	1642	109.0
S. Africa	662	590	15.8
Spain	2069	2166	59.1
Switzerland	157	605	90.3
UK	2177	3491	61.0
USA	30106	27691	108.0

By far the greatest proportion of the industry, in terms of numbers of firms, numbers employed and product value, is in plastics processing rather than plastics manufacture and plastics machinery manufacture. Plastics processing involves medium-sized companies, the majority of which, in the industrialised countries, employ less than 100 people per firm.

Western Europe and North America, with a share of about 60% of production and consumption worldwide, are still the chief plastics producers and consumers Table 1.6. Western Europe's share of production, however, is in recognizable decline and the growth of North America's depends largely on the powerful growth in capacity in Canada (1986: >2 million t). Asia has been able to increase its production share considerably. With a production of only 10 million t/a in Japan, this growth depends primarily on the increase in

1.8. Development Trends

Table 1.6. Worlds plastics production by regions

Region	Production (in million tonnes)				Share (in %)	
	1885	1988	1991	1993	1985	1993
Western Europe	24.5	29.7	31.7	32.2	32.0	29.9
North America	23.6	29.4	30.8	34.1	30.8	31.7
Asia	13.9	18.3	26.7	30.0	18.2	27.9
Eastern Europe	9.7	9.9	7.3	5.8	12.8	5.4
Latin America	3.0	3.7	4.2	4.1	3.9	3.8
Australia/Oceania	0.8	0.9	0.7	0.7	1.0	0.65
Africa/Middle East	1.0	1.9	0.6	0.7	1.3	0.65
World total	76.5	93.8	102.0	107.6	100	100

production in China, Taiwan and Korea. Taking into account the recently developed production capacities in the Near East, Asia will in the future claim an even higher share of overall production volumes. The capacities being developed in the emerging economies, including Latin America, are for the most part production plants for thermoplastic standard polymers for domestic consumption and export.

Plastics have now become indispensable materials not only in the industrial countries but also in the emerging economies, where the proportion of plastics applications in the industrial field is also continually increasing. For demanding engineering applications an annual growth worldwide of 8 to 10% is predicted. For consumer applications, the increase in consumption in the industrial countries must be of the order of magnitude of the increase in the gross national product. But there is still considerable need for the Asian and Latin American countries to catch up in this sector.

Plastics production is expected to reach around 120 million t in the year 2000. The breakdown of demand by type of polymer is shown in Figure 1.7.

The most important base material for plastics production – ethylene – is constantly extended by the utilization of the gases occurring during petroleum extraction and other petrochemical fractions that could previously be used only with difficulty. Their low proportion (at present about 4%) of the consumption of petrochemicals raw materials, combined with the possibilities of a new coal-based chemical industry excludes in the long term the risk of a shortage of raw materials. In addition, the specific energy consumption for plastics production (Fig. 1.8) and processing is considerably lower than for other, especially metallic materials and this makes plastics particularly environmentally friendly.

1.8.1. World Production and Consumption 21

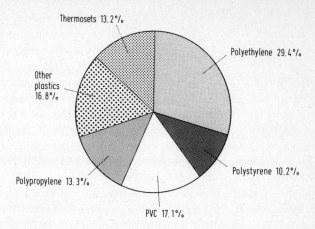

Fig. 1.7. Breakdown of Consumption in 1990 by Type of Polymer

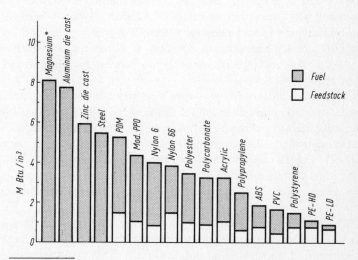

* Based on die-cast industry estimates of secondary metals usage: 20% magnesium, 46% aluminum, 5% zinc.

Fig. 1.8. Energy for the production of some plastics materials, including feedstock, in Btu (1 Btu = 1.06 kJ).

1.8.2. Materials Developments

The commodity *thermoplastics and conventional thermosets* account for about 85% of world consumption of plastics. A relatively small number of polymer groups (Table 1.7, 1 and 2a), cover the major applications: construction (profiles, especially for windows and claddings, translucent and transparent building components, pipes, roofing and constructional sealing sheets and felts, sealants and caulking materials, floor coverings, garbage barrels), packaging and transport (films, sheets, bottles, containers, closures, transport crates, pallets), automobile interiors, consumer goods and agriculture. There are a large number of different basic types, modifications and supply forms according to application.

Overlapping with the advances in the development of copolymers, alloys and composites (Section 1.3), the field of *engineering plastics* (Table 1.7, No. 2b to 4) comes next. These include a large number of polymers of different basic compositions, which are manufactured by complicated synthesis reactions (usually multi-stage) in low production volumes and command a higher price than standard polymers. Paramount, in this respect, are "specialties", which are characterized by properties such as very high thermal resistance ($\geq 200°C$), fire retardance with low smoke emission, etc. (Table 1.7, No. 4; Sections 4.1.9/11; 4.8), and include special phenolic resins. The growing demand for high-quality plastics permits production capacities to be developed, which in some cases is already leading to reduction in their extremely high prices, and therefore to extended applications.

The formulation of *plastics compounds* to meet special consumer requirements, requires a variety of compounding lines handling much smaller batch sizes than the processing units of the raw materials manufacturers. One example of this is the reaction extruder, which, together with its peripheral units, is suitable for (co-)polymerizing, alloying, and incorporating additives (Sec. 1.3.1, 3.1.4.1 and 4.1.3.1). Compounding companies are gaining increasing importance as intermediaries between large raw materials producers, plastics processors, and consumers.

Applications of engineering plastics are undergoing industry-wide development. They include electrical engineering, with electronics and instrumentation, as well as automobile construction, and aerospace. Their thermal resistance to temperatures greater than 100°C, combined with superior corrosion and permeation behavior, are utilized in chemical apparatus, for sanitary fittings and domestic heating appliances, (multilayer) packaging for industrial goods and foodstuffs (Sec. 4.3.2.1), sterilizable pharmaceutical requisites, sports equipment, etc.

In *electrical engineering*, with increasing power densities in the high and medium voltage range, tracking, arc, and flame resistant materials for service

Table 1.7. Main and Sub-Classifications of Polymers

No.	Main and Sub-Classifications, Section No. in Chapter 4[1]	Volume Proportion[2] %	Price Group Margins approx. DM/kg
1	*Standard thermoplastics*: basic types of polymers, modifications, copolymers, alloys: 4.1.2.1 PE–LD + PE–HD, 4.1.2.5 PP, 4.1.3 PS, 4.1.4 PVC	70–75	2.00–4.00
2	*Thermosets*		
2a	*Standard products*: binder resins, standard types of molding compounds 4.6.1.1/ 4.6.2.2 PF, 4.6.1.2/4.6.2.3 UF, 4.6.1.4 reactive resins (UP, PUR) in part, also varnish resins, adhesives	approx. 11	<1.50–4.00
2b	*Engineering reactive resins and molding compounds*: 4.6.1.3 furane resins, 4.6.1.5 in part UP, 4.6.1.6 EP, 4.6.1.8 PU, 4.6.2.2–4.6.2.5 special molding compounds (also BMC, SMC prepregs) based on PF, MF, UP, EP, DAP	approx. 9	
3	*Engineering thermoplastics*: multipolymers and blends of standard thermoplastics not included in 1, such as ABS/SAN and related polymers, PP composites, TPE, PE-UHMW, 4.1.2.2 PE-X, 4.1.2.3 CPE, 4.1.2.4 ethylene copolymers in part, 4.1.2.7 PB, 4.1.2.8 PMP, 4.1.5 fluoropolymers, 4.1.6 PMMA, 4.1.7 POM, 4.1.8 PA and TPU, 4.1.9.1/2 PC, PET, PBT, 4.1.10 modified PPE, PSU, PES, 4.1.12 CAB, CP	approx. 10	4.00–20
4	*Specialities*: 4.1.9.4 amorphous polyarylates, 4.1.9.5 self-reinforcing crystalline polyarylates, 4.1.10 PEK, 4.1.11 linear polyimides, 4.6.1.7 special casting resins, 4.8.3–4.8.8 high-temperature resistant polymers, 5.2 silicones	incl. remainder	20– >100

[1] also note: 4.1.1 preecursors and special products 4.2 thermoplastic semi finished products 4.6.3 semi finished thermoset products
[2] unclear definitions and possible double counting do not permit an exact total of 100%

1.8. Development Trends

temperatures of 150 to >210°C (300 to >410°F) are required for supports of current carrying parts of voltage transformers, switchgear units, plug connectors, lighting fixtures etc. Along with "classical" thermosets, which are still important, modified engineering thermoplastics and reactive resins are suitable. Such materials are also used for electrical power tools and domestic appliances. For halogen-free, fireproof, safety cable, elastomer-modified HT thermoplastics come into consideration.

For *microelectronic components* for data processing, communications, office and measuring technology, extremely precise miniaturization is also required. This is also needed for maintenance-free, low-wear, quiet-running gears and other precision mechanical moving parts of terminal equipment. These fields, whose proportion of plastics consumption in the electrical market is estimated to increase to 40%, have resulted in an expansion of the market for specialities (Table 1.7, No. 4) at affordable prices. Low birefringence, highly transparent polymers (PMMA, PC) for compact disks and data media for contactless laser reading are one special development. The necessity of protecting data media and audiovisual entertainment and power electronics equipment against electrostatic charges and for electromagnetic shielding has promoted the growth worldwide of large markets for materials with antistatic to semi-conducting properties.

Increasing the electrical conductivity of polymers with additives to that of metals (Sec. 3.1.3.3) is limited by the associated consequences on other properties. This is where the heavily promoted development of polymers that are intrinsically conducting because of their chemical structure and specific pretreatment without any additional additives, becomes relevant (polyacetylene, polypyrrole and polyaniline).

The series production of bodywork components from engineering plastics in *automobile manufacturing* poses quantitatively, qualitatively, and technologically important development tasks for both partners. For the demanding range of requirements:
- high thermal stability and rigidity,
- outstanding impact and notch impact strength, even at low temperatures,
- resistance to fuels and oils,
- good processability adapted to the requirements of rational mass production in automobile construction,

solutions are offered by: from the field of thermosets, BMC or ZMC injection-compression moldings, SMC compression moldings (Sec. 4.6.2.4.2/3) and PU(R)RIM, from the field of thermoplastics PP/EPDM, impact modified PC/PTP-, PPO/PA, and mat laminates. They are extensively used in the series production of bumper bars, headlamp supports, spoilers, side strips, wing mirror surrounds, underfloor protection, as well as for removable fenders,

1.8.2. Materials Developments

hoods, tail gates and door leaves for individual automobile models. Work has been done on plastics side windows and headlamp glasses and on over-coming problems in the in-line stove enamelling of plastics components with vehicle paints, improving the crash behavior of BMC and SMC front ends. For small series production of GF or hybrid reinforced front sections and other superstructures for road and track vehicles, that problem is of less importance. Engineering plastics are being increasingly used not only for fuel tanks and other containers but also for functional components which are resistant to fuel, oil and heat.

Spring leaves of advanced composites are in use for goods vehicles. Safety steering columns, rear axles, propeller shafts, connecting rods and other engine components of similar polymers are under development.

An essential aim of all this development work is saving of weight to reduce fuel consumption. It is expected that the proportion of plastics in motor vehicles, including interior fittings of the passenger compartment – which from 1970 to 1985 rose from 40 kg to over 100 kg – will double again by the end of the century (Fig. 1.9). All-plastic bodywork and plastics-ceramic engines, however, will remain the reserve of special models of sports cars and prototypes.

In *aeronautical engineering* a reduction of weight of up to 40% to save fuel is possible with lightweight composites as structural materials, and a primary development aim for interior fittings. Landing flaps, horizontal tail and rudder units, exterior doors and floor supports of C-fiber composites and increased fire safety through mainly phenolic resin bound honeycomb, foam and GF composites for interior fittings are state of the art for commercial airplanes, with sandwich C-fiber shells and long-life GF rotor blades for helicopters. Smaller touring aircraft with fuselages, wings and propeller shafts of APC (Sec. 1.3.2) are available. The composite wing center cell of 11.3 m span and 2260 kg weight for a military transport plane demonstrates the importance of lightweight materials in this field. The permeability of this material to radio and radar waves is of special significance here.

For components of *space vehicles*, very high quality special purpose polymers and APC materials which withstand conditions in outer space – temperature variations between $-180°C$ and $+150°C$ in vacuum and short wave UV radiation – without significant loss of weight and strength are indispensable.

In *precision instruments*, conventional materials have been replaced to a very great extent by engineering plastics for maintenance-free, wear-resistant, quiet-running mechanisms and bearings, operating elements, backing plates and cases.

Other important areas of application include chemical process plant and environmental protection where these materials play an important role in helping to meet the high standards required in these industries.

1.8. Development Trends

Prosthetics and clinical instruments for *medical and veterinary use*, made of very high performance plastics, are a special application that, although low in volume terms, is of great importance.

Dialysis equipment (artificial kidneys) utilizes products of membrane-technology, which is emerging with the new process of plasma polymerization for thin separating layers that are selectively permeable at a micro-level. A strong growth of applications is to be expected for other biomedical uses and for industrial separation and ultra-filtration processes.

2. LITERATURE ON PLASTICS TECHNOLOGY

The plastics industry is relatively young. Its main growth began in the 1950s and 1960s. However, it had its origins at the turn of the century, when the first polymeric materials based on natural products were used industrially. At about this time the first journal covering these products was published in Germany under the title "Kunststoffe". The US journal "Modern Plastics" appeared about 13 years later. Today there exists a large number of journals on plastics technology and polymer science in many languages published throughout the world.

In the following listings first of all the most important English-language and some of the leading foreign-language journals on plastics technology are given. These journals cover new developments in the total spectrum of plastics technology; e.g. materials, machinery, processing and applications, as well as aspects of marketing. Journals devoted to specific fields of application; e.g. packaging, construction, etc., or to the basic science of plastics (polymer physics and chemistry) are not included.

In addition to those journals concerned with current developments, there is a wide range of technical books, e.g. encyclopedias, handbooks, monographs, textbooks, etc. on most aspects of plastics technology. The list concentrates on some recently published English-language books giving an introduction to certain fields of plastics technology and to surveys of special topics. This can be only an incomplete listing but an attempt has been made to give references to the most important publications in recent years; in addition it shows the leading publishers in the field of plastics technology.

For the internationally operating plastics industry the different languages involved, as well as differing national standards, create a distinct barrier to the international exchange of know-how and products. The inclusion of some dictionaries on plastics technology in this compilation is aimed at reducing these obstacles. Information on publications of the national and international standard organizations is given in Chapter 6.

In addition to technical literature produced by publishers there are many publications in the form of books, brochures and leaflets produced by companies within the plastics industry. Most of them deal with technological aspects of company products but some also provide very fine introductions to certain fields of the technology. These publications can only be obtained by contacting the relevant companies directly.

2.1. Journals

2.1.1. English

USA/Canada

Canadian Plastics
Published 11 times/year by: Southam Business Publications,
1450 Don Mills Rd., Don Mills, Ont. M3B 2X7, Canada

Formulating & Compounding
Published bymonthly by: PTN,
Melville, N.Y. 11747

Injection Moulding
Published monthly by: Abby Communications, Inc.,
18 Fuller Ave., Chatham, NJ, 07928

Modern Plastics
Published monthly by: McGraw-Hill Inc.,
1221 Ave. of the Americas, New York, N. Y. 10020, USA

Plastics Engineering
Published monthly by: The Society of Plastics Engineers, Inc.,
14 Fairfield Drive, Brookfield Center, Conn. 068054-0403, USA

Plastics Technology
Published monthly by: Bill Communications, Inc.,
355 Park Avenue South, New York, N. Y. 10010, USA

Plastics World
Published monthly by: PTN Publishing Co.,
445, Broad Hollow Rd., Melville, N.Y. 11747, USA

Polymer Engineering and Science
Published semi-monthly by: The Society of Plastics Engineers, Inc.,
14 Fairfield Drive, Brookfield Center, Conn. 06805, USA

Europe
British Plastics and Rubber
Published monthly by: MCM Publishing, Ltd.,
37, Nelson Road, Caterham, Surrey CR3 5PP

European Plastic News
Published monthly by: Emap Maclaren,
Maclaren House, 19 Scarbrook Road, Croydon, CR9 1QH, Great Britain

International Polymer Processing
Published quarterly by: Carl Hanser Verlag,
Kolberger Str. 22, D-81679 München, Germany

Kunststoffe/Plast Europe
(bilingual edition English/German)
Published monthly by: Carl Hanser Verlag,
Kolberger Str. 22, D-81679 München, Germany

Modern Plastics International
Published monthly by: McGraw-Hill Publications Overseas Corp.,
14 avenue D'Ouchy, CH-1006 Lausanne, Switzerland

Plastics, Rubber and Composites – Processing and Applications
Published monthly by: Elsevier Applied Science Publishers, Ltd.,
Langford Lane, Kidlington, Oxford OX5 1GB,
Great Britain

Plastics Rubber Weekly
Published weekly by: Emap Maclaren, Ltd.,
Maclaren House, 19, Scarbrook Rd., Croydon CR9 1QH, Great Britain

Asia/Australia/Africa

Asian Plastics News
Published bimonthly by: Emap Maclaren,
Maclaren House, 19, Scarbrook Rd., Croydon CR9 1QH, Great Britain

Australian Plastics & Rubber
Published monthly by: IPC Business Press Australia Pty. Ltd.,
3–13 Queen St., Chippendale, N.S.W. 2008, Australia

China Plastics & Rubber Journal
Adsdale Publishing Company,
14/F Devon House, Taikoo Place
979 King's Road, Quarry Bay, Hong Kong

Plastics News International
Published bimonthly by: The Editors Desk Pty Ltd.,
P.O. Box 546, Mt. Eliza, Victoria, 3930, Australia

Plastics Southern Africa
Published monthly by: George Warman Publications, Ltd.,
Warman House, 77 Hout Street, Cape Town 8000, South Africa

2.1. Journals

Popular Plastics
Published monthly by: Colour Publications Pvt.,
126-A Dhuruwadi, Prabhadevi, Bombay 400025, India

2.1.2. Other Languages

German

Gummi Fasern Kunststoffe
Published monthly by: Dr. Gupta Verlag,
P.O. Box 104125, D-40852 Retingen/Germany

Kautschuk + Gummi Kunststoffe
Published monthly by: Hüthig Verlag,
Im Weiher 10, D-6900 Heidelberg/Germany

Kunststoffe
Published monthly by: Carl Hanser Verlag,
Kolberger Str. 22, D-81679 München/Germany

Kunststoffe Synthetics
Published monthly by: Vogt-Schild AG,
Zuchwilerstr. 21, CH-4501 Solothurn/Switzerland

Kunststoffberater
Published monthly by: Giesel Verlag,
P.O. Box 120 161, D-30907 Isernhagen/Germany

Österreichische Kunststoff-Zeitschrift
Published bimonthly by: Verlag Lorenz,
Ebendorfer Str. 10, A-1010 Wien/Austria

Plastverarbeiter
Published monthly by: Hüthig Verlag,
P.O. Box 102869, D-69018 Heidelberg/Germany

K Plastic & Kautschuk Zeitung
Published 21 times/year by: Giesel Verlag,
P.O. Box 120161, D-30907 Isernhagen/Germany

Swiss Plastics
Published 10 times/year by: Verlag Dr. Felix Wüst AG,
Seestraße 5, CH-8700 Küsnacht/Switzerland

2.1.2. Other Languages

French

Caoutchoucs & Plastiques
Published monthly by: E.D.I./S.E.T.E.,
5, rue Jules Lefebvre, 75009 Paris, France

Plastiques Flash
Published monthly by: S.E.P.E.,
142, rue d'Aguesseau, F-92100 Boulogne, France

Plastiques Modernes
Published monthly by: CEP,
42, rue des Jeuneurs, F-75002 Paris/France

Spanish

Plast 21
Published monthly by: Izaro S.A.,
C/Mazustegui, 21-3^a, 48006 Bilbao/Spain

Plásticos universales
Published quarterly by: Plastic Comunicacion,
La Llacuna, 162-Edf. Barcelona Activa, 08018 Barcelona/Spain

Revista de Plásticos Modernos
Published monthly by: Revista de Plásticos Modernos,
(FOCTTEC), Calle Juan de la Cierva 3, 28006 Madrid 6, Spain

Italian

Plastics
Published bimonthly by: Tecniche Nueove,
Via Ciro Menotti 14, I-20129 Milan/Italy

Macplas
Published monthly by: Promaplast Srl.,
Casella 24, I-20090 Assago (Milan), Italy

Materie Plastiche ed Elastomeri
Published monthly by: O.VE.S.T. S.r.l
Via Simone d'Orsenigo, 22, I-20135 Milan, Italy

Plast
Published monthly by: ERIS S.p.A.,
Vis Pola 15, I-20124 Milano, Italy

Czech

Plasty a Kaučuk
Published monthly by: Plasty a Kaučuk,
Priovská 2488, 760 01 Zlín, Czech Republic

Polish

Polimeri
Published monthly by: Instytut Chemii Przemyslowej,
ul. J. Rydygiera 8, pol. Warszawa 86/Poland

Croatia

Polimeri
Published monthly by: Društov Plastičara i Gumaraca,
Kaptol 22, 41001 Zagreb/Croatia

Hungarian

Müanyag és Gumi
Published monthly by: Delta Szaklapiado és Müszaki
Szolgaltato Leanyvallalat, Garay u. 5, 1441 Budapest 7

2.2. Books on Plastics Technology (in English)

2.2.1. Reference Books, Handbooks, Encyclopedias

Baijal, M. D. (Ed.): *Plastics Polymer Science and Technology*
1982, 992 p, Wiley, New York/USA

Berins, M. L.: *Plastics Engineering Handbook*
1991, 855 p, Chapman & Hall, New York/USA

Bisio, A. L., and Xanthos, M: *How to Manage Plastics Waste*
1994, 254 p, Hanser, Munich/Germany

Brandrup, J., and Immergut, E. H. (Ed.): *Polymer Handbook*
1989, 1904 p, Wiley, New York/USA

Brydson, J. A.: *Plastics Materials*
1989, 839 p, Butterworth, London/England

Charrier, J.-M.: *Polymeric Materials and Processing*
1990, 655 p, Hanser, Munich/Germany

2.2.1. Reference Books, Handbooks, Encyclopedias

Coleman, M. and Painter, P.: *Fundamentals of Polymer Science*
1994, 441 p, Technomic, Lancaster/USA

Domininghaus, H.: *Plastics for Engineers*
1992, 798 p, Hanser, Munich/Germany

Edenbaum, J. (Ed.): *Plastics Additives and Modifiers Handbook*
1992, 1113 p, Chapman & Hall, New York/USA

Ehrig, R. J.: *Plastics Recycling*
1992, 303 p, Hanser, Munich/Germany

Gächter, R., and Müller, H. (Ed.): *Plastics Additives Handbook*
1990, 1001 p, Hanser, Munich/Germany

Gum, W. F., Riese, W., and Ulrich, H.: *Reaction Polymers*
1992, 866 p, Hanser, Munich/Germany

Harper, C. A.: *Handbook of Plastics, Elastomers and Composites*
1992, 910 p, McGraw Hill, New York/USA

Hofmann, W.: *Rubber Technology Handbook*
1989, 651 p, Hanser, Munich/Germany

Katz, H. S., and Milewski, J. V.:
Handbook of Fillers for Plastics
1987, 467 p, Chapman & Hall, New York/USA

Klempner, D., and Frisch, K. C. (Ed.): *Handbook of Polymeric Foams and Foam Technology*
1991, 454 p, Hanser, Munich/Germany

Legge, N. R., Holden, G., and Schroeder, E. E. (Ed.):
Thermoplastic Elastomers
1987, 400 p, Hanser, Munich/Germany

Lubin, G. (Ed.): *Handbook of Composites*
1982, 786 p, Van Nostrand Reinhold, New York/USA

Mark, H. F., Bikales, N., Overberger, C. G., and Menges, G. (Ed.):
Encyclopedia of Polymer Science and Engineering
Vol. 1 to Vol. 19, 1985/89, each Vol. approx. 850 p, Wiley, New York/USA

Mark, H. F. Bikales, N., and Menges, G.: *Concise Encyclopedia of Polymer Science and Engineering*
1990, 1200 p, Wiley, New York/USA

2.2. Books on Plastics Technology (in English)

Nass, L. I., and Heiberger, C. A.: *Encyclopedia of PVC*
Vol 1: 1985, 680 p; Vol. 2: 1987, 675 p, Vol. 3: 1992, 608 p, Dekker, New York/USA

Oertel, G.: *Polyurethane Handbook*
1994, 688 p, Hanser, Munich/Germany

Progelhof, R. C., and Throne, J. L.: *Polymer Engineering Principles*
1993, 935 p, Hanser, Munich/Germany

Rosato, D.V., DiMattia, D. P., and Rosato, D. V.: *Designing with Plastics and Composites, A Handbook*
1991, 977 p, Chapman & Hall, New York/USA

Rosato, D. V., and Rosato, D. V.: *Plastics Processing Data Handbook*
1989, 408 p, Chapman & Hall, New York/USA

Rosato, D. V.: *Rosato's Plastics Encyclopedia and Dictionary*
1993, 884 p, Hanser, Munich/Germany

Rosen, S. L.: *Fundamental Principles of Polymeric Materials*
1982, 364 p, Wiley, New York/USA

Rubin, I. I.: *Handbook of Plastics Materials and Technology*
1990, 1745 p

Schwarz, S. S., and Goodmann, S. H.:
Plastics Materials and Processes
1982, 965 p, Van Nostrand Reinhold, New York/USA

Troitzsch, J.: *International Plastics Flammability Handbook – Principles, Testing, Regulations & Approval*
1990, 517 p, Hanser, Munich/W.-Germany

Waterman, N. A., and Ashby, M. F.: *Materials Selector* 1991, 2228 p, Chapman & Hall, New York/USA

Wright, R. E.: *Molded Thermosets*
1991, 213 p, Hanser, Munich/Germany

Modern Plastics Encyclopedia
(annual ed.) McGraw Hill, New York/USA

Chemical Resistance
Vol. 1: Thermoplastics, 1994, 1100 p;
Vol. 2: Thermoplastic Elastomers, Thermosets and Rubbers, 1994, 977 p
Plastic Design Library, Noirwich, N.Y./USA

2.2.2. Polymer Science Textbooks

Agassant, J.-F, Avenas, P. Sergent, J.-Ph., and Carreau, P.: *Polymer Processing*
1991, 499 p, Hanser, Munich/Germany

Billmeyer, F. W.: *Textbook of Polymer Science*
1985, 598 p, Wiley, New York/USA

Birley, A., Haworth, B., and Batchelor, J.: *Physics of Plastics*
1992, 549 p, Hanser, Munich/Germany

Chanda, M., and Roy, S. K.:
Plastics Technology Handbook
1992, 840 p., Dekker, New York/USA

Elias, H. G.: *Macromolecules*
1984, 1200 p (2 vol.), Plenum, New York/USA

Gedde, U. W.: *Polymer Physics*
1995, 352p, Chapman & Hall, New York/USA

Kaufman, H. S., and Falcetta, J. J.:
Introduction to Polymer Science and Technology
1977, 613 p, Wiley, New York/USA

Michaeli, W.: *Plastics Processing – An Introduction*
1995, 221 p, Hanser, Munich/Germany

Progelhof, R. C., and Throne, J. L.: *Polymer Engineering Principles*
1993, 935 p, Hanser, Munich/Germany

Sperling, L. H.: *Introduction to Physical Polymer Science*
1985, 592 p, Wiley; New York/USA

Ulrich, H.: *Introduction to Industrial Polymers*
1992, 188 p, Hanser, Munich/Germany

Young, R. J., and Lovell, P.: *Introduction to Polymers*
1991, 456p, Chapman & Hall, New York/USA

2.2.3. Engineering, Processing

DuBois, J. H., and Pribble, W. I.: *Plastics Mold Engineering Handbook*
1995, 776p, Chapman & Hall, New York/USA

O'Brien, K. T.: *Applications of Computer Modeling for Extrusion and other Continuous Polymer Processes*
1992, 551 p, Hanser, Munich/Germany

2.2. Books on Plastics Technology (in English)

Crawford, R. J.: *Rotational Molding of Plastics*
1992, 215 p, Wiley, New York/USA

Dym, J. B.: *Injection Molds and Molding*
1987, 416 p, Chapman & Hall, New York/USA

Gastrow, H. (Ed.): *Injection Molds – 108 Proven Designs*
1993, 258 p, Hanser, Munich/Germany

Griskey, R. G.: *Polymer Process Engineering*
1995, 496 p, Chapman & Hall, New York/USA

Gordon, J. M.: *Total Quality Process Control for Injection Molding*
1992, 618 p, Hanser, Munich/Germany

Hensen, F.: *Plastics Extrusion Technology*
1988, 738 p, Hanser, Munich/Germany

Johannaber, F.: *Injection Molding Machines*
1994, 328 p, Hanser, Munich/Germany

Kia, H. G.: *Sheet Molding Compounds*
1993, 267 p, Hanser, Munich/Germany

Macosko, C. W.: *RIM – Fundamentals of Reaction Injection Molding*
1989, 257 p, Hanser, Munich/Germany

Manas-Zloczower, I., and Tadmor, Z.: *Mixing and Compounding of Polymers*
1994, 896 p, Hanser, Munich/Germany

Manzione, L. T. (Ed.):
Applications of Computer Aided Engineering in Injection Molding
1987, 318 p, Hanser, Munich/Germany

Margolis, J. M. (Ed.): *Decorating Plastics*
1986, 152 p, Hanser, Munich/Germany

Menges, G., and Mohren, P.: *How to Make Injection Molds*
1992, 588 p, Hanser, Munich/Germany

Mennig, G. (Ed.): *Mold Making Handbook*
1995, 500 p, Hanser, Munich/Germany

Mennig. G.: *Wear in Plastics Processing*
1995, 452 p, Hanser, Munich/Germany

Michaeli, W.: *Extrusion Dies for Plastics and Rubber*
1993, 340 p, Hanser, Munich/Germany

Moore, G. R., and Kline, D. E.:
Properties and Processing of Polymers for Engineers
1984, 240 p, Prentice Hall, New York/USA

Pearson, J. R. A., and Richardson, S. M. (Ed.):
Computational Analysis of Polymer Processing
1983, 342 p, Elsevier, Essex/England

Pearson, J. R. A.: *Mechanics of Polymer Processing*
1985, 690 p, Elsevier, Essex/England

Powell, P. C.: *Engineering with Polymers*
1983, 360 p, Chapman and Hall, London/England

Pye, R. C. W.: *Injection Mould Design*
1983, 496 p, George Godwin, London/England

Rao, N. S.: *Design Formulas for Plastics Engineers*
1991, 142 p, Hanser, Munich/Germany

Rao, N. S.: *Computer Aided Design of Plasticating Screws*
1986, 142 p, Hanser, Munich/Germany

Rauwendaal, C.: *Polymer Extrusion*
1994, 568 p, Hanser, Munich/Germany

Rees, H.: *Understanding Injection Molding Technology*
1994, 140 p, Hanser, Munich/Germany

Rees, H.: *Mold Engineering*
1995, 652 p, Hanser, Munich/Germany

Rosato, D. V. and Rosato, D. V.: *Blow Molding Handbook*
1989, 1037 p, Hanser, Munich/Germany

Rosato, D. V., and Rosato, D. V.: *Injection Molding Handbook*
1995, 1145 p, Chapman & Hall, New York/USA

Schenkel, G.: *Kunststoff-Extrudertechnik*
1963, 540p, Hanser, Munich/Germany

Stevens, M. J., and Covas, J. A.: *Extruder Principles and Operation*
1995, 496 p, Chapman & Hall, New York/USA

Tadmor, Z., and Gogos, C. G.: *Principles of Polymer Processing*
1979, 736 p, J. Wiley, New York/USA

Throne, J. L.: *Plastics Process Engineering*
1978, 904 p, Marcel Dekker, New York/USA

Throne, J. L.: *Thermoforming*
1986, 400 p, Hanser, Munich/Germany

Tucker, C. L.: *Fundamentals of Computer Modeling for Polymer Processing*
1989, 642 p, Hanser, Munich/Germany

Whelan, A.: *Injection Moulding Materials*
1982, 398 p, Elsevier, Essex/England

White, J. L.: *Twin Screw Extrusion*
1990, 350 p, Hanser, Munich/Germany

White, J. L.: *Rubber Processing*
1995, 512 p, Hanser, Munich/Germany

Wright, R. E.: *Molded Thermosets*
1991, 213 p, Hanser, Munich/Germany

Wright, R. E.: *Injection Molding of Thermosets*
1995, 128 p, Hanser, Munich/Germany

Xanthos, M.: *Reactive Extrusion*
1992, 319 p, Hanser, Munich/Germany

2.2.4. Material Properties, Design

Baer, E., and Moët, A.: *High Performance Polymers*
1991, 335 p, Hanser, Munich/Germany

Belofski, H.: *Plastics Product Design and Process Engineering*
1995, 688 p, Hanser, Munich/Germany

Brostow, W., and Corneliussen, R. D. (Ed.): *Failure of Plastics*
1986, 486 p, Hanser, Munich/Germany

Domininghaus, H.: *Plastics for Engineers*
1992, 798 p, Hanser, Munich/Germany

Dym, J. B.: *Product Design with Plastics*
1983, 288 p, Industrial Press Inc., New York/USA

Gent, A. N: *Engineering with Rubber*
1992, 340 p, Hanser, Munich/Germany

Gordon, G. V., and Shaw, M. T.: *Computer Programs for Rheologists*
1994, 344 p, Hanser, Munich/Germany

Gruenweld, G.: *Plastics – How Structure Determines Properties*
1992, 371 p, Hanser, Munich/Germany

Ku, C. C., and Liepins, R.: *Electrical Properties of Polymers*
1987, 400 p, Hanser, Munich/Germany

Margolis, J. M.: *Engineering Thermoplastics – Properties and Applications*
1985, 393 p, Marcel Dekker, New York/USA

Mallick, P. K., and Newman, S.: *Composite Materials Technology*
1990, 400 p, Hanser, Munich/Germany

Malloy, R. A.: *Plastics Part Design for Injection Molding*
1994, 472 p, Hanser, Munich/Germany

Nagdi, K.: *Rubber as an Engineering Material*
1993, 322 p, Hanser, Munich/Germany

Nielsen, L. E. and Landel, R. F.: *Mechanical Properties of Polymers and Composites*
1993, 544 p,

Richardson, T.: *Industrial Plastics – Theory and Application*
1989, 521 p, South-Western Publishing Company, Cincinnati/USA

Tres, P. A.: *Designing Plastic Parts for Assembly*
1994, 256 p, Hanser, Munich/Germany

Utracki, L. A.: *Polymer Alloys and Blends*
1990, 367 p, Hanser, Munich/Germany

Ward, I. M.: *Mechanical Properties of Solid Polymers*
1983, 488 p, J. Wiley, New York/USA

Woodward, A. E.: *Atlas of Polymer Morphology*
1988, 541 p, Hanser, Munich/Germany

Woodward, A. E.: *Understanding Polymer Morphology*
1995, 128 p, Hanser, Munich/Germany

2.2.5. Polymer Science, Polymer Analysis

Brunelle, D. J.: *Ring-Opening Polymerization*
1993, 361 p, Hanser, Munich/Germany

Braun, D.: *Simple Methods for Identification of Plastics*
1987, 110 p, Hanser, Munich/Germany

Chan, C. M.: *Polymer Surface Modification and Characterization*
1993, 295 p, Hanser, Munich/Germany

2.2. Books on Plastics Technology (in English)

Cowie, J. M. G.: *Polymers: Chemistry and Physics of Modern Materials*
1991, 448p, Chapman & Hall, New York/USA

Dautzenberg, H. et al.: *Polyelectrolytes*
1994, 343 p, Hanser, Munich/Germany

Flory, P. J.: *Statistical Mechanics of Chain Molecules*
1989, 457 p, Hanser, Munich/Germany

Fouassier, J. P.: *Photoinitiation, Photopolymerization, and Photocuring*
1995, 250 p, Hanser, Munich/Germany

Garton, A.: *Infrared Spectroscopy of Polymer Blends, Composites, and Surfaces*
1992, 287 p, Hanser, Munich/Germany

Gelin, B. R.: *Molecular Modeling of Polymer Structures and Properties*
1994, 184 p, Hanser, Munich/Germany

Gilbert, R. D.: *Cellulosic Polymers, Blends, and Composites*
1994, 256 p, Hanser, Munich/Germany

Hummel, D. O.: *Atlas of Polymer and Plastics Analysis*
Vol. 1: Defined Polymers (1991, 1250 p)
Vol. 2: Plastics, Fibres, Rubbers, Resins; Staring and Auxiliary Materials, Degradation Products (Part a: 1984, 1035 p; Part b: 1987, 400 p)
Vol. 3: Additives and Processing Aids (1981, 696 p)
Hanser, Munich/Germany; VCH, Weinheim/Germany

Kausch, H.-H.: *Advanced Thermoplastic Composites*
1992, 380 p, Hanser, Munich/Germany

Kennedy, J. P. and Ivàn, B.: *Designed Polymers by Carbocationic Macromolecular Engineering*
1992, 255 p, Hanser, Munich/Germany

Lin, S.-C., Pearce, E. M.: *High-Performance Thermosets*
1993, 336 p, Hanser, Munich/Germany

Mathot, V. B. F.: *Calorimetry and Thermal Analysis of Polymers*
1994, 376 p, Hanser, Munich/Germany

Mitchell, J.: *Applied Polymer Analysis and Characterization*
Vol. I, 1987, 550 p, Hanser, Munich/Germany
Vol. II, 1992, 480 p, Hanser, Munich/Germany

Mobley, D. P.: *Plastics from Microbes*
1994, 288 p, Hanser, Munich/Germany

Shalaby, S. W.: *Medical Polymers*
1994, 272 p, Hanser, Munich/Germany

Spells, S. J.: *Characterization of Solid Polymers: New Techniques and Developments*
1994, 384p, Chapman & Hall, New York/USA

Vieth, W. R.: *Diffusion In and Through Polymers*
1991, 330 p, Hanser, Munich/Germany

Wool, R. P.: *Structure and Strength of Polymer Interfaces*
1995, 656 p, Hanser, Munich/Germany

2.2.6. Dictionaries

Glenz, W.: *A Glossary of Plastics Terminology in 5 Languages – English, Deutsch, Français, Español, Italiano*
1995, 350p, Hanser, Munich/Germany

Welling, M.: *Dictionary of Plastics and Rubber Technology*
Volume 1: German–English (323 p)
Volume 2: English–German (193 p)
1994, Pentech Press Ltd., London/England

Wittfoht, A. M.: *Plastics Technical Dictionary*
Part 1: English–German, (550 p)
Part 2: German–English, (534 p)
Part 3: Reference Volume, Illustrated Systematic Groups (512 p)
1992, Hanser, Munich/Germany

3. POLYMER SYNTHESES; PLASTICS PROCESSING AND FINISHING

3.1. Polymers and Compounds

3.1.1. Polymerization

The linear or branched macromolecules of primary thermoplastics and the cross-linked ones of thermosets and elastomers (Chap. 5.1) can be synthesized by either (ASTM D 883–83a):
 (1) *Addition polymerization*, in which monomers are linked together without the elimination of water or other simple molecules
 (2) *Condensation polymerization*, in which monomers are linked together with the elimination of water or other simple molecules.

3.1.1.1. Double-bond polymerization

The most widely used process for the synthesis of thermoplastic polymers is double-bond polymerisation (Table 1.3,). The gaseous or liquid monomeric reactants combine to form long-chain molecules by the "opening" of the unsaturated double bonds. These reactions, after having been *initiated* (by radical or ionic catalysts), are *propagated* until the growth of the macromolecular chains is brought to an end by *termination* reactions. Chain termination in radically-initiated reactions at a desired degree of polymerization is effected by mutual saturation of growing chains, or radical scavengers added to the reaction mixture as *regulators*. The absorbed radical can start a new chain (*chain transmission*). If it attaches itself intramolecularly to a finished chain, then a chain branch occurs (b, in Fig. 1.1).

More material-specific than radical initiators are initiations by ion-formers. Chain termination by mutual saturation of growing chains is not possible. Anionic polymerization results in "*living polymers*" without saturated end groups. These can be incorporated into chain segments in *block copolymerization*.

Mass (or bulk) polymerization to produce clear products (e.g., PS, Sec. 3.6.5.3, cast acrylics, Sec. 4.2.6.1) and *gas-phase polymerization* in a fluidized bed reactor (PE, PP, Sec. 4.1.2.1 and 4.1.2.6) are carried out in the absence of a diluent. In *solution polymerization* (PIB, Sec. 4.1.1.2.3) removal of residual solvent by evaporation is difficult. If the polymer is insoluble in its own monomer or in the monomer solution, *precipitation polymerization* occurs. An example is the two-step mass polymerization to very pure PVC (Sec. 4.1.4.1). In *suspension polymerization* the monomers are dispersed by mechanical agitation in a liquid phase (e.g., water for PVC, initiator slurries in diesel oil for low-pressure PE, Sec. 4.1.1.10). The resulting polymers may be rather coarse pearls, beads or irregular granules, which are easily separated when agitation ceases. In *emulsion polymerization* the monomers are emulsified in an aqueous medium by a soap-like surfactant. Stable *polymer dispersions* are produced that can be used as such (Sec. 4.1.2.1). Emulsion polymers produced by evaporating the water are finely dispersed and easily processible but contain residues of the emulsifier, which reduce the clarity and water resistance of the product.

Special metallo-organic *coordination catalysts* called *Ziegler-Natta catalysts* influence the steric order of the side groups in the macro-molecular chains, a phenomenon called *tacticity* (Fig. 3.1). Only stereoregular polymers with side groups can partially crystallize (Sec. 1.5.2).

Copolymers can be prepared with two or more different chemical monomer units or parts of polymer chains (e.g., terpolymers with three components). The properties of copolymers depend on the arrangement of the components. This

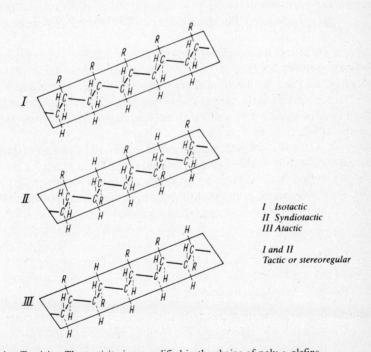

Fig. 3.1. Tacticity. The tacticity is exemplified in the chains of poly-α-olefins.
I Isotactic polymers: In a planar conformation of polymer $[CH_2-CHR]_n$ the substituents, R, are either all above or all below the plane of the backbone chain.
II Syndiotactic polymers: The substituents, R, alternate below and above the backbone chain.
III Atactic polymers: The substituents, R, show an irregular, random arrangement about the main chain.

3.1. Polymers and Compounds

is demonstrated for two components in Fig. 1.1. Copolymers as well as *polyblends* or *alloys* containing microphase entanglements structured in a different way, are discussed in 1.3.2. Intra- and inter-molecular combined polymers are considered in Section 1.3.1.

Network or *cross-linking* polymerization is a process available for the curing of products at atmospheric pressure, as for example in the processing of unsaturated polyesters (Sec. 4.6.1.5).

3.1.1.2. Condensation polymerization

Various monomers or intermediates are linked together through their reactive end groups by the elimination of volatile by-products such as water or ammonia. The molecular size of the polycondensates is regulated by the relative proportions of the feedstock and other reaction conditions.

A variant of polycondensation is *oxidative coupling*. Single monomers that at the outset have hydrogen in certain positions are linked by oxidation in such a way that water is eliminated. Polyphenylene oxide (PPO or PPE) is an example (Sec. 4.1.10).

Examples of linear molecules containing materials synthesized by condensation reactions are polyamides (Sec. 4.1.8), linear polyesters (Sec. 4.1.9) polyethers and polysulfones (Sec. 4.1.10.4), and short-chain unsaturated polyester resins (UP) to be cross-linked with styrene (Sec. 4.6.1.5), such as polyols as reactants for epoxy resins (Sec. 4.6.1.6) and polyurethane (Sec. 4.6.1.8). These types of plastics are summarized in Table 1.4.

The stepwise condensation polymerization of materials with more than two reactive positions leads to thermosets. Phenolics and aminoplastics (Table 1.2, p. 17) are examples. Most volatile by-products are lost in the first stages of the reaction. On hardening by condensation polymerization, non-porous products are achieved by the selection of appropriate processing conditions (high pressure).

3.1.1.3. Rearrangement polymerization

Rearrangement polymerization (or "polyaddition," in a restricted sense) is the joining of monomers or intermediate products with different reactive end groups without formation of volatile by-products. In a way "condensation" has been anticipated by the chemical structure of these end groups. The process is suitable for atmospheric pressure curing of plastics. Epoxy resins and polyurethanes are typical products (Sec. 4.6.1.6 and 4.6.1.8).

3.1.2. Characteristic Values for Polymers and Plastics

Molding compounds are described by a selection of characteristic values which are held in data banks of material properties such as CAMPUS and POLYMAT (see chapter 7).

The overall behavior of a plastic within the temperature range of its application is determined by the molecular weight (M), the chemical structure and the form and reactivity of the monomers involved in the synthesis of the polymer (dealt with generally in Chapter 1 and in detail in Chapter 4) as well as by the compounding of auxiliary materials.

The degree of polymerization, DP, is defined as[1]

$$DP = \frac{M\,polymer}{M\,monomer}$$

For non-cross-linked polymers this is generally in the range of 10^4 to $> 10^5$ and has a significant influence on flow, on T_g and T_m (Chap. 1.5) and on mechanical behavior. All synthetic polymers contain a size distribution of large molecules (Fig. 3.2). Without "fractionation," the expense of which would only be justifiable for scientific purposes, one can only determine the average molecular weight. This refers to the overall distribution, and depending on the mode of determination will be the *weight average* M_w or the *number average* M_n or some intermediate value, e.g. M_η in the case of viscosity measurements. The more

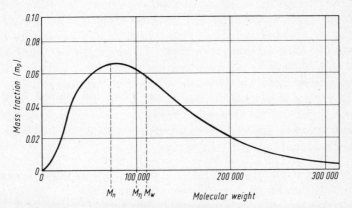

Fig. 3.2. Typical molecular-weight distribution curve.

---

[1] The symbol M_r is that for relative molecular mass in the SI system. U.S. literature and standards still refer to this as the molecular weight (MW). The molecular weights quoted for polymers are averages: M_n is the number-average molecular weight, M_w is the weight-average molecular weight and M_η is often used to indicate a viscosity-average molecular weight.

3.1. Polymers and Compounds

M_w/M_n or DP_w/DP_n is greater than 1 the wider is the molecular weight spread. This enhances processibility on the one hand but impairs the mechanical properties on the other.

In practice the following determinations are used to make comparative judgements on the behavior of thermoplastics and on any deterioration resulting from processing:

(1) *Viscosity number $[\eta]$:* The term for the relative increment of viscosity (specific viscosity, Chap. 1.4) for a dissolved polymer as concentration (c) $\to 0$ in ml/g or 100 ml/g measured under standard conditions (ASTM-D 2857-77, ISO/R 1628, DIN 53726/8 and others), specific for each individual group of polymers: for PE, PP, PVC, PMMA, PA, PETI, PC, CA[2]) (Sec. 7.1.2).

For the relation between viscosity numbers 50 to 190 and K values 45 through 84 relating to PVC see Table 4.16. Viscosity numbers and K values are related by the equation $[\eta] = KM^a$ and increase with rising MW. They can be measured meaningfully only for soluble polymers without the presence of additives.

(2) *Melt flow index MFI, melt volume index MVI, or Flow rate FR* (ISO/R 1113, ASTM D 1238, DIN 53735), covers the measurement of the rate of extrusion of molten thermoplastics – generally measured in grams per 10 minutes or MVI in $cm^3/10$ min – through a standardized die under prescribed conditions of temperature and load in a standardized piston-barrel plastometer. In accordance with the varying flow of single plastics groups, special combinations of temperature and load are applied, which must be stated (e.g., for 190°C and 2.16 kg: MFI 190/2.16 or by code letters for the conditions, see Section 7.1.2 and, Table 7.1). The Flow Rate Ratio (FRR) of values measured under different conditions has been found to be useful for the characterisation of materials like PE.

The method is generally applied to PE, E/VA, PP, PS, SB, SAN, ABS, ASA and PC. Melt flow of polymers decreases with rising MW. The method can be applied to compounds containing fillers or other auxiliaries that influence the MFI values.

For each thermoplastic there is a particular correlation between the viscosity number (η_{rel}) and the melt-flow index (MFI). The relationship is semi-logarithmic and empirical straight lines are shown in Figure 3.3. For example, for PA6, the following relationship holds:

$$\log(\text{MFI}) = 2.923 - 0.00851 \eta_{rel}$$

---

[2] For these abbreviations see p.XII. The arrangement of the polymers always corresponds with the arrangement of those in Chapter 4.

3.1.3. Auxiliaries and Additives for Plastics

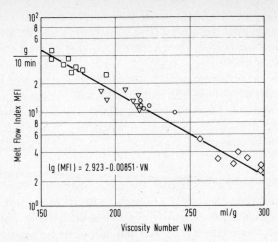

Fig. 3.3. Correlation between melt flow index (235°C, 2.16 kg) and viscosity number (25°C, m-Cresol, 10g/l) for PA 6

and for a particular PC the correlation is:

$$\log(\text{MFI}) = 3.0132 - 0.03566 \eta_{\text{rel}}$$

The MFI was determined at 300°C using 1.2 kg and the viscosity number was measured at 25°C at a concentration of 5 g/dm^3 in methylene chloride.

(3) There are some other methods for measuring *flow behavior of thermoplastic and thermosetting molding materials* under molding conditions (e.g., the spiral flow test) that as yet have not been generally standardized (Sec. 7.1.3).

3.1.3. Auxiliaries and Additives for Plastics

Although they have not achieved precise definitions that are universally accepted, the terms "auxiliaries", "additives", and "compounding ingredients" are used with the following meanings in the ensuing pages:

Auxiliaries are materials, such as catalysts, emulsifying agents, etc., which are used in the manufacture of polymers. In the final products, generally only small residues of the auxiliaries remain.

Additives include lubricants, stabilizers, flame retardants, etc. They are included for their processing functions and to influence the behavior of the compound in use. They too are added in relatively small amounts. For reasons of hygiene and better dosage, they are offered in dust free, readily dispersible, pourable formulations, or preprocessed into master batches. In the microencapsulation technique, additives (e.g., catalysts and flame retardants) enclosed in a solid casing can be released when required by mechanical destruction of the microcapsules.

Compounding ingredients are materials added in relatively large proportions (10–70% or more of the polymer) to make significant changes in the properties of the plastic compound. They include plasticizers, fillers and reinforcements. There are no rigid divisions between these categories. A number of types of additives will now be considered.

3.1.3.1. Lubricants or parting agents are metal soaps, montan or paraffin waxes, wax-like polymers (Sec. 4.1.1.11), higher fatty alcohols, fatty acid esters and silicones. They reduce the viscosity of molding compounds (internal lubricating action) and/or are effective as external lubricants between plastic melts and metal walls. Parting agents sprayed into molds for moldings which will after-treated must be silicone-free.

3.1.3.2. Stabilizers protect against thermal degradation during processing and in use and/or guard the products against oxidation and breakdown by weathering agents, particularly UV radiation. Many chemically different stabilizer systems are necessary for almost all plastics applications, and they must be appropriate not only for the polymer involved but also for the requirements dictated by each use. Stabilizer systems for PVC, manifesting "synergistic effects" between their components are considered on in Section 4.1.4.4. Carbon black is an almost universally effective UV stabilizer.

3.1.3.3. Antistatic agents and conducting additives are hydrophilic substances (e.g., amino derivatives and polyethylene glycol esters) that reduce the surface resistivity of plastics (normally 10^{13}–10^{15} ohm), so much that problems due to friction-induced electrostatic charges are prevented. *Conducting additives* that can, in small amounts, reduce specific resistance to 1 ohm cm are of particular importance in the *screening of high-frequency emissions* from electrical and electronic equipment. Various materials, such as aluminium flakes coated with coupling agent, steel microfilaments, silvered glass fibers and spheres, nickel surface coated textiles (Baymetix), special carbon blacks and carbon fibers are applied to meet the legally required (EMI) electromagnetic interference and radiofrequency damping in the 10 kHz to 140 GHz range with moderate to good screening at 60 dB. For surface-screening, conducting lacquers (Electrodag) or vacuum metallizing (Sec. 3.9.3.4, Elamet) is used. (For more about conductive plastics see Sec. 1.8.2).

3.1.3.4. Flame retardants are necessary to reduce the flammability of plastics products so that they can satisfy the specifications for electrial applications, for transport vehicles and for building (Sec. 7.3.5). Chlorine- or, more effectively, bromine-containing organic compounds evolve products when exposed to flames that inhibit the admission of oxygen and chemically retard the flame reactions. Phosphorus-containing compounds favor carbonization and encrustation. Both types can be incorporated chemically or combined with the monomer. The effect of halogens is strongly and synergistically reinforced by the presence of antimony trioxide. Relatively large concentrations of additives can have an unfavorable effect on the behavior of the product in use. Public health considerations militate against several additives. The formation of corrosive substances and the generation of smoke in the event of fires pose further fire-protection problems. Recent work is aimed at halogen-free flame retardants, e.g. 25–30% intumescent combinations of nitrogen-containing oligomers with ammonium polyphosphate (Spinflam).
Inorganic fillers proportionately reduce the flammable element within the plastic and favor encrustation during burning. Fire protection, without the development of smoke, is conferred by aluminium hydroxide, $Al(OH)_3$, or the microfibrous Dawsonite, $NaAl(OH)_2CO_3$, which at temperatures of about 200°C split off steam or steam and carbon dioxide. Zinc borate, $Zn(BO_2)_2 \cdot 2H_2O$, $Mg(OH)_2$ or $MgCO_3 \cdot H_2O$ at $\geq 300°C$ work in a similar way.

3.1.3.5. Colorants are specially prepared insoluble organic and inorganic pigments and plastics-soluble dyestuffs. Inorganic pigments provide temperature and light stability (Sec. 7.4.3), organic the most demanding color requirements. Polymer-soluble dyestuffs are used to a limited extent for transparent products and fluorescent products (Fuco, Lisa, DE) for luminous directional reflectance and absorption. Artificial horn and PA are surface colored with an aqueous solution. Optical-brightening agents are also soluble colorants.
Discolorations produced by reactions between sulfur-containing dyestuffs and lead or tin stabilizers, interference with the curing of resins by metal complexes, bleeding as a result of additives, the influence of pigments on crystallization and the electrical properties of plastics are all considerations that must be taken into account in the choice of colorants.
Masterbatches consisting of pigments and the plastics to be processed and concentrated pigments with higher pigment content in an inert bonding agent are used to color plastics in a dust-free manner.

3.1.3.6. Flexibilizers are chemically reactive or non-reactive additives for toughening thermosetting or cold curing resins. However, some thermoplastic compounds containing small amounts of plasticizers (cellulosics Sec. 4.1.12, PA. Sec. 4.1.8.3) are also called flexibilized.

Impact modifiers are non-rigid elastomer-like polymers (Sec. 1.3.1) used for enhancing the impact resistance of rigid thermoplastics by alloying. They are contained in most HI-modified thermoplastics as spherically microdispersed phases in the rigid matrix (Sec. 1.3.1), but there are also network-like distributions of the phases (e.g. PVC/EVA, PVC/PE-C).

Plasticizers are low molecular or oligometric additives compatible with rigid thermoplastic polymers solvating them to semi-rigid or leathery or rubbery plastics materials (Sec. 1.5.3). The most important PVC-plasticizers are discussed in Section 4.1.4.6.

Random or alternating copolymerisation (Fig. 1.1 a_3 and a_4) can result in some degree of "internal" plastification.

3.1.3.7. *Coupling agents* are molecular bridges at the interface of an inorganic filler and an organic polymer matrix. They contain hydrolyzable groups binding the inorganic filler and organofunctional groups in the same molecule. Semi-organic silanes and titanates have long been used for glass-fiber reinforced plastics (GFRP, Sec. 3.1.3.8.4). With specific organofunctional groups they are also very important for all kinds of polymer composites. Some of them can be added in normal additive blending instead of pretreating the fillers (e.g., "coating" with stearates) in a separate operation.

Interface layer binding is needed for multi-layer films, sheets and blow moldings combining non-compatible polymers. It can be produced by reactive gases blown between, but mostly co-extruded adhesive bonding layers are used (Sec. 4.1.2.4.3/4 and Sec. 4.1.3.1.2).

3.1.3.8. *Fillers, enhancers and reinforcements*

From the classic Bakelite thermosetting PF molding compounds to the newest thermoplastic range of composites, fillers have been used not only as "extenders," i.e., to reduce the resin content of the plastic and so reduce costs, but also to improve production rates (by increased thermal conductivity) and properties of the finished products. These include modulus, impact strength, dimensional stability, heat resistance and electric properties. Such components of plastics compounds can be differentiated approximately by the aspect ratio (a. r.) of length (or length and width) to thickness as follows:

Fillers: irregularly shaped granules or spheres, a. r. ≥ 1
Typical property enhancers: short fibers, e.g., wood flour, milled or chopped glass fibers, wollastonite, whiskers, talc or mica flakes, a. r. 10 to $\gg 100$.

3.1.3.8. Fillers, enhancers and reinforcements

Reinforcements: filaments, non-woven or woven textile products, a. r. very large.

This list does not cover the special surface-reinforcements used principally in rubber and elastomers, such as carbon blacks, pyrogenic highly dispersed silica (Cab-O-Sil, Aerosil) or precipitated, ultrafine, carboxylated rubber-coated $CaCO_3$ (Fortimax).

1. Natural organic materials: Wood flour has traditionally been used for thermosetting compounds (Sec. 4.6.2.1) and recently for thermoformable PP sheets (Sec. 4.1.13.1) and wood-like PVC moldings (Sweden). For cellulose fibers, chips and fabrics see thermosetting molding compounds (Sec. 4.6.2.2). Such materials as sisal, jute, etc. are attracting interest, especially in developing countries. Masterbatches containing hydrophobed starch are marketed as performance additives, in particular for biodegradable thermoplastics (Sec. 4.1.15).

2. Mineral fillers and property enhancers: Ground chalk, limestone or marble ($\emptyset \leq 3\,\mu m$, specific surface $6-7\,m^2/g$) and precipitated calcium carbonate ($\emptyset \leq 0.7\,\mu m$, specific surface $\geq 30\,m^2/g$) are generally used as a cost effective means of improving temperature resistance, impact strength, surface quality and stiffness. In PVC, $CaCO_3$ improves fire performance by binding HCl. "Snow White" is anhydrous calcium sulfate approved for contact with food. Barium sulfate is a heavy filler for sound and radiation protection. Silica flour is highly corrosion resistant, but is rather abrasive (mold wear). Calcined kaolin is used in high-voltage technology. Feldspar is of significance in transparent products on account of its refractive index. For alumina and other flame retardant fillers see Sec. 3.1.3.4.

Talc, naturally found as platelets (widely used with PP, Sec. 4.1.2.6.3), delaminated mica flakes (up to $1\,\mu m$ in thickness) or mixtures of such minerals with short glass or polyester fiber, is used for well balanced improvements of modular, flexural and impact strength. Asbestos is a first-rate heat-resistance improver, but, because it is a health hazard, it is likely to be replaced by mineral-fiber mixtures, such as wollastonite or $CaSO_4$-crystal microfiber ($\emptyset = 4-6\,\mu m$, a. r. >100; Franklin fiber, US). Foamed clay (Norpil, Tecpril, $\lambda = 0.045-0.1$ W/mK) is used for high temperature thermal insulating products.

Small amounts of finely dispersed $CaCO_3$, silica or special silicates (Sipernat) reduce sticking and improve the paper-like feel of PE films (Sec. 4.2.3.3).

3. Spheres: As spheres are fully isotropic, compounds incorporating large amounts of them flow easily in processing. Sphere-filled products are less

3.1. Polymers and Compounds

susceptible to shrinkage and distortion. Solid glass microspheres ("ballotini", $\emptyset < 50\,\mu m$) improve modulus, compressive strength, hardness and surface smoothness, "Cenospheres" are mixed solid or hollow spheres manufactured from the fly ash of coal-burning power plants. Hollow borosilicate or silica spheres, called microballoons ($\emptyset = 5$ to $250\,\mu m$; $d_R = 0.28$–$0.50\,g/cm^3$) (Q-cel, US; Fillite, JP; Microcel, BE) and used for boat hulls, aircraft, and automobiles are described under "Syntactic Foams," Section 3.1.6.

4. Fibrous reinforcements: Fibers, fabric cuttings and fabric web are used to reinforce thermosetting molding powders and laminates Sec. 3.2.4 and 4.6.3.5). Milled or chopped short fibers $<1\,mm$ to $0.1\,mm$ long are incorporated in structural foams (Sec. 3.1.6 and in granulated thermoplastic compounds in disordered but homogeneous distribution. Long-fiber types are prepared by the incorporation of flowing chopped glass strands up to 6 mm in length or by coating of rovings and then cutting into granules. Reaction resins (Sec. 4.6.1.4) incorporating random or oriented fibrous reinforcement are processed into high strength parts.

Textile glass reinforcements are predominantly made of continuous glass filament, which is drawn from molten glass (Type C), 5 to 25 μ in diameter, mostly from low-alkali glass. Glass strand combining 100 to 200 monofilaments, and filament products such as mats varying from 250 to 900 g/cm^2 (68.9–248 oz/in^2) (Table 3.1), have been further developed. Blown-drawn short staple fibers are used for resin-bonded light mats (30–150 g/cm^2) or fleece, for smooth, resin rich surfaces. For this, the more acid-resistant C-glass is preferred. Textile size, which is necessary for the pretreatment of the glass filament (Sec. 3.1.3.7) becomes effective as coupling finish for reaction-resins when used in combination with coupling agents. Noncoupling textile sizes must be burned off before further processing (e.g., finish 112 for woven glass fiber: $<0.1\%$ residue). Mats or fleeces are strengthened by means of synthetic resin binders based on PVA, EP, for GMT also on PUR.

High modulus and high-strength fiber reinforcements made of R- or S-glass (trade names) are more expensive than those of E-glass fibers. Polyester, acrylic or PVC synthetic fibers are used mainly in cover fleece. Synthetic fiber reinforced plastics (SRP) have significantly lower E-moduli than GRP in the direction of the fiber, but there is a higher resistance to damage by distortion. For mechanically and thermally highly stressed parts for the aerospace sector, in transport vehicles and sports equipment, fibers made from an aromatic polyamide (Aramid, Sec. 4.8.8) with a high proportion of 1.4 chemical bonds directly between aromatic rings (Kevlar, Twaron) are used; fibers made by carbonizing pitch or polyacronitrile fibers as 'precursor' are also employed. The commonly used carbon fibers, produced at about $1000°C$, have a low modulus and high strength. Those made by progressive graphitization at

3.1.3.8. Fillers, enhancers and reinforcements

Table 3.1. Filament Textile Glass Products according to ISO 472 (DIN 61850)

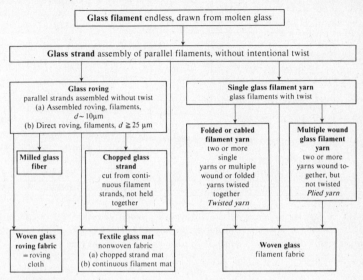

Special non-woven fabrics:
Scrim: an open mesh, several layers of parallel yarns are stuck together at an angle by adhesives.
NUF mat: Continuous non-woven unidirectional Fiberglass mat, cross laminated every 3 in;
Bigelow mat: Chopped strand mat without binder entangled by needling.

Special weaves of woven fabrics:
Unidirectional weaves strong warp threads and weaker filling threads or vice versa – gives maximum strength in one direction.
High-modulus weave: Strong warp and filling threads fixed by weaker warps and wefts without undulation – gives greater stiffness.

2000°C are the high-modular graphite fibers (Table 3.2). There are IM (Intermediate Modulus)-types exhibiting σ_B-values 4000 to 5000 MPa, E 300 GPa (Celion G40). All these fibers have lower densities and a specific strength and/or elastic modulus many times higher than that of glass fibers. They are also much more expensive. They improve the properties of the product even when present as a limited proportion of the reinforcing fibers.
Aramid-carbon, Aramid-glass, Aramid-carbon-glass and carbon-glass hybrid fiber reinforcements are all commercial products. Extremely high mechanical properties and temperature resistance (> 1000°C) are manifested by inorganic fiber-forming single crystals (whiskers) or crystalline long fibers from aluminum (Fiber FP, US) or calcium titanate silicon carbide. "Cobweb-whiskers" (Xevex, NO) containing micro-monofilaments result in better elongation values than single crystal whiskers. The very costly boron fibers on a tungsten-core are used for e.g. spacecraft. For high strength PE-fibers see Sections 1.3.2 and 4.1.2.2.

3.1. Polymers and Compounds

Table 3.2. Comparative Values for Reinforcing Fibers, in SI Units

Group	Fibers[1]	Density g/cm^3	Elongation at break (%)	Tensile strength (MPa)	Tensile modulus[2] (GPa)
GF	E-glass	2.54	~2.5	2000–3500	~70
	S-glass	2.49	~2.8	4570	86
AF	Aramid, normal types	1.44	3.3–3.7	3000—3150	63–67
	HM types	1.45	2.0–3.0	2700–ca. 4000	120–130
CF	Carbon,				
	HT/ST types (PAN)	1.74	1.4–1.8	3000–>4000	230–240
	IM types (PAN)	1.81	1.4–2.3	4000–(7000)	290–300
	HM types (PAN)	1.91	0.3–0.6	1900–2500	500–570
	HM types (MPP)	2.18	0.27	2200	827
SC	Silicon carbide, cryst. (JP)[3]	~3.2	<0.5	9000	640

[1] Types: HT/ST high tensile strength standard type (for aircraft industry); IM intermediate modulus; HM high modulus; in brackets precursor, see Sec. 4.88. [2] $E_\parallel$. For GF, $E_\perp$ is about the same, for the other highly oriented fibers much less (5·15 GPa). [3] not commercial.

3.1.3.9. Blowing agents for expandable plastics are considered in Sec. 3.1.5.2/3.

3.1.4. Plastics Compounding

3.1.4.1. Thermoplastic compounds

For molding and extrusion, most thermoplastic resins require admixture of auxiliaries and additives (Sec. 3.1.3) and processing to *compounds* in the form of pellets or granules.

Continuously operating, extensively automated compounding lines are used to homogenize and plasticize powdered and liquid components in heated roller mills, kneaders, or, most commonly, high output extruders (Sec. 3.6.1); at the same time reinforcements are incorporated and the materials degassed (Fig. 3.4). The hot compound extruded through a perforated die plate is granulated either by the action of a dieface cutter followed by cooling the lens-shaped granules in water or by cooling first and then cutting into thick cylindrical pellets about 2 to 5 mm in length. Newer technology aims at direct extruder lines that simultaneously compound the materials and produce semi-finished products. Processing scrap can be prepared for reprocessing by slitter-type granulator mills.

Powder coating and molding compounds (Sec. 3.2.1 and 3.2.3) are reduced with impact grinders. Special compounding lines are needed for *dry blends*, i.e.,

3.1.5.1. Some generalities on cellular plastics

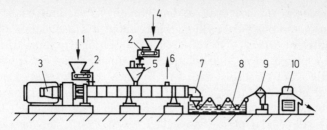

Fig. 3.4. Installation for incorporation of chopped reinforcing fibers into the plastic melt.
(1) Thermoplastic and additive, (2) Metering balance, (3) Twin screw kneader, (4) Reinforcing material, (5) Feeding hopper, (6) Degassing by vacuum, (7) Extrusion die, (8) Water bath, (9) Blowing air drier, (10) Strand pelletizer.

coarse-grained, free-flowing powder compounds. Large-scale processors and converters of rigid and plasticized PVC prepare such blends in the plant. The polymer powder and additives are metered and are batch mixed in a programmed sequence, using the frictional heat generated by high-speed agitators to slightly pre-gel the compounds. Open rolling mills contra-rotating at different speeds, which were formerly used for mixing and plastifying, are today used only as intermediate holding units in continuous processes.

3.1.4.2. Thermosetting compounds

To prepare standardized heat-curing molding compounds in specific installations in the resin plant, fluid or molten, low-molecular-weight resins are mixed with fillers or enhancers and condensed to a predetermined processing stage. Finally grinding and fractionated screening produces dust-free granules of a specific particle size. Compounds containing fibrous reinforcement are made by soaking the fibers in resin solution, drying and, in some cases, chopping. Webs for laminates (Sec. 4.6.1.4) are resinified on spreading machines (for UP-resin mats and reaction-resin prepregs see Sec. 4.6.2.4.3).

3.1.5. Expanded Plastics

3.1.5.1. Some generalities on cellular plastics

Cellular plastics are, according to ASTM D 883-83a and ISO, "plastics containing numerous small cavities (called cells) interconnecting or not, distributed throughout the mass." They can be made in various ways. One way is the generation of bubbles ($\varnothing < 0.5$ to > 3 mm) within a polymer melt or in potentially cross-linking resins. Then as a result of the high viscosity of the resin the bubbles do not coalesce but remain viable until cooling or the completion of the polymer reaction brings about a consolidation of the cell walls. A different process involves mechanical beating (frothing) with air – as in making whipped cream – of solutions (UF foam, Sec. 4.6.1.2) to be cured subsequently or of plastisols (Sec. 4.1.4.9). In a *two-step process* mechanical or chemical cream-like frothing of fluid reaction-resins prepolymers is followed by blowing the finished foamed product in a second stage (PUR, Sec. 4.6.1.8.2.6). All cellular plastics are called *foamed plastics* or *foams*. Foamed plastics made under pressure contain mainly *closed cells*, that is, the cells are not interconnecting. Many foams made at atmospheric pressure are *open celled*. Their cells are not totally enclosed and hence interconnect. Test methods for open cell content have been given by ASTM D-2856 – DIN ISO 4590.

3.1.6. Foam-plastic manufacturing

During processing to molded material, plastics pass through a state in which they are moldable. In this state, finely distributed gas or vapour bubbles of about 0.2 to 3 mm diameter can be introduced either by direct aeration (3.1.6.1) or using blowing agents (3.1.6.2) and nucleation (3.1.6.3). This produces foams, cellular materials with densities a fraction of that of their framework materials. They are extremely versatile as their properties differ according to their structure.

Foams are produced with uniform cell distribution over the complete cross section in densities ranging from $300 \, kg/m^3$ $(0.3 \, g/cm^3)$ to under $10 \, kg/m^3$ $(0.01 \, g/cm^3)$ which is less than 1% of the density of the unfoamed material. In closed-cell foams, cell walls completely enclose all the cells and gas and fluid interchange is only possible by diffusion. The cell walls of 'mixed-cell' foams are partly perforated. They, and even more the cell walls of 'open-celled' foams, are open to gas and liquid exchange between the cells and (in as far as no foam skin is present) the surroundings. In the extreme case of reticulated foams (MF p. 341, PUR p. 368) only the cross-pieces of the cell walls remain. Structural or integral foams are manufactured with non-homogeneous cell distribution over the cross section so that there is a steady (often parabolic) change from an almost solid outer skin to a minimum density in the middle of the material (Figure 3.5). This distribution of material and a tensile outer skin leads to optimum strength.

Syntactic foams are polymeric materials whose gas-filled, closed cells are produced by hollow spheres as fillers in the polymer matrix. This produces highly compression resistant foams with densities of up to $800 \, kg/m^3$.

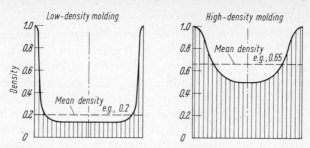

Fig. 3.5. Density profile of structural foam moldings.

3.1.6.1. High and low pressure blowing

1. Permanent gases (usually nitrogen) under high pressure (200 to 600 bar) are incorporated in an extruder with accumulator using the Airex slabstock method for PVC melts at about 170°C and the USS method for thermoplastics (PE, PP). In the Airex two-step process the cooled slab, under pressure in the compression mold is foamed in the open at 50–90°C. In the UCC method the material expands by flowing out into the mold previously placed in the accumulator.

2. In the mechanical foam blowing method, air, nitrogen or carbon dioxide under a small pressure is whirled with low-viscosity plasticized PVC pastes (p. 000) or urea intermediate/hardener solutions and surfactants in special mixing equipment to give wet foam which is conveyed to the application by the pressurised gas. UF foam cures chemically, e.g. as insulating material in pipe ducts, while plasticized PVC wet foam gels cure on heating as a backing layer for artificial leather or floor coverings.

3.1.6.2. Physical and chemical blowing agents

3. Physical and chemical blowing agents are either low-boiling organic solvents (preferably hydrocarbons) whose vapor pressure under the processing conditions causes foaming or finely distributed solids which decompose yielding a gas. By-products of polymer formation reactions can also serve as blowing agents, for example carbon dioxide from reaction of isocyanate with water (PU foam) or methanol or ethanol vapor from reactions forming imides (PMI and PI). Suitable combinations of different blowing agents are freely used. Chemical blowing agents are required to decompose (usually exothermically) to form inert gases and odorless, tasteless, non-toxic residues at temperatures around the processing temperature. Examples are azo-compounds, N-nitroso-compounds and sulfonyl-hydrazides which at tempera-

tures between about 90 and 275°C yield 100 to 300 cm^3 of nitrogen per gram of compound. The much-used azodicarbonamide has a decomposition temperature of 230–235°C which can be lowered to 155–200°C by use of 'kickers' such as metal compounds, e.g. the Pb and Zn stabilisers found in PVC mixtures. For PE-X foam there are similar gas-forming cross-linkers. Because it has been shown that the destruction of the ozone layer in the stratosphere is caused mainly by chlorofluorohydrocarbons (CFCs), these are being increasingly replaced as physical blowing agents in the manufacture of plastic foams (especially PU foams) by CFC-free blowing agents. The development of further chlorine-free, hydrogen-containing fluoroalkanes and fluoroethers, e.g. as insulating gases in PU rigid foams, is not ruled out at the present time.

3.1.6.3. Nucleating agents and pore size regulators

Nucleating agents and pore-size regulators are important additives, especially for physically blown thermoplastic foams. The polymer melt is supersaturated with blowing gas and the nucleating agents act as 'boiling stones' causing a uniform foam structure. They are fine particled solids or mixtures which release CO_2, e.g. sodium bicarbonate with solid organic acids such as citric acid. Gases or easily vaporised liquids formed as by-products on foaming (such as CO_2 in PU and ethanol in PI) can also act as nucleators.

3.1.6.4. Manufacturing information

1. Special techniques for closed thermoplastic foams are
   – the particle-vapor impact method for polyolefines and polystyrene with physical blowing agents incorporated during polymerization. This is applicable to the manufacture of sheets, webs and moldings;
   – the high-pressure mechanical blowing method for rigid and plasticized PVC slabstock foam.

   High temperature-resistant thermoplastics (PES and PEI) are slab-foamed by reducing the pressure on hot gels swollen with solvent.

   The mechanically blown foam method for plasticized PVC and UF (see above) leads to mixed and open celled foams.
2. Alongside the short-run, non-continuous, block foaming in separate molds with floating cover, continuous, (double-)web foaming and slabstock foaming (Fig. 3.6, p. 59) are the preferred methods for liquid thermoplastic materials containing curing agents and chemical or physical blowing agents (PF and MF, Sec. 4.6.1.1; rigid and elastic soft PU foams, Sec. 4.6.1.8). Whether closed or open cells are formed depends on the method. Sandwich sheets are also made by this method by continuously adding

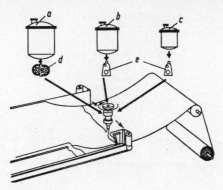

Fig. 3.6. Continuous foaming plant
(a) component I, (b) component II, (c) activator, (d) gear pump, (e) piston pump for less viscous liquids.

outer layers on both sides. Elastic soft foamed moldings are mass produced from PU pre-products by low-pressure casting.

3. Refrigerators are foamed in-situ using stationary plant; UF and PUR foams are used in-situ in building. Mobile plant is used to produce acoustic and thermal insulation as well as joint seals.
4. The extrusion method is used with physical blowing agents to form XPS light foam webs and with chemical blowing agents to form structurally foamed thermoplastic profiles, tubes and sheets (PS).
5. Thermoplastic structural foam moldings are produced by the thermocasting method and by blow-molding.
6. Rigid and semi-rigid PU integral foam moldings up to very large sizes are produced in two-component reactive resin casting (RIM and increasingly RRIM methods). The pivoted mold carriers for large parts are moved and closed hydraulically, causing the filled mold to be brought into the filling and foaming position (Fig. 3.7). Smaller PU structural foam moldings (e.g. for the shoe industry) are sprayed automatically from a central feed head into moving molds with ejectors; cycle times are about 2 min.
7. Thermoplastic products from material with 0.25 to 0.5% w/w of chemical blowing agent contain gas or filler. Uses of such materials include cable insulation and PE and PP decorative tape and cord which, because of the light cellular structure, gives a satin sheen on stretching. 0.01 to 0.03% w/w of blowing agent is added to injection molding materials for thick-walled parts in order to avoid collapse and voids caused by shrinkage on cooling.

3.2. No-Pressure and Low-Pressure Molding Processes

Fig. 3.7. Large mold carrier for reactive resin integral foam molding (PU-RIM) SFT range (Krauss-Maffei): clamping force 1000 to 10 000 kN; platen area 1500× 1000 mm to 2750 × 1700 mm.

3.2–3.7. Plastics Processing

3.2. No-Pressure and Low-Pressure Molding Processes

3.2.1. Dipped and Powder-Coated Articles

1. *Dipped hollow articles* (e.g., protective boots and gloves, dispatch covers) are made from plastisols (Sec. 4.1.4.9), dispersions (Sec. 4.1.1.10) or solutions of such viscosity that a uniformly gelled layer attaches to dip formers made of ceramic or metal for subsequent gelling or drying by hot air. Sensitive materials are dipped to encapsulate them in pull-off protective covers.

2. In *powder-coating technology* flame spraying to protect metallic objects is only of limited significance. Fluidized plastic powder is sprayed and melted onto the surface in a reducing blow flame.

Plastic powder with a particle size of 50 to 300 μm is used in *fluidized bed coating* or *sintering*. It is placed in the sintering equipment and air or nitrogen is blown in through a porous base plate (pore size < 25 μm) to fluidize the bed. On dipping preheated metal objects a layer of plastic is melted on, taking from 2 to 5 seconds. This is rendered pore-free where necessary by passage through a second oven. Profiles and tubes are coated by pneumatically conveyed powders (the Traflux skin process).

PE, E/VA, PVC (dry blend and extruder mix), PA11, PA12, CAB, CP, EP and UP are available as sinter powders. Objects to be coated must be metallically pure and, for several plastics, be treated with primer. Film thicknesses of 250 to 500 μm are customary; 75 μm is the minimum.

For *electrostatic coating*, powder with a range of particle sizes between 40 and 100 μm is used. It issues from a pressure spray gun into a (safe) high-voltage field of 50 to 90 kV so arranged that the particles adhere to one side of earthed metal parts in the area of spray for a long time. Waste powder (up to 70%) is recovered. Thermoset plastics are melted and cured in hot-air ovens at about 200°C (EP, UP and acrylic resin combinations). Thermoplastics (PE, plasticized PVC, PA11 and CAB) are melted.

The metallic surfaces to be coated must be smooth for the thin (40–150 μm) layers applied by electrostatic coating; they must be sand-blast roughened for the thicker layers produced by fluidized sintering.

In large installations rotating steel tubes up to 2 m in diameter are coated in a hot-melt process with PE to between 2 and 4 mm thick. Pipelines preheated to 250°C are coated electrostatically with PE to a thickness of 250 μm. Textiles are stiffened and laminated by flat ironing in PE or PA powder.

3.2.2. Casting and Spraying of Preproducts

1. *Open casting* is a process whereby polymerizable monomers (Sec. 4.1.1.1), gelling PVC pastes (Sec. 4.1.4.9) and hand-mixed reaction resins (Sec. 4.6.1.4) are used in the reproduction of organs and the embedding of preparations of interest in medical and scientific work, in model making and in arts and crafts where only small numbers of products are required. For the reproduction of objects with undercuts, elastomeric molds cast from the original are used.

2. *In batchwise processing*, e.g., reaction-resin joint filling compounds or mortars (Sec. 4.6.1.4.1), metered multicomponent resins are mixed, by weight or volume, with additives by a hand-drill or in a concrete forced mixer. For this

3.2. No-Pressure and Low-Pressure Molding Processes

technique measured packages and cartridges containing predetermined quantities of components are available.

3. *Reaction-resin casting equipment for electrical components* comprise of heatable mixing vessels and casting molds with automatic clamping mechanism. For air-free parts the equipment is placed under vacuum. The continuous drip impregnation of motor windings and bobbins is a variation of the process.

4. *Multicomponent casting and spraying machines* with mixing heads deliver reaction-resin components, and additives if required, for exact mixing and metering to the mixing head. They are designed to eject the mixture into open or locked molds or to coat surfaces by casting or spraying continuously or to fill hollow spaces.

   For coating and filling on building sites movable machines are fitted with hoses, to convey the components under pressure, and hand-mixing heads. In polymer concrete (PC) machines with output of 30–300 kg/min, the metering for the reaction-resin components is connected to the feed and mixing screws for the aggregates.

   Machines for intermittent running are operated on a cyclical basis. This is possible with high-pressure and low-pressure equipment. High-pressure equipment (Fig. 3.8) with a greater range of variables is more generally applicable (Table 3.3).

   Casting equipment for molded articles with feeder flows of 0.5 to 400 l/min and shot weights of 0.5 g/s to 100 kg/10 s is widely automated for the handling of solid or foamed PU and other reaction-resin castings. In machines with a stationary mixing head, many molds circulate over filling, curing, withdrawing and preparation stations.

   In *RIM reaction injection molding* (or **RRIM** if reinforced) for the in-situ polymerization of catalysed resin mixes to large articles (PU Sec. 4.6.1.8.2, EP Sec. 4.6.1.6, acrylic systems Sec. 4.1.1.1, cast nylon Sec. 4.1.8.2), the mold is intermittingly filled through the mixing head (Fig. 3.9). RIM has achieved a similar degree of productivity, precision and automatic operation to injection molding (Sec. 3.3.2.6). The largest machines can accommodate molds up to $3 \times 1.25\,\mathrm{m}^2$ ($10 \times 4\,\mathrm{ft}^2$) weighing up to 15 tons. Opening strokes of 2 m and clamping forces up to 2500 kN (250 tons) are common. Four column design of the mold carriers is preferred to the C-form design shown in Fig. 3.7. Mold curing times vary from 1 to 5 min, and devices are available to reduce material and mold changing times.

   In the continuous foaming of rigid or flexible PU foamed blocks (Sec. 4.6.1.8.2.6) or of PF rigid foam, mixing heads traverse to and fro over a longitudinally moving track. Sandwich moldings are made continuously by

3.2.3. Rotational Molding and Centrifugal Casting

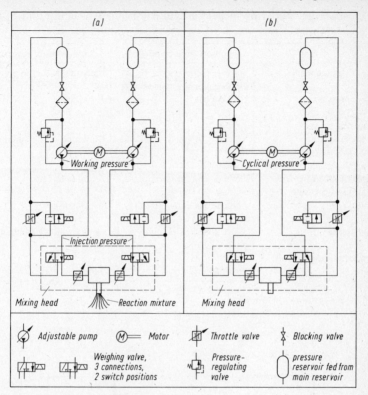

Fig. 3.8. High-pressure reaction-casting machine.
(a) Working position, (b) cyclical operation.

machines with feeder units for outer layers and where necessary for lateral and cross limitation devices.

5. *Multicomponent spraying equipment* without mixing heads works on the principle of the spray gun, with air pressures of 1 to 3 bar, or as a high-pressure plant with piston pumps. The components are mixed partly on impact and partly by blending on the sprayed surface. They are used mainly in GRP processing in applying the gel coat and in combination with glass-fiber cutters in the resin-fiber spray process (Sec. 3.2.4).

3.2.3. Rotational Molding and Centrifugal Casting

1. Hollow articles are *slush molded* by filling open, hollow molds with paste or powder and, after a sufficient layer of materials has gelled or melted at the hot mold wall, pouring out the remainder. In *rotational casting* measured

3.2. No-Pressure and Low-Pressure Molding Processes

Table 3.3. High- and Low-Pressure Multicomponent Casting and Spraying Plants

	High-pressure equipment	Low-pressure equipment
Supply of components	Special (–series-, axle-, dipping–) piston pumps	Precision gear type pumps
Working pressure	100–350 bar	3–40 bar
Viscosity range	3–2500 m Pa s[1]	> 50 m Pa s[2]
Max. mixing ratio	100:30	100:1
Filler processing	Special equipment	Continuation possible
Pressure loss on prolonged working	Slight	Large
Principle of mixing	Counter-current injection	Agitator mixing
Regulation of pre- and post-operation	Forced regulation unnecessary	By pressure
Working cycle	Short using self-cleaning mixing head	Long – with cleaning at intervals
Automatic scrap runoff at molding	Possible	Technically feasible

[1] With feed pumps up to 5000 m Pa s.
[2] Normally up to 5000 m Pa s, possible up to 60,000 m Pa s.

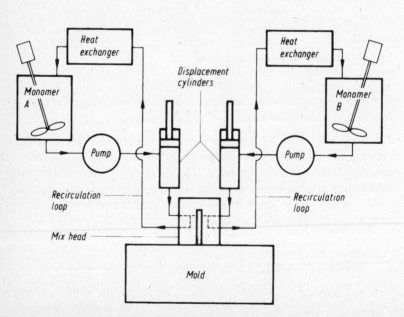

Fig. 3.9. Basic components of a RIM processing system.

quantities of material are rotated against the inner walls of molds alternately heated by hot air and cooled by spray cooling. As a result of moving the hot molds on two axes, the plastic melts or polymerizes and gels in a uniform layer. The rotation speeds are low (in contrast with centrifugal casting), and so no centrifugal forces arise. The mostly carousel-type automated units are subdivided into feeding, heating, cooling and removal stations.

Mold carriers, which can accommodate up to 50 molds, are symmetrically arranged about the main axis of rotation. They rotate around this and the second axis at right angles to it, according to the shape of the molding, at different speeds from 10 to 40 rpm.

Tools for large receptacles are made of steel plate 1.5 to 2 mm thick. Simple multiple tools are made from aluminium castings. Complex molds are made by electroforming from wax or reaction resin patterns. Mold parting lines can be arranged in such a way that undercuts are no trouble. To equalize pressure the molds are provided with venting tubes, which can also serve to introduce inert gas in the case of oxidation-sensitive melts or of compressed gas as a supplementary molding medium. Balls, dolls, toy animals, garden figures and items such as traffic cones or arm rests in vehicles are made from plasticized PVC pastes or dry blends and EVA. Large toys and vessels with capacities up to 80,000 l are formed by rotation molding from PE powder; cross-linking PE (Sec. 4.1.2.2) is used for packaging of heavyweight goods, cable drums, multi-layered boats and surfboards (80 kg). The process is suitable for irregularly shaped receptacles, such as motor vehicle gasoline and cooling-water tanks. PA 6 articles are made of dry powder or by anionic polymerization of a lactam melt (Sec. 4.1.1.1.4).

2. *Centrifugal casting*, with molds rotating quickly on one axis, gives rise to thick-walled, bubble-free and nearly stress-free rotational articles from plastic preproducts. Long-tube centrifugal machines are used to manufacture pipes and poles from GF-UP and GF-EP; the pipes have quite smooth external and internal surfaces and can also contain fine-grained fillers (Sec. 4.6.3.7). Disc casting machines produce compact gearwheels, rings and the like by the polymerization of capro- or lauryl- lactam (Sec. 4.1.1.1.4). Cylindrical centrifuging drums can produce sheets of plasticized PVC, PMMA, PU, UP and SI rubber, which are removed before the reaction is complete and cut up. The reaction is brought to completion by laying the sheets flat in a heating cabinet.

3.2.4. Low-Pressure Processes for GRP Moldings

Low-pressure processes for GRP moldings from fluid resins and reinforcing materials are set out in Table 3.4. For cold-curing resins, i.e., those processed at

Table 3.4. Low-Pressure Processes for Reinforced Moldings

Process	Hand lay-up	Fiber spraying	Vacuum and injection	Wet compression molding		Winding	Centrifuging
				Cold	Warm		
Materials: reaction resins (A) Acrylate resins (B) Butadiene resins	UP, EP A, B, SI	UP	UP, EP	UP, EP	UP, EP B, SI	UP, EP A, B	UP, EP
Glass-fiber reinforcement (weight percent) — Mats	20–30	–	30–40	25–40	30–50	Ribbons 50–70	25–35
Fabrics	35–50	–	40–65	50–65	50–65	50–80	30–40
Rovings	locally > 50	20–30	–	–	–	–	25–35
Preforms	–	–	–	30–50	30–50	–	–
Gel coat possible	Yes	Yes	Yes	Conditionally	Conditionally	Yes	Yes
Molds: W = wood; G = gypsum; P = plastic; M = metal; S = steel	W, G P, M	W P, M	M, P (O)	P, M	S	S P liner Core	M, (P)
O = open, one or more parts C = closed, two parts	O	O	C	C	C		Various
Processing: Temperature	Room	Room	< 60°	< 60°	~ 150°C	Aftercuring at higher temperatures	Centrifugal
Working pressure (bar)	By hand	By hand	ca. 1	< 10	< 50	Tension	min to h
Molding times	30 min	Up to a day	1–10 h	5–30 min	2–10 min	–	Little
Finishing needed	Yes	Yes	No	Yes	Little	Yes	
Molding:							
Smallest radius (mm)	5	5	5–10	3	3	10	–
Draft, degrees (approx.)	2	2	2	1	1	–	–
Undercuts possible	Yes	With parted molds	No	No	No	Yes	No
Variation in wall thickness possible	Yes	Yes	Limited	Limited	No	Yes	Limited
Wall thickness (mm)	2–10	2–10	0.5–10	1.5–10	1–10	1–10	3–10
Thickness tolerances (%)	20–50	Up to 80	20	20	10	< 20	< 20
Smooth walls	One side	One side	All around	Both sides	Both sides	One side	Varying
Investment	Light	Moderate	Medium	Medium	High	High	High
Working costs	High	High	Medium	Medium	Medium	Light	Light
Application range	Individual parts, small runs, large articles	Small runs, linings	Small runs of larger moldings	500–5000 articles	> 10 000 articles	Pipes, containers	Pipes, containers

3.2.4. Low-Pressure Processes for GRP Moldings

room temperature, light molds, often made from GRP itself, are adequate. Complete cold curing of moldings requires several hours of subsequent heating in a warm-air oven (tempering) in conditions appropriate to the type of resin. To achieve a smooth surface that obscures the glass fibers of underlying layers and protects against external influences, a 0.3 to 0.6 mm (0.012–0.025 m) fiber-free *gel coat* of resin is painted or sprayed onto the mold, which has been treated with separating pastes and a parting lacquer. This layer must be allowed to gel for 5 to 20 min before proceeding further. A variation on this process is the embedding of fiber fleece in a resin-rich surface layer.

A build-up, characterized as a chemically resistant layer, consists of a 10% C-glass overlay, and two or three glass fiber mats (900 g/m^2, 12.8 lb/in^2) with 25% highly resistant UP resin, laid on before further laminate build-up. This applies also to the winding process for GF-UP tanks and pipes.

In the *hand-lay-up* process mat or cloth laminate is assembled one after the other and impregnated without air bubbles by portion-wise addition of cold-curing resin. The resin is incorporated by means of lambswool rollers or brushes. The hand process is essential for making large moldings, with the incorporation of stiffeners (Dekotubes), and for jacketing for plant.

In the *fiber-spray process* the resin mixture and finely cut glass fiber are introduced simultaneously with multicomponent spraying equipment (Fig. 3.10), equipped with a cutter for rovings. For low-pressure spraying two feeds are used, one with resin-containing hardening agent and the other with an

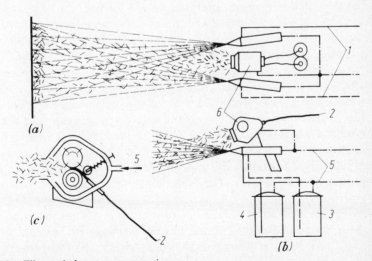

Fig. 3.10. Fiber-resin low-pressure spraying.
(a) Plan, (b) Side view of equipment, (c) Section through cutting head (6).
(1) Resin channel, (2) Glass rovings, (3) Resin with hardener, (4) Resin with accelerator, (5) Compressed air, (6) Cutting head.

accelerator. The sprayed fiber-resin layers must be consolidated manually with a roller. With high-pressure spraying, hardener can be introduced with the catalyzed resin in the spraying head through a linked metering pump. Units with movable spray heads are used for small series production of moldings, for reverse-side reinforcement of large moldings made of other plastics and also for GRP lining of building structures. The sprayed resin-fiber coat must be densified by hand with sheepskin rollers, which requires experienced operatives.

Steps towards pollution-free mechanization of the fiber spraying process are the manufacture and coating of round containers using rotating spray units or workpieces. Other steps include compacting with downstream ribbed rollers and robot-operated, computer-controlled closed systems for series production of moldings such as surfboards, engine covers and shower bases. The molding is densified by laying on a film under pressure and exhausting the air, and after curing and removal, deflashed with computer-controlled hard metal or polycrystallite diamond cutters.

In *resin transfer molding processes* reinforcing mats or fabrics are impregnated in a pre-closed mold. In *vacuum injection molding processes*, after laying-up the reinforcement, the yielding and elastic thin-walled top of the GRP tool is sealed shut. The mold is filled under the pressure of a 1 m column of resin and then evacuated until resin overflows. The application of the top of the mold densifies the product. Such processes are suitable for large parts and for sandwiches with inserted cores. Special highly reactive, low-viscosity resins ($\eta = 0.2\,\text{Pa}\,\text{s}$ instead of the $2\,\text{Pa}\,\text{s}$ of normal resins) are needed for this.

In the vacuum bag *autoclave process* for heavy duty *composite components*, e.g. for aerospace applications with $\geq 60\%$ fiber content, laminates laid up on the mold with resin impregnation, or EP resin fabric prepregs (Sec. 4.6.2.5.2), possibly with honeycomb sandwich cores, are first evacuated under a flexible cover with extraction of the excess resin. The evacuated bundle is cured at about 200°C in the autoclave at ≤ 7 bar. The molding is subsequently tempered in the open.

In the *wet compression molding process* the reinforcing material with the resin mixture poured over it is placed on one half of the mold. This is done so that as the press closes with a slowing during the last 5 to 10 mm of travel the resin is uniformly distributed through the molding. Preforms made by sucking fibers onto a screen may be suitable. They can be bound provisionally by aqueous resin dispersions.

Cold compression molding is suitable for runs of up to several thousand using molds made of filled UP and EP resins. For hot pressing with hot-curing resins (UP < 150°C, EP 105–200°C) steel molds are necessary, and these are economical only for long runs. The tools must have pinching or cutting edges. Ejecting and multipart molds are not feasible since fluid resin penetrates into the joints; ejection is by compressed air. Low-pressure compression presses are

3.2.4. Low-Pressure Processes for GRP Moldings

constructed with platen surfaces of up to $30\,m^2$ (Sec. 3.3.4.2). For reaction-resin high-pressure *molding compounds* and preresinified mats (*prepregs*) see Section 4.6.2.4.2.

In *filament winding*, rovings, tapes or threads bath-preimpregnated with resin are wound onto a core under tension, if needed, with subsequent curing in an oven or a heated autoclave. Fig. 3.11 illustrates the design principles for machines that wind at any angle. Large tubes are continuously wound onto a running spiral, an endless steel band acting as the core. Molded convex ends

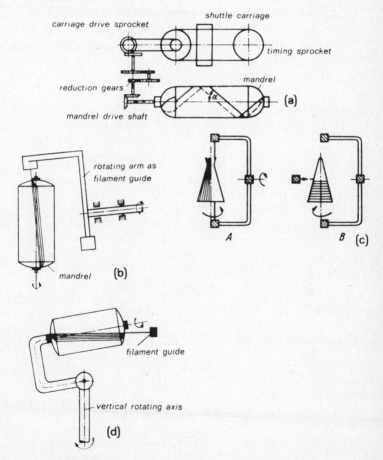

Fig. 3.11. Winding machines.
(a) The type or traversing carriage machine for helical angle α 15° to approximately 90°.
(b) Planetary machine with rotating arm for polar winding.
(c) Strickland winding of cones: (A) Longitudinal, followed by (B) circumferential.
(d) Tumbling machine: the mandrel is tumbled end over end in a horizontal or vertical plane.

3.2. No-Pressure and Low-Pressure Molding Processes

and junction fittings are attached with adhesives to cylindrical containers. Mobile on-site installations are available (Table 3.5) to manufacture tanks or other chemical equipment that is not transportable on account of its size. Closed containers are wound onto a lost liner or onto a core that can be melted, washed out or split.

Large-scale machines are used for manufacturing reinforced helicopter rotor blades, deep sea submarine hulls, or airplane fuselages, to diameters of about 10 m and length of 50 m. *Winding robots* permit reproducible filament winding in any desired direction. Their use for series production of non rotationally symmetrical, long-fiber reinforced, and therefore highly stressed motor vehicle components, such as propeller shafts, transverse control arms, leaf springs and torsion springs, pistons and connecting rods is under development. Dry winding with subsequent vacuum impregnation produces high quality laminates free of air pores.

Table 3.5. Processes for the Manufacture of GRP Tanks

Process	Investment cost	Run size	Dimensions	Field of Application
Hand lay-up	Very low	Small to medium	Up to 3000 l	Agricultural storage and transport tanks
Fiber spray	Low			Wine storage tanks
Wet compression, cold	Medium	Medium	2000 l to 5000 l	As above – hemispherical shells for heating oil tanks and container bottoms
Wet compression, hot	High	Large		
Centrifugal	Medium	Only for series production	120–2200 mm diam.	Tanks, pipes
Rotation-fiber spray			> 1000 mm diam.	Silos and storage tanks of large diameter
Parallel and helical winding	High	Large or small	Tanks up to 150 m³	Storage and transport tanks, above- and underground heating oil tanks, pipes
Drostholm process with endless steel band core	Very high	Continuous, large	Up to 4500 mm diam.	Large-run tanks, pipes
Planetary winding and other three-dimensional winding processing	Very high	Special process		High-pressure tanks (aerospace)
Mobile on-site winding		Individual manufacture	> 1000 m³	Large storage tanks and chemical plants

Production processes for *GRP profiles and corrugated webs* are described in Section 3.4.2/3.

3.3. Compression, Injection and Blow Molding

3.3.1. Introduction

3.3.1.1. The following *high-pressure molding techniques* (Fig. 3.12) are available for mass production of moldings (ASTM D 883):

- (A_1) Compression molding – the method of molding a material in a confined cavity by applying pressure and usually heat.
- (A_2) Transfer molding – a method of forming articles by fusing a plastic material in a chamber and then forcing the mass into a hot mold where it solidifies.
- (B) Injection molding[1] – the process of forming a material by forcing it, in a fluid state and under pressure, through a runner system (sprue, runner gate(s)) into the cavity of a closed mold.

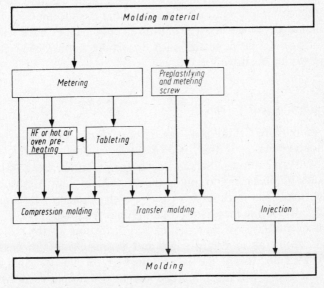

Fig. 3.12. Flow diagram for compression, transfer and injection molding.

---

[1] The definition includes RIM (p. 62).

(C) Blow molding – a method of fabrication in which a heated parison is forced into the shape of a mold cavity by internal gas pressure.

(A_1) and (A_2) are the classic methods for thermosets.
Preweighed amounts of thermosetting material, generally preheated, are placed in each cavity or antechamber in the fixed half of an opened and heated tool on the press platen. After closing the tool and building up the pressure the material melts, flows and solidifies on curing. The moldings are then ejected from the re-opened tool while hot.

In methods (B) and (C) melted or fluid material is forced into the closed (B) or closing (C) mold. Blow molding (C) is used for thermoplastics only. Injection molding, a one-stage process, is widely used for all kinds of plastics and rubbers. In curing thermosetting material the tool must be heated to curing temperature. Thermosets or vulcanized moldings are ejected hot from the molds, whereas thermoplastics are cooled to an optimum temperature below the glass transition of the polymer (Chap. 1.5) before ejecting.

An interesting variant of injection molding is injection-compression molding. In this, the measured molding compound is sprayed into a slightly opened mold. The clamping unit moves both halves of the mold together just before injection is finished and the item obtains its final form by compression. The molds are more expensive than normal injection molds because vertical flash faces are necessary to prevent material escape and flash formation. Thermoset and thermoplastic molded items produced in this manner have less orientational anisotropy and hence scarcely any distortion. *Extrusion* is a continuous process and will be dealt with in 3.6.

3.3.1.2. Designing plastics parts

The walls of moldings should, as far as possible, be uniformly thin. With injection-molding materials that can be readily melted, wall thicknesses of 0.4 mm (1/64 in) and a flow path-wall thickness ratio of 250:1 can be produced. However, such products tend to split in the direction of flow. A ratio of 100:1 is generally feasible. Wall thicknesses of very massive moldings are limited to 10 to 20 mm because of the increased molding times required.

Localized concentrations of material lead to distortions of the molding with variations of temperature. Sharp edges and abrupt changes of cross section must be avoided because of the notch sensitivity of plastics. Shape-induced stress concentrations are most marked in glassy-amorphous polymers and least in reinforced molding materials. Figure 3.13 indicates various shapes.

For ejection, *wall tapering*, in the direction of the parting of the mold, should be at least 0.5°. For polyolefins, reinforced thermoplastics, PF, UF, UP and EP molding compounds in particular it should be more than 2°.

3.3.1.2. Designing plastics parts

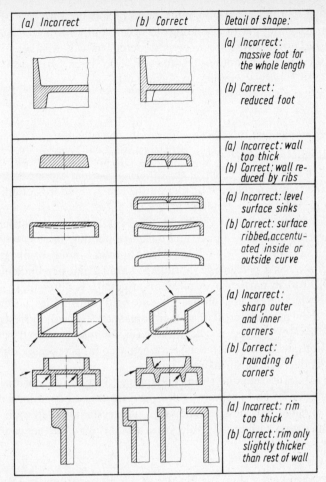

Fig. 3.13. Shapes of plastics moldings.

Larger demolding angles facilitate trouble-free, fully automatic operation. Low stiffness moldings with undercuts of approx. 1 mm (0.04 in) in the case of thermoplastics and even more in the case of elastomers can be demolded elastically when warm. Otherwise moveable mold sections are required. As far as possible undercuts should be so designed as to minimise the complexity and hence cost of the mold.

Metal inserts can be sufficiently firmly anchored in the molding by the plastic flowing around them. Adhesion, pressing-in or ultrasonic embedding (Sec. 3.8.6.2.2) of metal parts may often be preferable and cheaper on account of their variable thermal expansion. Embedded or locking nuts are better than threads for attaching plastics moldings to other components. For sufficiently elastic plastics, snap fittings can be considered, as can welding for joining thermoplastics to one another. PP-injected film hinges are dealt with in Section 4.1.2.6.2; *outsert* molding in Section 4.1.7.

3.3.1.3. Shrinkage and tolerances

The dimensions of moldings after cooling are less than the corresponding dimensions of the mold because of the differing thermal expansion of plastics and steel, and they can decrease further on storage in standard conditions for a given period. Material-dependent values of *mold shrinkage* are generally expressed, according to ISO 2577 and ASTM D 955, as the differences in length between a molded test piece and the corresponding mold cavity, divided by the latter, as "initial", "24-h" and 48-h "normal" mold shrinkage in mm/mm (= in/in) or as a percentage. Shrinkage parallel and perpendicular to the direction of flow, in particular of injection moldings, is different.

Thermosets which cross-link on heating and rigid amorphous thermoplastics shrink less than partially crystalline thermoplastics. Table 3.6 gives *shrinkage values*. In DIN 16901 these provided the basis for *tolerances* and permissible mass variations, which can be reconciled when account is taken of the differing requirements due to tool shrinkage.

Large amounts of differential shrinkage are one of the causes of the undesirable deformation of an injection molded part; this must be minimised by gate choice and molded part design, especially with crystalline thermoplastics and thermoplastics used with organic fillers (e.g. wood flour). Further causes of the

Table 3.6. Shrinkage values for Plastics Materials according to DIN 16901

Shrinkage values (%)	Thermoplastics		Thermosets
	Partially crystalline	Amorphous	
0–1	GF–PA*, GF–POM*	PS; SAN; SB; ABS; rigid PVC; PMMA; PPO mod. PC; PET, amorphous	PF and MF types with inorganic fillers
1–2	PP + inorganic filler*; PA*, POM, <150mm long*; PET, crystalline	CA, CAB, CAP, CP	PF and MF types with organic fillers, UP types
2–3	PE*, POM >150 mm*; PP*, PE/PP		
3–4	PB molding compound	Plasticized PVC with low plasticizer content 1–3	

*At wall thicknesses >4 mm, the next-higher shrinkage value group.

3.3.1.3. Shrinkage and tolerances

deformation can be anisotropic shrinkage of sheet (caused, for example, by differing glass fiber orientation) or dissimilar cooling of both sides of the molding either in the mold or after demolding.

The effect of the type of plastic and processing conditions on the shrinkage are complex and it is difficult to predict accurately processing shrinkage and shrinkage difference from, for example, pvt-diagrams and processing parameters. Semi-empirical models, in which measured values are incorporated into the shrinkage calculation, are more promising.

The tighter the tolerance specification, the higher the manufacturing costs and costs generally, as for example in the making of precision parts that require extremely high molding pressures (Table 3.7).

Chemical after-reactions and post-crystallization or relaxation can effect persistant *post-shrinkage* for a lengthy period, which can lead to distortion of the moldings. It is accelerated by raising the temperature; therefore the product can be dimensionally stabilized by relatively short tempering (24–168 h).

Mineral-filled moldings give the smallest post-shrinkage ($\leq 0.2\%$), measured as relative length differences of the molding before and after tempering. Changes in amorphous thermoplastics are considerably less than those in partially crystalline polymers, and those in phenolics are less than those in similarly filled aminoplastics. With unfavorable shapes and processing conditions post-shrinkage of organically filled moldings can be of the order of $< 1\%$.

The application temperature and moisture absorption from the atmosphere (swelling) must be taken into account in the determination of *operating tolerances* in individual instances. Polyamides and UF in particular can swell after years of storage in damp conditions.

Table 3.7. Comparison of Allocated Manufacturing Costs

Cost categories	Mass-produced article (%)	Technical parts (%)
Electrical	35	15
Water	20	4.5
Auxiliary and factory	1	0.5
Fixed plant	44	80
(a) Machine	100	100
Material	80	45
Machine	10	25
Tool	5	20
Personnel	2	5
Special and general	3	5
(b) Total article-specific cost	100	100

3.3.2. Injection-Molding Machines

3.3.2.1. Assembly and working cycle

Injection-molding machines (Fig. 3.14, for elements of the process see Sec. 3.3.2.4) include, as a rule:

1. The *injection unit*, including the nozzle for the batchwise output of plasticized molding material (for machines with more than one unit see Sec. 3.3.6.3).
2. The *locking unit* with the stationary (cavity) platen close to the nozzle and the moving platen on which the corresponding parts of the two- or multipart split molds are clamped.

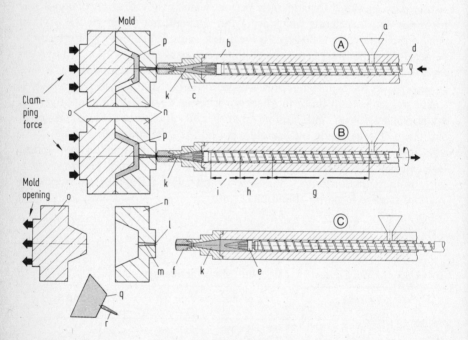

Fig. 3.14. Principle of Injection Molding of Thermoplastics.
Injection unit components
a) Hopper, b) Injection cylinder with heating bands and measuring sensors, c) Cylinder head, d) Screw, e) Non-return valve, f) Heated nozzle.
The screw is divided into three zones: g) the feed section, h) the compression section, and i) the metering section. k) is the residual material melt cushion in the cylinder head.
The following features of the mold are shown: l) gate, m) feed channel, n) injection half of the mold, o) closing half of the mold, p) melt, q) molding, and r) sprue.
A) Screw moves forwards, pressing the melt through the jet into the mold cavity.
B) The screw continues to press the melt at a lower pressure until the melt in the jet and gate are solid (holding pressure). This compensates for the volume shrinkage while the molding cools.
C) The solidified molding is ejected.

3.3.2.1. Assembly and working cycle

3. Those parts that are necessary for carrying out the injection cycle, including those running and controlling the machine itself and the associated temperature control units.

Injection and locking units are generally arranged horizontally in a line. In several cases they are placed at right angles. There are also circular installations with several locking units. Figure 3.14 illustrates the working cycle of *reciprocating screw molders*. Transfer two-stage screw plunger units are special types, for example for precision molding of small thermoplastic parts, structural foam molding (Sec. 3.3.6.4) or thermoset injection (Sec. 3.3.7). The main *performance data* for injection molding machines are

(1) *Clamping force*, expressed in metric tons or – according to EUROMAP rules[1] – in kN (1 ton = 10 kN)
(2) *Calculated swept volume* in cm^3 (EUROMAP), which is approximately equal to the calculated capacity for general-purpose polystyrene ($d = 1.05$) in ounces or grams. This theoretical value is about 10 to 30% higher than that derived practically for the volume of the injection molding.

Many machines are fitted with interchangeable parts, e.g. three cylinder plus screw sets of different diameters. Table 3.8 gives further technical data for the

Table 3.8. Approximate Productivity of Normal Injection Molding Machines, 100–400 MPa or Tons Clamping Forces Range

	International size:			
Calculated swept volume	cm^2	200–2000	in^3	12–120
Injection capacity, calc. for GPPS	g	200–2000	oz	7–70
Plasticating capacity, calc. for PS	kg/h	45–320	lb/h	100–700
Max. injection flow	g/s	100–1000	oz/s	3.5–35
Dry cycle time (Fig. 3.17)	s	3–10	s	3–10
Max. injection pressure	bar	1000–2500	ton/ft^2	1000–2500
Screw set diameters	mm	35/50–70/100	in	1.4/2.5–2.8/3.9
Max. screw speed	min^{-1}	200–300	min^{-1}	200–300
Max. open daylight	cm	25–150	in	10–60
Practical shot weight range, approx.	g	100–1500	oz	3.5–53

---

[1] From 1969 EUROMAP (European Committee of Producers of Rubber and Plastics Machinery) has adopted a designatory system involing
(1) Calculated swept volume in cm^3 and (2) clamping force in MPa = metric tons. In 1982 it modified the sequence mentioned above by supplementing it by H for horizontal, V for vertical and L for side-feed horizontal machines. For example: Designation of a horizontal machine with 2000 kN clamping force and 500 cm^3 swept volume: new 2000 H-500, obsolete 500/200.

3.3. Compression, Injection and Blow Molding

general-purpose machines in the 100 to 400 tons clamping force and 200 to 2000 cm^3 swept volume range. The commercial limits for normal injection machines are about 10 times these values. Individually produced machines of 30 000 to 100 000 kN clamping forces and corresponding high swept volumes or shot weights are in use for, e.g., boats, large automotive parts or garbage containers with capacities of up to 1000 l. For the mass production of e.g. cups or closures fast cycling units with 500 to 1000 kN and screw diameters of about 25 mm are on the market.

3.3.2.2. Elements of the injection unit

Screws as processing and plasticizing instruments will be dealt with individually in Sec. 3.6 on extruders. *Universal screws* used for thermoplastics injection molding, with a length:diameter ($L:D$) ratio 22 to 25 are mostly *three-stage screws*. The screw stroke for each molding material feed-cycle is limited to 2.5 to 4 D. *Degassing screws* of greater length allow the removal of moisture and monomer residues at melt temperatures, sometimes more economically and in a cleaner way than by lengthy pre-drying at lower temperatures.

Many machines can be provided with screws suitable for dealing with thermoplastics that are difficult to process (e.g., unplasticized PVC) and with low compression (0.9–1.1:1) thermoset or elastomer screws. With larger machines there is the danger that reactive plastics will cross-link while on the screw. For such processing, special machines with short screws are appropriate.

DIF (Direct Incorporation of continuous Fibers) using injection molding machines was introduced in 1989; the fiber strands of roving spools are drawn into a degassing oven. This technology combines the hitherto usual compounding during manufacture of long fiber granules and manufacture of the molding in one step.

The *screw cylinders* of thermoplastic machines and the injection nozzles are heated in zones by means of heater bands; thermoset and elastomer machines use circulating fluid. The feed zone can be cooled and designed for granule, powder or band delivery feed.

The *drive* for screw rotation is generally provided by a continuously regulated hydraulic motor. Electric motors are used for high-speed screws. The drive for the axial screw stroke is mostly hydraulic, but there are fully electromechanical machines. High-performance machines are required to provide programmable screw speeds to correspond with changing *shot speed* in the course of injection. The after-pressure (*dwell* or *holding pressure*) must also be controllable, and the rotating screw must be able to build up a *back pressure* on its return to promote mixing and improved plastification of the incoming feed. Regulatable pumps or several connected pumps are used for this purpose; they may be supplemented by a maximum efficiency oil hydraulic accumulator.

For thermally sensitive materials (PVC, thermosets) the *screw head* runs into an extended unflighted tip with an advancing spiral. There must be a cushion between the cylinder head with the inserted nozzle and the head of the screw to avoid damage and to maintain pressure until the completion of the after-pressure phase (Fig. 3.17). In processing low-viscosity melts the screw head must be provided with an effective non-return valve. When working with a non-return valve, one must not inject with the screw rotating. In the *intrusion* or *flow-casting process* for thick-walled parts the mold is filled by means of a rotating screw.

With rigid PVC *open injection nozzles* must be used; with other viscous plastics open nozzles can be used. For more easily flowing materials *closed nozzles* (Fig. 3.15) are needed. The simplest in design are the sleeve- and needle-locking varieties. Needle-locking nozzles permit free injection, in contrast with the sleeve-locking ones.

In processing fiber-reinforced and mineral-filled molding materials the injection unit is subject to considerable *wear*. For that reason bimetal cylinders with an inner lining of Co-Ni-Cr alloy (Reiloy or Xaloy) and laminated or nitrided wear-resistant screws are used.

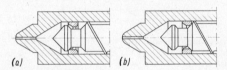

Fig. 3.15. Non-return valve in the screw head.
(a) Plastifying, (b) Injection and afterpressure.

3.3.2.3. Elements of the locking unit

The *end frame* at the outermost end of the locking unit is anchored to the machine bed. It is connected to the *mounting platen on the nozzle side (the fixed platen)* by massive *tie bars* that guide the *moving platen*. Both platens have an alignment system for each half of the mold. In mechanical mold locking (usually by hydraulically actuated toggles), which can generally be used in machines of up to 400 tons, the tie bars take up the locking force like tension springs (Fig. 3.16a). The kinematics of the toggle automatically slows the mold movement before it reaches the final position. For bigger machines the hydro-mechanical clamp (Fig. 3.16b) utilizes a toggle system similar to that of a standard mechanical clamp to open and close the mold only. A large, short-stroke clamp ram is then hydraulically pressurized to lock the mold. In fully hydraulic systems, long-stroke lower pressure hydraulic units are also used to

3.3. Compression, Injection and Blow Molding

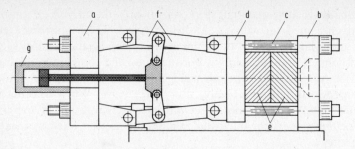

Fig. 3.16a. Clamping unit with double toggle mold clamping mechanism.
a Backing plate, b Platen on the fixed half of the mold, c Columns, d Movable platens, e Mold halves (male and female), f Toggle, g Hydraulic cylinder

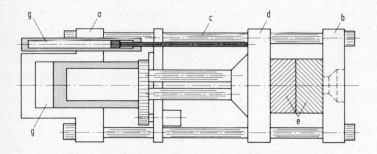

Fig.3.16b. Fully automatic mold closure with separate operating cylinder.
a Backing plate, b Platen on the fixed half of the mold, c Columns, d Movable platens, e Mold halves (male and female), f Hydraulic cylinder

close and open the mold. Switching to a low end speed on opening and closing protects (*damping*) the tool and the ejector plate from too sudden an end stop. In *injection-compression* moldings of items with large surfaces, the material is allowed to flow several tenths of a millimeter into the "breathing" mold and is compressed in the final mold by high locking pressure.

3.3.2.4. The injection-molding process

The following phases, which overlap and must be coordinated, occur in the injection cycle (Fig. 3.14): the tool locking and opening movement (Fig. 3.17a), the forward and backward motion of the injection unit (Fig. 3.17b), the movement of the screw in the injection process and the material entry (Fig. 3.17c). The injection pressure and dwell pressure must be adjusted so that after solidification of the melt (mold sealing point, Fig. 3.17d) the molding cools in such a way that it can be ejected without pressure. It is possible to calculate this

3.3.2.4. The injection-molding process 81

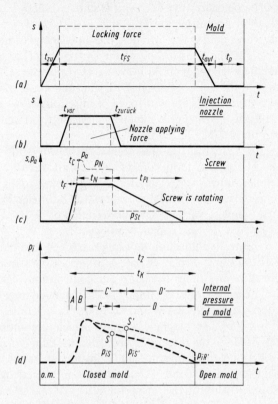

Fig. 3.17. Operating cycle during injection molding.
(a) to (c): Strokes (s, ——— graphs), forces or pressure (p, - - - - graphs) of
    (a) The moving mold half on the moving platen,
    (b) The injection unit to apply the nozzle to the speed bush,
    (c) The screw,
    (d) The sequence of internal pressure in the mold during (A) filling, (B) molding material compression, (C) formation of the molding and (D) molding cooling down at the correct afterpressure, C' and D' at too high afterpressure.

Times: (a) $t_{zu} - t_{auf}$ = mold closing-opening time,
    t_{FS} = mold stand time, t_p = change of time
  (b) $t_{vor} - t_{zurück}$ = *injection unit applying-returning time*
  (c) t_F = mold filling time, t_C = compression time
    t_N = dwell time, t_{pl} = plastifying time
  (d) t_Z = cycle time, t_K = cooling down time
    Dry cycle time: $t_{zu} + t_{vor} + t_{auf}$, i.e. time of the mold closing + screw advancement + mold opening movements.

Pressures: (c) p_a = maximum injection pressure, p_N = dwell pressure,
    p_{ST} = back pressure
    p_{iS}, P'_{iS} = internal mold pressure at the sealing point S, S'.
    P'_{iR} = residual pressure on the molding following too high dwell pressure: molded part blocks in the mold, danger of stress cracking.

3.3. Compression, Injection and Blow Molding

"isochore" of the process if the dependence of the specific volume of the plastic on pressure and temperature (pvt diagram) is known exactly.

In actual production operation the *predominant influences* – plastifying and injection temperatures across the cylinder, the mold temperature through the temperature cycle and the temperature of the hydraulic oil for the axial screw drive and, where necessary, for mold movements – must be set optimally and held constant within the limits of normal machine and material values (Tables 3.9, 3.11). Temperature, pressure and speed of injection are decisive factors in determining mechanical properties, dimensional accuracy, freedom from warpage and the surface quality of the molding.

In injection-molding machines currently in use, *temperatures* are normally *feedback controlled*. The closed loop control circuits are fitted with sensors to measure the temperatures of the cylinder, oil, melt and tool (for measuring the latter there are also built-in heating pipes) and with electronic measuring and switch gears. Stepwise variable time preselection of *pressures* and *speeds* is mostly effected by *feed forward* (open loop) *control*.

In the computer controlled injection molding machines common today, input of all machine settings and display of values of processing parameters such as travel, force, pressure and times occur on one control unit. The machine settings for the mold can be stored on floppy discs, magnetic tape or in program modules or battery-backed up RAM and read into the machine. There are also tolerances for most of the machine settings which are monitored and if exceeded are reported via the integrated error diagnostics. For important processing steps such as metering, injection and hold-time there are up to 16 preset values available. Closed loop control of the important processing parameters is now performed in software by the microcomputer system rather than through hardware components. Thus the optimum parameters for the control circuit can be determined and optimum control conditions can be obtained.

Graphic display of the hydraulic pressure in the injection cylinder, screw stroke, injection velocity and internal pressure of the mold allows automatic process monitoring using software comparison with ideal values. Up to 16 process parameters per cycle can be programmed in quality tables and can be evaluated by integral or external software.

Standardised communications interfaces for CPUs of injection molding machines (EUROMAP 15) and peripheral equipment (EUROMAP 17) form the basis for CIM (Computer Integrated Manufacturing).

Fig. 3.18 is a schematic of a fully-electrical injection molding machine. Advantages of this type of machine are:

- it is free from oil and hence leakages,
- the starting up and running processes are invariant as there is no longer a dependence on varying oil temperature (less waste),

3.3.2.4. The injection-molding process

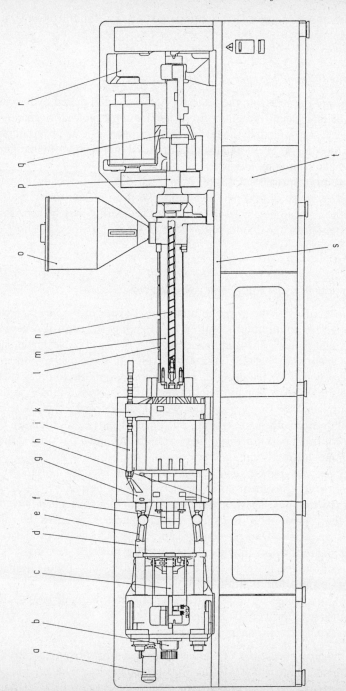

Fig. 3.18. Schematic of a fully-electric injection molding machine. Drive is provided by ac servo-motors in which linear motion is transferred via ball screws. Mold locking and clamping forces and linear and rotational screw movement are performed by separate drives.
a Mold height adjustment motor, b Servo motor to closing, c Ball screw, d Guide pins, e Double toggle, f Supporting rollers, i Mechanical lock, k Solid platen, l Heating bands, m Injection cylinder, n Screw, o Material hopper, p Adjustment plate in front (Injection unit), g Moveable pressure plate with motor for turning the screw, r Adjustment plate behind with drive motor for injection and carriage movement, s Machine frame, t Switching compartment

3.3. Compression, Injection and Blow Molding

– it is noticeably more energy efficient compared with machines working with hydraulic oil.

3.3.2.5. Ejection of moldings and their aftertreatment

Small moldings are pushed from the mold by ejectors and gathered by moving receptacles or belts. Delicate moldings are sucked from the mold. Movement of hydraulically activated bi-axially linear movable grippers can be programmed into the machine control. Multi-axial robots are used on fully automated lines to remove and place large moldings.

With the exception of sprueless hot runners and self-degating injection systems sprues must be removed with separating tools. To prevent deformation after ejection the moldings can be placed on a former. Tempering may be necessary to deal with after-shrinkage due to strains (Sec. 3.3.1.3). For the deflashing of moldings see Section 3.3.3.3.

3.3.2.6. Automation of the injection molding factory

Microprocessor control of the injection molding sequence in the machine can be complemented for continuous operation by extending the control system to the peripheral equipment. This requires, first, flexibly switchable pneumatic *conveying equipment* and – usually mounted directly on the machine – *mixing and metering equipment* for the injection molded compound. Second, for quality control (in addition to display monitoring of the process status), on-line measuring equipment for dimensions and weight can be connected, with an integrated switch mechanism to separate out rejected parts. Additionally the computer can control a *post processing line* for removing and separating sprues, as well as drilling, cutting, welding, assembly and deposition stations.

Mold and injection unit quick-change systems with facilities for simultaneously connecting all supply, control and measuring lines permit retooling of machines in a few minutes. They are used for *fully automatic injection molding*, which, under the control of an overriding master computer, removes the components from storage via a transport control and conveys them to a pre-heating station for heating to the operating temperature of the machine. In conjunction with similarly automated material feed and change-over, new production can continue without any additional tooling time.

For rapid mold exchange for small parts manufacture in small batch sizes, rotary magazines have proved their worth.

There are two types of mold changing systems:

1. *Integrated clamping systems.* Clamping cartridges are integrated into the platens of the machine which grip the bolts projecting from the mold platen

with claws. Tension is exerted by springs and the tension released hydraulically. Transferring the system to other types of machine can be difficult.
2. *Adaptive clamping systems.* Clamping units with positive clamping and holding of the tool by means of toggles, deflecting levers, spring tension, wedging, friction or permanent hydraulic pressure are mounted on each platen. Aligning aids and centering pins are used to position the mold correctly.

An automatic coupling provides a single-block connection for temperature control medium, hydraulic oil for core pullers, electrical current for hot runners, control signals and mold codes, and air. This allows the set-up times to be markedly reduced and provides the preconditions necessary for unmanned operation.

3.3.3. Presses and Ancillary Equipment

3.3.3.1. Types of molding presses

Molding processes in which the mold is filled while open require vertical clamping units. *Hand presses* with unfixed molds (50–800 kN clamping force) are used for trial molding.
Toggle presses with clamping forces of up to 1.5 MN and *downstroke hydraulic presses* of the frame or four-column type with locking forces in the range of 150 kN to 100 MN (15 to 10^4 tons) operate in a semi-automated way for the production of moldings. A second platen can be fitted with a pressure piston working from below (Fig. 3.19). In presses for transfer molding activated from below, the upper ram serves only to keep the mold closed under pressure (Fig. 3.20). Upstroke presses, in which several pressure cylinders are placed in a pit, are used with multi-daylight platens for producing laminates and for polishing sheets of thermoplastics.

3.3.3.2. Metering and preheating equipment

Bulk material can be metered volumetrically or by weight. *Pre-pelleting* of the powder can be effectively carried out in eccentric presses or hydraulic pelleting machines. *High-frequency preheating* equipment adjacent to the press is preferable to hot-air ovens. Molding powder pellets are taken up to the beginning of softening (110–120°C) in about 1 minute. This shortens the time in the mold for materials, which then flow readily under reduced pressure.

3.3. Compression, Injection and Blow Molding

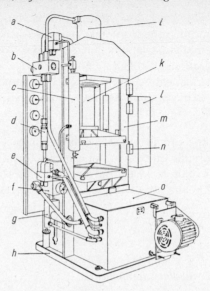

Fig. 3.19. Hydraulic press with a force of 1 MN with supplementary hydraulic ram to the platen (0.4 MN), which can be used as a transfer ram or an ejector. (Drawing by Hahn and Kolb, Stuttgart.)
(a) Magnetic valve, (b) Upper sleeve valve, (c) Pressure accumulator, (d) Back-Pressure valve, (e) Double magnetic valve, (f) Lower sleeve valve for lower ram, (g) Pressure-regulating valve for lower ram, (h) Base plate, (i) Upper platen cylinder, (k) Upper ram, (l) Switchboard for electrical controls, (m) Press frame, (n) End stop, (o) Pump box and drive.

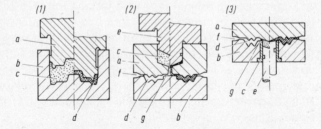

Fig. 3.20. Ram movement in compression and transfer molding.
(1) Compression molding, (2) transfer molding from above, (3) transfer molding from below. (a) Tool: upper part in (1) and (3), middle part in (2); (b) Tool: lower part; (c) Loading chamber or injection cylinder; (d) Mold cavity; (e) Injection ram: in (2) in upper part of tool; (f) Venting channel; (g) Transfer channel.

3.3.3.3. Arrangements for finishing moldings

Because thermosets do not harden immediately at the mold walls, their entry as fluids between the parting surfaces of the mold is unavoidable. Flash of > 0.4 mm thickness or from fiber-filled molding material must be removed by hand. Small parts made by transfer or injection molding with thinner flash can be treated mechanically by barrel or jet deflashing. Soft materials, such as apricot nutshells or polyamide granules, are used for this purpose, and the water used to moisten them contains antistatic agents. Very thin flash (<0.1 mm) is blown off with compressed air. For machining of moldings see Section 4.1.4.1.

3.3.4. Automated and Large Area Presses

3.3.4.1. Automated presses

Among available units for molding are *vertical automated presses* for moldings of up to 100 g. They include precompression and infrared preheating. They are further equipped to remove flash and clean the mold with compressed air. Automatic feed of several molds with metal inserts and mechanical unscrewing of threads is possible. Rotary machines with up to 20 stations are used for small moldings.

Horizontally operating transfer molding machines are similar to plunger injection-molding machines but material is taken in for a single shot. It is plasticized hot while under pressure and injected into the activity. A plunger-type injection-molding machine for doughy molding substances injects cold material into the prepared hot mold.

Metering and plasticizing screws for molding materials save several stages in the molding process (Fig. 3.12). They are used as mobile units for charging molds and transfer chambers, or they are built into automated presses. In the tubular ram injection-molding process, screw plasticized plastics are fed into a heated annular space at the head of the screw. The piston designed as a tubular ram thrusts forward and injects the mass through the nozzles into the receiving mold. The nozzles are subject to wear, but the screw is not stressed by the injection pressure. Automatic production equipment programmed completely from material feed to post-processing for thermosets are analogous to that for thermoplastics. They are used for manufacturing products such as clutch rings and brake linings.

3.3.4.2. Large-area presses

For *low pressure* molding of large GRP parts (Sec. 3.2.4), hydraulically-operated downstroke or four-column presses with 1–30 m^2 platen area are usual (Fig. 3.21a). For cold pressing, they are designed for 5–10 bar, for hot pressing, 10–50 bar pressure.

3.3. Compression, Injection and Blow Molding

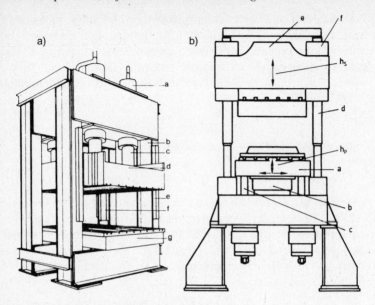

Fig. 3.21. Large area presses for GRP mouldings.
a) Frame press: hydraulic downstroke GRP press with four work cylinder (model SHG)
(a) work cylinder, (b) frame, (c) main piston, (d) movable platen (ram); (e) ram guide, (f) return piston (to raise the slide), (g) stationary platen
b) Construction of an upstroke SMC short-stroke press in cross-section (model Müller)
(a) hydrostatically mounted lower platen, (b) main cylinder, (c) return piston, (d) press column, (e) closing unit, (f) clamping unit, (h_p) press stroke, (h_s) closing action

Closing rates and the precision of parallel travel of the ram and the stationary platen are inadequate for mass production of large-area SMC parts, e.g. for vehicle construction. Upstroke, short-stroke presses with hydrostatically mounted backing plates (Fig. 3.21b) permit a positioning accuracy of up to 0.5 mm for the mold parts. The press ram in Fig. 3.21b has a quick-action closing system, which is interlinked with the columns. All functions of such *high-performance presses* are programmed to be self-regulating. They are conceived for integration into production lines that, at least from the prepreg feed to the finished part, are fully automatic (Fig. 3.22), such as are required by suppliers to the automobile industry: the necessary rate of production and uniform quality cannot be achieved if manual stages are interposed. They also permit in-mold coating of SMC parts (Sec. 4.6.2.4.3), molding of components from glass fiber mat reinforced thermoplastics (Se. 4.1.1.3.2) and mass production of high-strength, fiber-reinforced structural elements for aircraft construction.

3.3.5.1. Materials and production process

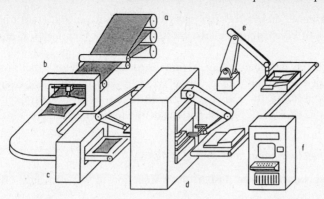

Fig. 3.22. Automated SMC processing by the IKV concept (German Institute for Plastics Processing)
(a) SMC rollers, (b) cutting station, (c) preheating station, (d) press, (e) milling robots, (f) process control

3.3.5. Molding Tools

3.3.5.1. Materials and production process

Molds and plastics machinery are the key equipment required for processing molding compounds into moldings (compression-, injection- and blow-moldings). Because of the high pressure in the mold, special steels are used in their construction to provide sufficient mechanical integrity. Distortion of the individual mold components at the highest molding pressure (300 to 1000 bar and above) and corresponding closing and locking forces should usually be limited to about 0.01–0.02 mm (depending on the tolerances demanded for the molding).

For the cavity of the mold, *case-hardened steel* alloy with a glass-hard polished surface and a toughened core is usually used. Tools made of nitrided steels satisfy the most demanding requirements in dimensional stability. Corrosion-resistant steels with a high chrome content are necessary for such plastics as PVC and are recommended for tools cooled to very low temperatures where rust is formed as a result of condensed water. Molds are often highly chromed. Cast molds made of *zinc alloys*, which can be exposed to temperatures of about 100°C, are suitable for simple injection moldings and for blow moldings. *Light metal castings* can, on account of their surface porosity, be used for blow moldings. Alloys of *copper, cobalt* and *beryllium*, which are heat conducting, are used as injection nozzle inserts and for mold parts that will be intensively cooled.

For moldings faithfully reproducing fine structures of a master model (e.g., for technical toys) *hard nickel* mold cavity walls copied galvano-plastically from the model are fitted into steel molds.

3.3. Compression, Injection and Blow Molding

Methodical mold design requires specifying the number and layout of cavities, the method of construction of the mold and the cooling, ejection and venting systems.

A mold can be designed either 'from inside outwards' or (and this is to be recommended for small and medium sized molds) by choosing standardised mold units suitable for the size of the molding or (for multi-cavity molds) moldings and the cavities in the platens of the mold, i.e. from outside inwards.

3.3.5.2. Injection molds and gating systems

Fig. 3.23 shows the basic design of an injection mold with only one parting surface.

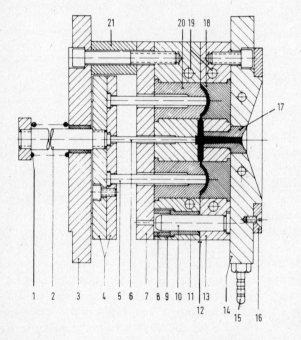

1. Compression spring
2. Ejector rod
3. Platen on moving mold half
4. Ejector plate
5. Ejector
6. Central ejector
7. Intermediate platen
8. Intermediate bushing
9. Mold plate
10. Guide bush
11. Guide bush
12. Mold parting surface
13. Mold plate
14. Platen on fixed mold half
15. Nipple for cooling hose
16. Locating ring
17. Gate
18. Mold insert
19. Cooling channel
20. Mold insert
21. Support bushing

Fig. 3.23. Injection mold.

The tool contacts the injection nozzle through the *sprue bush*. The radius of the nozzle seat must be slightly greater than that of the nozzle itself. The bore of the sprue must be somewhat larger than that of the nozzle. The sprue opens up at the entrance to the mold cavity at the *gate*, at which the molding is separated from the feed system. Various types of gating are shown in Fig. 3.27. On a single cavity mold with a conical sprue the molding must subsequently be separated at the gate. A pin gate (Fig. 3.24.2) with a continuously heated pre-chamber permits sprueless injection. Some flat or conical moldings are injected at the separating surfaces (Fig. 3.25).

In multi-cavity molds (Fig. 3.26) the material is distributed through runners from the nozzle to the cavities. The individual moldings are cut from the runners and then separated by sprue pulling. In three-plate molds (Fig. 3.28) with insulated runners the material remaining in the interior of the sprue and runners is fluid enough to allow injection. The same situation is achieved in the hot-runner mold by heating these channels. Plates 1 and 2 are separated only on the interruption of a run. For undercuts or holes in walls sliding cores, which can be jaws or blocks, are used; these are actuated by swallow-tail guides that slant upward. For internal threads unscrewable cores are used. In the

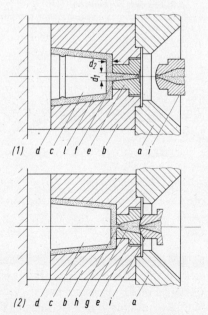

Fig. 3.24. Central sprue for a tub-shaped molding
(1) Conical sprue, (2) Pin point gating.
(a) Mold mounting plate, (b) Cavity, (c) Core, (d) Mold parting line, (e) Sprue bush, (f) Sprue gate, $d_1 > d_2$, (g) Hot well, (h) Pin gate, (i) Injection nozzle.

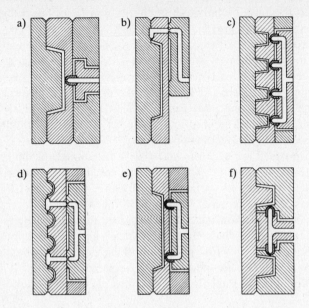

Fig. 3.25 Examples of melt passages with hot runners.
a) Central gating of a cavity b) Side gating of a single impression mold c) Central direct gating of a multi-cavity mold d) Indirect side gating of a multi-cavity mold e) Multi-gating of a cavity f) Side direct gating of a multi-cavity mold.

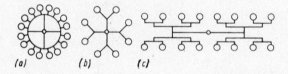

Fig. 3.26. Designs of multi-cavity molds.
(a) Ring spider, (b) Star spider, (c) Multilayered series spider, also used in multilayers.
a) is unfavourable because of the differing flow path lengths. b) and c) are better.

simplest designs the mold is unscrewed by hand after it is opened; otherwise it is treated in the same way as other moveable parts.

For the injection molding of hollow bodies, such as 180° tube bends, cores made of alloys melting at 50 to 130°C can be fixed in the center of the mold. These maintain higher injection temperatures for plastic materials and can be melted out from the molding without loss.

3.3.5.3. Compression and transfer molding tools

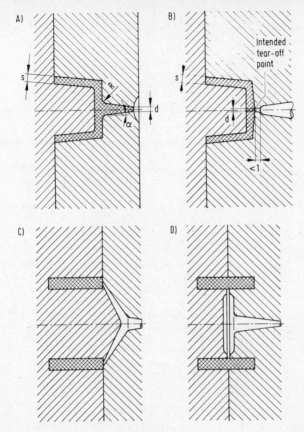

Fig. 3.27. Various types of gating and sprue.
(A) Sprue gate, α = demolding taper, s = wall thickness, d = sprue gate, diameters d ⩾ s, d ⩾ 0.5.
(B) Pin gate, d ⩽ 2/3 s,
(C) Umbrella gate,
(D) Diaphragm gate,

For *In-mold Decoration (IMD)* printed sheets (Sec. 3.9.3) are inverted, for *Powder Mold Coating (PMC)*, used for BMC parts, melting powder is electrostatically sprayed into the cavities of the open mold before the change over to injection is made.

3.3.5.3. Compression and transfer molding tools

The basic construction of the tools for compression and transfer molding is shown in Fig. 3.20 (p. 86). To ensure consolidation of the material in the

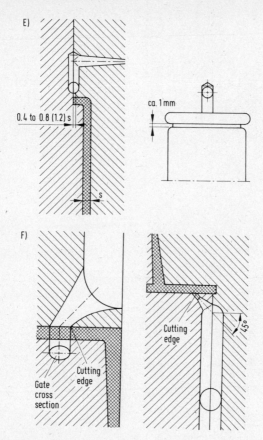

Fig. 3.27. Various types of gating and sprue (*cont'd*).
(E) Film gate, mainly for large area moldings,
(F) Tunnel gate.

mold cavity one has to work with excess of powder. In the simple flash mold (Fig. 3.29a), with horizontal parting surfaces, the excess appears as flash of varying thickness between the outer squeezing surfaces.

Semi- or *full-positive molds* (Fig. 3.29b) permit more exact metering and with levelled exit surfaces between punch and die allow the formation of an easily separable vertical flash collar.

Transfer molds resemble injection-molds in their construction beyond the injection channel which is designed as a pinpoint or film gate (Fig. 3.24, 3.25, p. 91).

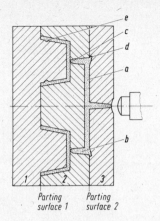

Fig. 3.28. Three-plate mold. Sprue with runners and connecting channels is separated from molded part with double daylight opening of the mold.
(a) Runner, (b) Connecting channel, (c) Gate, (d) Undercut for gate, (e) Undercut of core.

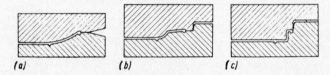

Fig. 3.29. Sealing surfaces of compression molds.
(a) Flash mold with horizontal land, (b) Flash-transfer mold, (c) Transfer mold with vertical flash.

3.3.6. *Molding Techniques for Thermoplastic Materials*

3.3.6.1. General injection technique

In the molding cycle of the screw-plunger injection-molding machine (Fig. 3.14, p. 76) the injection-molding time is generally not much more than one second, whereas the dwell period to solidification of the sprue or in the gate requires several seconds. The dominant element of the cycle is the cooling time to ejection, which can take several minutes with large moldings. During this period the rotating and withdrawing screw draws additional material into the cylinder. The demolding and nozzle contact times (Fig. 3.17, p. 81) are also short.

Table 3.9 gives guide values for the *processing conditions* of standard polymers. For newer engineering thermoplastics the injection units must be designed for temperatures up to 500°C (930°F) (Fig. 3.30, see also Table 4.47, Table 4.49). The material temperature as measured at the nozzle with a penetration

3.3. Compression, Injection and Blow Molding

Table 3.9. Data for the Injection Molding of Thermoplastics

Materials	Temperatures		Injection pressure (bar)	Non-return valve	Locking nozzle	Comments
	Material °C	Tool °C				
PE 0.92 thin walled 0.96 thick walled	220–260 180–220 260–300 240–280	20–60	600–1500	(+)	+	Pressure dependent on flow behavior (melt index)
PP TPX	200–300 270–300	20–60 70	800–1800	(+)	+	Low viscosity, >270°C.
PS SAN ABS ASA	170–280 220–260 200–280 230–280	5–70 30–85 40–80 40–85	600–1800	+	(+)	(A)BS slowly at higher temp. but do not overheat.
PVC: rigid flexible	180–210 170–200	20–60 15–40	1000–1800 300–1500	–	–	Slowly, possible intrusion, injection unit corrosion resistant.
PCTFE PFA, FEP	200–280 340–360	80–130 120–180	approx. 1500 300–700	(+)	(+)	Injection unit corrosion resistant.
PMMA: VST 80 VST 110	150–200 180–230	50–65 60–90	700–1000 800–1200	(+)	(+)	For high quality injection unit is chromed. Highest pressures for optical equipment.
PA: 6 6.6 6.10	230–290 280–300 230–290	40–60 40–30 30–60	900–1400	+	+	Fast injection, wide gates, fine crystals at high mold temps.
PPO: modified	250–300	80–100	1000–1400	(+)	(+)	Slowly, wide nozzle
POM and copolymers	180–230	60–120	800–1700	+	+	Crystallize, as for PA.
PC	280–320	85–120	800–1500	+	(+)	Preheat to 120°C.
PET PBT	260–280 235–270	120–140 60–80	1200–1400 1000–1200	(+) +	(+) +	Particular types: with cooled molds (20–40°C) glass-clear, amorphous.
CA, CAB	180–230	40–50	800	+	(+)	Chromed injection unit.

+ = Recommended. (+) = Can be useful. – = Not possible.

3.3.6.1. General injection technique 97

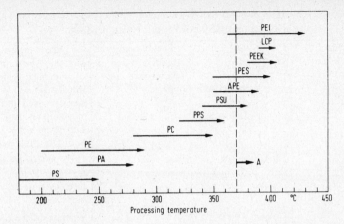

Fig. 3.30. Processing temperature ranges for various plastics. Beginning of steel transition at 370°C (A).
PS: polystyrene, PA: polyamide, PE: polyethylene, PC: polycarbonate, PPS: polyphenylene sulfide, PSU: polysulfone, APE: aromatic polyester, PES: polyethersulfone, PEEK: polyaryletherketone, LCP: liquid crystal polymer, PEI: polyetherimide.

pyrometer or a thermocouple at the nozzle end is not the same as that of the cylinder wall. For thin-walled moldings of high viscosity, higher material temperatures and pressures are required than for thick-walled parts. Plunger machines in general require about 200 bar extra injection pressure.
Dwell pressure to avoid sink marks (Fig. 3.17, p. 81) is usually fixed lower than injection pressure and 10 to 20% of the injection pressure is suitable as *back pressure*. With the intrusion process (Sec. 3.3.2.2) the mold is slowly filled by the rotating screw at almost no pressure. The dwell pressure is high. With *injection-compression* about half the injection pressure is sufficient.
The higher the *mold temperature* (within given limits), the better will be the surface gloss and if possible the crystallinity of the molding. If crystallinity is required, the cooling time is extended. The large-scale production of small parts is carried out in molds cooled to low temperatures, about 5°C, which can suppress crystallization (shock cooling).
The *position of the gate* must be so chosen and the injection procedure so carried out that the material fills the cavity preswelled rather than as a jet. Otherwise "sausage" injection (Fig. 3.31) occurs with separation from the after-swelling mass. Table 3.10 lists typical faults in injection-molded parts

Fig. 3.31. "Sausage" injection.
(a) Sausage, (b) After-swelling melt.

3.3. Compression, Injection and Blow Molding

Table 3.10. Faults in Injection Molding of Thermoplastics, Possible Causes and Recommended Solutions (Source: Bayer AG)

Defect	Possible appearance	Possible causes	Recommendations
Contamination of the granules	Grey foreign particles which sparkle and reflect light	Abrasion of feed tubes, vessels and hoppers	No tubes, vessels or hoppers of aluminium or tin-plate, instead use steel or VA tubes (cleaned internally) and steel/VA plate. Conveying sections should be as straight as possible
	Dark fish eyes, discolored streaks	Dust or dirt particles	Keep the drier clean and regularly clean the air filter; carefully close broken-open sacks and containers
	Colored streaks, flaking off of skin particles in the area of the sprue gate	Mixing with other plastics	Separate different plastics, never dry different plastics together, clean the plasticizing unit, check subsequent material for purity
Contamination of the regrind	As for granules (see above)	Mill abrasion	Regularly check mills for surface abrasion or damage and keep in good condition
		Dust or dirt particles	Store scrap in a dust free environment, clean dirty articles before grinding, reject items from moist processing (PC, PBT) as well as thermally damaged items
		Other plastic regrind	Always store different plastic regrinds separately
Moisture streaks	U-shaped, long drawn streaks which are open against the direction of flow	Residual moisture in the granules too high	Check drier and drying process, measure the temperature of the granules, observe the drying time
Silver streaks	Long drawn out silver lines/streaks	Thermal loading of the melt too high: temperature too high, melt dwell time or screw speed, nozzle and flow channel cross section too small	Check melt temperature, choose a more favorable screw diameter, lower the screw speed, enlarge nozzle and flow channel cross sections

Table 3.10. (Cont'd)

Defect	Possible appearance	Possible causes	Recommendations
Streaks (with incorporated air)	Long drawn, linear streaks spreading out over a wide area. Bubble formation is also often visible with transparent plastics	Injection speed too high. Air drawn in, back pressure too low	Increase injection time Raise the back pressure within the allowed limits
	Line and nozzle shaped, concentrated black coloration (diesel effect) at weld/flow lines or on flow paths	Air enclosed and compressed in the mold unable to escape via the parting surface or ejector	Improve mold evacuation especially in the region of the melt flow line at the end of the flow path and at depressions (threads, pins and characters), correct flow front variation (wall thicknesses, position of the gate, flow aids) or lengthen the injection time somewhat
Grey streaks	Grey or dark-colored stripes distributed unevenly	Wear on the plasticizing unit	Exchange the complete unit or individual parts, use corrosion and abrasion resistant plasticizing unit
		Dirty plasticizing unit	Clean plasticizing unit
Cloudiness	Extremely fine fish eyes or metal particles forming clouds	Wear on the plasticizing unit	As above
		Dirty plasticizing unit	Clean plasticizing unit
	Cloudy dark discolorations	Too high a screw rate	Reduce screw speed
Dark (usually black) fish eyes	Size less than 1 mm^2 to microscopic	Wear on the plasticizing unit	As above
	Size greater than 1 mm^2	Tearing and flaking of the boundary layer formed on the screw and cylinder surfaces	Clean plasticizing unit and use corrosion and abrasion resistant plasticizing unit. For PC and PC blends: heat the cylinder through at 160–180°C when production is interrupted
Charred streaks	Brown discoloration with streaks	Melt temperature too high	Check melt temperature and lower, examine regulator
		Melt dwell time too long	Shorten cycle time, use a smaller plasticizing unit

3.3. Compression, Injection and Blow Molding

Table 3.10. (Cont'd)

Defect	Possible appearance	Possible causes	Recommendations
	Periodically occurring brown discoloration with streaks	Temperature control in the heating channel unsatisfactory	Check heating channel temperature, examine regulator and temperature sensors
		Plasticizing unit worn out or dead corners on sealing faces	Check elements such as cylinder, screw, non-return valve and sealing faces for wear and dead corners
		Unsatisfactory flow area in plasticizing unit and heating channels	Eliminate unfavorable flow transitions
		Injection speed too high	Increase injection time
Exfoliation or delamination	Flaking off of skin particles in the sprue gate region (especially with blends)	Contamination by other, non-compatible, plastics	Clean plasticizing unit and check subsequent material for purity
Dull spots	Matt spots around the gate, on sharp edges and wall thickness)	Disturbed melt flow in the sprue gate, on transitions and bypasses (shearing, tearing of the already solidified surface skin)	Optimise gate, avoid sharp edges especially in the transition from gate to mold cavity, round off transitions on feed channels and wall thickness changes and polish, stepwise injection (slow-fast), so that laminar flow is maintained throughout the mold and thus the wetting of the contours can occur from gate to mold end systematically without disrupting the laminar flow
Concentric grooves	Extremely fine grooves on the surface (e.g. with PC) or dull grey rings (e.g. with ABS)	Too high flow resistance in the mold so that the melt stagnates, melt temperature, mold temperature, injection speed too low	Raise melt and mold temperatures, increase injection speed

3.3.6.1. General injection technique

Table 3.10. (Cont'd)

Defect	Possible appearance	Possible causes	Recommendations
Cold grafts	Cold melted particles enclosed in the surface	Nozzle temperature too low, nozzle cross section and aperture too small	Choose a suitable heater higher power, fit thermo sensors to nozzle and controller, increase nozzle cross section and aperture. Avoid cooling of the sprue bush. Remove nozzle from sprue bush earlier.
Voids and sink marks	Air-free voids in the form of round or long drawn out bubbles only visible in transparent plastics; depressions in the surface	Volume contraction during cooling not compensated	Increase holding pressure time, increase holding pressure, lower melt temperature and alter mold temperature (for voids raise and for sinks lower), check melt cushion, increase nozzle aperture
		Molding not suitable for the particular plastic (e.g. too great a difference in wall thickness)	Design appropriately for the plastic in use (e.g. avoid wall thickness changes and material accumulation), use flow channels and gate cross section appropriate to the molding
Bubbles	Similar to voids but much smaller diameter and many more of them	Moisture content in the melt too high, residual moisture in the granules too high	Intensive drying, sometimes replace degassing screw by normal screw and work with pre-drying, check drier and drying process and possibly install air drier
Melt jetting	Visible strand formation of the initial inflowing melt on the surface of the article	Unsuitable sprue location and dimensions	Avoid jetting by moving the gate (inject against a wall), increase gate cross section
		Injection speed too high	Increase injection time and inject stepwise (slow-fast)
		Melt temperature too low	Raise melt temperature
Incompletely shaped molding	Incomplete filling, especially at positions on the flow route and where walls are thin	Flow properties of the plastic inadequate	Raise melt and mold temperatures

3.3. Compression, Injection and Blow Molding

Table 3.10. (Cont'd)

Defect	Possible appearance	Possible causes	Recommendations
Flow weld strength unsatisfactory	Clearly visible notches along the flow line	Injection speed too low	Shorten injection time and/or raise injection and/or hold pressures
		Wall thickness of the molding too thin	Increase the wall thickness of the molding
		Nozzle not sealed against the mold	Increase nozzle contact pressure, check radii of nozzle and sprue bush, check centering
		Gate system has too small a cross-section	Enlarge sprue, flow channel and gating into the molding
		Mold evacuation inadequate	Optimise mold evacuation
		Flow properties of the plastic inadequate	Raise melt and mold temperatures, possibly reposition gate to improve the flow properties
		Injection speed too low	Shorten injection time
		Wall thickness too small	Adjust wall thickness
		Mold evacuation inadequate	Improve mold evacuation
Warped moldings	Moldings are not flat and show angle warpage, propeller warpage or convex shape	Wall thickness and shrinkage differences too great, strong transverse orientation from glass fiber in the interior of the article	Design according to the plastic, alteration of the gate position
		One-sided stronger orientation in the machine direction by external glass fibers	Alter the position of the wall thickness so that the melt flows so as to give a symmetrical glass fiber distribution
		Mold temperature unsatisfactory	Differentially heat mold halves
		Change over point from injection to holding pressure unsatisfactory	Alter change over point

Table 3.10. (Cont'd)

Defect	Possible appearance	Possible causes	Recommendations
Article sticks in mold	Dull spots and finger or cloverleaf shiny depressions on the surface (mostly near the sprue gate)	Local temperature on the mold wall too high	Lower mold temperature
		Demolding too early	Increase cycle time
Molding is not ejected or is deformed	Molding sticks. Ejector pins deform the molding or punch through it	Mold over charged, undercuts too strong, insufficient polish on screw threads, ribs and pins	Increase injection time and lower holding pressure, remove undercuts, re-machine mold surfaces and polish longitudinally
		Vacuum between article and mold on demolding	Improve mold evacuation
		Elastic mold deformation and core misalignment caused by injection pressure	Increase the stiffness of the mold, check core plate
		Demolding too early	Increase cycle time
Flash formation	Formation of plastic skins on mold gaps (e.g. parting surfaces)	Mold pressure too high	Somewhat lengthen injection time and shorten holding pressure, advance change over point from injection to holding pressures
		Mold parting surfaces damaged by overfeeding	Re-machine the mold in the region of parting lines or contours
		Closing or locking force not adequate	Raise closing pressure, possibly use the next largest size of machine
Rough and dull molding surfaces (with GF reinforced thermoplastics)		Melt temperature too low	Raise melt temperature
		Mold too cold	Raise mold temperature, fit thermal insulation to mold, use a more powerful heating unit
		Injection speed too low	Shorten injection time

with indications of their possible causes. For pre-drying of materials or processing with a vented barrel see Section 3.3.2.2 and 3.6.3.3.

Cleaning injection-molding machines is particularly difficult after their use with high-melting materials. The injection unit is cleaned with an easily removable material, the flow temperature of which is related to the melt temperature of the preceding mass. Such cleaning materials are commercially available. Finally wood chippings are run through the machine, which is rubbed clean with brushes and polishing rags. These are fastened to a rotating head that fits the cylinder.

3.3.6.2. *Injection molding of filled compounds*

Mineral-reinforced thermoplastics (Sec. 4.1.13) require injection temperatures at least 10°C higher and higher pressures than corresponding unfilled polymers. Because the melts solidify quickly they must be injected as fast as possible and the mold temperature must be maintained as high as possible in order to obtain stain-free moldings with smooth surfaces. To obtain fault-free ejection of the rigid moldings tapering of 1 to 2° is necessary.

3.3.6.3. *Multi-component injection molding*

Multi-component injection-molding machines have two or three independently controlled injection units working to one locking unit. Multi-colored products are made with locking units whose nozzle-sided platens have an opening for each nozzle. Such products include car rear-light assemblies and typewriter keys, double-walled pots and closures with molded-on soft gaskets. The molded part is presented to the following injection unit after each cycle for completion. Simultaneous regulation of the feed of a variety of colored materials into the mold enables colored patterned moldings to be produced.

In *sandwich injection molding* the laminar flow of the material at the inflow into the mold cavity is used: within the solidifying skin of the outer material adhering to the wall, the core material flows forward, so that it is possible to inject a different type of core within the peripheral material. In the process two injection units operate together to feed the same sprue.

The feed is regulated by valves or multi-cut-off nozzles, which make it possible for material from either of the injection units to flow first or second, as preferred (to seal the sprue), or to flow simultaneously. To avoid breaking through to the core, the outer layer must be at least 1 to 1.5 mm thick. The sandwich injection technique is suitable only for moldings with total wall thicknesses ≥ 5 mm. With very thick walls, which require long cooling times, two alternating locking units are used. Sandwich injection molding permits the commercial production of moldings, the outer skin of which can be highly stressed with the core made, for example, from cheap, filled or recycled material. Further, a variety of plastics, such as PP-PE, PMMA-PS, CA-ABS and PC-ABS, can be combined with one another. Shear-sensitive plastics, such as rigid PVC, are not suitable.

In *foamed-core sandwich injection* the quality of the surface is approximately equal to that of the solid molding, so that foamed-sandwich parts do not generally require any finishing (see the following section).

3.3.6.4. Structural foam and controlled internal pressure molding

Thermoplastic injection molding of foam moldings with a dense outer skin over a foamed core (Fig. 13.5) has expanded the field of application of injection molding to thick-walled rigid moldings. Wall thicknesses are about 5 to 15 mm (2–6 in), lateral dimensions more than 1 m (> 3 ft) and weights up to 40 kg (90 lb). These are used for housings, containers, palettes, sports equipment made of HI-PS or PE. It competes with reaction injection molding of PU (Sec. 4.6.1.8.2). Foam injection processes present a problem in that the foam forming at the mold walls collapses, resulting in rough, streaky surfaces. Then moldings must not only be lacquered but also ground and polished. Processing developments aim at improvement of the surface structure as in foamed core sandwich moldings (v. a.).

In low pressure methods sufficient material is projected into the deepest part of the mold at 100 to 200 bar back pressure above the gas pressure generated by the blowing agent to fill it to 60 to 80%. The mold needs to accommodate only the foam pressure of the expanding mass, about 10 to 20 bar. The locking units of the machines and the molds can be made correspondingly light. So that air pressure does not counteract the foam pressure, degassing channels (0.08 mm, approximately 3 mil) are introduced. With long flow paths, multicasting through hot runners is useful and, for large moldings, parallel operation of several injection units is used.

Molding compounds with chemical blowing agents for structural foam articles up to 1.5 kg can be plasticized and injected by conventional screw injection. The machines are fitted with non-return valves and needle shut-off nozzles and fast-shot arrangements. In transfer machines dealing with greater volumes, screw plasticized material is collected in an accumulator for a shot, which is projected into the mold by gas pressure (Fig. 3.32).

The cooling time of the moldings is significantly longer than that of normal solid injection moldings. They must be thoroughly cooled otherwise they continue to expand after ejection. Hence, multi-station machines with changing locking units are often used.

In the Controlled Internal Pressure *Cinpress process*, compressed gas is injected at the same time as the molding compound, computer controlled in such a way

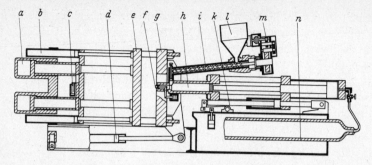

Fig. 3.32. Structural foam transfer molding machine. Stroke volume, 19,000 cm^3; material pressure, 1100 bar; locking force, 600 MP; Daylight, 1200 mm; platen dimensions, 1900 × 1900 mm. (a) Locking force cylinder, (b) Locking cylinder, (c) Locking, (d) Hydraulics for reciprocating locking unit, (e) and (g) Platens, (f) Nozzle lock, (h) Accumulator, (i) Plastifying unit, (k) Vertical adjustment, (l) Hopper, (m) Screw drive, (n) Compressed gas cylinder.

that, instead of irregular bubbles, well defined hollow channels in reinforcing ribs are formed beneath solid and smooth surfacing walls.

3.3.6.5. Transfer-injection stretch molding is a two-stage process for densified moldings of semi-crystalline plastics. The mold contains two cavities, which are placed in turn against an injection unit by a hydraulic slide or turntable. A blank is injection molded in a cavity of simple design, for example a hollow cylinder. After a suitable holding time, the mold is opened, the part is ejected from the second, shaping cavity, e.g. for a gear wheel, and the blank remaining on the core on the stationary side is transferred into the second cavity. There it is compression molded into a finished molding by the force applied through the closing unit. If the temperature is properly controlled, the moldings densified by crystallite orientation have the same service temperature range as injection molded parts.

3.3.6.6. UHF and *resistance sintering of extremely viscous melts.* If PTFE, PE-UHMW and highly filled compounds, which because of their high viscosity and shear sensitivity cannot be injection molded, are coated with a UHF-sensitive medium (carbon black, Zeolith T, Trigonoc C) and heated in the "UHF Applicator" chamber of a press, the heat absorbed by the coating is conducted to the interior of the granules. The compound is then impact sintered to a homogeneous blank, which can be shaped by a transfer process in a second mold. If such molding compounds are made sufficiently electrically conductive by addition of about 2% carbon black, they can be heated by direct resistance heating in a process that is programmable for high manufacturing rates, followed by impact sintering.

3.3.6.7. Blow molding

In *blow molding*, hollow articles are made by a two-step process. The first step is extruding an open tube or injection molding tubes closed at one end as preforms or *parisons*. In the second step, parisons heated to an appropriate flowability are clamped in cooled outer molds and molded to the finished shape mainly by gas pressure. If necessary auxiliary mechanical expanding and calibrating devices are used. For symmetrically shaped articles *blowing* of parisons conditioned at optimal stretching temperature can be combined with *biaxial stretching*.

The two steps can follow each other immediately in a single heating phase, called *one-stage blow molding*, or separately in a *two-stage blow molding* process. For blow forming of sheets see Section 3.8.4.2. Blow molding is used for the manufacture of hollow articles of nearly any shape from < 10 ml to 10 000 l (< 0.003–2600 gal) in volume. They include regularly shaped packing and storage units as well as complicated shapes, e.g., pallets, heating pipe manifolds, bellows and gasoline tanks for motor vehicles, large toys (rocking horses) and sporting goods (surfboards, toboggans).

Because the *blow molds* (mostly two part) are external to the molding they seldom give rise to ejection problems. Blowing pressures do not exceed 10 bar, and the cavity walls are not subjected to wear, because the moldings only contact them. Blow molds can therefore be lightly constructed and can be made by casting from master molds using zinc or light-metal alloys without excessive cost. They are therefore suitable for short runs. They are amply provided with channels for de-gassing and have a finely sand-blasted internal surface so that demolding is facilitated by the penetration of air between the cavity and the molded article. Cooling times determine cycle times. External cooling by means of cooling channels in the mold is augmented by purging with air or water sprays or cryogenic gases (N_2 or CO_2).

Extrusion blow molding

Euromap 40 (1940) is a model specification for extrusion blow molding machines. This document also contains a Euromap parameter summary. This consists of closing force in tons (1 t = 10 kN), the maximum separation of the tool platens, the maximum distance in mm between the platens with the clamping unit open, the length of the mold platens in mm (measured in the direction of the parison delivery), the screw diameter in mm or the accumulator volume in cm^3. An example is 15/700 × 600-2500. Additionally information should be given about the size, weight and quantity of blow moldings required. *Extruders* with screw of 30 to 160 mm in diameter (1–6.4 in), usually 20 *D* long, resemble those used in the production of pipes (see Sec. 3.6.3.2 and 3.6.4).

3.3. Compression, Injection and Blow Molding

Depending on circumstances, parisons of up to 1 kg (2 lb) for blow moldings of between 10 and 30 l (2.5–8 gal) can be produced in less than 30 seconds from molten tubes emerging vertically and continuously from *side-fed dies* (for PE) *or mandrel support heads* (for PVC, see Sec. 3.6.2.3. These are taken off by grips or directly by *alternating or rotating molds* and delivered subsequently to finishing stations. When, because of its size or material characteristics, a parison remaining in free suspension for too long stretches or cools too much, the melt is extruded hydraulically by a continuously running screw through an *accumulator* during a preset phase of the plastifying period. Tubular ram accumulator heads (Fig. 3.33) are used for stroke volumes of 2 l (0.5 gal) and more. For heating-oil tanks with a capacity of 10 000 l (2660 gal) blow molding machines fed by several large extruders have reservoirs of up to 300 kg (660 lb) PE and extrude at the rate of 12 kg/s.

To regulate the wall thickness of the products for varying requirements and/or blow-up ratios, conical parison head dies are generally fitted with electronically controlled adjustable cores (Figs. 3.33 and 3.34, a,b) to alter the die gap during extrusion of the parison. *Multilayer* coextrusion (e.g., for containers with barrier layers) is possible (Fig. 3.35). On locking the blow mold the parison is pinched off by replaceable parts of the mold. In the production of containers one of these is constructed as the welding and pinch-off edge for the base, the other as the calibrating setting for the opening. Blown air is led through a mandrel incorporated in the calibrating unit.

Flash is pinched off outwards; in the case of bottles, this can account for 20% of the material used and for over 50% in asymmetric moldings. The proportion of flash or its reworkability for production can affect the economics of the process. High-viscosity PE-HD grades (Sec. 4.1.2.1.5.3, also with intermediate barrier layer) are used predominantly for large extrusion blow moldings because they offer a favorable combination of good flow, thermal stability in the run

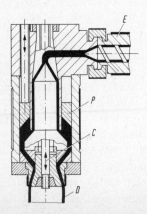

Fig. 3.33. Storage or accumulator head.
(E) Extruder
(P) Annular piston
(C) Wall thickness control
(D) Parison

3.3.6.7. Blow molding 109

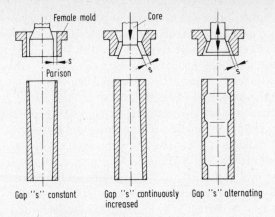

Fig. 3.34a. Principle of wall thickness regulation.

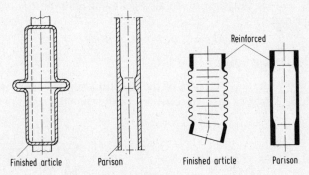

Fig. 3.34b. Examples of wall thickness regulation.

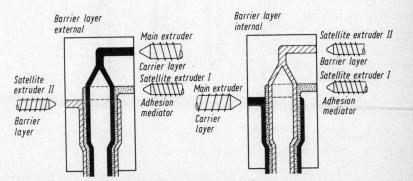

Fig. 3.35. Diagram of process for three-layer coextrusion.

through the accumulator, re-use of flash and properties of the product. PE-LD, PP, and ABS are also very suitable for extrusion blowing. PVC-U is used for bottles of up to 2 liters. Impact-resistant PS with physical blowing agents is also used for double-walled insulated beakers.

A variety of ancillary equipment is associated with extrusion blow molding machines including blowing and ejection stations, flash removal, weighing and density determination, discharge stations, filling, locking, printing and labeling of containers, welding or molding of supports, etc. In mass-production lines extrusion blow molding machines are largely automated, often electronically regulated and controlled and hydraulically or pneumatically activated.

Extrusion blow molding is the most generally useful process applicable to products of all sizes and shapes.

Dip blow molding is a particular form of extrusion blowing (Fig. 3.36). The process delivers flash-free hollow moldings without weld lines with slightly drawn walls and is suitable for small packaging and bottles up to 1 gal (approx. 4 l) capacity.

Injection blow molding

In injection blow molding the core of the injection mold is constructed as the blow mandrel. Parisons are injected cyclically on the mandrel with an

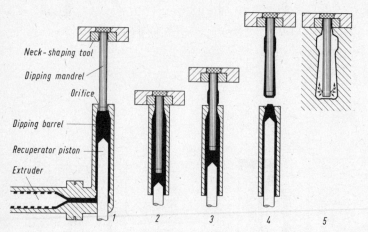

Fig. 3.36. Dip blow molding.
(1) dipping mandrel starts plunging into the dipping barrel
(2) mandrel is dipped completely, neck is formed in the neck-shaping tool
(3) preform thickness is controlled by the relative speed between mandrel and piston
(4) between piston and mandrel a solid melt strand is formed; it is cut after both have reached their upper position, the melt strand, now completely covering the mandrel, yields a closed-end preform
(5) preform is blown into shape

3.3.6.7. Blow molding

accurately shaped neck and a variable wall thickness depending on the hollowing out of the mold halves. The mold platens with the attached blow mandrel or the injection and female blow molds operate intermittently up to the ejection of the molding. Where appropriate, a conditioning station for biaxial stretching is switched between injection and blowing (Fig. 3.37). Smaller machines with only two stations and rotating injection and blow models are used for flash-free production of, for example, accurately molded pharmaceutical containers without bottom weld lines.

In flow compression molding, melt in a mold cavity flows upward around the blowing mandrel which dips in the mold to produce the parison.

Collapsible tubes, ampoules and roll bottles are press blow molded by:

(1) Injection molding of the closed-off part,
(2) Drawing of the closed off part in the injection mold while the parison is formed from the nozzle mass;
(3) Expansion of the parison inside the locked blow mold;
(4) Ejection of the hollow molding while pinching off the bottom flash.

Fig. 3.38 shows the manufacturing sequence for a shaft sleeve of TPE. In the vertical, self-operating injection mold there is a three-jawed gripper (with in-built blow mandrel) in which the external contour (here the small flange of the sleeve) is formed. Drawing up of the mold, jet-cone setting to regulate the wall

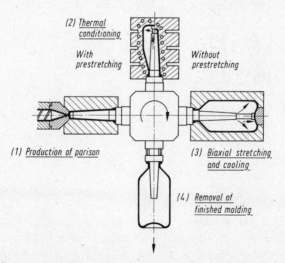

Fig. 3.37. Principles of injection stretch blow molding with rotating injection molding and blowing mandrel. Station 2 is absent from plants for injection blow molding without stretching.

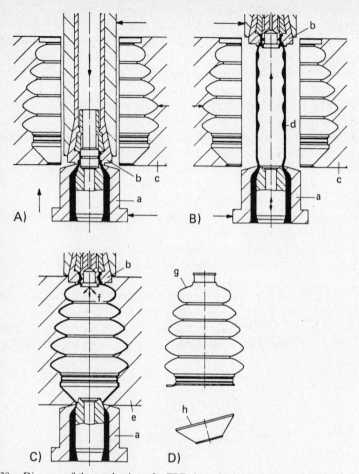

Fig. 3.38. Diagram of the production of a TPE sleeve by injection blow molding.
A: Injection molding of head, B: Press-drawing of parison controlled by wall thickness and speed, C: Blow molding of sleeve, D: Demolded article with base cut off
a: Melt feed head, b: Injection mold for the head, c: Opened blow mold, d: Extruded parison with different wall thicknesses, e: Closed blow mold, f: Blowing air inlet, g: Collar, h: Sprue

thickness and pressing out of the melt from the accumulator of the die head is controlled by a suitable program so that the shape of the preform can exactly match the desired wall thickness distribution of the blow molded part.

This technique allows very accurate blow molded parts to be produced which, at least on the head of the article, are scrap-free. The end of the sleeve with the large opening is produced after automatically cutting off the body. The following dimensional tolerances can be obtained: 0.03 mm on the small sleeve (the

3.3.6.7. Blow molding

accuracy of an injection molded part) and 0.1 mm in the folded region which would not be possible with extrusion blow molding with freely hanging parison.

Stretch blow molding

Stretch blow molding is the mass-production process for plastic bottles replacing the glass variety. In particular these are used for oil, wine, spirits, milk, still and carbonated mineral water, soft drinks and beer. It should be borne in mind that the situation will vary in different countries according to their consumer habits and the legal situation in relation to food handling. To meet steeply rising demand, stretch blowing plants have been developed to produce more than 10,000 bottles per hour.

Stretch blowing involves mechanically extending parisons produced by one of the processes already described at optimal stretching temperature (by means of a stretching mandrel internally and tongs externally) longitudinally and radially either simultaneously or consecutively by blowing. This is regulated to give a blow-up ratio of 5 to 10. Such biaxial stretching results in:

(1) A several-fold increase in stiffness and impact resistance, particularly at low temperatures, and thus in a reduction in the weight of the products;
(2) Significant reduction of the specific permeability to oxygen, CO_2 and water vapor;
(3) Improvement in transparency and surface gloss.

The optimum stretching temperatures for an amorphous plastic such as PVC-U lie within its broad thermoforming range (Fig. 3.70) and can be achieved by cooling the parisons formed from the higher melt temperature. Amorphous thermoplastics are especially suitable for *single-stage stretch molding*. Partially crystalline plastics are susceptible to stretching only within certain crystal content limits (5–25%). For completely transparent products the crystallites must not grow and must not, for example, aggregate into spherulites. In bottle blowing with PET (Sec. 4.1.9.3), T_m of 255 to 260°C, spherulite formation is reduced by force-cooling the melt from 285°C down to the region of the glass transition temperature of the amorphous component (T_g approximately 80°C) and conditioning at the stretching temperature of 90 to 100°C. Such plastics lend themselves to *two-stage stretch molding*, in which the heating, conditioning and blowing equipment for parisons cooled to below the T_g and drawn in other positions is linked with the stations at which the bottles are filled with liquid. In the single-stage extrusion blow molding (Fig. 3.39) of biaxially stretched glass-clear bottles from polypropylene copolymer with slow crystal nucleus formation, an extruded tube is run through three calibration and cooling tanks for controlled crystallization at about 95°C. Then, for the required stretching and

3.3. Compression, Injection and Blow Molding

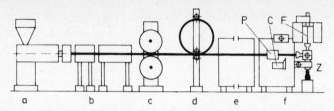

Fig. 3.39. Principle of single-stage-stretching plant for PP containers (Bekum/Hercules).
(a) Extruder with pipe die, (b) Pipe calibration and cooling, (c) Hand-off, (d) Equalizing loop, (e) Oven, (f) Stretch-blow machine: P Puller, C Cutter, Z Claw system, F Mold locking unit and calibration

blowing, it is run through a long oven at a temperature of 150 to 155°C, the lower limit of the crystallite melting range.

Typical bottle weights are 23 g for a 0.33 l PVC bottle, 32 g for a 0.625 l PP bottle and 140 g for a highly rigid 2 l bottle made of polycarbonate (PC). A major consideration in the commercial viability of plastic bottles is their ability to be re-used or re-cycled, see section 00.

3.3.7. Molding Techniques for Thermosets and Elastomers

The cycle time for molding of thermosets is largely determined by the *time required for curing* the molding before removing it from the mold (Fig. 3.40). Thermosetting molding compounds are supplied in a variety of grades. They are characterized by their flow behavior (soft, medium-hard, hard) and their cure speed. Injection molding grades must be so adjusted that they can be plasticized at the lowest possible cylinder temperature and withstand a cylinder residence time of 3 to 6 min and also cure quickly at the elevated mold temperature. Very soft materials with high flow are required for moldings with high flow-path wall thickness ratios, for complicated parts with cores around which the plastic must flow and for semi-automatic production of moldings with inserts that must be placed in the tool by hand. In such cases a slower cure must be accepted.

For uniform feeding in compression, transfer and injection molding, dust-free and free-flowing granules are required. Special pelletizing is necessary for impact-resistant molding powders with fiber or chopped cloth reinforcement. Long-fibered or coarsely structured materials cannot be easily injection molded.

Thermosets can contain or develop gas-forming volatile substances in the course of processing. In compression molding of inadequately preheated materials, the punch is raised ("breathing") for 2 to 3 s just prior to completion of compression. Inadequate or delayed breathing causes blistering or porosity. In transfer molding the tools must be provided with venting channels and, in

3.3.7. Molding Techniques for Thermosets and Elastomers

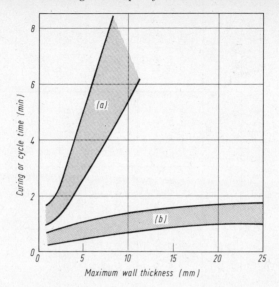

Fig. 3.40. Dependence of hardening or cycle time on wall thickness of phenolic moldings in compression and injection molding.
(a) Phenolics pressed without prewarming at 160°C curing time, (b) Thermosets injection molding cycle time.

injection molding, volatiles are for the most part lost from the injection cylinder through the feed hopper.
Temperature and pressure values for compression and injection molding of thermosets are given in Table 3.11. In transfer molding a pressure of 400 to 500 bar higher than that for injection molding is necessary. The mold is usually heated by electrical cartridge heaters, whereas the injection cylinder is heated by circulating heating fluid. *Back pressure* is important in the injection molding of thermosets because of its effects on plastification and the equalization of variations in shot volume. With mineral-filled powders careful metering is necessary because they produce high frictional heat and increased wear of the injection unit. With hot runner molds (Sec. 3.3.5.2) thermosets are injected without a sprue. In the transfer-injection process the feed plate is hydraulically exchanged directly after injection for a gate-locking plate which makes for fast curing in strongly heated molds. Processing shrinkage of thermosets is higher in injection than in compression molding. Table 3.12 surveys types of faults in thermoset moldings and their possible causes.
Cylinders heated by oil circulation are used in the injection molding of thermoelastic peroxide cross-linking polyethylenes and for vulcanizable elastomers. Raw rubber mixes are mostly fed as strip feed on a drum, from which it is

Table 3.11. Data for Processing Cross-linking Molding Compounds

Process	Compression		Injection					
			Cylinder Temp.*					
Molding compounds	Temp. (°C)	Pressure (bar)	Feed zone (°C)	Nozzle (°C)	Mold temp. (°C)	Back pressure (bar)	Injection pressure (bar)	Holding pressure (bar)
PF, filler:								
mineral	150–165	150–400	60–80	85–95	170–190	to 250	600–1400	600–1000
wood-flour	155–170	150–350	70–80	90–100	170–190	300–400	600–1400	800–1200
fiber flock	155–170	250–400	70–80	95–110	170–190	to 250	600–1700	800–1200
chopped fabric	155–170	300–600	–	–	–	–	–	–
MF, filler:								
cellulose	135–160	250–500	70–80	95–120	150–165	300–400	1500–2500	1000–1400
wood flour	145–170	250–500	70–80	95–120	160–180	to 250	1500–2500	800–1200
asbestos twine	145–170	300–600	65–75	90–100	160–180	to 150	1500–2500	800–1200
UP mineral + GF	130–170	50–250	40–60	60–80	150–170	without	200–1000	600–800
EP mineral + GF	160–170	100–200	ca. 70	ca. 70	160–170	without	to 1200	600–800
PE, cross-linking	120°C melt to 200°C cross-linking – temperature		135–140	135–140	180–230			

*The material temperature cannot be determined exactly because of frictional heating.

unwound by the screw. Several vulcanizing presses are fed from one injection unit.
For manufacturing large surface area moldings of SMC and BMC (Sec. 4.6.2.4.2), special presses with relatively short cycle times (Sec. 3.3.4.2) produce the highest strengths, because the respective fiber reinforcement structures are largely retained. In injection molding, extensive uncontrollable reorientation of the fiber structures, including shearing of the fibers by the screw must be avoided if possible. Special machines with metering feed units and large nozzle cross-sections are used for BMC, while ram feed is used for ZMC processes, see Sec. 4.6.2.4.2. Injection compression molding with "breathing" nips up to 40 mm can produce an orientation of the fibers, increasing the strength (Sec. 3.3.2.3). Strips with lengthwise reinforcing fibers are readily introduced and conveyed by the screw.

3.4. Production of Semifinished Thermosets

3.4.1. Casting

Solvent-free liquid PF resols (Sec. 4.6.1.1.1b), formulated for slow curing at elevated temperature, and liquid, cross-linking reaction resins (Sec. 4.6.1.4) (which can also contain fine granular mineral filler) are cast in open molds to produce sheets, tubes or other profiles of limited dimensions, which are then further machined. Depending on the formulation, curing may take place in the hot or in the cold.

3.4.2. Laminating

The supporting webs of industrial and decorative laminates are immersed in hot solutions of curable resins in continuously operating plant, then dried and cut up into appropriately sized sheets. Plies of the impregnated web in the required numbers are placed one on top of another between polished plates with cushioning pads between the plates and platens. These assemblies are fed into the individual daylights of multi-daylight high-pressure presses. At 130 to 180°C and pressures of 7 to 20 N/mm^2 (70–200 bar) the laminae are consolidated by the curing resin into laminated sheets, which are cooled under pressure (Sec. 4.6.3.5).
Flexible decorative laminates (0.3 to 0.6 mm thick) and flat or corrugated *GRP continuous webs or sheets* (Sec. 4.6.3.6 are continuously produced with solvent-free liquid resin (Fig. 3.41). For GRP filament-wound pipe see Sec. 3.2.4, for centrifugally cast pipes Sec. 3.2.3.2. Pipes of high rigidity and high bursting strength can be cast in centrifugal casting molds from braided roving tubes.

3.4. Production of Semifinished Thermosets

Table 3.12. Faults in Processing Thermosetting Plastics

Location of fault	Molding material								Injection or compression unit							Tool and clamping unit								
			Flow setting		Curing speed		Material temperature		Metering		Injection speed		Injection pressure			Tool design				Tool temperature		Hardening time		
Reasons for fault	Too moist	Too much mold lubricant etc.	Too rigid	Too soft	Too high	Too low	Too high	Too low	Too high	Too low	Too high	Too low	Too high	Too low	Holding pressure too low	Runners too narrow	Otherwise unfavorable	Ejector not right	Degassing inadequate	Too high	Too low	Too long	Too short	Locking pressure too low
1. Material faults																								
1.1 Bad mixing ⎫ With coarse fillers 1.2 Porosity ⎬ heterogeneous mass 1.3 Cloudiness and striation ⎭				+												+	+							
1.2 Porosity	+			+				+		+		+		+	+	+			+		+		+	+
1.3 Cloudiness and striation				+			+				+		+											
1.4 Large blisters – moldings matt and deformed	+			+		+		+																
1.5 Small blisters, burst, moldings smooth	+						+				+		+			+				+		+		

Table 3.12. (Cont'd)

Fault											
2. Surface faults										+	+
2.1 Not smooth (orange peel)		+				+		+		+ +	+ +
2.2 Inadequate shine				+					+	+ +	+ +
2.3 Matt spots			+			+ + +		+			
2.4 Light spots				+		+ + +		+			
2.5 Burn marks			+		+					+	
2.6 Sticky								+			
		+ +		+		+			+	+ +	+
				+	+	+	+				
						+ + +		+		+	
						+ + +		+		+	
			+	+			+		+		+ + +
				+	+					+	+ +
			+			+					
						+					
3. Appearance faults	+	+				+ + +	+				
3.1 Sink marks			+ +	+		+					
3.2 Bubbles				+			+		+		
3.3 Short moldings				+		+ +			+		
3.4 Flow marks	+	+				+ +		+	+ +	+	+
3.5 Grooves show through		+		+	+	+					
3.6 Sticking to mold			+	+							
3.7 Binding in the mold	+										
3.8 Excess flash						+					
4. Structural faults											
4.1 Molding warped	+ +										
4.2 Moldings torn	+ +										
4.3 Metal inserts damaged or bent											
4.4 Plastic in metal inserts											

3.4. Production of Semifinished Thermosets

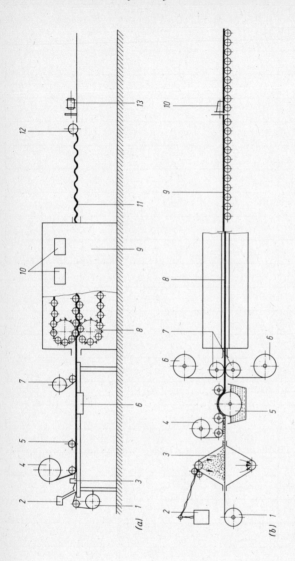

Fig. 3.41. Production of continuous strand UP-GF webs and sheets.
(a) Laterally corrugated

(1) Lower separating film
(2) Resin coater
(3) Metering unit
(4) Glass mat
(5) Impregnating roll
(6) Visual screening in transmitted light
(7) Upper separating film
(8) Chain links with former
(9) Forming and curing zone
(10) Ventilating flaps
(11) Corrugating sheet
(12) Lateral trimming
(13) Separating disc

(b) Flat or corrugated longitudinally (Filon plant) with correspondingly profiled resin content

(1) Perlon spool
(2) Rovings
(3) Chopped glass strand deposit
(4) Perlon
(5) Immersion tank
(6) Film rolls
(7) Double contacting rolls
(8) Curing zone
(9) Cooling zone
(10) Cutter

3.4.3. Extrusion and Pultrusion

Pheno- and amino-plastics are extruded to make decorative profiles and stay bars of limited cross section. Special phenolics are employed in the extrusion of corrosion-resistant chemical tubes. Extruders are used in which the profile hardens in the die and generates the necessary back pressure.

Profiles reinforced with continuous longitudinal fibers with reactive resin binders are manufactured by the pultrusion process. Roving strands fed tangentially by rollers, and possibly also tapes or strips of fabric are bundled through a resin impregnation bath and subsequently squeezed to remove excess resin. This mass is then passed through a high frequency preheating zone and fed to the profile steel mold, which may be up to 70 cm long. The electrical band heaters should be adjusted so that the profile is completely cured in the mold. Friction with the walls of the profile in the mold requires considerable tensile forces to be applied by the downstream caterpillar or hinged drawing device. *"Pulforming"* is a further development of the pultrusion process. The resin coated and HF-preheated strand is drawn in continuous, successively circulating shaping tools of changing cross-section for straight moldings (safety steering columns) or with any desired cross-section for curved products (automobile leaf springs). For more information regarding drawn and wound composite profiles, see Sec. 4.6.3.3/4.

3.5. Semifinished Thermoplastics: Older Processes

3.5.1. "Celluloid" Process

Cellulose nitrate and camphor (cellulose acetate and plasticizer in the case of acetylcelloid) are compounded with a solvent at a relatively low temperature. Flat or rolled sheets are made from these compounds on lightly heated rolls and these are pressed and layered one on top of another in presses into thick flat or round blocks. Various patterns can be produced in this way by repeated layering of differing color components.

Full profiles are continuously sliced or cut longitudinally from the flat or round block with a profile cutting tool. Sheets or laminae are cut from blocks on a horizontal cutting machine in the same way as cut veneers. Celluloid and acetyl celluloid must be freed from solvent by lengthy air drying. The cut films and sheet are finally polished by pressing.

3.5.2. "Astralon" Process

In the *Astralon process* rigid vinyl copolymers are first worked on heated mixing rolls (Sec. 3.1.4.2), plasticized solely by heating and finally drawn into sheets on the calender (Sec. 3.7.3).

3.5.3. Pressing Sheets and Blocks

Cuttings from calendered sheets of soft plastics and thin celluloid sheets are pressed between metal plates in heated multi-daylight presses. They are removed from the press after cooling under pressure. To obtain thicker sheets several thinner laminae are consolidated together in the press. The surface of the polishing plates is reproduced on the plastic. For pressing thicker blocks, the material is enclosed within a detachable heat-insulated framework between the heating and cooling platens.

Polishing presses are used to produce plastic sheets of uniform anisotropic high quality and surface finish such as is needed for drawing equipment, windows in soft top cars and high-quality flooring.

3.5.4. Casting Films

Solutions of polymers are processed through a slit die and cast, either onto a continuous belt made of polished metal, for example, nickel, or onto a casting wheel, and then heated to remove the solvents. Such cast films are distinguished by their great clarity and parallel surfaces. Continuous webs are also cast in this way from dispersions and pastes. For "casting" from melt see Sec. 3.6.5.7.

3.6. Extrusion of Semifinished Thermoplastics

3.6.1. General

Extrusion differs from molding in that it is a continuous process forming products, such as pipes, tubing, profiles, filaments, sheets and films, by forcing plasticated materials through a shaping orifice, called the die. It is used for thermoplastics as well as for elastomers and thermosets (see Sec. 3.4.3.).

The plasticizing extruder has a threefold purpose:

(1) To draw in powder or granules through the feed mechanism and to convey the material while compressing and on occasions devolatilizing it;
(2) To mix and produce a homogeneous melt;
(3) To develop the pressure required to overcome the flow resistance of the open die so that the profile emerges continuously from the die.

Screw extruders are universally available for this. In basically the same way as for thermoplastic semi-finisheds they are also used for mixing and granulating compounds (Sec. 3.1.4.1), for the production of parisons in blow molding machines (Sec. 3.3.6.7) and for feeding other machines, e.g., calenders (Sec.

3.7.3). Some extruder types (Sec. 1.8.2 and 3.1.4.1) are designed to combine polymerizing, compounding and production in one unit.

Hydraulic extruders are now used only for thermoplastics with extremely high melt viscosities. The ram extrusion of tubes from PTFE is considered in Sec. 4.1.5. The Engel process uses twin ram extruders with two pistons working in cycle under pressures of $1000\ N/mm^2$ to plasticize the mass adiabatically in the shortest time. They then feed the injection head alternately. The process is used for cross-linked PE tubes and also for profiles of very high molecular weight PE.

Raw materials melts are transferred from polymerization vessels with *gear-type pumps*. They are interposed between extruder and mold, relieve the extruder of the task of building up the pressure and deliver a steady stream of melt. They are used for accurate metering, e.g. for extruding strands or fine film for photographic film backings. Gear-type pumps with 20–5000 kg/h throughput and 350 bar maximum pressure are available for pipe, profile, cable sheathing and flat film production lines and those with up to 1000 kg/h for film blowing.

3.6.2. Construction and Operation of Screw Extruders

3.6.2.1. Types and sizes

Screw extruders can be fitted out with one or with more screws working in coordination. Construction and size are designated internationally by number codes. Thus, according to Euromap 20, 1-90-25 signifies:
  1–The number of screws
  90–Screw diameter D in mm
  25–Effective screw length as a multiple of the screw diameter ($L = 25\ D$)
Degassing extruders are indicated by the supplementary letter V (vented).

3.6.2.2. Extruder components

Multi-screw extruders require more gearings and thrust bearings in the gearbox and a cylinder cross-section appropriate to the type of screw. Otherwise they have the same components as in the single screw extruder diagram, Fig. 3.42. Infinitely variable commutator motors (1:6) or thyristor regulated dc motors (1:10) power the screw drive and control its speed. Hydraulic drive is seldom used.

The *feed hopper* is a separate component with water cooling around the feed section. Single or two-stage heatable vacuum feeds for pneumatic delivery of powders provide significant increases in output and quality. The flow of mate-

3.6. Extrusion of Semifinished Thermoplastics

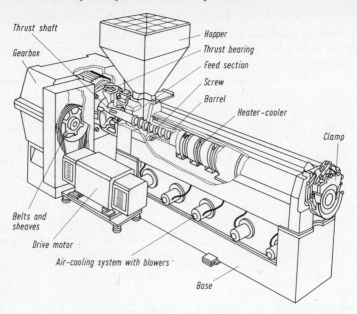

Fig. 3.42. Typical single-screw extruder.

rial to the screw can be regulated by the conveying spirals, baffle elements, the stirrer or the metering screw in or after the feed hopper.

The *barrel* and *screw*, made of high-strength steels, are nitrided. Hard chroming protects against corrosion. Protection against wear is provided by coating the (bimetal) cylinder with carbide and by case-hardening of the screw.

Barrel heating-cooling is zoned into from three to seven sections, each section electrically heated and cooled by compressed air or circulating (distilled) water or steam. At the inner surface of the cylinder, induction heating is more effective, though more expensive than resistance heaters at the cylinder jacket. Screws with internal cooling can be additionally temperature controlled with air, oil or water.

3.6.2.3. Auxiliary parts between clamp and die

Flow restriction bushes, adjustable nip widths at the screw tip or perforated *breaker (strainer) plates* are used to adapt the melt pressure to the die resistance. The breaker plates are also used as carriers for the screens that are exchangeable during operation. Flight-less (Fig. 3.55), side feed (Fig. 3.43)

3.6.3.1. General 125

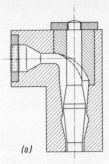
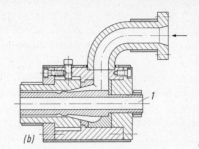

Fig. 3.43. Types of side-feed extruder head.
(a) Side-feed cross-head tube die with mandrel held from above and heart-shaped groove insert.
(b) Spiderless offset tube head, simple calibrating air flow and fixing of calibration devices at 1. Not suitable for PVC pipes on account of shearing of the melt by the double bend (danger of burned spots).
For (c) flight-less head for PVC pipes see Fig. 3.55, p. 000.

extruder heads or *melt distributors* (Fig. 3.58) are introduced to obtain a uniform distribution of the melt in the die land.

3.6.2.4. Extruder operating parameters

The most important processing parameters are screw speed and torque, the temperature profile along the length of the screw cylinder up to the die and the metering rate and temperature of the feedstock. These determine the melt pressure and temperature at the screw head and, together with the downstream equipment, the quality of the product which can be monitored by non-contact testing of contours, wall thicknesses and weights.

The temperatures are held constant in each zone by means of individual electronic control circuits. High performance machines use computers for process control with sensors for melt pressure and viscosity.

3.6.2.5. Processing faults and their causes

See table 3.13.

3.6.3. Single-Screw Extruders

3.6.3.1. General

Single-screw extruders used in the processing of thermoplastics are commercially available with screw diameters of 25 to 200 mm (1–8 in), lengths between 20 and 40 D and drive powers of 5 to 500 kW.

3.6. Extrusion of Semifinished Thermoplastics

Table 3.13.

Defect	Cause	Remedy
Small blisters	Volatile ingredients	Predry and/or degas
Large blisters	Air inclusions Insufficient feed	Increase feed
Small surface blisters	Working temperature too high	Reduce temperature
Discoloration or small fish eyes (mostly with PVC)	Decomposed material through overheating	Reduce temperature. If necessary, disassemble machine and check for dead spots
Striations, fish eyes	Contamination of melt with dust	Ensure dust-free conditions throughout
Rough surfaces	Surface temperature and/or pressure too low	Increase temperature and/or pressure (screw speed/filter). If necessary use larger machine
Variation in output ("pumping" of the screw)	Fluctuations in screw drive or haul-off efficiency. Voltage fluctuations in the heating circuit. Irregular feed. Screw running too fast and overloading (melt fracture). Insufficient flow resistance in the die.	Stabilize voltage Stabilize feed Reduce screw speed

In single-screw extruders the front face of the flight conveys the melt to the cylinder wall which, because of adhesion friction, is hotter than the screw. Mechanical energy is converted into heat sufficient in the *autogenic or adiabatic working* of fast runners (250–1000 min^{-1}) to continuously plasticize the plastic after start-up without external heating. Short fast runners (D 25–80 mm, L/D 16) are used in single-purpose machines for uniform large-scale production of PE tube. For the greater part of production operations *polytrope* machines (i.e., neither purely isothermal nor adiabatic) with variable speeds of between 200 min^{-1} and 400 min^{-1} are technically and economically advantageous. For shear- and temperature-sensitive plastics (PVC-U, fluoroplastics) slow-running twin screw extruders (Sec. 3.6.4) with positive conveying action and, usually, external heating are used.

The throughput of the conventional single screw extruder is determined by the difference between the "drag flow" by the melt and the opposite "pressure flow" from the die which is needed to homogenise the melt. From screw characteristic and die resistance diagrams, optimum operating conditions can be calculated.

3.6.3.2. Extruders with a forced conveying feed section

Multisection screws of extruders consist of the following sections in order:

Feed section,
Compression section,
Metering section.

The *compression ratio* is significant with respect to the type of machine and screw. This ratio relates the channel volume of the first full turn of the screw at the feed end to that of the last full turn at the discharge end (Fig. 3.44). In the multi-stage screw (Fig. 3.44b, c, d, f) this variation occurs by the diminishing depth of thread, or less commonly by diminishing channel width with variable pitch. (Fig. 3.44a).

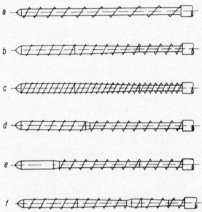

Fig. 3.44. Some forms of extruder screws.
(a) Single start screw, reducing pitch, (b) Single start screw, flight depth in the middle zone gradually reducing (3-zoned screw), (c) Twin start screw, otherwise as (b); (d) Single start screw, stepwide reduction of flight depth (short compression zone screw), (e) Single start screw with torpedo head (smooth or fluted, e.g., as "Leroy" shear section), (f) Two single start screws with reducing flight depth in series: venting screw (p. 128).

3.6.3.2. Extruders with a forced conveying feed section

The development of pressure in a *conventional extruder* with smooth feed zones and compression ratios of 5:1 to 3:1 is essentially due to friction in the shallow flighted metering section of the screw or to a smooth torpedo with a narrow annular space at the screw head (Fig. 3.44e). At this end the pressure in the sheared melt is therefore at a maximum. The plasticizing and homogenizing capacity of such an extruder is limited. Its output, beyond a given mass, cannot be increased and is subject to fluctuations as a result of external disturbances. *Extruders with grooved feed zones* exert a positive conveying action throughout. In the strongly cooled, wet feed zone several deeply cut axial grooves are introduced around the screw. Granulated material is – as in the movement of a counterrotating safety washer on a rotating spindle – moved on to the

3.6. Extrusion of Semifinished Thermoplastics

plasticizing and homogenizing sections under a pressure of 800 to 1300 bar. The subsequent zones use the maximum pressure developed at the end of the conveying feed zone. They are overridden by a small overall compression ratio of about 1.2:1 and can therefore be freely fitted with mechanical and thermal homogenizing shear sections and viscosity-equalizing mixing sections (Figs. 3.45, 3.46).

The introduction of the feed zone with effective conveying capacity and of supplementary shearing and mixing elements of various design, contributes significantly to increased output of single-screw extruders. Because such extruders hardly shear heat-sensitive plastics, they are especially suitable for the manufacture of high-quality products.

3.6.3.3. Special types of construction

In *vented extruders*, screws with several compression sections are so arranged that, in the intervening decompression zones, the melt can be degassed and

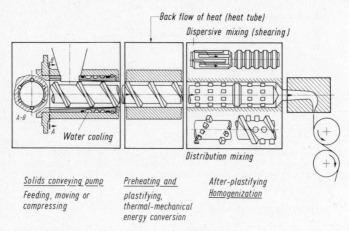

Fig. 3.45. Single screw with solid transport *(IKV Aachen)*.

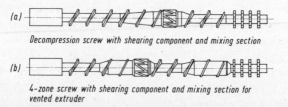

Fig. 3.46. Examples of screw types with shearing and mixing sections *(Reifenhäuser)*.

dried under vacuum (Fig. 3.44). Extruders with flow-restricting zones are used for the addition of physical blowing agents and other additives. The *cascade extruder* is a variant serving the same purpose. It has separate and consecutive plasticizing and hot-melt screws (Fig. 3.47).

In the *planetary gear extruder* the melt is finely layered and plasticized between the positively driven central planetary screw, the barrel and the planetary screws rotating around the central one (Fig. 3.48). It can be used for PVC-U single screw extrusion.

In the *Maillefer screw* with its two superimposed flights (Fig. 3.49) melt and unmolten compound are separated in a way that prevents the movement of the latter into the metering zone. The same effect is achieved in the *solids draining screw* (SDS) (Fig. 3.50) by feeding back the granules through a recycling hole in the screw core. Such "barrier" designs and others like them result in increases in output through high screw speeds.

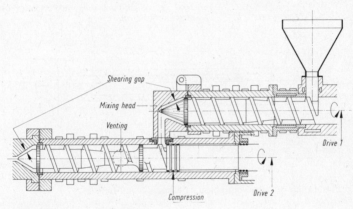

Fig. 3.47. Schematic of a cascade extruder.

3.6.4. Multi-screw Extruders

Twin-screw extruders with tightly intermeshing contra-rotating screws are different in their action from single-screw extruders working through adhesion drive and drag flow and from co-rotating twin-screw extruders used for compounding. The former are used as plastics processing machines with closed individual chambers constituting pumping systems with forced conveying. Because of the twin-screw drive and support they are technically preferable to the single-screw extruder. They are generally used as special purpose machines for PVC processing because of the positive conveying, high output and short residence time, the possibility of self-cleaning of the screws, and

3.6. Extrusion of Semifinished Thermoplastics

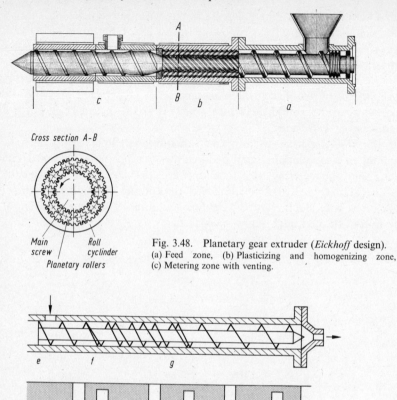

Fig. 3.48. Planetary gear extruder (*Eickhoff* design).
(a) Feed zone, (b) Plasticizing and homogenizing zone, (c) Metering zone with venting.

Fig. 3.49. Plasticating-extruder with phase separation *(Ch. Maillefer)*.
(a) First screw channel, (b) Second screw channel, (c) Barrier flight, (d) Overflow flight, (e) Granules, (f) Melt, (g) Cylinder.

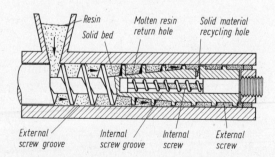

Fig. 3.50. Simplified drawing of construction of solids draining screw *(J. and R. Klein)*.
Through the drain hole adjacent to the trail side of screw flight, unmolten material escapes to enter the internal screw groove so as not to give unnecessary shear force to the molten resin. The barrier flight situated immediately after the drain hole prevents solid material from entering the die.

3.6.4. Multi-screw Extruders

gentle plasticization of thermally sensitive powder compounds which are difficult to feed. Such slow-running (40–50 rpm) vented extruders with deep screw channels, 50 to 170 mm (2–7 in), L/D 16 to 22 and motors of 7 to 200 kW deliver 80 to 150 kg/h (180–3300 lb) of rigid PVC pipes and about 60% of this output in profile, sheet or film.

The multi-screw extruders from different manufacturers have different arrangements of the screws (Fig. 3.51a, b) and build up of pressure under compression of the melt up to the screw head. Of the possibilities indicated in Fig. 3.51, the following have been produced: c,d,e (extended delivery screws with the purpose of facilitating the transition to circular dies), f, i and k. Cylindrical screws are arranged axially (in tandem). The tapering arrangement allows more room for the drive gears.

Twin screw extruders normally function with a full hopper like single screw machines. Metering units provide a wider range of operation with the same set of screws with PVC dry blends of different bulk densities.

PVC powder is "agglomerated" in the feed zone under compression by baffles. Before venting the entrained air, it is detached by breaker plates as baffles (Fig. 3.52) or by the removal of flights (Fig. 3.53). The surging caused by forced feed via the spaces between the flights of the intermeshing screws is smoothed out by at least three-channel plasticizing and discharge sections of the twin screws each delivering an overlapping output.

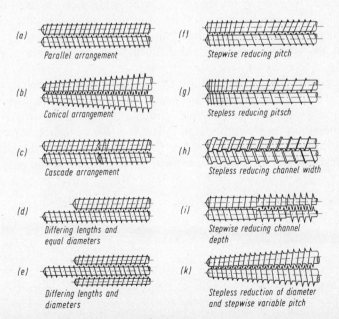

Fig. 3.51. Arrangements and geometries of twin screws.

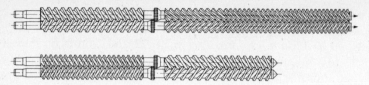

Fig. 3.52. Paired screws breaker plates prior to venting zone. (*Thyssen*-Plastik Machines).

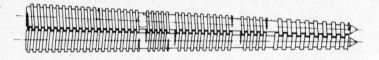

Fig. 3.53. Tapered twin screws *(Cincinnati)*.

3.6.5. Extrusion Plant for Various Products

3.6.5.1. Pipes and hollow sections

Figure 3.54 illustrates elements of production lines for rigid pipes or single- or multi-chambered hollow sections, (e.g., window frames, double- or triple-walled sheets. For flexible pipes (PE up to medium grades) wind-up units are used in place of take-off mechanisms.

A typical *hollow profile tool* is the straight-through die for pipes as shown in Fig. 3.55.

If the mandrel-support spider legs in a center-fed die disturb the melt flow, then axial breaker plate dies, side-fed die heads (Fig. 3.43) or smear devices (Fig. 3.58) can be used. Cores must be well centered by automated devices because the volume of flow is proportional to the third power of the die cross sections. The subsequent *calibration and cooling zone* (4 and 5 in Fig. 3.54) determines the dimensional stability of the products. Sizing plates are combined with a

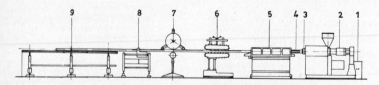

Fig. 3.54. Layout of plant used for the extrusion of pipes and hollow sections.
(1) Heating and cooling equipment, (2) Extruder, (3) Extruder head and die, (4) Calibration head, (5) Cooling zone, (6) Caterpillar take-off, (7) Marking device, (8) Saw, (9) Discharge. Automated lines are equipped with metering and control units from 1 to 9.

3.6.5.2. Solid profiles

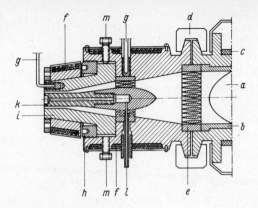

Fig. 3.55. Extruder head with pipe die.
(a) Screw head, (b) Cylinder lining, (c) Cylinder heating, (d) Attachment of the die head with bolt, (e) Breaker plate; drop-shaped holes, favor good flow, (f) Electrical heating of the die or tool, (g) Temperature-measuring pockets with copper contacts, (h) Die screw coupling, (i) Die ring, (k) Mandrel, die land 30–40 × nip width, (l) Compressed air inlet, (m) Centering screws.

water bath or spray cooling and vacuum calibration. Compressed air calibration with a floating plug is used for bigger pipes. The take-off units overcome considerable frictional resistance. For large pipes up to 1600 mm (400 inch) diameter, 10 to 100 kN take-off tension is required. Lines are available for foamed and/or multilayered pipes (e.g. PVC-U with PVC foam core) and other hollow sections with ancillary units dealt with principally on p. 000.
Polyolefin and PVC wound tubes made in lengths of 5 to 6 m with diameters of 500 to 3000 mm (20–120 in) are produced by spirally winding hot plasticized strips emerging from the die onto a core and then welding them together. Good stiffness/weight ratios can be achieved with ribbed or hollow profiles.

3.6.5.2. Solid profiles

In the *Technoform process* for solid profiles with tight tolerances, the still-molten material enters the calibration die and is formed by it and, as in the drawn-metal process, is drawn down onto the end product. For partially crystalline plastics the mold is subdivided by a cooling die into a heated and a sharply cooled area. The melt continues to flow in a solidified jacket, which arises at the cold die and thus prevents the formation of voids following the non-linear reduction of specific volume in the crystalline freezing temperature range.

3.6. Extrusion of Semifinished Thermoplastics

3.6.5.3. Foamed profiles

Structural foams made with chemical blowing agents are produced on conventional extruders, the temperature program of which causes them to expand at the die outlet. Circular profiles are made by free-foaming into the calibrating die, so that the expansion ratio is correspondingly greater than that of the die lips. With difficult profile cross sections, die-restrictor grids and/or torpedos are used to achieve even foaming (Fig. 3.56).

Structural foaming with chemical blowing agents can decrease the density to 50 to 70% of that of solid plastics. *Light foam profiles*, e.g., "boards" (Woodlite, density $= 256 - 356 \text{ kg/m}^3$) or beading profiles (density $= 45 - 75 \text{ kg/m}^3$) made from PS and heat-insulating tubes of PE (density $\sim 35 \text{ kg/m}^3$) with inner and outer solid skin are all made by directly gassing the melt with physical blowing agents (Sec. 3.1.6.1). Solution of the blowing agent in the hot melt under the required high pressure of 300 to 400 bar lowers its viscosity so much that it must be cooled considerably for extrusion. Especially long extruders ($44\,D$) with nine temperature zones are required for this difficult process. The blowing agent is fed in after the short melting section, in suitably equipped cascade (Fig. 3.47) or twin-screw extruders (Sec. 3.6.4). PS-foamed film and insulating panels (Sec. 4.2.3.4) are extrusion foamed with physical blowing agents.

3.6.5.4. Bristles, filaments and nets

Plastic filaments (monofilaments) and bristles are extruded through multi-holed plate spinnerets, with 10 to 40 holes for monofilaments and up to 150 holes for the subsequently bundled bristles. They are then cooled in thermostated baths to improve strength, drawn stepwise and wound on many spools.

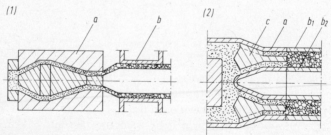

Fig. 3.56. Tubular structural foam tools.
(1) Calibration 20 to 40% wider than special pipe tool (*Armocel* process, Soc. Armosig).
(2) Slit die for double-walled tube, which produces a foam core between the outer and inner calibrated tube walls (*Celuka*-process, Ugine-Kuhlmann).
(a) Die, (b_1) and (b_2) outer and inner calibrating devices, (c) Torpedo in double-slit die.

The diameters of the products range from several millimeters to 0.08 mm (3 mil). Finer filaments are associated with the production processes and technology of chemical fibers. For filaments made by slitting film tape see Sec. 3.6.5.10.
Plastics net tubes are produced with tools in which the die and mandrel rotate continuously in opposite directions or alternate periodically.

3.6.5.5. Wire and cable covering

Coating with cross-head dies is used (Fig. 3.43, p. 125) mainly for the insulation of electrical conductors – telephone wiring at the rate of 2700 m/min – and cable covering. The preheated wire or the already insulated cable are introduced through a central bore hole in the mandrel.

3.6.5.6. Blown films from annular dies

Vertical film blowing extruders with 60 to 250 mm (2.4 to 10 inch) diameter, 25 to 33 D screws, horizontal annular dies 100 to 1800 mm (4 to 72 inch) diameter and 0.5 to 2 mm (20 to 80 mil) die gap are widely used for mono- and multilayer tubular films of all PE grades (Sec. 4.1.2.1.5), also combined with barrier plastics (Sec. 4.3.2.1), 10 to 300 μm (0.4 to 10.5 mil) thick and of 1.2 m to 2.5 m (3.0 to 8.2 ft) doubled lay-flat tube width. Take-off speeds are 150 m (500 ft)/min, output rates 100 to 1200 kg (220 to 2600 lb)/h.
Such machines are highly automated; some lines include ancillary units for cut-to-size reels, printing, welding and stacking, e.g. 50,000 refuse sacks (40 μm film) to 200,000 tear-off perforated bags in one shift.
Figure 3.57 illustrates the principle of a blown film plant. The flow toward the annular die gap is regulated by an adjustable gap in the conical die. Emerging tube is hauled-off longitudinally and transversely stretched by air injected under pressure. For external calibration a calibration basket is placed around the blown film. High output necessitates intensive cooling of the tube by blowing from the outside, especially in the bubble expansion zone where the material is still soft, and by the circulation of air or another cooling medium within the tube. The lay-flat tube is taken through guide boards and through nip and deflecting rolls. If uncut, it is wound onto one roll; if cut at the edges, it is picked up on two winding rolls.
The normal upward working direction, up to 20 m high in large plants, avoids a marked gravity effect on the still-plastic tube. To even out wall thickness differences along the length of the tube that give rise to hoops in winding, one part of the plant, usually the extruder or the blowing die, is allowed to rotate slowly in one direction or reversibly. Smear devices or spiral mandrel distributors in *blown film dies* obliterate spider marks in the melt (Fig. 3.58). To

3.6. Extrusion of Semifinished Thermoplastics

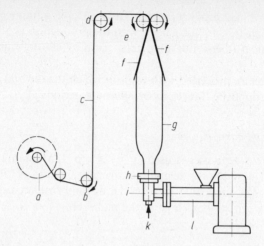

Fig. 3.57. Blown-film plant, operating vertically upward.
(a) Winding mechanism with tensioning roll, (b) Deflection roll, (c) Lay-flat tubing (uncut), (d) Deflecting roll, (e) Nip rolls, (f) Guide boards, (g) Film bubble, (h) Cooling, (i) Tubular film die head, (k) Air inlet, (l) Extruder.

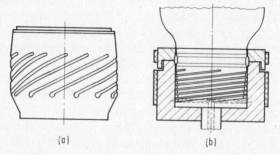

Fig. 3.58. Melt manifold tubular film dies: (a) Smear device in die mandrel, (b) Spiral melt distributor die.

make multi-ply composite films, melt streams from several extruders are brought together by adapters (feedblocks) or immediately on leaving the die (Sec. 3.1.2 and 3.6.5.9). Two to three layer films, e.g. of different PE grades, need two extruders, five layered (with intermediate tie layers, Sec. 3.6.5.7) three or four, seven layered (with inter-layers of reprocessed materials) five extruder units.

Annular dies grooved along their length with rotating cores give spirally circulating mechanically bonding between different film layers (Japanese process). A monofilament die (Sec. 3.6.5.4), as part of a rotating system, is used to produce composite films with an internal support lattice (criss-cross Baroflex

process). Two-layer dies, fed by one extruder yield a pore-free two-ply blown film.
Horizontal working is suitable for rigid PVC tubular film and for PS foam films with blow ratios of 1:3 to 1:6, densities of 50 to 200 kg/m^3 (3–13 lb/ft^3), thicknesses of 0.1 to 4 mm (4–120 mil) and a double width of 1500 mm (4.9 ft). By laminating the still-warm foamed film or using afterfoaming heaters in the haul-off, thicknesses of up to 8 mm can be produced. (For details see Sec. 3.6.5.7 and 3.8.4.4).

3.6.5.7. Sheets, films and coatings with slot dies

Slot die extrusion lines for sheets, sheeting and "cast" extrusion flat films have been developed to the same high degree of automated control as tubular film plants dealt with in the preceeding paragraph – that is, up to completely unmanned operation. Plants for making and converting multi-ply products (with up to 10 layers) are of growing importance for films as well as for sheeting. As an example, Fig. 3.59 shows a fully automated food can plant.

Horizontal working slot dies of adjustable widths (Fig. 3.60) are used to extrude continuous webs and sheets from 0.3 to 30 mm (0.08–1.2 in) in thickness and 3.5 m (11.5 ft) wide. Cast films (see Sec. 3.5.4), up to 4 m (13 ft) wide from plastics of low melt viscosity, are extruded obliquely downward with die

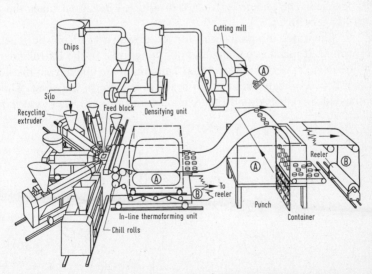

Fig. 3.59. Coextrusion and thermoforming of plastics food cans (diagram courtesy of ER-WE-PA).

3.6. Extrusion of Semifinished Thermoplastics

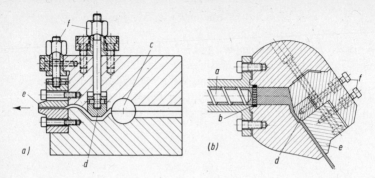

Fig. 3.60. Slot die.
(A) Sheet die. (B) Film die.
(a) Extruder, (b) Breaker plate, (c) Manifold, (d) Flex adjustable restrictor bar, (e) Variable lip, (f) Adjusting screws.

gaps between 0.3 and 0.6 mm (Fig. 3.6b). They are finally drawn by the chill-roll process to 8 to 15 μm (0.3–0.6 mil) thickness (Fig. 3.61). Vertically downward operating dies are used for extrusion coating. The manifold channel of *slot dies*, with heating bands arranged transversely to the run of the web and also arranged in several sections for the die lips, can be fed centrally or from the side. The die gap section width across the slot die is regulated by thermal expansion pieces or electrically driven screws actuated by non-contact web thickness gauges. The geometry for a centrally fed die often takes the form of a coat-hanger (Fig. 3.62).

Figure 3.63 illustrates the plant used for the production of plastic sheets for thermoforming. It includes cooling and polishing rolls. In the Extrulander and Calandrette devices, webs between 1 and 2 mm thick are taken from the slot die and drawn between heated rolls down to 0.1 to 0.8 mm in thickness. They are then embossed or polished and finally cooled and wound. These mechanisms

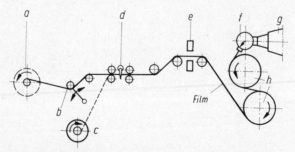

Fig. 3.61. Principle of the chill-roll-process.
(a) Film wind-up, (b) Free roll, (c) Waste wind-up, (d) Edge cutting, (e) Thickness measurement, (f) Slot die, (g) Extruder, (h) Cooling rolls.

3.6.5.9. Multi-layer and multi-colored extrusion 139

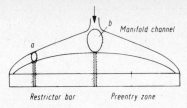

Fig. 3.62. Coat-hanger or fish-tail die for rigid PVC.
(a, b) Cross section of flow channel.

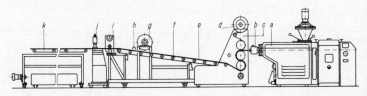

Fig. 3.63. Line for the production of plastic sheets.
(a) Extruder, (b) Slot die, (c) Three-roll polisher, (d) Runoff for laminating film, (e) Continuous web, (f) Guide rolls, (g) Idler roll, (h) Edge cutter, (i) Haul-off device, (j) Transverse cutter, (h) Discharging and stacking mechanism.

combine the working principles of the calender (Sec. 3.7.3) with those of the extruder. For foamed sheets see Sec. 3.6.5.3.

3.6.5.8. Stretch process for flat films

Biaxial orientation of fine films of < 10 to $25\,\mu m$ (0.4–1 mil) thickness, particularly those made from PP, PS, PAN, PA and PET, gives a several-fold increase in strength, high transparency, low permeability and high cold strength. They are stretched by a two-stage process in which power-driven heated rolls draw the web in the machine direction and clamping tools stretch it transversely (Fig. 3.64). Alternatively they are biaxially oriented simultaneously using clamping tools only. Raising the temperature considerably for a short time in the closing heat setting section of the stretching machine gives stress-free recrystallization in PS, PC and PET films. For example, this occurs in $2\,\mu m$ (0.08 mil) thick drawn condenser films with $< 1\%$ shrinkage on heating (Sec. 4.2.8.2). PP and PE-HD packaging films (Sec. 4.2.2.3) but not extruded PVC film can be heat set.

3.6.5.9. Multi-layer and multi-colored extrusion

Fig. 3.65 is a drawing of the different coextrusion systems (see also Fig. 3.35, p. 109). In conveying the melt streams outside the die it is possible to facilitate adhesion between plies by blowing reactive gas into the die gap.

3.6. Extrusion of Semifinished Thermoplastics

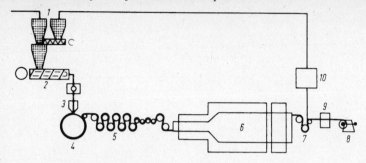

Fig. 3.64. Diagram of two-stage stretching installation.
(1) Granule feed system, (2) Extruder, (3) Die, (4) Film haul-off, (5) Longitudinal stretching, (6) Transverse stretching, (7) Haul-off equipment, (8) Wind-up, (9) Corona pretreatment for PP, (10) Film waste recyling.

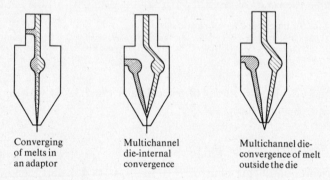

Converging of melts in an adaptor

Multichannel die-internal convergence

Multichannel die-convergence of melt outside the die

Fig. 3.65. Coextrusion system.

Coextruded products include multi-layer films and sheeting, PMMA-coated PVC window profiles, rigid profiles with a flexible sealing edge extruded onto them, and multi-colored spirally patterned coating of electrical wiring, tubes and patterned profiles made with a rotating manifold.

3.6.5.10. Filaments and fibers from films

High-density PE and PP film extruded from an annular or slot die is processed by passing over cutting equipment with fixed knives, double hot-air stretching to a draw ratio of 10:1 and hot-air fix stretching to give high-strength flat filaments or film tapes 10 to 30 μm (0.4–1.2 mil) thick and 2 to 10 mm (0.08–0.4 in) wide. Fibrillated filaments and yarns with titers of between 0.8 and 8 tex are made with fibrillating and needle rolls. Applications include binding yarn, polypropylene rope and weather- and corrosion-resistant industrial and decorative fabrics, webs and nets. In the weaving process, filaments cut from stretched film at the loom can be used to replace the warp bem. For PE reinforcement fibers see. Sec. 4.1.2.2.10.

3.7. Webs Made of Thermoplastics

3.7.1. Films Made from Solutions

Films are made from solutions by casting (Sec. 3.5.4) and, in the case of cellophane, by spinning into precipitating baths. Such films are generally thin (0.01–0.1 mm) and glass-clear.

3.7.2. Extrusion of Webs

Extrusion of webs (details Sec. 3.6.5.6/7) using annular dies produces blown film 0.005 to 0.3 mm thick in any widths while films from 0.01 mm up to 4 m in width and sheets of rigid plastics 0.6 to 30 mm thick and 2 to 3.5 m wide are produced using slot dies.

3.7.3. Calendering of Plastic Film

Rigid and plasticized PVC films are manufactured on a large scale by calendering from continuously supplied material that has been preplasticized by heat (for the compounding and plasticizing processes see Sec. 3.1.4).
The drives, bearings and heating (up to 200°C) of four- or five-roll calenders, which might be arranged in Z-, I-, F-, or L-form, can be individually regulated (Fig. 3.66). Embossing rolls and printing follow directly on the calender or in

Table 3.14. Calendered Film

Products	Calender rolls		Output	Normal film	
	Length, max. (mm)	Diameter, max. (mm)	(m/min)	Width (mm)	Thickness (mm)
PVC-E rigid film made by the "Luvitherm" process, after-treated at 220–250°C and semi-stretched drawn stretched to >200%	≤ 1700	500	6–25	Semi-finished product	
				to 1400	>0.18
				to 1200	0.18–0.135
					0.08–0.02
Other PVC-U films	≤ 2600	600	10–20	to 1500	0.8–0.06
Flexible PVC-P films	(3500)	≤ 900	8–60	≤ 2300	0.6–0.06

3.7. Webs Made of Thermoplastics

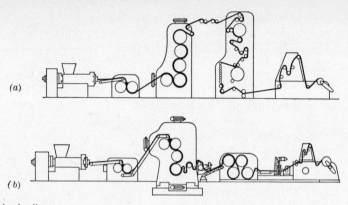

Fig. 3.66. Calender lines.
(a) A five-roll calender installation with an L configuration for the production of rigid PVC film by the Luvitherm process. (b) A four-roll calender system in an F configuration for the production of plasticized PVC film.

separate operations. For combinations of extruders and small calender for rigid films see Sec. 3.6.5.7.

3.7.4. Multilayered Webs

3.7.4.1. Bonding several layers

For the extrusion of multi-layered plastic webs see Sec. 3.6.5.6. For sealing pores in films, they are laminated with an embossing calender or a vacuum-embossing machine. The Auma machine is used particularly for multi-layered flooring. Preheated webs are pressed against a driven heated cylinder by means of an endless circulating steel pressure band (Fig. 3.67).

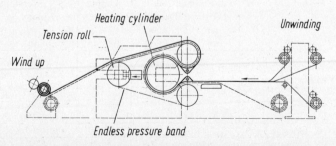

Fig. 3.67. *Berstorff* Auma for doubling and laminating several film layers.

3.7.4.2. Spreading, coating and laminating

The usual processes for coating substrates with plastics are
- Extrusion coating plastic melts (Sec. 3.6.5.5)
- (Air) knife application or reverse-roll coating (Fig. 3.68) solutions, dispersions (Sec. 4.1.1.10), organosols (Sec. 4.1.4.9), PU pre-products, (Sec. 4.6.1.8.2.5)
- Hot melt roller application (Fig. 3.69) of plastics powder
- Calender laminating paper, non-woven, knitted and other industrial fabrics for many applications, e.g. leather cloth, sometimes with post-foamed plastic core
- Laminating webs from the reels by adhesives on the laminator
- Coil-coating sheet metal in large plants, using hot melt roller application of powders as well as laminating plastics film.

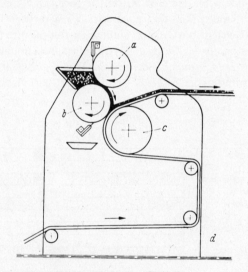

Fig. 3.68. Reverse roll coater, *Olbrich* system.
(a) Doctor roll, (b) Furnishing roll, (c) Backing roll, (d) Conveyor belt.

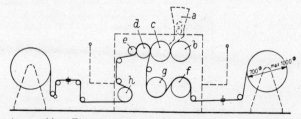

Fig. 3.69. Coating machine, *Zimmer* system.
(a) Material feed, (b, c) Melt rolls, (d) Rubber roll, (e) Embossing or polishing roll, (f, g) Preheating rolls, (h) Cooling rolls.

3.8. Forming and Joining Semifinished Plastics

3.8.1. Forming Techniques for Thermoplastics

3.8.1.1. Temperature ranges (Fig. 3.70)

The viscoelastic behavior of long-chain polymers leads to irreversible flow and more or less reversible energy and/or entropy-elastic deformation in all changes of form of thermoplastics under mechanical stress. These states of quasi-elastomeric flexibility or viscous flow, depending on the amorphous or partially crystalline structure, are influenced very differently by each of the parameters of time, temperature and load:

1. All rigid amorphous glassy thermoplastics creep under long-term stress ("cold flow") but some are too brittle for forming at room temperature. Less brittle-rigid amorphous thermoplastics below their glass transition temperature, T_g, as well as partially crystalline thermoplastics below their crystalline melt temperature, T_m (Fig. 1.4), are ductile enough, within certain limits, to allow *cold forming* (Sec. 3.8.2). This involves three-dimensional permanent forming under forces acting for short periods in the production of simple moldings.

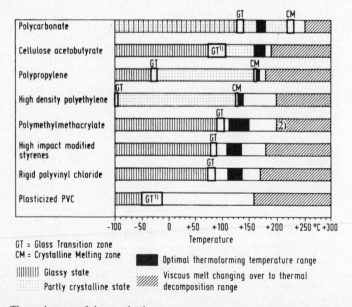

Fig. 3.70. Thermal states of thermoplastics.
(1) Depends on plasticizer content. (2) Only acrylic molding materials; cast acrylics cannot be molded.

3.8.1.1. Temperature ranges

2. In the most important *two-dimensional stretching* of films and sheeting and *one-dimensional longitudinal drawing* of monofilaments, wires and webs, the surface or length measurements are increased by more than 100%, with a corresponding reduction in thickness or cross-section. These processes aim to improve the properties of the product by orientation of amorphous macromolecules or of their crystalline regions. In particular they bring about a significant strengthening in the direction of stretch. They are combined with molding and extrusion, as for example in stretch blowing (Sec. 3.3.6.7), the extrusion of tubular film (Sec. 3.6.5.6), cast film (Sec. 3.6.5.7) and filaments (Sec. 3.6.5.10). Amorphous thermoplastics can be stretched to a limited degree below their T_g. Stretch in the thermoelastic state must be frozen in (see 3 below). Partially crystalline plastics are stretched by more than 100% by cold drawing at room temperature above their yield point (Sec. 7.6.2.1). With associated *heat setting*, to relax the stress used to orient the amorphous regions, the effects of stretching become resistant to high temperatures. For stretch blowing in the crystalline region at high temperatures see Sec. 3.3.6.7.

3. Amorphous thermoplastics become quasi-elastomeric thermoelastic over quite wide temperature ranges above T_g (Fig. 1.4; Fig. 3.70). Hand and industrial *thermoforming* (Sec. 3.8.3., 3.8.4) uses this range in forming semi finished materials into finished parts while using only limited force. Partially crystalline plastics that have a sharp transition at T_m into a melt can be formed to a limited degree. Only some grades of PE-HD and PP can be handworked in this way.

    At the lower limits of the optimal thermoforming region, relatively stiff elastic material can be formed with more than 100% flow and stretch extension without danger of crack formation. Articles formed at higher temperatures are more temperature resistant because of increasingly isotropic flow in the forming process. With further elevation of temperature, elongation and strength at break of the increasingly fluid materials are reduced so far that they can no longer be formed because of their "heat brittleness". All thermoformed objects are anisotropic and reversibly entropy-rubbery elastic at forming temperature. Their shape must be frozen in by cooling the molding under tension down to well below T_g or T_m. On re-heating they revert in part below these temperatures; above them they revert almost completely.

4. *Rubbery plastics* whose T_g lies below the service temperature cannot be thermoformed. Semi-stiff leather-like plastics such as PVC with 20 to 30% plasticizer are borderline cases and are considered in the following section.

3.8. Forming and Joining Semifinished Plastics

3.8.1.2. Embossing, stretching, roll mill pressing, shrinking, annealing

In the transition from the flowing to the rubbery-elastic state, high-gloss surfaces or embossings can be impressed onto plasticized PVC products. Films, pipes and semi-rigid tubes are strengthened by uni- or bi-axial stretching in the thermoelastic state and the stretching effects are frozen in under tension. Their capacity to revert on re-heating is used in "shrink tubing" to coat passenger support bars and tool grips and in "shrink films" for tight packaging. Roll mill pressing is a technique for producing self-reinforced partly crystalline thermoplastic slabs by three dimensional orientation of the crystallites between pinch rolls. The products are cold post-formable (see 3.8.2).

Residual orientation strains generated by molding tools or take-off units in semi-finished thermoplastics are annealed by sections of the plant or special annealing equipment. For example, plasticized PVC flooring is heated at 70 to 100°C before cutting. In pre-heating for thermoforming, extruded sheeting contracts about 3% along its length and about 1% or more across its width while increasing in thickness. Cast acrylic shrinks when first heated about 2% both lengthwise and across.

3.8.2. Cold-Forming Thermoplastics

Very high ultimate tensile strengths can be conferred on monofilaments, wires and ribbons of crystallizable polymers by *cold drawing* several hundred percent at room temperature. Crystallite orientation remains stable up to melting temperatures.

Cold forming of thermoplastic semi-finished products by techniques and machines used for cold rolling, deep drawing, forging sheet metal and bar makes possible very fast and economical mass production of relatively simple shapes such as cups, covers, flanges, castors, gears, screws or cartridge shells. Optimal working temperatures (mostly near the T_g or T_m) and other conditions for the production of such very strong and highly anisotropic parts have to be ascertained for each production line. They are different from those appropriate for cold-flowing metal. The solid phase pressure forming (SPPF) method for PP packaging and drinking cups is an example. For solid phase extrusion of high strength fibers see Sec. 4.1.2.1.10.

3.8.3. Hand-Fabricating Thermoforming

Semi finished goods for forming are heated in the workshop in hot-air ovens or with locally applied hot air in assembly work or in hot fluid baths. They must then be formed in one operation and finally cooled with water or air. Local heating with a gentle luminous flame demands much experience. *Pipes* are bent

3.8.3. Hand-Fabricating Thermoforming

by hand after they have been filled with hot sand and subjected to local heating. Acrylic tubes can also be carefully bent by hand in matched dies while inflated with compressed air. Pipe ends are calibrated in heated tools. Shrink-on sleeves are fixed in this way or formed onto the complementary pipe. Collars are heat formed onto polyolefin pipes. Wide-diameter pipes are made by first hot-bending sheets and welding with a longitudinal seam and then rounding them out in a second heating on a wood former.

Sheets of up to 10 mm thickness can be folded or bent against templates when a bending zone at least six times the sheet thickness is heated, most effectively by radiant heaters.

In *drawing processes* involving three-dimensional forming with draw ratios of up to 1:1, semi finished products must be heated to the upper limits of the forming temperature range (Fig. 3.70). These processes are (Fig. 3.71):

(a) Matched mold forming in cold, clamped tools on a hand press without a female mold;
(b) Deep drawing using tools heated on both sides and a holding plate. Mechanical slip forming with sliding holding plates provides a more uniform thickness than do drape and vacuum forming with clamping

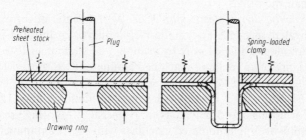

Fig. 3.71a. Plug and ring forming.

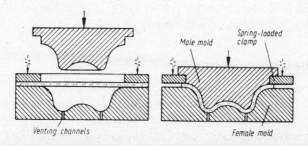

Fig. 3.71b. Mechanical deep drawing.

frames. These are more suited to industrial thermoforming. Spherical shells can be blown freely without molds.

3.8.4. Industrial Stretch Thermoforming

3.8.4.1. Principles of thermoforming by machines

In thermoforming machines for *pneumatic forming* by vacuum, all edges of the sheet or sheeting to be formed are firmly fixed airtight around the tool. Therefore, the material cannot slide (as in deep drawing, see 3.8.3.), but is stretched during sucking in or on the tool. Those parts of the material, at the margins and flat areas, which are first sucked down adjacent to the tool take no further part in the forming. As a result the wall thickness at the corners and edges and in deeply stretched places can be unacceptably reduced. For a more equal wall thickness distribution, the final vacuum forming is combined with free pre-blowing by hot air and/or plug-preforming (Fig. 3.72).

Using vacuum alone one can obtain depths of 40% of the tool diameter. In combination with blowing and/or mechanical stretch forming, 100 to 250% is generally obtainable and, at the limit, 400 to 500%. In the "snap-back" variation of the process, male vacuum formers are moved onto pre-blown hemispherical preforms. The surface lying adjacent to the vacuum tool is always smoother than the other.

For *flowing in the thermoelastic state* (Sec. 3.8.1.1 material for each forming cycle is preheated, as a rule by radiation from quartz- or infrared-heaters. The unheated tool blocks, loaded only by vacuum and sometimes fitted with cooling channels, are often made of machined laminates (Sec. 4.6.3.5.1) or molded concrete. Tools made of light metals or brass castings are good heat conductors, as are EP-resins (Sec. 4.6.1.6) with a 60% aluminium filling. Hydraulically driven pre-form plugs need only 20 N/mm^2 pressure. Ignoring the costs of the peripheral units of fully automated machines with pre-programmed controlled heating, thermoforming equipment costs about 20% of comparable injection molding equipment. Sheeting is, however, more expensive than molding compounds, and cost considerations require separation and reuse of edge waste. The aim is to produce parts without edging.

3.8.4.2. Areas of application

In their thermoelastic stage (Sec. 3.8.1.1) semi-finished materials made of ionomers, biaxially drawn and impact-resistant PS, ABS, rigid PVC, PMMA, PC, PET, also multilayer sheeting, GFR-thermoplastics and cellulose esters can easily be thermoformed. PS foam films expand on forming at 95 to 100°C to double their thickness, provided they have not been prefoamed. PE-HD and

3.8.4.2. Areas of application   149

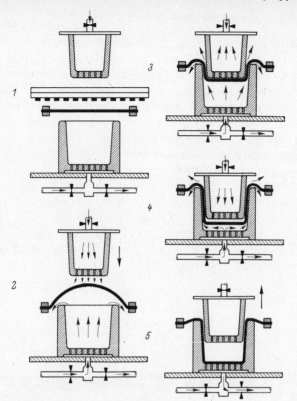

Fig. 3.72. Combined process for thermoforming.
(1) Heating (radiant heater or clamping frame, displaceable).
(2) Clamping frame moved downward, finally airtight. Preblowing with hot air.
(3), (4) Preforming with plug. Material slides on the simultaneously blown hot air (airslip process).
(5) Forming by exhausting at the mold wall.

PP above the "clear point" must be heated from both sides, air supported and slowly cooled to recrystallization after molding.

Thermoformable semi-finished goods range from 10 mm sheets to about 0.1 mm films. Typical applications include large-surfaced and complex shaped parts (e.g., facade parts, sanitary cells, garden seats, refrigerator housings, containers, vehicle parts and boats). Hollow articles (e.g. surfboards, fuel tanks) are made in two-part tools. Air is blown between two edge-welding sheets while outward suction is applied. Other applications include mass produced small parts, in particular for the packaging of very large numbers of items.

Dome-lights are produced by free blowing without a mould with a uniform wall thickness that is strengthened by biaxial stretching.

3.8. Forming and Joining Semifinished Plastics

3.8.4.3. Thermoforming machines

Cut-sheet thermoformers for large parts, which can be economical for small production runs of up to 5000 items, are similar to large area presses (Sec. 3.3.4.2). Platens up to an area of $18\,m^2$ ($2000\,ft^2$) are combined with suction and blow equipment; the clamping frame and down-stroke assistant plugs move vertically, the upper and lower heaters displace horizontally. Thermoforming as shown in Fig. 3.72 can produce items up to about 250 cm (100 inches) in height, using the many possibilities of program control, including automated feeding/demolding, regulated cooling by spray water or cold gases, high speed tool changing systems.

Roll-fed thermoformers for the mass production of packaging cups are highly automated lines, combined sometimes with the extruder, with punching and trimming and also with filling and sealing stations (Fig. 3.73). There are also rotational roll-fed thermoforming lines. Polystyrene or PVC cups are also formed by mechanical stretching only, similarly to those from PP using the SPPF-process (Sec. 3.8.2) just below its T_m.

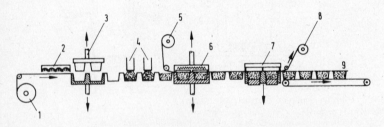

Fig. 3.73. Schematic drawing of a thermoforming, filling and sealing machine.
1 material roll 2 heating station 3 forming station 4 filling station 5 sealing film 6 sealing station 7 trimming station 8 web scrap roll 9 removal

3.8.4.4. Laminating, skin packaging and blister packaging

When former-laminating with plastic films, the object to be coated itself serves as the tool. It must be able to withstand the vacuum effective on the edges and through suction holes for negative sections and must have a contact adhesive layer.

In skin packaging, heated film is taken to the commodity to be packaged on heat sealable board and is formed into a skin around the article by the application of vacuum. In blister packaging, the article is placed in preformed hemispheres or shells of film, which are then heat sealed or welded at the interface to a substrate.

3.8.5. Forming Materials Other Than Thermoplastics

1. *Vulcanized fiber* (Sec. 4.4.3) of limited thickness is bent, after moistening when necessary, on metal plate or hot bending machines. Steamed material can be printed and drawn (embossed, grained, edged, recessed).
2. *Industrial laminates* (Sec. 4.6.3.5.1) up to a thickness of 1 mm can be bent after heating. The curve must be fixed. Inscriptions can be made with a punch heated to 150 to 200°C (the high-relief letters being cut out bar shaped and somewhat wedge shaped) with colored embossing film worked in. Thin laminated paper is available in special heat-forming qualities.
3. *Decorative laminated materials* (Sec. 4.6.3.5.2) need to be bent in the cold with radii of > 300 mm and to be glued under stress. Heat bendable types are heated on both sides for 3 min to 140 to 150°C and bent on formers with radii of up to 35 mm. After cooling on the former the curve tends to revert a little.

3.8.6. Welding of Thermoplastics

Semi-finished parts and moldings of identical (and, to a limited degree, differing but mutually compatible) thermoplastics are welded or heat-fusion jointed by heating the areas to be welded until viscous, pressing the parts together and allowing them to cool under pressure. The creep rupture weld efficiency should be > 0.6 of the strength of the basic material and may exceed it. ISO/R 194 and 472, DIN 1910, T.3, DIN 19960 and ASTM D 2675 contain specifications of welding processes and construction guidelines. The International Institute of Welding (IIW Com. XVI Welding of Plastics) in London publishes documents in English based on the codes of the national trade organizations. In the *Multilingual Collection of Terms for Welding*, Part 8, 667, is a collection of plastics welding expressions in English, French and German.

3.8.6.1. Shop and field welding techniques

Manual hot gas and heated tool welding are the most important joining processes in pipework and equipment construction and for moisture- and damp-proof plastic membranes.
Hot gas welding torches or guns (Fig. 3.74) heat the welding region with hot compressed air (200–300°C, with 0.2 to 0.8 bar excess pressure). The air must be free of oil or water vapor. Single or double V butt joints between sheet or pipes are hot welded using round or triangular coiled filler rods of the base material. This has to be preheated by hot gas, as do the grooved edges beveled exactly to a given angle. Torches with a flat nozzle are used to lap weld overlapping sheets manually without the need for filler rod.

3.8. Forming and Joining Semifinished Plastics

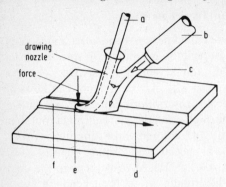

Fig. 3.74. Hot gas high speed welding (from DVS 2207, part 3).
a welding rod
b welding torch
c hot air
d welding direction
e welded join
f welded seam

Welding pressure and pressure during cooling may be applied by a skilled welder using the torch directly (Fig. 3.75). For semi- or non-rigid plastics (PE-LD, PVC-P) pressure rollers are used. Semi-automatic, self-propelled machines are in use for welding long simple seams, e.g., between the beveled edges of floor-covering material or the overlapping joints of moisture-proof sheeting.

In extrusion welding material is delivered by a small portable extruder to fill preheated grooves or form molten beads in lap welding.

In hot plate welding electrically heated platens are used to make butt welds. The square faces of the joints to be fused are heated on both sides of the platen and, after removal from it, they are pressed together until flash has been formed and the weld cooled. To obtain a good joint all welding parameters, temperatures, times, heating and welding pressures, must be accurately controlled; manually or semi-automatically operated welding devices containing jigs and positioners are used. Special heating tools are available for a variety of purposes. Heated tool welding is generally used for polyolefins. Welding sleeves, incorporating electrical heating elements, are also available for PE-HD pipes.

Electrical *heating wedges* in the form of soldering irons provide simple heating tools for minor work.

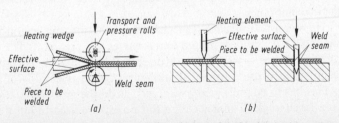

Fig. 3.75. (a) Mechanically heated edge welding, (b) heated tool welding with cutting edge.

3.8.6.2. Industrial welding techniques

1. Heated tool welding

Roller seam welding machines (Fig. 3.75a) operating continuously like industrial sewing machines are used to fabricate plastic garments, tarpaulins, etc. A wedge-shaped heating tool is introduced between the overlap ahead of a pair of pressure rollers. Envelopes, bags, etc., made of double foils are cut and welded in one operation by *knife-edged heater bars* (Fig. 3.75b).

In *indirect heat welding* of sheets using welding presses or tongs with movable bars, the heat is conducted through the sheets to the interface. There is a difference between *heat sealing* with continuously heated tools and *impulse sealing* with a small resistant element in the sealing surface that is strongly heated for only a small portion of the sealing cycle. PTFE slip sheets on the surface of the bars prevent blocking.

Automated *butt welding machines* are used for joining injection moldings or other parts as a cost-effective technique for mass production. The automated manufacturing of bags and sacks by heat tool welding equipment immediately connected to extrusion lines is considered in Sec. 3.6.5.6.

2. Ultrasonic (US) welding

In ultrasonic welding equipment
– the *generator* produces high-frequency electrical oscillations (20 or 40 kHz);
– the ceramic *converter* converts these into mechanical ultrasonic oscillations;
– the *sonotrode* transmits the vibrations to the parts to be welded, which are fixed between it and the anvil, and thus produces frictional heat, raising the temperature at the interface to that required for welding in between 0.02 and 2s.

The maximum output of large area 20 kHz ultrasonic equipment is 3 kW and the lower-noise 40 kHz machines produce 1.5 kW. Only a few sonotrode materials (titanium, possibly carbide protected) withstand vibrational stress over long periods. A welding pressure of 2 to $5 N/mm^2$ is necessary to ensure that the sonotrode does not destroy the material by separating from it.

In *contact or near-field ultrasonic welding* filled packing tube, or similar, is welded transversely with spider-shaped sonotrodes irrespective of the nature of the filler: powder, paste or fluid. The major area of application is *remote welding* of moldings. The vibrations are conducted from sonotrodes through the adjacent part of the article to welding edges that are arranged to avoid overlapping the molten welding parts (Fig. 3.76). PS, SAN, ABS, PMMA, POM, PC, PET/PBT and dry PA are rigid plastics which can be welded in this way at 40 kHz, and they are also partly mutually weldable. The presence of fillers makes leakproof welding more difficult.

3.8. Forming and Joining Semifinished Plastics

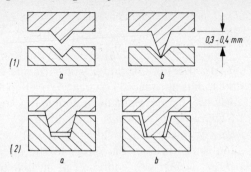

Fig. 3.76. Ultrasonic welding.
(1) Flash seam, (2) Groove-spring combination.
(a) Incorrect, (b) Better.

Ultrasonic welding at 40 kHz is suitable for mass production, for example for car rear lights, film cassettes, slide mounts, fountain pens and ballpoint pens. Welding tools do not wear out, so that automatic production of millions of items in the shortest of time cycles is possible at low cost. Other ultrasonic applications include the embedding of metal parts, riveting, and combining thermoplastics with porous materials. Large items (e.g. car spoilers) are welded at 20 kHz.

3. Friction welding

In *spin welding* circular objects heat is generated by mechanically rotating one component against another for a short time (< 0.5 s) and then cooling the weld under a dwell pressure of 0.1 to 2 N/mm^2. The process is used in the workshop in the welding of collars on pipes or fittings and industrially for the joining of hemispheres to form hollow articles.

In *vibration welding* the parts to be joined are vibrated against each other at about 100 or 240 Hz. With 240 Hz machines welding areas can be up to 150 cm^2. Machines with sufficiently rigid holding arrangements oscillate longitudinally for long narrow parts and at an angle for rectangular or rounded parts.

Friction welding is used in the rapid manufacture of articles made of almost any material including materials which cannot be ultrasonically welded including, for example, moist PA. Items produced include fuel tanks and bumpers of reinforced material.

4. Dielectric (RF or HF) welding

Heating thermoplastics which have a dissipation factor ≥ 0.01 by dielectric loss in a radio-high frequency (RF or HF) field between cold or constant temperature linear electrodes can raise the interface between two films or sheets in contact with the electrodes to welding temperature, while the surrounding material remains cool. Plastics with a limited polarity, such as polyolefins, polystyrene and PTFE, are not weldable by HF. The major field of application is flat weldings with plasticized PVC sheet and also PVC combined with board, wadding, foam and textiles. CA, PA, TPU and many thermoplastic elastomers can also be welded by HF. HF welding plants operate at electrical loads of 0.1 to 100 kW with pneumatic or hydraulic welding presses (locking forces of 0.1 to 1000 kN (0.01–100 ton)) and welding tables with areas from several hundred square centimeters up to dimensions of 2×3 m (80×120 in), the largest for inflatable rescue rafts. HF welding equipment is often coupled with feeding and after-treatment stations to form automatic production lines. As a rule the frequency of 27.12 ± 1.6 MHz, internationally permitted for RF welding, is used. The upper electrode is of brass rods 1 to 2 cm high and several millimeters wide, with which decorative seams and stampings can be produced. The (usually) flat counter electrode is in the press platform. Intermediate insulating layers and solid-state safety circuits that are activated in microseconds protect against dangerous electrical breakdown.

The major applications of small- and medium-sized welding machines are coverings, bookbinding, portfolio goods, medical wares, garment manufacturing and inflated articles. Larger machines produce automotive equipment such as car door coverings, seat fittings, armature boards and vehicle canopies. Concave- or convex-curved electrodes can also be used, for example for the manufacture of rainwear in one operation from the film roll. Rigid PVC fan blades are welded into base plates with special electrodes.

5. Induction welding

This is a welding process especially applicable to moldings made of plastics that are difficult to combine, for example, because they are made of different plastics. The parts to be welded contain iron wires or band inserts or intermediate layers of magnetizable iron oxide-ceramic powders or these materials incorporated into the plastic. External induction coils using high frequency alternating current can produce melt temperature in a few seconds.

3.8.7. Bonding of Plastics

Adhesion strength when bonding plastics can be traced back to secondary valence, dipole and dispersion forces. The plastics to be bonded differ not only in their surface energy (wettability) but also in their chemical composition which makes possible the formation of these forces. The well-known resistance of non-polar polyolefins (polyethylene and polypropylene) to bonding shows

3.8. Forming and Joining Semifinished Plastics

the strong influence of polarity. Non-polar plastics can only be bonded with sufficient strength after surface treatment (oxidation by acids, corona discharge or flaming) – see Table 3.15.

The solution properties and diffusion behaviour of thermoplastics often makes diffusion bonding possible. The main solvents for dissolving or swelling individual thermoplastics are:

Plastics	Solvent(s)
polyvinyl chloride (PVC)	tetrahydrofuran, cyclohexanone
polystyrene (PS)	toluene, xylene
polymethylmethacrylate (PMMA)	methylene chloride, methylethylketone
polycarbonate (PC)	methylene chloride
cellulose acetate (CA)	methyl ethyl ketone, methyl alcohol
polyphenylene oxide (PPE, PPO)	chloroform, toluene
polyamide (PA)	formic acid
polyethylene terephthalate (PET)	benzyl alcohol

In contrast to solvent adhesives, solvent-free adhesives do not cause any change in the parts to be bonded as they do not contain any monomers which can dissolve them. In general, adhesives based on epoxide resins (EP), polyurethane (PU), methyl methacrylates (MA) and unsaturated polyesters are suitable. Likewise cyanoacrylates have found wide applicability in bonding small-surface plastics and elastomers.

Diffusion bonding by swelling or partly dissolving of the surfaces to be jointed is possible with thermoplastics with the exception of PE, PP, POM and polyfluorocarbons. It leads to weld-like combinations, but can give rise to stress cracking.

Adhesion bonding by physical and chemical processes at the interface between plastic surfaces and adhesives is possible with nearly all plastics, with the exception of PS and to a limited degree of plasticized PVC. It is the process by which plastics can be bonded to other materials.

1. *Articles made of the same thermoplastic* can be well bonded by the application of light pressure after swelling the surfaces with solvent. The action of the solvent can be controlled by the addition of non-solvents or solutions of plastics (*solvent adhesives*). The overlapping "*solvent-welding*" of PIB and plasticized PVC sheeting is a solvent adhesive process. POM and PTP can be swelled and made adhesive with hexafluoroacetone sesquihydrate or by chemically etching the surface with contact or reaction-resin adhesives and then bringing the surfaces together.

2. In *hand fabrication*, acrylics bonded with solution adhesives can withstand limited stressing. Polymerizable bonding agents are optically flawless and

3.8.7. Bonding of Plastics

Table 3.15. Bonding of some plastics

Bonding	Plastic	Polarity + polar − non polar	Solubility + soluble − insoluble or soluble only with difficulty	Ability to Diffusion bond	Ability to Adhesion bond
Good	Polyvinylchloride (unplasticised)	+	+	+	+
	Polyvinylchloride (plasticised)	+	+	+	+
	Polystyrene (also foamed)	+	+	+	+[1]
	Polymethylmethacrylate	+	+	+/−[2]	+
	Polycarbonate	+	+	+	+
	ABS copolymers	+	−	+	+
	Celluloseacetate	+	+	+	+
	Polyurethane (also foamed)	+	−	−	+
	Polyester (unsaturated)	+	−	−	+
	Epoxide resins	+	−	−	+
	Phenolic resins	+	−	−	+
	Urea Melamine resins	+	−	−	+
Limited	Polyamide	+	−	+/−	+
	Polyoxymethylene	+	−	−	+
	Polyethylene terephthalate	+	−	−	+[3]
	Rubber polymers	+	−	+/−	+
Difficult	Polyethylene	−	−	−	+[4]
	Polypropylene	−	−	−	+[4]
	Polytetrafluoroethylene	−	−	−	+/−[4]
	Silicone resins	+/−	−	−	+

[1] only for PS foams
[2] only for non-cross-linked PMMA
[3] after pre-treatment with caustic soda solution
[4] only after pre-treatment

weather resistant. Solutions of post-chlorinated PVC are used for PVC-U chemical equipment. With PVC pipes surface dissolving, gap-filling adhesives with tetrahydrofuran as solvent are used. PE and PP cannot be reliably bonded by adhesives. PE sheets are laminated with webs consisting of filled PIB. These serve as adhesion promoters in further processing with GRP as well as for lining. PIB webs are bonded to concrete with bitumen-plastic-melt adhesives and to metals with special contact adhesives.

3. *Contact bonding agents* based on natural or synthetic rubbers are suitable for bonding plastics, particularly in web or sheet form, to impermeable substrates (metal, concrete, stone, glass). They are generally applied to both surfaces. After evaporation of the solvent the materials are joined together by rubbing or knocking. Good contact adhesives can absorb considerable shear forces with lasting flexibility. Two-component adhesives forming elastic networks maintain this property even at elevated temperatures. Polychloroprene adhesives increase their strength by slow partial crystallization.

4. Solvent-free *dispersion adhesives* (Sec. 4.1.1.10) are suitable for bonding plastics, particularly films, to porous materials (paper, board, felt, textiles, leather, wood). Fresh, fluid adhesive can be removed with water, but this cannot be done after the adhesive has dried. The bonds are very moisture resistant.

5. *Thermosetting moldings* are bonded to one another or to other materials with the same type of hot- or cold-curing resins. The adhesives discussed in (3) and (4) above are used for decorative laminates. Vulcanized fiber and synthetic horn can be bonded to each other and to wood by all the normal wood adhesives.

6. *High-load-bearing joints* between fiber-reinforced plastic parts or between these and other materials can be produced with solvent- and pressure-free reaction-resin adhesives (Sec. 4.6.1.4). PU-preproducts that quickly cure in an electron beam have proved useful for laminating multilayer films.

3.8.8. Screw, rivet, snap-in joints

There are manifold possibilities of temporary, limited life and permanent joining of plastic parts with each other or with items of other materials. The type of plastic and the design of the molded parts can only be decided upon by the designer after the method of joining has been chosen. This decision often strongly affects the economic viability of the manufacture and assembly of the plastic components and consequently their cost-effectiveness and useability.

3.8.8.1. Joining with screws and rivets

Because of their low strength and the stiffness of the plastic, plastic nuts and bolts should only be used when there is no alternative. This may occur if electrical insulation or very high corrosion resistance is desired or the screws must be transparent and/or colored throughout (see VDI 2543 and 2544 Screws of thermoplastic materials).

Screw joints using metal screws however widespread with, for example, brass sockets with internal threads incorporated into injection molded parts. These inserts can be avoided and the thread produced by post processing by cutting the thread into the thermoset or thermoplastic (forming the thread during the injection molding involves expenditure in mold manufacture and, because of shrinkage, the thread pitch is insufficiently accurate). A thread depth of 2 to 3 times the screw diameter is chosen. Such thread cutting is only economic when it can be combined in one process with other post processing such as drilling, cutting, milling, etc.

Special self tapping screws have proved especially useful in screwing together thermoplastic parts.

In screwing together thermoset plastic parts, self tapping screws are useful; they do not cause distortion of the plastic (as in tough thermoplastics) but effect optimal cutting.

Riveting of plastics to each other or to other materials can be effected using copper, brass or aluminium rivets and also with pop rivets.

If one of the parts to be joined is a thermoplastic, the end of the pin of the molded part involved can be formed into a rivet head by ultra sound.

3.8.8.2. Snap-in joints

Snap-in joints are in many cases a technically- and cost-effective method of joining which can be made for single or multiple use. Because many plastics permit large strains without breaking or permanent deformation, they are particularly suitable for this type of joining. Figures 3.77 a and b show simple examples with snap hooks, Fig. 3.78 shows a chassis covering with two snap fishplates which can be released by thumb pressure in the direction of the arrow. Fig. 3.79 shows an opened snap ring joint.

The most important basic forms for such joints are: spring hooks, snap rings and twist snap joints. Raw material manufacturers give information on calculating hook thickness (h), deflection force (Q) and adhesion force even in complicated shapes (e.g. Bayer AG: Plastic Snap Joints, Formation and Calculation), order number: KU 46036, published 12/90).

3.8. Forming and Joining Semifinished Plastics

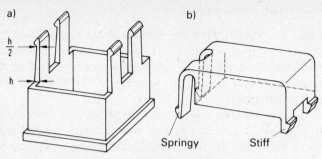

Fig. 3.77a. Component for control panel with four snap-in hooks.
Fig. 3.77b. Cover with two stiff and two springy snap-in hooks.

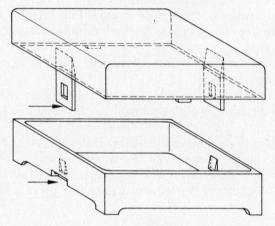

Fig. 3.78. Separable snap-in joints on a chassis cover.

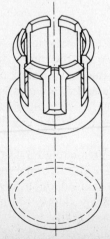

Fig. 3.79. Snap-in ring joint.

3.9. Other Fabrication Processes

3.9.1. Cutting, Punching and Separating

Most of the machines and types of tools normally used for cutting, blanking, flat edge trimming and hole punching of metal sheets are also applicable to semi-finished plastics.

Soft plastics are cut with heated knives and foams are cut by hand or by machines with hot wire tools. Thin acrylic sheet and decorative laminates can be scratched with a hard, metal tipped stylus and then broken like glass. The *CO_2-laser burning* process with fume extraction has proved effective for the cutting and drilling of PMMA and other thermoplastics but less so for thermosets. Prepregs and articles made from fiber-composite materials are edge trimmed by means of *high-pressure water jet* cutting (2500–4500 bar). The equipment is operated by hand or by a coordinate guiding machine fitted with electronic sensing. Both techniques provide precise cutting edges, small cutting width and a high rate without dust or shavings.

3.9.2. Machining

The specific cutting forces for plastics are small – about $200 \, N/mm^2$ ($300 \, lb/in^2$) for thermosets and $100 \, N/mm^2$ ($150 \, lb/in^2$) for thermoplastics. Fast-running, rigidly built, powerful machines for light metals and wood permit the high working speeds at reduced feeds and small rake angles that are necessary to control the cracking of brittle plastics and heat build-up. Excessive heating of the poorly conducting plastic can result in smearing, burning the product or annealing the tool. Because plastics expand on heating, the material being worked should be cooled. The best method of cooling is by an air jet, which is indeed the only method possible with thermoplastics that have long and cohering shavings. To ensure that such shavings do not remain hanging the tool tips are rounded off. The short, dusty chippings of thermosets must be extracted. Cooling of GRP by fluids can be useful. Lubricants (MoS_2) improve the quality of the surface, e.g., of PMMA.

Equipment has to be fitted with elastic chucks to avoid stressing the product. Chucks for metallic workpieces are not always suitable for plastics.

Tool steel is not generally suitable as the *material for these tools*. High-speed steels can have a cost-effective service life in use with unfilled thermoplastics. Ceramic oxide materials, when suitable, provide the necessary high cutting speeds. Diamond-tipped tools permit very tight tolerances over long runs.

Guiding principles for the working and the machining techniques for individual thermoplastics, thermosets, and reinforced plastics are based on current general national standards for these processes. Examples can be found in:

3.9. Other Fabrication Processes

VDI 2003: Machining of Plastics
VDI Publishers, D-40001 Düsseldorf, Germany

Machining Data Handbook
Metcut Research Assn.
3980 Roselin Dr., Cincinnati, OH 45209, USA

Machining of Plastics, Kabayashi
McGraw-Hill, Inc.
1221 Ave. of the Americas
New York, NY 10020, USA

3.9.3. Surface Treatment

1. *Polishing*: Thermoformed and machined parts are hand polished with buffing discs and special grinding and gloss waxes to deal with surface blemishes. Small parts are tumbled with antistatic granules and polishing material.

2. *Lacquering* of structural foams and FRP moldings (Sec. 4.6.1.8.2.1 and 4.6.2.4.2) with reactive-resin lacquers generally requires surface preparation by after-grinding and priming if thin layers of lacquer cannot be sprayed into the mold. Black-pigmented UV-filtering lacquer (Transfer-Electric) protects polyolefins from photochemical decomposition. Abrasion-resistant conducting lacquer is used for antistatic lining of petrol tanks, etc., and filled with silver, nickel or copper is used for high frequency shielding of electronic equipment. The scratch resistance of PMMA (Sec. 4.2.6.1) and PC (Sec. 4.2.8) is significantly improved by special clear lacquers. For surface effects on thermoplastics, such as pearlescence with fish silver or Iriodin pigments or two-toned coloring of artificial leather, lacquers with related raw material and special solvents are used.

3. *Printing and decorating*: Thermosets are hardly ever printed. Flat MF or GRP products are decorated by embedded printed or painted paper or patterned textile under a clear covering layer (Sec. 4.6.3.5.2). Ornamin decorative sheets are inserted into the mold for compression and injection moldings.
Polyolefins, polyacetals and polyfluorocarbons, which are not attacked by print ink solvents, can be printed on after flame corona or plasma treatment or pickling of surfaces. Similar effects produced by photosensitizers and UV irradiation (200–300 nm, $1000\,J/cm^3$) can be given a long shelf life by subsequent treatment with polyisocyanates.
Sheet from the roll is printed on conventional rotary machines. Cut sheet is printed (appropriately distorted for subsequent thermoforming) on sheet

3.9.3. Surface Treatment

printing machines modified for plastic materials. There is a multiplicity of special machines for multicolored printing of cylindrical, conical four-edged and flat moldings. Printing ink manufacturers market special types for plastics. Relief printing is used, with metal or plastic forms, in the normal printing process for small runs in sheet print and for printing on flat moldings. Aniline printing or flexoprinting are the main processes for printing from the roll with rubber printing forms, which can take up several colors. Engraved (intaglio or gravure) printing permits technically perfect reproduction on sheet or circular bodies, but its high capital and print-forms costs make it economical only for long runs. For the universally applicable screen printing, originally a hand process for short runs, there are automatic multicolored print machines that can be used for round articles.

In thermodiffusion printing for polyolefins, PA and POM, thickly applied printing ink diffuses into the surface by a brief treatment of the printed article at 100 to 150°C. The *Therimage process* consists of the transmission of an engraving on waxed paper onto a hollow thermoplastic article by hauling off and fixing at elevated temperatures. In the *Formprint process* for blown articles preprinted sheets of appropriate material that combine with the product are previously placed in the mold, as is the case in the *Ornamat process* for injection-molded parts, in which the Ornatherm or machine-fed sheet is likewise inserted in the mold. Irregular and large moldings that cannot be printed are decorated by *ink spraying* around a stencil. Local *blind* and *color stamping* is done with a heated plug. HF blind stamping is combined with the melting-in of stamping foils for PVC sheet. "Plastettes", which are multicolored pictures on printed sheets, are welded-in by heated stamping dies. There are special machines and sheets for reverse-side stamping of transparent sheet (counterstamping) and the stamping of plane surfaces and of moldings, e.g. writing equipment or bottles. Print color foils can also be raised (dry printing). For *velouring* of sheets and moldings by electrostatic flocking see Fig. 3.80. Velour effects can also be achieved with special two-component lacquers. Flock spraying is also used for noise reduction, insulation and damping purposes.

4. To *metallize*, individual parts are held on rotating supports and sheets on reeling mechanisms in high-vacuum tanks (10^{-4}–10^{-5} mbar). Here they are coated with a 0.1 to 1 μm thick layer of metal, which has been volatilized by electrical heating. From 0.2 μm up the metallic coating is opaque. Gas-producing (plasticizer-containing) plastics require a primer lacquer while mirror like thin metallic coatings need a protecting one. Thicker layers may be added by electroplating.

To key metal coatings of 50 to 100 μm thickness applied by *electroplating*, the parts are first immersed in etching baths to chemically roughen up the

3.9. Other Fabrication Processes

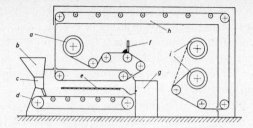

Fig. 3.80. Flock finishing of plastics.
(a) Unwinding, (b) Flock hopper, (c) Metering device, (d) Transport belt, (e) Screen electrode, (f) Adhesive doctor knife, (g) Electrostat, (h) Drying channel, (i) Wind-up.

surface. There are currently appropriate etching media for PE, PP, PVC, ABS, ABS/PC blends, mod. PPO, PF and EP resin molded parts. The roughened surfaces are activated in baths containing solutions of noble metal salts so that firmly cohering copper coatings can be precipitated from copper baths onto the plastic surfaces without current. These are then further coppered, nickeled or chromed to the desired layer thickness by electroplating. Moldings of special PP compounds (Caprez DPP, GB) can be electroplated without pretreatment. These processes lead to composite products with improved mechanical and thermal resistance (Table 3.16) but maintaining the other advantageous properties of plastics (light weight, no corrosion) which are so desirable.

Table 3.16. Effect of Electroplating

		ABS		PP	
		Not galvanized	Galvanized	Not galvanized	Galvanized
Flexural stress, yield	N/mm^2	39	52	45	83
E-modulus	N/mm^2	2200	6300	1200	6200
Ball indentation hardness	N/mm^2			54	86
Deflection temperature to	°C	90	130	148	>170

4. THE INDIVIDUAL PLASTICS

4.1. Thermoplastics: Raw Materials and Molding Compounds

4.1.1. Preproducts and Special Products

4.1.1.1. Monomer reaction resins

1. *Styrene* is a volatile, colorless liquid with a characteristic smell, boiling at 140°C, ignition temperature 35°C. Its vapor attacks mucous membranes. It is delivered with stabilizer, which is removed prior to processing by washing with soda solution, distillation in vacuum or adsorption on Al_2O_3. It should be stored in a cool, dark place. Polymerization to hard polystyrene (for properties, see Sec. 4.1.3.1) at 80°C takes 8 days. It can be catalyzed by the addition of benzoyl peroxide (0.1 to 2%). High-volume contraction can result in voids. This can be reduced by polymerizing as slowly as possible at not too elevated a temperature, by the solution of 10 to 25% polystyrene, or by the addition of paraffin, wax, bitumen or mineral oil. The additives can reduce the clarity of the polymerization product. Unfilled glass-clear or PE-, PTFE- or silica-filled cast resins for electronic applications (Stycast, US) are cross-linkable. For the hardening of unsaturated polyesters by cross-linking with styrene see Sec. 4.6.1.5.

2. *Vinyl carbazole* exists as colorless, light-sensitive stable crystals melting at 65 to 67°C. In the molten state with added surface-active fillers and catalysts it polymerizes to a heat-resistant thermoplastic, stable up to 150°C. The surface resistivity of polyvinyl carbazole is reduced by light. This photosensitivity is used, *inter-alia*, in xerography in the generation of a latent image on selectively illuminated plates coated with sensitive material.

3. *Methacrylates* are colorless liquids with characteristic odors. Methyl methacrylate boils at 100°C and is readily flammable (ignition point, 10°C). PMMA is soluble in the monomer. Peroxides act as polymerization initiators on warming. For polymerization in the cold, peroxides are supplemented by amino activators or other redox polymerization-initiating systems. Partially cross-linked final products are produced with the inclusion of multifunctional components in the hardening system. Diacryl 101 is a low-viscosity bifunctional liquid condensation product based on Bisphenol A (Table 1.4) with methacrylic end group. At temperatures of 100 to 150°C and with peroxide initiators it polymerizes to an acid- and alkali-resistant product, HDT:F_{ISO} A 140°C, that can be combined with unsaturated polye-

sters, etc. in Sec. 4.2.6.2 (see also Tables 4.29, 4.30). Oligourethane methacrylate resins can be cured in a similar fashion.

Among the applications of the cast resins are gap-filling transparent acrylic adhesives, laminated glass, wood impregnation, embedding of demonstration objects, decorated colored filled button sheets, highly transparent glass-reinforced sheets and artificial stone with inlaid marble. Two-component systems with mineral fillers, polymer powder and peroxide initiators on the one hand, liquid monomer and amine catalysts on the other, are used to make highly resistant floor screeds, quick-hardening patching cement and road markings. They are also used in surgery, orthopedics and dental prosthetics.

A low-viscosity dispersion with 72% solids containing finely divided mineral filler in MMA (Asterite, GB) and similar highly filled cross-linking two-component casting resins (Paraloid VII, US) can be cast and heat hardened to make very serviceable thick-walled marble (Corian, DE, US) or ceramic-type surfacing and basins for sanitary and kitchen equipment.

Low-viscous, cross-linking modified MMA-resins (Modar, GB; Acpol, US) cure completely and very rapidly to give an excellent surface finish and mechanical properties. They can be highly filled, even if reinforced, e.g. with flame retardant alumina trihydrate or 60% cristobalite. Grades are available for GRP-resin transfer injection and cold press molding (Sec. 3.2.4) as well as for pultrusion (Sec. 3.4.3).

4. *Caprolactam* (m. p. 70°C) and *lauroyl lactam* (m. p. 153°C) are polymerized on a large scale at 250 to 300°C to PA 6 and PA 12 which melt at these temperatures. Co-catalysts (acylation agents, particularly isocyanates) permit rapid, pressureless polymerization to strong and stiff cast PA 6 and PA 12 (Table 4.37) at 100 to 200°C, although with a 15% volume contraction. Large, thick-walled articles up to 1 ton in weight are cast by a two-component process using simply-produced molds such as used in metal casting. The two components are either caprolactam or mixed melts with materials which act as initiators or co-catalysts at reaction temperature. Rollers and pipes are made by centrifugal casting (Sec. 3.2.3.2), whereas containers of up to 10,000 l capacity are made by rotational casting (Sec. 3.2.3.1). RIM articles (Sec. 3.2.2.4) of nylon block copolymer (NBC) are produced by the Nyrim process with the caprolactam component on the one side and a catalyzed polyether elastomer component on the other. Scrap can be reused for injection molding or extrusion.

5. *Methyl-2-cyano acrylate* is a water-clear liquid that is polymerized directly in the presence of atmospheric moisture or weak alkalinity (e.g., from glass) at room temperature, almost without reduction in volume. One component cyanoacrylate adhesives are expensive but versatile. For example, methyl-

2-cyano acrylate is serviceable at up to 160°C for short periods in precision engineering applications.

6. *Triallyl cyanurate* (m. p. 27°C) contains three allyl double bonds symmetrically arranged around the stiff cyanurate ring. These can be polymerized simultaneously or successively. As a result it serves as an additive in directed cross-linking, e.g., of PE cable insulation and under-floor heating tubes (Sec. 4.1.2.2).

4.1.1.2. Olefin polymers

1. Ethylene (co-)polymers and chlorinated products

Low molecular weight PE: Epolene, Zetabon[1], (US) Hoechst-Wachs PA 190, Kuroplast KR 2175[1], Veba-Wachs, Le-Wachs, (DE) and PE with a hydroxyl content such as Elvon (US) are used as compounding materials in plastics processing.

Ethylene copolymers with vinyl acetate, acrylates and acrylic acid (see also Sec. 4.1.2.4) such as Bakelite DQR, Elvax, Ultrathene, and Zetafax (US); Kuroplast types (DE) are compounded with paraffins, waxes and synthetic resins to produce polishes, melt adhesives and coating compounds. For cross-linking synthetic rubber grades see Table 5.1.

By alcoholysis of the vinyl acetate and further reaction of the resulting OH groups and sometimes also graft polymerization, copolymers can be "denatured" in many ways (Dumilan, JP). Such products are used as the interlayer in safety glass and for laminating of glass bottles. Acrylamides containing copolymers are used as the adhesion agent between metals and polyolefins.

Chlorinated PE (PE-C) (Fortiflex, Tyrin (US); Daisolac, Elaslen (JP); Bayer CM, Hostapren, Lutigen (DE); Haloflex (GB); Solpolac (IT); Kelrinal (NL)) with a chlorine content of 25 to 40% is flexible to soft-rubbery and cold resistant. It is compatible with many plastics and is compounded with PE to reduce its flammability, with PVC to raise its impact strength (Sec. 4.1.4.3.3) and with PS, etc., for recycling waste-mix compounds. Applications of plastics grades are summarised in Sec. 4.1.2.3, and of chlorosulfonated PE (Hypalon, US) as a synthetic rubber in Sec. 5.1.13. Chlorinated polypropylene (PP-C) is a resin lacquer useful in corrosion protection. Adhesive layers on polyolefins are a further application of chlorinated polyolefins and special copolymers.

---

[1] also containing carboxyl groups.

4.1. Thermoplastics: Raw Materials and Molding Compounds

2. Atactic polypropylene (APP) and other α-olefin polymers

Low molecular weight APP and α-olefin polymers (A-fax, US; Vestoplast, DE) and atactic polybutene (APB) delivered in blocks, granules or as molten liquids in tank wagons, are formed as amorphous soft by-products in Ziegler-Natta polymerization (PP, Sec. 4.1.2.6; PB, Sec. 4.1.2.7) and are produced in special grades.

Applications include the coating of paper packaging and the back coating of carpets from the melt, car vibration dampings, corrosion protection wraps, road markings, melt adhesives, sealants, lubricants and bituminous blends for road making and for aging-resistant construction sealants and roof sheeting.

3. Polyisobutylene (PIB)

Trade names: Parapol (low molecular), Vistanex (US); Oppanol-B (DE); Hyvis, low molecular, (GB).
Commercial grades of Oppanol-B are given in Table 4.1.
Grade B.3 is used as an electro-insulation oil, whereas B.10 and B.15 are applied as viscosity-improving additives to mineral oils and in stabiliser-free form in the manufacture of chewing gum. All the types are used in adhesives and sealants, which have properties that do not vary from their glass temperatures (about −60°C) up to almost 100°C. B.100 is added to laminating waxes and polyolefins. The highest molecular weight grades are processed into linings and roof sheeting at 170 to 200°C (Sec. 4.2.2.2) with mineral fillers protecting against light.
The moisture and gas permeability of PIB (Oppanol B 200) is very small, similar to that of PE-HD. Addition of PIB to rubber mixes reduces their gas permeability and improves their weathering and aging characteristics as well as the adhesion of tire treads (cf. butyl rubber, Sec. 5.1.4). PIB (d_R 0.92 g/cm^3) has excellent dielectric properties (specific resistance, 10^{16} ohm cm; $\varepsilon_R = 2.2$; tan

Table 4.1. Commercial Grades of Oppanol-B

		Molecular weight	
		Number average	Viscosity average
B.3	Viscous oil	820	–
B.10		24 000	40 000
B.15	Sticky to soft plastic mass	40 000	85 000
B.50		–	400 000
B.100	Raw rubber-like	–	1 110 000
B.150	Short-term-stressed	–	2 600 000
B.200	Elastomer	–	4 000 000

$\delta = 0.0005$) and is resistant to aggressive chemicals, apart from nitric acid and halogens. It is soluble in aliphatic and aromatic hydrocarbons and insoluble in esters, ketones and alcohols. Oppanol-B conforms with the requirements of the food laws.

4. *Ethylene copolymer + Bitumen blends (ECB)*

Lucobit as a thermoplastic polyolefin material consisting of ethylene copolymers and special bitumen. The plastic forms the continuous phase in which bitumen droplets are embedded. Lucobit contains no plasticizers and is resistant to weathering and ageing. It is supplied as black granules (density 0.97 g/cm^3) and is extruded at temperatures between 140°C and 190°C into sheeting or injection molded between 160°C and 220°C. Thick-walled moldings are obtained by casting at atmospheric pressure at temperatures between 250°C and 280°C. The main areas of application of the various Lucobit grades are membranes for roofing (Lucobit 1210 and 1221) and tunnels and underground works (Lucobit 1233).

Roofing membranes made of Lucobit 1210 with single-sided or internal glass or polyester non-woven reinforcements are flexible (E modulus $< 15 \,\text{N/mm}^2$), exhibit high elongation at break $> 300\%$) and, like sheeting made of other Lucobit grades, resist root penetration and remain elastic at low temperatures. Lucobit 1221 roofing membranes have higher heat resistance and are used, particularly in warmer climates, for flat roofs.

Lucobit 1233 is used as a waterproof membrane below ground, e.g. in tunnelling to prevent the ingress of water under pressure. Like Lucobit 1210/1221, it can be welded easily and is distinguished by high toughness and biaxial extensibility.

Added to bitumen melt and asphalt compounds, Lucobit significantly improves elastic stability under load over a wide range of temperature.

4.1.1.3. *Styrene copolymers*

Styrene copolymers with 15 to 40% butadiene are used as compounding materials in rubber processing. Variously modified multi-copolymers (Sec. 4.1.3.1.2) are used as interphase coupling agents in polyblends, intermediate bonding layers for films, flexibilizers and other property enhancing additives.

4.1.1.4. *Vinyl chloride (co-)polymers*

PVC and copolymers with VAC, maleic acid, etc., or terpolymers containing vinyl alcohol with K values of 40 to 50 (Table 4.16) are soluble in ketones, esters, chlorinated hydrocarbons and solvent mixtures as well as in aromatics. They are used as lacquer resins. Terpolymers with free carboxyl groups and

4.1. Thermoplastics: Raw Materials and Molding Compounds

moderate adhesion to metals are used for corrosion protection dipping packages, can coatings and for cocooning irregularly shaped objects with weatherproof protective casing.

Mixtures of PVC of low K value and containing high levels of plasticizer are liquid at 100 to 140°C. They are used for tear-off dip packaging and for casting flexible molds.

PVC post-chlorinated in carbon tetrachloride is soluble in esters, ketones and mixtures of them with aromatic and chlorinated hydrocarbons. Such solutions serve as "PC adhesives" for PVC-U semi-finished articles (Sec. 3.8.7). Their resistance to inorganic chemicals is the same as that of the material.

4.1.1.5. Vinyl acetate (co-)polymers and their derivatives

1. Polyvinyl acetate (PVAC) and copolymers

Trade names: Daratak, Elvacet, Vinylite (US); Gohsenyl (JP); Mowilith, Vinnapas (DE); Rhodopas (FR); Emultex, Epok (GB); Neo Vac (NL); Vipolit (CH).

The many grades of PVAC ($d = 1.17\,\text{g/cm}^3$) with K values of 20 to 90 (MW 35,000–2,000,000) are glass-clear, soft-to-hard resins whose lack of adequate temperature resistance makes them unsuitable for molding materials. However, they are readily soluble in most solvents (apart from aliphatic hydrocarbons and water-free alcohols). They form light-fast films that are stable in oil, naphtha and water, although they swell somewhat in water (water absorption up to 3%). Chemically PVAC can be attacked by saponification reactions. PVAC has a high capacity for accepting pigment and is of limited compatability with plasticizers and nitrocellulose. Many grades are supplied only in solution or dispersion, (p. 173) for paints, adhesives, textile finishes and coatings.

2. Polyvinyl alcohol (PVAL)

Trade names: Elvanol, Monosol (US); Poval, Gohsenol (JP); Mowiol, Polyviol, Vinarol (DE) and Rhodoviol (FR).

PVAL is prepared by hydrolysis of polyvinyl acetate. It is a white-to-yellowish powder, insoluble in organic solvents. Partially hydrolyzed grades with about 13% residual PVAC are readily soluble in water, more so than fully hydrolyzed PVAL. PVAL gives clear, colorless, light-fast films. PVAL can be processed using polyhydric alcohols, such as glycerol, as plasticizer into thermoplastic leatherlike products with good aging qualities such as petrol- and solvent-resistant tubes, membranes, packing and release films, e.g., for wet prepregs (Table 4.73). Water-soluble packaging films are described in Sec. 4.1.1.11.

PVAL with the addition of dichromate becomes water insoluble after UV irradiation; applications of this substance are in graphics.

3. *Polyvinyl acetals (Butyral PVB, Formal PVFM)*
Trade names: Butacite PVB, Butvar, Formvar, Saflex (US); S-lec (JP); Mowital, Pioloform (DE), Rhovinal B (FR).

Polyvinyl acetals are prepared by condensation of polyvinyl alcohol with aldehydes. Formaldehyde and butyraldehyde acetals are solid resins soluble in organic solvents.

Applications of PVB, depending on the properties of the special resin, are coatings (film lacquers, heat-sealable primers, stoving enamels, gold lacquer), printing inks, impregnating agents, adhesives, shrink capsules and tear-off packaging. Interlayers for laminated safety glass are made from high molecular weight PVB. Moldable thermoplastic polyvinyl acetals are made into oil- and naphtha-resistant seals and tubes. Fuel-resistant lacquers made from polyvinyl formal are used for petrol containers; resin wire lacquers and heat-curing film adhesives for metals are manufactured using a combination of phenolic resins and PVB.

4.1.1.6. Polyvinyl ethers

Trade names: Lutonal, V-Wax (DE).

Polyvinyl methyl ether (Lutonal M) is soluble in cold water and in organic solvents except aliphatic hydrocarbons, ethyl ether (A), isobutyl ether (I) and decalyl ether (Z, ZI). According to their degree of polymerization PV ethers are oils, or soft or rigid solids and are not saponifiable and have good electrical properties. They are used as electrical insulants, contact adhesives, insole build-up in shoes, and as blended resins, e.g. for chewing gum and dentures (Lutanol AK). The octadecyl ether (V-Wax) is a basis of high-gloss polishing wax.

4.1.1.7. Polyacrylate resins

Homopolymeric acrylic acid esters are soft resins of little significance. Their stability in light to oxidative influences and to heat and its elastifying effect manifests itself in many copolymers and terpolymers with styrene, vinyl chloride, acetate, methacrylate, acrylic acid, acrylonitrile, etc. These are supplied as solid resins or solutions (e.g. Acrysol, US; Acronal F and L, Plexigum, Plexisol, DE), but mainly as dispersions (Sec. 4.1.1.10). Elastoplastic copolymers form the basis of sealing compounds. They are water soluble with $>20\%$ acrylic acid. Polyhydroxyethyl methacrylate (Hydron, US) swells in water.

Saturated with 40% water and softened, it is used in contact lenses. Further applications include coatings (e.g., of spectacles) and coating-controlled moisture uptake and permeability in medicine and technology.

Heat- or radiation-curing lacquer resins are prepared by copolymerizing acrylics with cross-linking components (e.g., polyisocyanate adducts, Sec. 4.6.1.7/8). Hard methacrylate copolymers are used as finishing lacquers for artificial leather and as fuel-resistant finishes. MMA-VC copolymers (e.g., Paraloid, US) confer elasticity on PVC (Sec. 4.1.4.3).

4.1.1.8. Polyamide copolymers and polyamine products

Polyamide copolymers such as polyamide 6/66 are soluble in aqueous-alcohol mixtures. They are used in the manufacture of petroleum-resistant electrically insulating lacquers, ready to use solutions of textile adhesives and coatings (Nylosol) and hot melt adhesives. (Vestamid T, DE; Gril-tex, CH).

Polyaminoamide and the similar polyaminoimidazoline, which are prepared by the condensation of dimeric or trimeric fatty acids with excess polyamine, are flexible cross-linking agents for epoxy sealant and adhesive systems, (p. 335 Versamid (US); Euredur (DE); Genamid, Wolfamid (GB), Casamid (NL). They are coupling agents for PVC-plastisol stoving enamels (Euretek, DE). High molecular weight thermoplastic polyamides of a similar nature (Versalon, US, Eurelon, DE) are used as solvents and hot melt adhesives and as printing ink binders.

4.1.1.9. Natural product derivatives

1. *Rubber conversion products*

Chlorinated rubber

Trade name: Pergut (DE).

Manufacture: Chlorination of pre-swollen or dissolved rubber up to a chlorine content of 65%.

Properties: Hard resin, readily soluble in benzene, chlorinated hydrocarbons and ketones; insoluble in naphtha, mineral oils, alcohol and water; resistant to concentrated acids and alkaline solutions and halogens up to 100°C. Not flammable.

Applications: Corrosion-resistant coatings, also in combination with other lacquer resins and in synthetic rubber adhesives.

4.1.1.9. Natural product derivatives

Rubber hydrochloride

Manufacture: Treatment of rubber with hydrochloric acid gas.
Properties: More or less flexible products; soluble in aliphatic hydrocarbons; resistant to many solvents, fats and oils, acids and alkalis; ages in light.
Applications: Lacquer resins, leather-hard moldings, most importantly cast packaging film (Pliofilm, US).

Cyclized rubber

Trade name: Pliolite NR (US), Alpex CK 450 (DE).

Manufacture: By the action of concentrated sulfuric acid (*interalia*) a horn-like thermoplastic hard rubber vulcanizable resin is produced, suitable for shellac-like materials and lacquers.

2. Cellulose conversion products

Cellulose nitrate (CN, collodion cotton)

Supplied as:
(1) Fibrous flakes moistened with 30 to 35% ethanol or butanol.
(2) Celluloid-like lacquer material in the form of chips gelatinized with 18 to 20% plasticizer.
(3) Paste with 22 to 40% collodion cotton (according to type) in mixed solvents.

Plasticizer-free colodion cotton should not be allowed to dry, as it assumes explosive properties. If it dries, it must be moistened.
Application: Alcohol soluble (nitrogen content below 11%) and ester soluble (nitrogen content about 12%) grades in various viscosities for nitro lacquers and adhesives.

Cellulose acetate, cellulose acetobutyrate and cellulose propionate

Distinguishable esterification grades: lacquer and adhesive resins with plasticizers that at 150°C give workable melt and dip materials, for tear-off packaging. For molding powders, tubes and sheets see Sec. 4.1.11 and 4.2.9.2.

Ethyl cellulose (EC) and benzyl cellulose are used in the same way as the esters. For water-soluble cellulose ethers see 4.1.1.10.

4.1.1.10. Dispersions

In emulsion polymerization, finely divided particles of polymers are produced in water. They are designated as coarse dispersions at sizes of <1–$10\,\mu$m (<0.4–0.04 mil) and as fine dispersions or latices at 1 to $<0.1\,\mu$m (0.04–<0.004 mil). They are stabilized with protective colloids (PVAL, water-soluble cellulose derivatives), and they are stable within defined pH limits. Some are also stable when frozen. Dispersions with 40 to 50% solids content, concentrated by creaming to 70% solids, have a low viscosity compared with polymer solutions, even after the incorporation of plasticizers. They differ from plasticized PVC pastes, with plasticizer as the liquid phase (Sec. 4.1.4.9), in that they can be diluted with water to any extent. They are extensively miscible with aqueous solutions and most are readily compatible with fillers. The minimum film-forming temperature from dispersions is depressed according to the fine-particle distribution of the resin and as appropriate by the addition of small amounts of volatile solvents. Dried film or precipitates are not redispersible. They are water resistant. If the emulsifiers and protective colloids are absorbed by a substrate, the films are somewhat swollen by water. *Hydrosols* are extremely finely divided dispersions (approximately $0.0045\,\mu$m) with similar penetrating capacity to primers based on solvents. Some types of dispersions, particularly those based on vinyl acetate (Sec. 4.1.1.5), and acrylic resins (Sec. 4.1.1.7) are available for solvent-free adhesives, paints and coatings and spreading impregnation of textiles and papers. Many contain cross-linking components, which increase resistance of the film to solvents. Dispersions are also used as binders for non-woven textiles and are precipitated electrolytically on fibrous artificial leather. Many conform with the requirements of the food laws. They are used not only for food-packaging material but also for the direct coating of hard sausage and hard cheese. Dispersions resistant to alkaline saponification are employed as polishes, in admixture with hydraulic hardening cement, for the modification of concrete, as the adhesive agent between old and new concrete and for industrial flooring.

Availability: Dispersions are supplied in paraffined wood drums, polyethylene casks, containers or tankers. For storage, insofar as they are not in supply containers, corrosion-resistant storage tanks (e.g., of stainless steel) or GRP containers can be used, but not iron vessels. They should be stored in sealed containers and in the cold. Evaporation encrustation cannot be dispersed, nor can dispersions that have coagulated as a result of low temperature.

Redispersible powders can give film-forming latices when stirred in water, with plasticizer if appropriate. They serve futher to increase the resin content of mastics and knifing fillers and as additives to hydraulic binding agents that can be delivered in powder form, e.g., for ceramic wall tiles. Stable polyethylene dispersions in water or organic solvents with up to 70% solids content are made from PE powder with a particle diameter of 10 to 20 μm (Microthene) and dispersing and thickening agents.

4.1.1.11. Water-soluble polymers

General applications of such solutions are: protective colloids for dispersion and stabilization, thickening agents, textile sizes and finishes, paper sizes, binders for the reservoirs of ball-point pens, paint and printing colors, adhesives and coating materials.

The following water-soluble polymers are available.
  (1) Polyvinyl alcohol (PVAL) (see Sec. 4.1.1.5.2).
  (2) Polyvinyl methyl ether, (see Sec. 4.1.1.6).
  (3) Polyacrylic acid salts and similar copolymers.
  (4) Polyvinyl pyrrolidone and copolymers. The blood replacement Periston is a solution of polyvinyl pyrrolidone.
  (5) Water-soluble cellulose ether, methyl cellulose (MC), carboxymethyl cellulose (CMC cellulose-glycol ether), hydroxy-ethyl and hydroxypropyl cellulose, and cellulose with grafted ethylene oxide: ethylose.
  MC is soluble only in cold water and is precipitated by salt. Higher methylated types are dissolved in a methylene chloride-methanol mix. CMC solutions are compatible with alkalis and alkali salts and are not precipitated on warming, as they are on the addition of acids and solvents. Applications: bookbinders' adhesives, carpet adhesives and painters' size.
  (6) *Polyethylene oxide (PEOX)* Glycol polyethers are made by the polyaddition of glycol and ethylene oxide. Because of the structure of the chain, they are also called polyethylene oxide. They are wax-like in the molecular weight range from 1000 to 10 000 and plastic at higher molecular weights. They are soluble in water, methanol, ethanol and several chlorinated hydrocarbons at room temperature and in many solvents at higher temperatures.
  Carbowax (US) and Oxidwachs, Polywachs (DE) are applied in dilute solution as mold lacquers. Polyox (US) is a tough thermoplastic as a high molecular weight product. Aqueous solutions are highly viscous to gel-like.
  Low molecular polyalkylene glycols are synthetic lubricants. For polyether polyols as PU preproducts see Sec. 4.6.1.8.1a.

4.1. Thermoplastics: Raw Materials and Molding Compounds
4.1.2.–4.1.12. Thermoplastics for Molding, Extrusion and Calendering[1]
4.1.2. Polyolefins
4.1.2.1. Polyethylene

1. Product groups

Polyethylene molding compounds are internationally identified and grouped (Table 4.2) according to the principal characteristics:

density range, 0.910 to $>0.960 \, g/cm^3$, and
melt flow index, the (reciprocal) indicator of the polymerization grade.

The characterization through these "designatory properties" is quite inadequate to describe the behavior and suitability for use of the many PE grades made commercially available by suitable combinations of "structural parameters" (Table 4.3). The numerical values of certain physical properties, according to ASTM 1248, 2952 and 2853 are summarised in Table 4.4 (p. 179). ASTM D 3350, Polyethylene Plastic Pipe and Fitting Materials, also includes tests for environmental stress crack resistance, thermal stability and hydrostatic design basis (800–1000 psi).

2. Processes for synthesis

High-pressure, low density (LD) (Imperial Chemical Industries, 1939): Ethylene gas under a pressure of 1000 to 3000 bar (14,000–40,000 lb/in^2) at temperatures between 150 and 300°C with 0.05 to 0.1% oxygen or peroxide as initiator yields 10 to 20% conversion to a strongly branched PE-LD with molecular weights up to approx. 600 000. With a modified catalyst system it can also be used for PE-MD and for PE-LLD (Atochem, Enichem, see below).

Medium-pressure, high density (HD) (Phillips, 1953): Solution polymerization in xylene at 150 to 180°C at a pressure of ≥ 35 bar (500 lb/in^2) with Cr-Al-silicate catalyst: PE-HD, mainly linear, molecular weight about 50,000.

Low-pressure, high density (HD) with Ziegler catalysts (1953/55): Ethylene introduced into dispersions of mixed catalysts ($TiCl_4$ + Al-alkyl, etc.) at pressures of 1 to 50 bar (15–700 lb/in^2) and temperatures of 20 to 150°C polymerizes into almost unbranched, linear PE-HD of medium to high molecular weight with only a few short side-chains.

---

[1] For different ASTM-specification and the ISO/DIN designation tables for the single product groups in the following sections see: Chapter 6, Standardization, p. 418.

4.1.2.1. Polyethylene

Table 4.2. Groups of PE Molding and Extrusion Grades

Nominal density (g/cm^3)	Common term	Symbol	ASTM D 1248[1] Types	Designation ISO DIS 1872[2,3]	DIN 16776-E82[1,3]
0.910-0.925	Low-density polyethylene	PE-LD	I	14, 18, 23	15, 20, 25
0.926-0.940	Medium-density polyethylene	PE-MD	II	27, 33, 40	30, 35, 40
0.941-0.959	High-density polyethylene	PE-HD	III	45, 50, 57	45, 50, 55
0.960 and higher			IV	62	60, 65
Nominal flow rate (g/10 min)	Degree of polymerization (i.e. molecular weight)		[4]	Category	[5,6]
> 25	Low molecular (LM)		1	700, 400	
> 10–25			2	200, 090	
> 1.0–10	to		3	090, 045, 022 012	
> 0.4–1.0			4	006	
0.4 max.	High molecular (HM)		5	003, 001, 000	

[1] PE-homopolymers.
[2] Homo- and co-polymers with a maximum content of other alkylene monomers of less than 50 mole % and $\leq 1\%$ of non alkenic monomer; further coded information is given in chapter 6.
[3] Last two digits are the medium values of density cells.
[4] MF cond. E, ASTM 1238 (190°C, 2.16 kp).
[5] 10-fold MF-medium values of cells.
[6] For materials having a value of less than 0.1/10 min measured at 190°C/21.6 N (ISO 1133, cond. D), 190°C/50.0 N (cond. T), or 190°C/216 N (cond. G) shall be applied. The code letter (D,T,G) must be inserted in front of the MFR-cell designation according to ISO 1872. (Code letters of MF-cond. ISO 1133 are not identical with the code letters for ASTM D 1248.)

Special products are: PE-HD-HMW (MW: 200,000 to 400,000); copolymers with C_4 to C_8 for paper-like films (pp. 188, 302) and pipes (p. 300), but not for injection molding, and PE-HD-UHMW, MW $> 3 \times 10^6$ to 6×10^6 (see p. 189).

Low-pressure linear low density (LLD) polymerization with high-efficiency catalysts (halogen-free metal complexes on carrier bases or as hydrocarbon solutions), is up to a 100-fold more efficient. Developed since 1960, such processes have increased the production capacity of new and converted old plants. They make possible the production of PE of the desired density between 0.913 and 0.940 g/cm^3 in a wide range of molecular weights, but with a tight molecular

Table 4.3. Structural Parameters and Properties of PE

Structural parameter	Density (g/cm³)		Molecular form		Molecular weight average		Molecular weight distribution (Mw/Mn)	
	0.915	0.97	Strongly and much branched	Linear, only short side chains	Low, 20000–60000	High, 200000–400000	Narrow	Broad
Limiting values	−/+	++	−−	++	−	+	+	−
	+	−		−	++	−−		+
	+		+		+	−		
Degree of crystallization		↑		↑		↑		↓
Elongation at break		↓	↑			↑		↓
Stiffness and hardness		↑	↑			□		↓
Impact strength		↓	↑			↑		↓
Stress-cracking resistance		↓	↓			↑		↓
Crystalline melting and heat distortion temperature		↑	↑			↑		↓
Low-temperature brittle range		↑	↑			↑		↓
Chemical and solvent resistance		↑	↑			↑		□
Water vapor and gas permeability		↑	↑			□		□
Transparency		↓	↓			□		□

+ −: High or low values. →: Increasingly favorable influence in direction of arrow. □: Without effect on property.

4.1.2.1. Polyethylene

Table 4.4. ASTM Specifications for Polyolefin Molding and Extrusion Materials

	PE and copolymers	PP and copolymers	PB	Reinforced polyolefins[1]
ASTM	D 1248[2], D 2952	D 2146	D 2581	D 2853
Density	(IVa): 0.910–0.960[2]	(I): Homopolymer, 0.902–0.904 (II): Copolymers 0.898–0.900	(Ia): 0.905–0.909 (IIa): 0.910–0.920	–
Melt flow	D 1248: 5 categories[2,3] > 25–0.4 max. (2952): (7b), < 0.4–100	(7a): 0.3 max.–100 cond. L: 230°C, 2160 g	5 categories: < 0.25–25, cond. E: 190°C, 2160 g	–
Temperature characteristics	(9c): Brittleness temp., −20 to −100°C	(9d): Deformations under load, 50°C, 690 kPa 25–1.5% max.	–	(8d): Deflection temp. at 264 psi: 63–157°C (145–315°F)
Tensile strength	–	(9b): Yield 14–34 MPa (1920–4840 psi)	(IIb): Tensile strength, 20.7 MPa (3000 psi) (d) Elong. break, 300% (IIc): Yield strength, 10.3–13.8 MPa (1500–2000 psi)	(8a): 22.4–105 MPa (3200–15 000 psi)
Secant flexural modulus	(7a): < 35–1100 MPa, (< 5–160 10³ psi)	–	–	(8b): 1000–5600 MPa (150–800 × 10³ psi)
Izod impact strength notched	–	(9c): At 23°C: 15–300 J/m (0.3–5.5 ft–lb/in) (7d): at −18°C: 250–1500 J/m (4.6–28 ft–lb/in)	–	(7c): 27–216 J/m (0.5–4.0 ft–lb/in)
Others	(8d): Diel. const., 1 MHz 2.25–2.80 (9c): Diss. factor 1 MHz 0.00005–0.01 (6): Intrinsic viscosity 1–20	*Classes*: According to 1248 (PE) and 2581 (PB): A, natural color; B, colored; C, weather resistant containing not less than 2% carbon black. According to 2146 (PP) D dielectric: Diel. const. max. 2.30; diss. factor max., 0.001; insulation resistance, min. 1×10^{15} ohm; water immersion stability		

Roman numerals (PE, PP, PB) indicate types; the Arabic numbers indicate the quantity of cells, which may be combined for the designation of a product. The values are the minimum values for the lowest and the highest number of cells. The letters a to f mark the respective order of the cell in D 2952, e.g. 516140: Flexural modulus cell, 5(276–552 MPa); flow rate cell, 1(< 0.4); brittleness temp. cell, 6(−70°C max.) dielectric constant cell, 1(2.25–2.30); dissipation factor cell, 4(0.0005 max.), and intrinsic viscosity, 0, i.e., unspecified.

[1] A complete product designation must include reinforcement content and type, polymer type and classification numbers.
[2] Only homopolymers. In addition, 1248 gives specifications of 11 electrical, 4 jacketing and 4 pipe PE-grades.
[3] Both cond. E., 190°C, 2160 g. The order of the categories D 1248 is the reverse of that of the cells D 2952.

4.1. Thermoplastics: Raw Materials and Molding Compounds

weight distribution. The process is carried out in the gas phase (BASF, Union Carbide) in a slurry process (BASF, Phillips) or in the liquid phase (Dow, DSM, Du Pont Canada, Rexon-Montedision, Mitsui) with a smaller expenditure on energy and equipment than in the older processes.

Linear low density polyethylene (PE-LLD) with statistically distributed short side chains and high crystallinity is produced by these processes using 5 to 10% C_4- to C_8- olefins as co-monomer. It is stronger, tougher, more serviceable down to low temperatures and more heat resistant than conventional PE-LD of the same density and melt index. Film thickness can be reduced by half for the same strength compared to PE-LD. The narrower molecular weight distribution of PE-LLD makes feasible processes requiring higher shear viscosities and the substitution of PE-MD blends for cross-linked pipes, for injection molding and large rotational molding with reduced tendency to stress cracking. Higher α-olefin HAO grades are long-chain binary or tertiary copolymers containing up to 10% octene, 4-tetramethylpentene-1, and in some cases propylene. With an excellent balance of overall toughness, they are premium grades for heavy-duty (stretch) films. PE-V or PE-ULD, i.e. Very or Ultra light HAO grades, density 0.910 to 0.880 g/cm^3, are of very low crystallinity and nearly rubber-elastic, with elongation at break up to 900%. They can be highly filled and processed like other grades of PE. They are also used as enhancers of elastic properties and crack resistance in various polyolefin compounds.

3. *Commercial types and trade names*

Polyethylene types of every variety: Alathon, Bapolene, Dylan, Escorene, Hifax, Marlex, Petrothene (US); Novatec, Sholex, Sumikathene (JP); Lupolen (DE); Lacqtene (FR); Rumiten (IT); Carlona (GB); Stamylan (NL); Eltex, Finathene (BE); Ertileno (ES); Sclair (CA).

PE-LD, $d_R \sim 0.918$–0.935 g/cm^3: Fortiflex, PolyEth, Rotothene, Tenite (US); Mirason, Rexlon, Yukalon (JP); Baylon (DE); Lotrene (FR); Novex (GB); Bralen (CS); Ropol (RO); Tipolen (HG); Hipten, Okiten (YU).

PE-HD, $d_R \sim 0.935$–>0.96 g/cm^3: Paxon, Super Dylan (US); Hi-zex, Suntec (JP); Hostalen, Vestolen, (DE); Rigidex (GB); Hiplex (YU).

PE-LLD: Dowlex, Marlex TR 130 (US); Ulzex (JP); Lupolen (DE); Innovex (GB); Evaclear (IT); Lotrex (FR); Stamylex (NL); Novapol, Sclair (CA); Ladene (SA).

PE-VLD: Escorene alpha (US); Tafmer (JP), Norsoflex (FR) Stamylex (NL):

Delivery: natural or pigmented granules (diced, 3–4 mm cubes, pellets about 3 mm in diameter), PE-HD, PE-LLD also as coarse powder (grit), fine-grained, milled or precipitated powder for powder techniques (p. 000). For low-molecular weight PE and chlorinated PE (PE-C) see Sec. 4.1.1.2 and 4.1.2.3.

4. Properties

The property dependence on structural parameters (density, linearity, molecular weight and molecular weight distribution) is summarized in Table 4.3 (p. 178). For injection-molding, PPE grades, a narrow molecular weight distribution provides superior shrinkage, shock and cold strength characteristics. Wider molecular weight distribution is more suitable for extrusion and blow molding. The advantageous dielectric properties of non-polar polyethylene are practically independent of density and degree of polymerization.

The mechanical and thermal properties (Tables 4.5, 4.6) change approximately linearly with density. Figure 4.1 shows the temperature dependence of the shear modulus. Figure 4.2 indicates the change of tensile strength with time for various types of PE at different temperatures. Products from all types of PE are not deformed in hot water if they are not mechanically stressed. PE-HD and some PE-LLD grades resist boiling water.

PE withstands water, salt solutions, alkalis and acids, but strongly oxidizing agents, such as fuming sulfuric acid, concentrated nitric acid, nitration acid, chrome-sulfuric acid, chlorosulfonic acid and halogens, do attack PE. PE can be protected against oxidative chain degradation under the long term influence of heat by the use of antioxidants. Addition of 2% of special carbon black affords a 10- to 15-fold, light-colored light stabilization and a 4- to 6-fold improvement in UV stability. Below 60°C, PE is insoluble in all organic solvents, but it does swell in aliphatic and aromatic hydrocarbons and chlorinated hydrocarbons. The lower its density, the more it swells. Individual types of high-density PE and PE-LLD have been tested and approved as containers for heating oil and petrol and for vehicle fuel tanks. In some cases the internal surfaces have been fluorinated (using dilute F_2-gas) or sulfonated (by SO_3). Such tanks are practically impermeable to all kinds of fuels and hydrocarbons. The water-vapor permeability of PE is minimal. For oxygen, carbon dioxide and many odiferous and aromatic substances permeability is considerable, but it decreases with density. PE is stable to ionizing radiation if the dose is not too high. PE does not itself become radioactive on radiation. It is useful in the shielding of nuclear reactors and for laboratory installations in nuclear technology.

Surfactants (wetting and washing agents) can induce environmental stress cracking in mechanically stressed parts.

Stress cracking resistance increases with molecular weight and is enhanced by a slightly disrupted crystallinity, as e.g., in blends of PE-MD-HMW with PE-LLD grades or these grades (Sec. 4.1.2.1.2) alone. Blending with copolymers (Sec. 4.1.2.2) offers further possibilities of improving stress crack resistance.

PE is odorless, tasteless and physiologically inert. Most of the PE grades available commercially meet the current official guidelines for the use of

Table 4.5. General Properties of Polyolefin Molding and Extrusion Materials according to ISO/DIN/VDE

Generic types:		Polyethylenes			Polypropylenes							Polybutene
Special groups:	Unit	PE-LD (low density)	PE-HD (high density)	PE-UHMW (ultrahigh molecular wt.)	Homopolymers PP-H			Cop. PP-R MG~4×10³ 1,2	Reinforced compounds			MG 1-2×10⁶
					highly MG 2.4×10⁶ 1	isotactic MG 5-6×10⁵ 2	less isotactic		Talc 20%	Talc 40%	Glass beads 20%	Long glass fiber chemically bonded 30%	
Principle properties													
Density at 23°C	g/cm³	≤0.920	≥0.954	0.94	0.908-0.905	0.903	0.900-0.898	0.91-0.90	1.04	1.22	1.03	1.14	0.915
Melt flow rate 190/2	g/10 min	0.1-1.22	0.4-8	<0.01	3-20	0.5	<0.1-2	—	2-3	~3	—	~2	—
190/5	g/10 min	0.4-88	0.2-30	—	1.5-15	0.35	—	4.5-6.5	1.5-2	1.5-2	—	~0.8	0.3-20
230/2	g/10 min	—	—	<0.1	7-60	1.5	<0.1-6	14-16	6-10	7-9	~4	~4	—
230/5	g/10 min	—	—	—	—	—	—	—	—	—	—	—	—
Melting temp.	°C	105-110	130-135	135-138	158-166	158-165	157-162	130-164	158-167	158-167	164-167	164-167	125-130
Mechanical													
Tens. str., yield: 23°C (73°F)	N/mm²	8-10	20-30	20	30-37	32	22	27-30	31-32⁶	30⁶	28	71⁶	15-25
80°C (176°F)	N/mm²	~2	4-6	8	15-18	14	—	—	—	—	—	—	9
Elongation, yield	%	20	12	20	~12	16	—	—	—	—	—	—	10
Elongation, break	%	~600	400-800	>600	400-700	>700	700	450-650	15-20	10	170	6	150-400
Flexural str., yield	N/mm²	7-10	30-40	30-40	40-60	45	22-27	40-45	38-42	42-47	32	85⁴	15-25
Stiffness in torsion (Clash-Berg)	N/mm²	60-90	~400	250	450	380	450	300-400	550	600	370	720	—
Shear modulus⁵: 23°C (73°F)	N/mm²	700->1000	700->1000	~300	650-850	700	450	—	1200	1600	—	—	~200
50°C (122°F)	N/mm²	100-200	400-900	~150	~450	~400	200	~350	23°C:2300	3300	1500	5000³	—
100°C (212°F)	N/mm²	30-100	80-200	—	~250	~100	65	~140	80°C:900	1700	—	4300	—
Hardness: Ball indentation (30s)	N/mm²	<10	~50	38	75-85	65	36-45	57-67	80-85	82-88	70	110	39
Charpy imp. str.: unnotched +20°C	kJ/m²	~15	N.B.	N.B.	80-N.B.	N.B.	N.B.	N.B.	35-40	15-20	33	15	N.B.
−20°C	kJ/m²	N.B.	~5	N.B.	10	35	11-25	N.B.	15	10	—	—	N.B.
notched +20°C	kJ/m²	N.B.	6-N.B.	N.B.	4-7	8	6-24	10-20	5	4	4	6	N.B.
−20°C	kJ/m²	N.B.	>5	N.B.	1.5	2.5	2.4	>5	2	2	—	—	15-40
Thermal													
Vicat soft temp., B/50	°C	<40	75-70	74	105-100	95	××	85-80	105	105	96	—	70
Defl. temp., ISO R/75: A (1.8 N/mm²)	°C	~35	~45	95	60-65	55	—	~50	80	87	65	—	60
B (0.45 N/mm²)	°C	~45	75-80	—	90-100	85	~80	~80	125	132	110	—	110
Linear therm. exp. (20-80°C)	K⁻¹	2.5×10⁻⁴	2×10⁻⁴	2×10⁻⁴	1.6×10⁻⁴	1.6×10⁻⁴	1.5×10⁻⁴	1.5×10⁻⁴	~1×10⁻⁴	0.8×10⁻⁴	1.2-0.6×10⁻⁴	0.5×10⁻⁴	1.5×10⁻⁴
Thermal conductivity	W/mK	~0.35	~0.50	0.42	0.22	0.22	0.17	0.22	0.41	0.51	0.25	0.27	0.17

Guide values for all polyolefins, this table and Table 4.9.: Electrical properties: surface resistivity, >10¹³ ohm; volume resistivity, >10¹⁶ ohm cm; dielectric constant, approx. 2.3; dissipation factor, tan δ 0.0002-0.0007; dielectric strength, 700 kV/cm, (D 149:450-500 V/10⁻³ in); tracking resistance, KA 3c, KC>600; water absorption, 96h, <0.5 mg, reinforced, 1.5-2 mg.
¹ Injection molding. ² Extrusion (pipes) blowing. ³ Flexural creep modulus, 1 min. ⁴ Flexural strength at break. ⁵ $G = E/2(1+\mu) \approx E/2.7$. ⁶ Ultimate tensile strength.

4.1.2.1. Polyethylene

Table 4.6. General Properties of Ethylene and Propylene Polymers according to ASTM

Properties	Unit		Polyethylenes						
			Low density			Medium density	High density	Ultra high molecular weight	
	SI	U.S.	branched	linear					
Density	g/cm³	lb/ft³	~0.920	0.918-0.935	57.1-58.4	0.926-0.941	0.941-0.965	58.7-60.2	0.94	58.7
Melting temp.	°C	°F	~57.4	122-124	252-255	120-130	130-137	266-275	120-125	248-257
			107-115							
			221-239							
Mechanical										
Tensile strength, yield	N/mm²	10³lb/in²	4-16	13-28	1.9-4.0	22-32	3.1-4.5	22-27	3.1-3.9	18-39	2.5-5.6
Elongation, break	%		0.6-2.3							
Tensile modulus	N/mm²	10³lb/in²	≦650	≦900	≦600	≦1200	~500	
Flexural strength, yield	N/mm²	10³lb/in²	70-280	280-560	40-80	140-420	20-60	420-1300	60-180	140-770	
Flexural modulus	N/mm²	10³lb/in²	10-40			35-49	5-7			
Izod, notched	J/m	ft lb/in	—	~420	~60	~700	~100	~1000	~150	~1000	
			~280	53-106	1-9	27-800	0.5-15	27-1060	0.5-20		
Hardness	Shore (D) or Rockwell (R)		N.B.							
			D 40-50			D 50-60	D 60-70	D 60-70		
Thermal										
Linear therm. expansion	10⁻⁴ × K⁻¹	10⁻⁴ × °F⁻¹	2.5	—	—	2.2	1.2	2.0	1.1	2.2	1.2
Defl. temp. 1.8 N/mm² (264 psi)	°C	°F	—	—	—	—	—	—	—	43-49	116-120
Defl. temp. 0.45 N/mm² (66 psi)	°C	°F	40-44	104-112	—	—	210-240	80-91	175-196	74-82	166-180
Vicat soft. temp. (D 1525)[1]	°C	°F	90-100	195-210	—	100-115		120-130	250-265	—	—
Thermal conductivity	W/mK	BTU in / h ft² °F	0.30	2.1	—	0.36	~2.5	0.48	~3.3	—	—

For electrical and water absorption values see Table 4.5.

[1] For corresponding Vicat softening temperature methods see ISO 306 method A (9.81 N) and for values for ISO 306 B (49.06 N) see Table 4.5.

Table 4.6. General Properties of Ethylene and Propylene Polymers according to ASTM (cont.)

Properties	Unit		Polyallomer		Polypropylenes							
					Homo-polymer		High impact copolymer		40 % max. talc filled		40 % max. glass fiber reinforced	
	SI	U.S.										
Density	g/cm³	lb/ft³	0.896-0.899	56.0-56.2	0.900-0.910	56.2-56.8	0.890-0.905	56.2-56.5	1.00-1.30	62.4-81.2	1.05-1.25	65.6-78.0
Melting temp.	°C	°F	120-135	248-274	168	334	160-168	320-334	–	–	–	–
Mechanical												
Tensile strength, yield	N/mm²	10³lb/in²	21-28	3.1-4.1	32-42	4.5-6.0	20-33	2.8-4.5	30-39	4.3-5.5	42-105	6.0-15
Elongation, break	%		400-500	–	≤600		≤700		5-20		2-4	
Tensile modulus	N/mm²	10³lb/in²	–	–	1050-2100	150-300	700-1300	100-180	2800-5600	400-800	3500-7000	500-1000
Flexural strength, yield	N/mm²	10³lb/in²	–	–	42-56	6-8	42-49	6-7	56-63	8-9	70-155	10-22
Flexural modulus	N/mm²	10³lb/in²	490-980	70-140	1400-2100	200-300	910-2100	130-300	2100-4900	300-700	2800-7000	400-1000
Izod, notched	J/m	ft lb/in	91-203	1.7-3.8	27-106	0.5-2	53-1060	1-20	21-160	0.4-3	53-270	1-5
Hardness	Shore (D) or Rockwell (R)		R 50-85		R 80-110		R 50-90		R 94-110		R 102-110	
Thermal												
Linear therm. expansion	10⁴ × K⁻¹	10⁴ × °F⁻¹	1.3	0.72	1.6	0.89	1.5	0.83	0.8	0.44	0.5	0.28
Defl. temp. 1.8 N/mm² (264 psi)	°C	°F	49-51	120-130	49-60	120-140	46-60	115-140	82-132	180-270	149-154	300-310
Defl. temp. 0.45 N/mm² (66 psi)	°C	°F	74-88	165-190	107-121	225-250	85-104	185-220	129-143	265-290	166	330
Vicat soft. temp. (D 1525)[1]	°C	°F	130	266	150-155	300-310	140-145	285-295	–	–	–	–
Thermal conductivity	W/mK	BTU in / h ft² °F	0.13	0.87	0.10	0.7	0.12-0.17	0.8-1.2	0.32	2.2	0.36	2.5

For electrical and water absorption values see Table 4.5.

[1] For corresponding Vicat softening temperature methods see ISO 306 method A (9.81 N) and for values for ISO 306 B (49.06 N) see Table 4.5.

4.1.2.1. Polyethylene

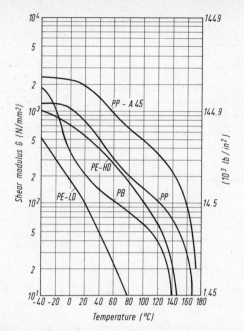

Fig. 4.1. Shear modulus temperature curves (ISO 537/DIN 53 445) for polyolefins: PE-LD: Density 0.92 g/cm³, PE-HD:Density 0.96 g/cm³, PP: normally stabilized, PP-A 45: reinforced by 45% asbestos, PB: Polybutene.

plastics in connection with food-stuffs. The outstanding dielectric properties of PE, which remain unchanged in water, are also quite independent of temperature and frequency. Antistatic treatment of molding compounds remains effective in the long term and does not deteriorate significantly on washing the product. PE continues to burn once ignited. It can, however, be flame retarded.

5. *Processing and application*

Abbreviations are as follows:
i_2 correponds to MFR 190°C, 2160 g, ISO cond. D, ASTM cond. E
i_5 corresponds to ISO cond. T, ASTM cond. P, 190°C, 5000 g
Abbreviation for grades: e.g., 0.918/2 = density 0.918 g cm⁻³/i_2 = 2

1. In *injection molding* (for process conditions see Table 3.9, p. 96) easy-flow PE grades ($i_2 > 25$) are processed into mass-produced articles. The shrinkage and density of the molding are dependent on the temperature history of the melt up to its cooling. Shock-cooled parts with a high amorphous component exhibit minimal processing shrinkage. There is,

4.1. Thermoplastics: Raw Materials and Molding Compounds

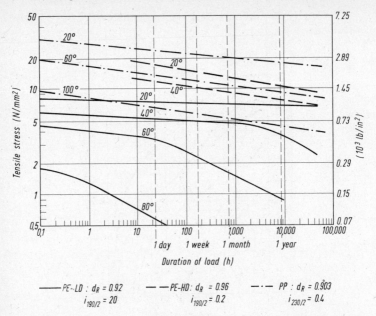

Fig. 4.2. Time to fracture curves of polyolefines under unidirectional tensile stress in air (see also Fig. 7.13). (a) PE-LD: easy-flow-molding compound, (b) PE-HD: blow molding compound, (c) PP: strongly heat stabilized.

however, marked shrinkage at higher temperatures with progressive crystallization. Warpage and cracking can be a consequence.

Mold temperatures of 40 to 70°C are required for high-density PE with low melt indexes to achieve moldings with a high gloss. Grades with an i_2 of 2.5 to 4 almost always help to deal with warping due to the freezing-in of stresses. *Gating brittleness*, resulting from strong molecular orientation, can be combatted by raising the injection temperature and using the grade with the highest applicable melt index. The range of applications of PE injection-molding compounds is indicated in Table 4.7.

2. PE grades with higher melt viscosities ($i_2 = 2$ to 0.2) must be selected for *blow molding* so that the parison does not tear under its own weight. PE-LD is used for flexible containers, bottles and canisters up to 60 l (13 gal) and vessels up to 200 l (40 gal) capacity. In the density range 0.94 to 0.96, $i_5 = 1- < 0.1$, there are special grades with increasing stiffness for petrol canisters, vehicle fuel tanks (Sec. 4.1.2.1.4), barrels of up to 50 gal, large toys and sports equipment, bottles for domestic chemicals, cosmetics and pharmaceuticals, milk bottles, heating-oil cellar tanks and other large containers of up to 10 000 l (2000 gal) capacity.

Table 4.7. Processing Behavior and Applications of PE Injection-Molding Compounds

Density (g/cm^3)	0.92	0.93	0.94[1]	0.95	0.96
MFI 190/2.16: >25–15[2])	Easiest flow; mass articles without particular stressing	Easy flow, moldings with large surfaces, little warping, good gloss	Easy flow, shock-proof moldings without particular stiffness	Easy flow, low warping, difficult to injection mold, household goods	Easy flow, hard, stiff, basins, sieves, dishes, transport cases[3], protective headwear
15–5	Articles with greater strength, less surface gloss	Little-stressed moldings with good surface gloss	Good shock resistance, little stress corrosion, highly stressed technical moldings	Easily processed, resistant to shock, screw caps, closures, technical moldings	Resistant to shock and creep, mechanically strongly stressed moldings, e.g., rubbish bins[3], seating shells
~1.5	Very good mechanical strength and resistance to stress corrosion		Good resistance to creep, little tendency to stress corrosion, particularly stressed closures	Resistant to stress cracking, good surfaces, highly stressed technical moldings	
<1				High molecular weight, most highly stabilized, pressure armatures, pipe bends, fittings	

[1] Mostly blends of PE-LD/HD, now also low-density PE-LLD.
[2] Super easy flow grades up to MFI >100.
[3] Bottle crates and rubbish bins with admitted amounts of tested recycled material.

4.1. Thermoplastics: Raw Materials and Molding Compounds

Because of its narrow molecular weight distribution, PE-LLD $(0.928/i_2 = 2.5)$ is less suitable for bottle blowing than for *rotomolding* of containers of up to 6000 gal and of other large hollow bodies (canoes, surfboards).

3. Carbon-black stabilized special purpose high molecular weight PE-LD and PE-HD grades ($i_2 = 0.5$ to 0.1) are available for the *extrusion* of sheets for thermoforming and for standard-sized pipes. Special grades of PE-HMW and HD are used for highly impact-resistant and creep-resistant pressure pipes and membranes for land-fills.

4. Heat-stabilized material is used mainly in the range of 0.918 to 0.930/0.15 to 0.35, for the insulation of long-distance cables and trunk lines. For foamed insulation it is extruded with tumbled blowing agents. Telephone lines, medium and high voltage cable (10 kV to 400 kV) are insulated with high purity cross-linkable grades. PE is used for cable sheathing with a small amount of carbon black for light stabilization. Materials with 30% conducting black have a resistance of 10 ohm cm. They serve as electrical current smoothers and as screening materials.

5. *Blown and cast films* (for manufacture, see Sec. 3.6.5.6/7) are made mainly from PE-LD. For tough, highly mechanically stressed products (sacks, bags, carrier bags) 0.918/0.2 to 0.5 grades are preferred. Those with an i_2 of 1.5 to 2 are transparent. One can produce thin glass-clear and somewhat stiffer films with higher takeoff speeds. Thin, paperlike HM or high-tenacity, HT, film < 10 to 30 μm (0.4–1.2 mil) with a dry touch is made from PE-HD-HMW, approximately 0.95/0.3. Similar PE grades, but not of such high molecular weight, are used as the starting film for fibrillated yarn. PE-LLD films can be blown down to a thickness of 0.5 mil (approximately 10 μm) (Sec. 4.1.2.1.2). Incorporation of ca. 400 ppm fluoroelastomer (Viton) enhances flow characteristics and output of PE, especially PE-LLD films.

6. Easy-flowing PE-LD ($i_2 = 4$ to 80) is necessary for *extrusion coating*.

7. *In powder techniques* (rotomolding, Sec. 3.2.3; fluidized bed sintering, Sec. 3.2.1) PE powder with a particle diameter of 30 to 800 μm (e.g., Micropol) of grades with a density of 0.92 to 0.95 and an i_2 of 1 to 5 is used. Those with an i_2 of 17 to 22 serve for the undercoating of carpets and smoothing materials. Precipitated PE powders with uniform particle size (Microthene) of about 50 μm are suitable for electrostatic coating of metals or fabrics. Still finer powders (8–30 μm) are dispersed in the beater in papermaking or in printing inks.

4.1.2.1. Polyethylene

8. *Ultra high molecular weight polyethylenes (PE-UHMW, MW 4 to 6×10^6 g/mol)* (Tables 4.5, 4.6) demand special processing.
Blocks and simple moldings (filter press plates) are preformed, then pressed at 200°C under pressures of 2 to 5 N/cm² and slowly cooled. Newer processes operate with melting compounders for feeding into molds of 1 to 110 l capacity, special grades can be injection molded through flow channels 2 to 4 times larger than for other PEs at 8 to 15 mm/sec into molds heated 40 to 60°C. The material temperature is held at 200 to 250°C and at an injection pressure of 1100 bar, and moldings are 10 to 300 g (0.35–10 oz) in weight.
High-pressure plasticization (2000–3000 bar) in a cyclic twin extruder (ram extruder, p. 000) is used to make *Teledynamic* injection moldings. Extruded profile is also made with slowly co-rotating twin-screw extruders at 180 to 200°C.
The material is impact resistant over a temperature range from -200 to 90°C and is not liable to stress cracking. It has outstanding abrasion and lubricating properties. Cross-linking with 0.2 to 0.5% peroxide (Sec. 4.1.2.2) increases wear resistance and reduces thermal expansion to 50%. There are antistatic and UL V-0 grades that contain carbon black. Lining of self-discharging ore and coal ships is a major field of application. Similarly stressed slides, etc., weaving shuttle gear wheels, seals and gland packings for electrical engineering and low-temperature technology and cable railway rollers are machined or injection molded. Filter materials are sintered. A particular area of application is in surgical implants for hips, knees and fingers.

9. *In the post-treatment* of PE articles, their resistance to attack is of importance. PE surfaces can be permanently printed on or lacquered only after local oxidizing pretreatment by high-voltage discharge plasma or glow discharge, oxidizing flame, ozone or chromosulfuric acid (special equipment). Mutual adhesive joints of PE products or, for example, adhesives on labels are not very strong mechanically even with special adhesives. Welded connections are secure.

10. *PE fibrides* (Lextar, US; Hostapulp, DE; Ferlosa, I) are fine cellulose-like fibers that are precipitated from solutions under shear. Because they are water repellent and have an affinity for hydrocarbons, etc., they are used as additives for mastics, in aqueous suspension for the adsorption of oil in environmental pollution and for moisture-resisting boarding and paper. Their application as reinforcing fibers in place of asbestos is being developed. *Ultra high-strength PE-fibers* prepared by the Melt-drawn Orientation Method (MOM) can be used for "self-reinforced" pultruded (Sec. 3.4.3) PE-profiles.

4.1. Thermoplastics: Raw Materials and Molding Compounds

Details on *semifinished products* including *foamed* PE are given on Sec. 4.2.3.

11. *PE foam sheet and webs* (Chap. 5.2). In the Neoplen process, pre-foamed particles are produced from a melt containing a blowing agent by hot-cut pelletisation after the extruder and are sintered using steam into blocks or moldings in the particle foam molding process.

4.1.2.2. Cross-linked polyethylene (PE-X)

Light cross-linking of polyethylene (about 5 cross-links per 1000 carbon atoms) improves time-dependent strength properties, cold impact strength and stress-cracking resistance very considerably. Because the material softens only in a rubbery-elastic way above the crystalline melting point, it can be exposed to high temperatures and for short periods even over-exposed.

According to ASTM D 2647, specified properties for cross-linked PE are

(1) Ultimate elongation (5 cells): < 25 to $> 450\%$
(2) Minimum of 75% retention of ultimate elongation after aging 168 h at either 121°C or 150°C
(3) Apparent modulus of rigidity (3 cells): < 70 to > 275 MPa; $< 10 \times 10^3$ to $> 40 \times 10^3$ psi
(4) Brittle temperature (4 cells): -75°C to -20°C
(5) Gel content, measured (D 2765) by percent extract (degree of cross-linking: from < 10 to $> 30\%$ for mechanical (M) and electrical (E) grades. For the latter a dielectric constant of 2.5 to 8.0, a dissipation factor of 0.001 to > 0.1 and volume resistivity of $< 10^4$ to 10^{15} ohm cm are also specified.

Higher density PE, cross-linked with a particular peroxide, is mainly processed by injection molding within a precisely defined temperature range between 130 and 160°C and then cross-linked in the mold at 200 to 230°C. The moldings, used in electrics, plant construction and cars, can withstand short exposure to temperatures of up to 200°C.

PE-LD extrudate, in particular insulation of medium- and high-voltage cable (10–110 kV) is cross-linked continuously in-line as extruded by the following methods:

(a) With dicumyl peroxide at 150 N/cm² and 150 to 200°C in a steam-heated pressure tube, for pipes also by special one-step cross-linking extrusion processes (Engel, Pont-à-Mousson).
(b) In the Sioplas (Dow Corning) or Monosil (Maillefer) technique, silane-grafted PE compounds are treated with a silane cross-linking catalyst

which "kicks-off" in the following hot water-pressure treatment and forms Si-O-Si bridges.
(c) By irradiation with electron beams (also for shrink films and insulation tubes, Irracure, Irrathene, US). For thick-walled electro-insulating shrink tubes the irradiation is applied so that only the outer layer is cross-linked and the shrunken, molten inner layer is closely applied to the conductor in use.
(d) By addition of azo-compounds before extrusion to introduce reactive side groups for N-cross-linking in a following hot salt bath (Lubonyl process).

Cross-linked warm-water pipes are made by all the preceding processes and also by other chemical processes. They are increasingly used as underfloor heating and warm water installation pipes. Pipes shrink-fitted to cover conveyor rolls and cylinders and shrink sleeves are all commercially commonplace. PE-X-HD, auto-crosslinkable after molding by a new modified organosilane process, has UHMW characteristics.

4.1.2.3. PE-C and chlorinated thermoplastic elastomers

For PE-C trade names see 4.1.1.2.1. The types listed include compounds for melt processing into unplasticized waterproofing sheets, industrial hoses, cable sheathings, etc. For peroxide or radiation cross-linkable PE-C as synthetic rubber, see Sec. 5.1.15. Amorphous thermoplastic elastomers on a base, not individually disclosed, of chlorinated, chemically cross-linked polyolefins (Alcryn, US) in the Shore hardness range A 60–80 have replaced oil resistant vulcanized nitrile or chloroprene rubber for the service temperature range −40°C to 121°C for static applications (hoses, seals, conveyor belts, wire and cable sheathing, including in automobile construction). It is not suitable for automobile tires, subject to dynamic bending, because its energy absorption leads to overheating. The granules are processable at about 170°C by all the processes used for thermoplastics, including on plants from the rubber industry.

4.1.2.4. Ethylene copolymers

1. *Copolymers with vinyl acetate (E/VA, E/VAC) with >50% VA content; also VAE*

    Trade names (see also p. 487):
    Elvax, Ultrathene (US); Evaflex, Evatate, Soarlex, Soablen (JP); Acralen, Hostalen LD-EVA, Levapren, Lupolen V, Wacker VAE (DE); Evaclene (IT); Evatane (GB).

4.1. Thermoplastics: Raw Materials and Molding Compounds

E/VA-copolymers for molding compounds are made by high-pressure polymerization in a manner analogous to PE-LD. When they contain 20 to 40% comonomers they exhibit combinations of enhanced tear and impact strength, extension at break, stress-cracking and UV resistance, and transparency, flexibility and resistance to cold compared with PE, changing to rubber-like behavior as the temperature is raised. They can accept a high proportion of fillers. For E/VA with a VA content of 1 to 50% (7 cells) and an MFI 190/2 of <0.1 to 100 (11 cells, code, 000 to 700), properties according to ISO DIS 4613 or DIN 16778 are given in Fig. 4.3 and Table 4.8. Many grades of E/VA are suitable for use with food. Films of high strength and toughness have also been made of slightly hydrolyzed E/VA grades (see following paragraph).

E/VA grades that are compatible with PE and PVC are used as elastifiers. Levapren 450 P and N (DE) grafted with VC is applied for improved impact-resistant PVC (Sec. 4.1.4.3.3).

2. *Copolymers with vinyl alcohol (E/VOH, E/VAL)*

Trade names: Eval (US, JP), Selar-OH (US), GL-resins, Soarnol (JP), Clarence (BE), Levasint (DE)

E/VOH resins are made by hydrolysis of E/VAC Copolymers.

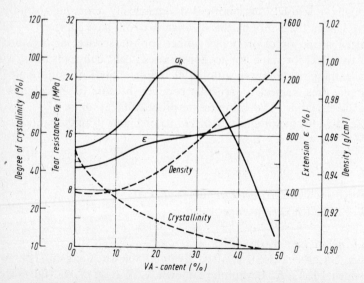

Fig. 4.3. Properties of E/VA copolymer in relation to its VA content.

4.1.2.4. Ethylene copolymers

Table 4.8. Properties and Applications of E/VA Copolymers with Varying VA Content

VA content of copolymer (% by weight)	Properties and Applications
1–10 (03, 08)*	In comparison with homo-PE-LD, more transparent, flexible, tougher (heavy sacking film, refrigerated packaging), can be sealed more easily (bags, laminated films), less liable to stress cracking (cable covering), higher shrinkage at low temperature (shrink films), smaller relaxation of pre-stretched films (stretched films).
15–30 (18, 25)*	Still melt processable, very flexible and soft, rubber-like (applications like those of plasticized PVC, especially for closures, gaskets, carbon black filled thermoplastics for the cable industry).
30–40 (35)*	High elastic elongation, softness with capacity to accept fillers, wide softening range. Polymer with great strength for coatings and adhesives.
40–50 (45)*	Products with even more pronounced rubbery qualities (cross-linkable with peroxides and radiation), e.g., for cable, for grafting reactions, modifiers e.g., for high impact PVC with very good weathering resistance. Highly filled "VAE" roofing membranes.
70–95	Applications in the form of latices for emulsion paints, paper coating, adhesives.

* cell number acc. to ISO/DIN, see text; for further coded information chapter 6, Standardization.

E/VAL types are rather hydrophilic, crystalline, random copolymers containing 50 to 70 mol% VAL. Levasint contains much more PE, but polar OH-groups in such an amount that it is well suited for impact resistant anticorrosive powder coating of metallic surfaces.

E/VOH with higher VAL content is used mainly in the coextrusion of packaging materials on account of its outstanding barrier properties against oxygen, nitrogen, carbon dioxide and fragrances. It resists oils and organic solvents; most grades are permitted for packaging foods. However, its water vapour transmission rate is not so good and, if it absorbs water (3 to 8%), the gas barrier properties deteriorate. Hence it is most suitable for coextrusion layers appropriately protected by PE- or PP-layers. With T_g ca. 66°C, T_f 160 to 180°C, E/VOH is extruded in the 200 to 250°C temperature range. It will withstand short contact with hotter resin streams (PP, PA, PTP) during extrusion. The oxygen transmission rate of EVOH increases reversibly with temperature. When EVOH composite films are subjected to boiling, the barrier properties are completely restored on cooling.

3. *Binary copolymers and terpolymers with acrylics (E/EA, E/MA, E/AA etc.)*

Trade names: Bynel, Nucrel, Paxon, Plexar, Primacor, Zetafin (US); Admer (JP); Lotadur (F), Lucalene (DE)

4.1. Thermoplastics: Raw Materials and Molding Compounds

Various different resin groups are marketed under trade names and are not always clearly disclosed or strongly demarcated.

Resins containing *E/EA ethylene ethyl acrylate copolymers* (also with butyl acrylate) and *E/MA methacrylate copolymers*, or both are used for rubber-like stress crack resistant blown or cast packaging film with low temperature impact strength. These can be highly filled. They are also used for heat sealable coatings. Contract with food is permitted for E/MA generally and for E/EA containing up to 8% EA. Semiconductive carbon black filled E/EA films and tubes are used for microchip packaging, explosives packaging and medical applications where static electricity is a hazard.

E/EA Ethylene acrylic acid copolymers, also terpolymers containing (meth-)acrylate groups or modified by acryl amide, are used for intermediate bonding layers in multilayer film, e.g. between PA/PE, and also for metal coating. For vulcanized high temperature resistant synthetic rubbers see Chap. 5.1.

Special grades of the resins named above containing acrylic acid and Zn^{++} or Na^+ ions transform into ionomers.

4. *Ionomer resins*

Trade names: Surlyn A, for films Sur-Flex, for tie layers Excor (US); Himiran, Copolene (JP).

Synthesis and behavior: Ionomer resins are copolymers of ethylene and 11% acrylic acid, of which half is as the sodium salt. The polar bonds tend to suppress crystallization completely and lead to "ionic" cross-linking in the temperature range of applications −40 to +40°C. The glass-clear products are tough and stiff. At high temperatures the polar bonds relax to the extent that the plastic can be worked by all the ordinary processes for thermoplastics at 290 to 330°C (554–626°C); mold shrinkage is 0.7 to 1.9%. At 120°C (248°F) it can be heat sealed. The tenacity of the melt leads to pore-free products, even in very thin layers. Films can be readily heat stretched. For physical properties see Table 4.9.

Surlyn A is resistant to alkalis, but less so to acids. Organic solvents merely swell it. It is resistant to alcohols, ketones, fats and oils. Its weathering properties are comparable to those of PE, and UV stabilization with carbon black is possible. Its permeability to water vapor, oxygen and nitrogen is similar to that of PE but it is less permeable to carbon dioxide. Its stress-cracking propensity is very slight.

Applications: Transparent coatings (5 to 12 g/m^2); transparent films for fat-containing foods; intermediate bonding layers; skin and blister packaging; bottles for vegetable oils, shampoos and liquid fats; corona and stress-

4.1.2.4. Ethylene copolymers

Table 4.9. Guide Properties/Values of Ionomers, PB and PMP according to ASTM

Properties	Unit SI	Unit U.S.	Ionomers	Polybutene	Polymethylpentene
Density	g/cm^3	lb/ft^3	0.94-0.96	0.91-0.925 | 56.9-57.8	0.83-0.84 | 51.9-52.2
Melt flow rate	$g/10$ min		58.8-60.0 | 14-1.0	1.0-0.4	
Melting temp.	°C	°F	85-99 | 185-205	126 | 259	230-240 | 446-464
Mechanical					
Tensile stress, yield	N/mm^2	10^3 lb/in^2	—	12-17 | 1.7-2.5	—
Tensile strength, break	N/mm^2	10^3 lb/in^2	20-35* | 2.9-5.1*	27-31 | 3.8-4.4	25-28 | 3.5-4.0
Elongation, yield	%		—	24	—
Elongation, break	%		350*-520	300-380	10-50
Tensile modulus	N/mm^2	10^3 lb/in^2	140-420 | 20-60	210-260 | 30-40	1100-2000 | 160-280
Flexural stress, yield or break	N/mm^2	10^3 lb/in^2	—	14-16 | 2.0-2.3	28-42 | 4.0-6.0
Flexural modulus	N/mm^2	10^3 lb/in^2	260-380 | 37-55	310-350 | 45-50	770-1800 | 110-260
Izod, notched	J/m	ft lb/in	320-800 | 6-15	N.b. | N.b.	16-64 | 0.3-1.2
Hardness			Shore D 56-68	Shore D 55-65	Rockwell L 67-74
Thermal					
Linear thermal expansion	10^{-4} K^{-1}	10^{-4} °F^{-1}	1.1 | 0.61	1.3 | 0.71	1.2 | 0.67
Defl. temp., 1.8 N/mm^2 (264 psi)	°C	°F	—	54-60 | 130-140	41 | 105
Defl. temp., 0.45 N/mm^2 (66 psi)	°C	°F	38-50 | 100-120	91-112 | 215-234	100 | 212
Vicat soft. temp. (D 1525)	°C	°F	57-72 | 135-160	108-113 | 226-235	179 | 350
Thermal conductivity	W/mK	Btu·in / h·ft²·°F	0.24 | 1.68	0.22 | 1.50	0.17 | 1.16
Electrical					
Volume resistivity	Ohm/cm		5×10^{15}-10^{18}	10^{16}	$> 10^{16}$
Dielectric strength	kV/mm	V/10^{-3}in	40 | 1000	18-40 | 450-1000	
Dielectric constant			2.4 (10^6 Hz)	2.52 (10^6 Hz)	2.12 (10^7 Hz)
Dissipation factor			~ 0.003 (50-10^6 Hz)	0.0002-0.005 (10^6 Hz)	0.00015 (10^7 Hz)
Water absorption	% 24 h		0.5-0.3	< 0.02	0.001

* Machine direction blown film

cracking resistant insulation; snap-fitted car windows and structural foam bumper guards, ski-boot parts and assemblies, golf ball covers.

5. *Copolymers with tetrafluoroethylene, etc.*, are injection moldable and extrudable (details in Sec. 4.1.5.2).

4.1.2.5. Olefinic thermoplastic elastomers (TPE)

Trade names: Ferroflex, Geolast, Polytrope, Prolastic, Ren-flex, Rimplast, Santoprene, Vistanex (US); Milastomer, Soflex (JP); Levaflex, Vestopren (DE); Ferra-Flex, Kelpox (NL); Dutral, Dutralen (IT).

Most olefinic TPE are based on blends of a crystalline polyolefin (PP, less frequently PE) with EPM or EPDM rubber (p. 000), often lightly cross-linked. However, there are also block-copolymers with crystalline PE domains in amorphous PP, dispersed fully cross-linked EPDM particles in a PP matrix (e.g. Santoprene) or, with better oil resistance, a dynamically vulcanized blend of nitrile rubber in a PP matrix (e.g. Geolast). The character of such blends is not always clearly disclosed, and there are no sharp demarcations between these and rubber-like α-olefin copolymers (Sec. 4.1.2.1.3). A wide range of general purpose grades with relatively poor elastic recovery is used for wire and cable insulation, automobile bumpers, spoilers and trim while engineering grades with better recovery and high temperature performance are used for industrial hose, automotive under-bonnet application and sports goods.
Good ozone resistance, weatherability and aging properties are exhibited by these TPE. With Shore hardnesses of A55 to A95 they can be used at temperatures ranging from $-40°C$ ($-40°F$) to $90°C$ ($200°F$) and some up to $120°C$ ($250°F$).

4.1.2.6. Polypropylene (PP) and copolymers

1. *Syntheses and product groups*

Quantitatively the major polymer component of all molding compounds is *isotactic polypropylene* i.e., 60 to 70% crystallizable PP with 5000 to 10 000 units

$$\left[\begin{array}{c} H \quad H \\ | \quad | \\ C - C \\ | \quad | \\ H \quad CH_3 \end{array}\right]$$

in sterically regular and uniform order (Fig. 3.1). Pure isotactic PP (not commercially available) with $d = 0.91$ has a crystalline melting range above $170°C$.

4.1.2.6. Polypropylene (PP) and copolymers

The basic synthesis is low-pressure precipitation polymerization of propene gas at the surface of Natta (1954) organometallic catalysts. These are stereospecifically effective Ziegler catalysts dispersed in hydrocarbons. The development of high-efficiency catalysts analogous to those used for PE syntheses (Sec. 4.1.2.1.1/3) led to gas-phase high-pressure (BASF) and low pressure (Rexene/Novamont) processes that resulted in highly pure products. Many new plants use the economical, superactive catalytic "Spheripol"-process (Himont) under license.

According to the conditions of the reaction, more or less of the crystallization-inhibiting atactic PP arises as a by-product, which can be separated as a heptane-soluble component and used for other purposes (Sec. 4.1.1.2.2). The percentage which is not heptane soluble, the *Isotactic index*, is one of the designatory properties of PP-Types (Table 4.10).

Table 4.10.   ISO/DIN Designation of Propylene Polymers

ISO 1873 order a to d	ISO 1873	DIN 16 774 E82
a) Coding letters for types[1,5]		
All propylene homopolymers	H	H
Block copolymers[2]	C	B
Random copolymers[2]	R	R
Mixtures of types[2]	M	Q
c) Isotactic index (II)[3]:	Cell range	
95	>90 to 100	
85	>80 to 90	
75	>70 to 80	
65	>60 to 70	
55	>50 to 60	
d) Melt flow rate (MFR)[4] in g/10 min:	Cell designation	
0.10 to 0.25	002	
<0.1 to 0.2 to 0.4		000–001–003
>0.25 to 0.5 to 1.0	004–007	
>0.4 to 0.8 to 1.5		006–012
>1 to 2 to 4 to 8	015–030–060	
>1.5 to 3.0 to 6.0 to 12.0		022–045–090
>8 to 16 to 32 to 64	120–240–480	
>12 to 25 to 50 to 100		200–400–700
>64 to 100	820	

[1] Additional end-use coding letters in the order b (data-block 2) see Chapter 6.
[2] Thermoplastic types with ≤ 50% content of other aliphatic olefin monomers.
[3] Percentage by weight not soluble in boiling heptane, not applicable for mixtures (code: 00).
[4] ISO condition M (230°C/2.16 kg) or – when a condition code is to be added – cond. T (190°C/5 kg).
[5] In the order e of ISO 1873 "NAT" designates an unmodified "MOD" a modified material, for kind and amount of filler or reinforcing material in data-block 4 see Chapter 6.

4.1. Thermoplastics: Raw Materials and Molding Compounds

In CR (controlled rheology) grades, excessively long polymer chains have been cracked by post-reactor treatment to yield injection-molding grades with narrow molecular weight distribution that flow well.

Other processes include *random copolymerization* by admixture of 1 to 4% ethene to the monomer; stepwise graft or block copolymerization, with higher proportions of ethylene, in the readily crystallizable polyallomers (Table 4.6), thermoplastic-elastomer ethylene-propylene copolymers (Sec. 4.1.2.6.1), and finally, mixtures with other polyolefins. The group classification and designation system relating to these products with MFI as the second designatory property according to ISO 1873 or DIN 16774 are indicated in Table 4.10. ASTM D 2146-80 (Table 4.4) differentiates between

Type I: Polypropylene characterized by its rigidity and resistance to deformation under load and

Type II: Polypropylene copolymers and propylene elastomer compounds, characterized by improved resistance to impact, especially at low temperatures.

Both types have up to $0.915\,g/cm^3$ maximum density based on uncolored, unfilled plastics. For reinforced PP see ASTM D 2853, Table 4.4.

2. *Trade names, forms of delivery and grades*

Trade names for molding compounds of all grades:

Bapolene, Fortilene, Profax, Tenite (US); Noblen, Sunlet (JP); Hostalen PP, Novolen, Vestolen P (DE); Moplen (IT); Lacqtene P, Appryl PP (FR); Stamylan P (NL); Eltex P (BE); Daplen (AT).

Injection and extrusion compounds are generally supplied as coarse powders or pellets. With regard to the long-term uses of PP at elevated temperatures, the compounds are stabilized against oxidative and other harmful influences in the following grades:

(1) Normally heat-aging resistant, natural and in part in standard colors. Also with lubricant additive, non-toxic.
(2) High-temperature aging resistant, additionally light stabilized, compatible colors, partly non-toxic.
(3) Highly weathering resistant – stabilized with carbon black, with sterically hindered amines, colored for outdoor use, a 10-fold improvement in light stability.
(4) Highly heat-aging resistant in the presence of hot detergent solutions and in streaming hot water, natural, white and ivory colors, non-toxic.
(5) Highly heat-aging resistant in contact with copper and other metals, colored after being made compatible.

4.1.2.6. Polypropylene (PP) and copolymers

Special grades are formulated to be resistant to γ-radiation for medical sterilization, flame retardant for building applications (B.1 to DIN 4102) and electrotechnical applications (UL-94 H B to UL-94 V-O), antistatic or electrically conducting, directly electroplatable (Sec. 3.9.3.4) or structurally foamed to $d_R < 0.6$. Filled and reinforced compounds are described below under GMT and in Sec. 4.1.13.2.

3. Structure and properties

Tables 4.5 and 4.6 give data for PP molding compounds. PP has a smooth dry feel, is not susceptible to stress cracking and can be nailed. According to type and stabilization, it is usable for long periods up to temperatures between 80°C (176°F) and 120°C (248°F). Unstressed, it is resistant to boiling water under all circumstances. PP and its copolymers have a low density (< 0.9 g/cm^3 up to slightly more than 1 g/cm^3 for highly filled products) and with many grades on the borderline between standard and engineering plastics combine advantageous properties both technically and commercially. They have a greater hardness and stiffness than PE but a lower brittleness than standard PS, outstanding resistance to hot water and to chemicals, and very good electrical properties. Thus PP and its copolymers have a very wide range of applications in the form of moldings, profiles, tubes, films, fibrillated filaments and textile fibers.

Highly isotactic homopolymers, at $\sigma_B \geq 36$ MPa and $E \sim 2100$ MPa, exhibit very high strength, stiffness and temperature resistance, but low notched impact strength with a lower service limit of about 0°C. Reduction of the crystallinity by a proportion of atactic PP or random copolymerization produces flexible, shock resistant types with 10–20% lower values for σ_B and E, e.g. for packaging containers, stretch films or blow-molded bottles. To increase the transparency of thin walled products and improve the general processing characteristics of extruded ones, a fine grained crystallite structure is sought by the addition of nucleating agents. One such agent, 1% PP regrind, partly cross-linked by electron irradiation, has proved better than foreign substances (such as sodium benzoate), which may produce defects. Block copolymers become increasingly less rigid and less heat resistant with increasing ethylene content, although they are still satisfactory. Elastomer blends can be made impact resistant down to approximately −40°C.

For mineral-filled, reinforced PP molding compounds, specific bonding promoters, such as acrylic acid grafted PP (Polybond, GB) or organo-silicon based materials, permit, at filler proportions $\geq 60\%$, moldings with less processing shrinkage and deformation with good surface characteristics. Chalk-reinforced PP, having at the same time good impact resistance, is an alternative to polystyrene for mass-produced articles. PP compounds highly reinforced with talc, mica, aluminum oxide trihydrate, and also glass spheres, short glass fibers or

4.1. Thermoplastics: Raw Materials and Molding Compounds

aligned long glass fibers are engineering materials with correspondingly improved mechanical heat resistance and flame retardance. (Tables 4.4, 4.5). PP sheet molding compounds filled with glass fiber (Elastogran, Symalit), 40–60% wood flour or sawdust (Wood-Stock) or shredded waste paper (Papia, JP) are used for thermoformed car interior cladding. For GMT, glass mat thermoplastics with PP matrix, see Sec. 4.1.13.2.

4. Processing and applications

In *injection molding* (Table 3.9) normally stabilized compounds are used for domestic and kitchen equipment parts, sanitary equipment, tool and travel cases, shoe heels, photo accessories, housewares, toys, etc. Structural foam moldings are also used for such purposes. Base and lid shells and similar moldings can be injected together and joined by a film joint that is strengthened on bending by stretching. Garden furniture has been made of weather-resistant grades.

Heat-resistant PP grades stabilized against detergent solutions have an enormous number of applications as injection-molded functional parts in household washing machines as well as in the textile industry, e.g., for dyeing spools and other machine parts that are also copper resistant in electrical applications. In both fields reinforced components are also of importance.

In the automotive sector many reinforced and unreinforced PP grades are used for, e.g., injection molded battery cases, ventilator blades and frames, for air filter housings, dashboards, steering wheels, low temperature impact-resistant to elastomeric copolymers for front and rear bumpers, fenders etc.

Higher molecular weight PP grades are used for *blow molding* of large drums, hot water reservoirs in heating technology or motor-cycle trunks and claddings. Glass-clear bottles for hot bottled drinks are blown from random copolymers (Sec. 3.3.6.7).

There are several high-molecular, standardized PP grades for the *extrusion* of pipes and sheets or *press molding* of large parts such as filter press plates. The chemical industry needs highly chemical- and temperature-resistant homopolymers (Sec. 4.2.2.1). Random copolymers with toughness and flexibility across a wide temperature range are used for industrial hot water pressure pipes and for heating pipes under floors. Hot waste water PP installations in buildings are flame retarded. For extruded "cast" and biaxially oriented PP films see Sec. 4.2.2.3.5.

Fibers with coarse and fine titers are made from easy-flow, low molecular weight PP by melt spinning. For *woven film tape* and fibrillated fibers made from blown and flat film, preferred PP grades are those of relatively high molecular weight and limited crystallinity. The amorphous component contributes as internal lubricant in the stretching and splitting.

PP fibrids (Lextar, US; Ferlosa IT) are fine filaments similar to PE fibrids (see Sec. 4.1.2.1.10).

PP foam panels and webs. PP copolymer granules are impregnated under pressure in aqueous suspension with a blowing agent and expand after reduction to atmospheric pressure into foam particles which are used to manufacture moldings, such as bumper cores and containers.

4.1.2.7. Polybutene-1 (PB)

Trade name: Shell PB (US)
Form of delivery: Natural, colored and black granules for pipes, sheets and cable; $i_{190/5} = 0.5$ for blow moldings; for injection moldings and films $i_{190/5} = 2\text{-}150$. Flame retardant grades are also available.

Manufacture, properties and processing

Butene–1 is polymerized to a largely isotactic product of very high molecular weight by means of specific Ziegler-Natta catalysts. The polymer flows readily at 190°C. On cooling, 50% crystallizes first as a metastable flexible modification (d_R approximately 0.89), which is then spontaneously converted (with corresponding shrinkage) in 1 or 2 days into the stable form ($d_R = 0.915$).

Tables 4.4 and 4.9 summarize ASTM D 2581 specifications and properties. PB in its stable form has short-term properties at room temperature and reduction of the shear modulus with temperature between those of PE and PP. As a consequence of its high molecular weight PB exhibits high creep resistance even at high temperatures. It is also highly resistant to stress cracking. These favorable properties are unimpaired with 20% carbon black.

Above 190°C PB can be readily injection, blow, and compression molded as well as extruded. Thermoelastic aftershaping presents some difficulties. Because PB goes into the metastable condition, the shaping reverts after cooling. Pipe bends are made by winding the extruded tube onto a drum and allowing it to convert into the stable modification. PB is readily welded.

For relatively low molecular weight atactic PB see Sec. 4.1.1.2.1.

Applications:

Moldings for pipework are made by injection, and corresponding stressed hollow articles by extrusion blowing. Pipes (see Sec. 4.2.1 and 4.2.2.1) are used for underfloor heating, for hot water and for the large-pipe sector. Blown or cast films that are heat sealable at 195 to 245°C and are FDA approved are used for hot-filled food and meat packaging. Addition of from 1 to 5% of easy flowing PB to PE, PP and PS increases the extrusion speeds of these thermoplastics. PB is also suitable as a carrier for readily processable pigment concentrates and filler concentrates for polyolefins and PS.

4.1.2.8. Polymethylpentene (PMP)

Trade name: TPX Polymers (JP).
Delivery: unfilled, 15% and 30% GFR granules.
Properties: 4-methylpent-1-ene polymerizes to a highly branched very light glass-clear hard product with 90% transmission and good heat resistance. The chemical behavior resembles that of other polyolefins and it is not recommended for outdoor applications.

For general properties of unfilled PMP see Table 4.9, p. 195.
Properties of FR-TPX: σ_B 50 to 60 N/mm^2 (7 to 8.5 × 10^3 lb/in^2), E_f 30 to 55 kN/mm^2 (4.3 to 7.1 × 10^5 lb/in^2), HDT 175 to 190°C; flame retarding E_f 71 kN/mm^2, HDT 165°C, UL 94 V-O.
Processing: Injection molding is at 260 to 320°C, mold temperature is 60°C and mold shrinkage 1.5 to 3.0%. Extrusion presents some difficulties because of the narrow melting range. Blow molding is carried out at melt temperatures of 275 to 290°C. Pre-drying is not required before molding.

Applications: Spectacles, domestic lighting and laboratory equipment, sterilizable moldings for medical use and packaging, carriers for microwave cooking food, dyeing spools and Raschig rings.
A low flammability grade that is flexible at −10°C is used as an insulant for coaxial cable and electronic circuitry.
FR-TPX is used for injection molded electrical/electronic equipment components. Its excellent dielectric properties remain constant, nearly independent of frequency, temperature and humidity.

4.1.3. Styrene-Based Polymers

4.1.3.1. Homopolymers, copolymers and alloys

1. *Structures, properties and polymerization techniques*

Styrene (⌬–CH = CH$_2$) is synthesized by the addition of ethylene to (70% w/w) benzene and subsequent dehydrogenation. For 50 years it has been the raw material for relatively brittle "standard" polystyrene (PS) homopolymers. It has been the predominant monomer (weightwise) in

- Copolymers with improved thermal and mechanical properties such as SAN with acrylonitrile (CH$_2$=CHCN), SMS with α-methylstyrene (⌬–C(CH$_3$)=CH$_2$) and SMA with maleic anhydride
- Copolymers with improved impact strength (PS-HI) such as BS with butadiene (styrene) rubber or such with other elastomers

4.1.3.1. Homopolymers, copolymers and alloys

– Terpolymer engineering plastics with balanced thermal, mechanical and impact properties in many modifications, such as ABS with acrylonitrile + butadiene (acrylic) rubber

An overview of the system of designatory and specifying properties of PS, SAN, SB, ABS (AXS) used for the ISO/DIN- and ASTM-standardization of the styrene polymer group is given by Tables 4.11 and 4.12 (p.208).

Reactive polystyrene (RPS) with multifunctional monomers in the main chain (commercial products include RPS with oxazoline groups from Dow) can be grafted with side chains of polymers of the most diverse kinds in a second reaction stage, and thus otherwise incompatible polymers (e.g. PS and polyolefins) can be synergetically combined. Such reactions can be carried out on a small scale with relatively little effort as substance graft copolymerization in extruders designed for intense mixing and shear. They provide the compounder with the ability to offer new tailor-made combinations of styrene based plastics.

Standard polystyrene is made mainly by batch or continuous mass polymerization (Sec. 3.1.1). PS is taken off as a melt and then pelletized. Redispersible EMU powder (Sec. 4.1.1.10) is emulsion polymerized. Foamable PS containing physical blowing agent (pentane), SAN and other heat-resistant copolymers are suspension polymerized. For multistage chemical graft polymerization (Fig. 1.1) of rubber-modified products the prepolymerized rubber is either dissolved in styrene and the solution mass or suspension polymerized (SB process), or a rubber latex is used as the medium for emulsion graft polymerization of styrene and acrylonitrile or other components as appropriate (ABS process).

In addition to reactive double bonds containing butadiene (co-) polymers, saturated polyacrylic or EP(D)M rubbers serve as grafting sites for better weathering products. Two-phase copolymers with dispersed (1 to 10 μm) elastomeric particles in a disperse styrene (co-)polymer phase are opaque.

Newer developments in polymerization techniques and knowledge of the limits of compatibility have led to the optimization of products in terms of specific properties such as impact or crack resistance. They have also led to products with elastomeric particles < 1 μm, which are nearly glass-clear, and to block copolymers with defined segment lengths and/or comonomers with comparable refractive indexes, which are quite glass-clear.

"Teleblock" copolymers with blocks of up to 50% styrene in which longer rubber molecules are bound to the ends of the blocks in a cross-linked manner are general purpose *thermoplastic elastomers* (p. 0) which represent about 50% of the TPE-market (Cariflex, Elexar, Kraton, GB; Europrene, I; Solprene, BE). Technical rubberware is made from such SBS copolymers (Cariflex T R, Kraton D, Solprene). For example, shoe soles are injection molded. Other types, also containing isoprene or ethylene-butene as comonomers, are the

Table 4.11. ISO/DIN Designations of Styrene Based Polymers

	DIN	ISO	Block 1[1] Acrylonitrile content % w/w	Block 3[1]: Sequence of properties, is the same in all styrene group standards				
				Vicat softening point ISO 360, B/50 °C	Flow rate Condition indicated g/10 min	Charpy impact[2] ISO 129/1 kJ/m^2	Izod impact[2] ISO 180/1 kJ/m^2	Flexural Modulus ISO 178 GN/m^2
PS	7741-86	1622–85		(4) ≤ 80 to >100	(4.5) ≤ 2 to >16 200°C/5 kg			
SB	16771-86	DIS 2897–85		(6) ≤ 80 to >100	(4) ≤ 4 to >16 200°C/5 kg		(5) ≤ 3 to >12	(4) ≤ 1.5 to <2.5^3
SAN	16775-86	DIS 4894-85	(3) <10 to 50	(4) ≤ 90 to >110	(4) ≤ 5 to >20 200°C/5 kg			
ABS AXS[4]	16772–75 16777-E-86	2580–85 6402–85	(2) <10 to <30[5]	(4) ≤ 90 to >110	(4) ≤ 5 to >20 200°C/10 kg	(5) > 2 to >18	(5) >3 to >30	(4) ≤ 1.8 to >2.5^2

[1] See Sec. 6.2.3. Arabic numbers in brackets give the quantity of cells, followed by the total cell range corresponding to the cells' middle value code, expressed in the units given above. Number of cells and consequently special code and limits of single cells may change slightly in the future, because generally unified ISO/DIN-standardization of all styrenics is still in process.
[2] Specimens in stress-free condition, notched. C for Charpy or I for Izod have to be inserted in the designations, after the values.
[3] Only ISO, optional.
[4] X: Acrylic esters (ASA), ethylene propylene (AES), chlorinated polyethylene (ACS).
[5] In the continuous phase.

4.1.3.1. Homopolymers, copolymers and alloys

raw materials for (melt) adhesives and coatings. Styrene *teleblocks* allow the alloying of special SBS into a PS matrix to produce toughened products with about the same softening temperatures as PS itself. Graft-polymerized high-impact types are converted to super-high impact.

The same SBS types can be used as PP elastifiers. Styrene polymers are also components in many single-phase and two-phase alloys that are used for molding compounds or as modifiers, e.g., for PVC (Sec. 4.1.4.3.1). For details see Sec. 2 following and Tables 4.13 (p.209) and 4.14.

2. *Products and trade names*

PS and styrene polymer compounds in general:

Dylene, Fostarene, FyRid, Lustrex (US); Esbrite, Toporex (JP); Vestyron (DE); Edistir (IT); Gédex, Lacqrene (FR).

SAN copolymers (partly SAN and ABS):

Lustran, Magnum, Tyril (US); Litac (JP); Luran Sconarol (DE); Kostil (IT);

also reinforced compounds such as Absafil (ABS), Acrylafil (SAN), Colimate (SAN), Comalloy, Fiberfil, Fiberite, Plastalloy, Starasan (SAN), Thermocomp, Thermofil (SAN, ABS).

SB copolymers and other toughened polystyrenes:

Fosta Tuflex, Superflex (US); Luran (DE)
Flame retardant: Tuf-Flex 702, 707 (US). Transparent: K-Resin (US); Styrolux (DE).

ABS copolymers:

Abson, Cycolac, Kralastic (US); Diapet, Stylac, Toyolac, Tufrex (JP); Novodur, Terluran, Sconater (DE); Ronfalin (NL) Ravikral; Ugikral (FR); Forsan (CS).

Other terpolymer molding compounds:

AAS (methacrylate acrylic rubber styrene); Vitax (JP) AES (acrylonitrile ethylene-propylene rubber styrene), Rovel (US); Novodur AES (DE); ASA (acrylonitrile styrene acrylic rubber); Stauffer SCC 100 (US); Sunrex (JP); Luran S (DE); MBS (methyl methacrylate butadiene styrene), transparent; Cycolac CIT (US); Cyrolite (DE); Toyolac 900 (JP); Sicoflex (IT); Methacrylene (FR); etc. See also MMA-Cop, Sec. 4.1.6.1 and PAN-Cop, Sec. 4.1.6.2. SMA (styrene maleic anhydride(ter-)polymers): Cadon, Dylark, Rovel (US).

Table 4.14. Property Ranges for Styrene Terpolymers and Alloys

Properties[1]	Unit		ABS				ASA (European types)		Alloys		
	SI	U.S.	High impact		Impact				ABS/PVC[2]		ABS/PC[3]
Density, 23 °C (75 °F)	g/cm³	lb/ft³	1.02-1.04	63.8-65.0	1.04-1.05	65.0-65.6	1.05-1.07	65.6-66.9	1.07	66.9	1.16-1.21	72.5-75.6	1.12-1.16	70.0-72.5
Mechanical												
Tensile strength	N/mm²	10³ lb/in²	30-45	4.3-6.4	40-60	5.7-8.5	45-62	6.4-8.8	47-66	6.7-9.4	30-52	4.3-7.5	35-45	5.0-6.4
Elongation, yield/break	%		3.5/30		3.0/20		2.5/<20		3/<20		–/(125)		4.5/<50	
Tensile modulus	kN/mm²	10³ lb/in²	1.5-2.0	2.1-2.8	2.0-2.6	2.8-3.7	1.7-3.0	2.4-4.3	2.3-2.9	3.3-4.1	2.1-2.3	3.0-3.3	2.0-2.2	2.8-3.1
Flexural strength	N/mm²	10³ lb/in²	45-65	6.4-9.2	70-80	10-11	70-100	10-14	70-100	10-14	63-76	9.0-11	75-85	11-12
Hardness, ball indent.[4]	N/mm²	10³ lb/in²	65-75	9.2-11	85-110	12-16	90-135	13-19	72-127	10-18	–	–	80-90	11-13
Hardness, Rockwell	R-scale		80-90		105-115		107->	115			96-100		111-120	
Charpy impact, +23 °C unnotched	kJ/m²		n.b.		80-n.b.	n.b.	60-n.b.	n.b.	40-n.b.		–		n.b.	
−40 °C	kJ/m²		n.b.		50-60	n.b.	20-80		18-40		–		n.b.	
Charpy impact, +23 °C notched	kJ/m²		20-30		8-12		8-16		4-12		–		25-30	
−40 °C	kJ/m²		5-10		3-5		1-8		–		–		10-15	
Izod, notched 23 °C ⅛ in (3.2 mm) specimen	J/m	ft-lb/in	400-500	7.4-9.4	100-300	1.9-5.6	100-340	1.9-6.4	–		530-640	10-12	530-580	10-11
−40 °C	J/m	of notch	40-200	0.7-3.7	40-50	0.7-0.9	35-100	0.6-1.9	–		59-69	1.1-1.3	65-220	1.2-4.0
Thermal												
Vicat soft. temp., B/50	°C	°F	90-100	194-212	90-105	194-221	102-121	216-250	93-101	199-214	80-85	176-185	110-130	230-266
Defl. temp., ISO/R 75[5] A, 1.8 N/mm² (264 psi)	°C	°F	85-95	185-203	90-100	194-212	95-111	203-232	94-99	201-210	70-86	158-186	95-105	203-221
B, 0.45 N/mm² (66 psi)	°C	°F	93-101	199-214	95-103	203-217	100-116	212-241	98-104	208-219	77-93	170-199	100-126	212-259
Linear thermal exp.	10⁻⁵ K⁻¹	10⁻⁵ °F⁻¹	9-12	5-7	8-10	4-6	7-11	4-6	8-10	4-6	8.3	4.6	7-8	4-5

[1] Properties measured by ISO or ASTM methods agreed within 10%. [2] Flame retardant alloys: UL 94 V-1 and V-0, hardest types (R ≧ 110): Izod (23 °C). 5.0-2.0 ft-lb/in. [3] UL 94 HB to V-0. [4] H 358/30 [5] annealed to 10 K (18 °F) higher.

4.1.3.1. Homopolymers, copolymers and alloys

Alloys with styrene polymer components[1] (Examples)

Polyphenylene oxide (PPO)/PS see Table 4.48; Polysulfone (PPSU)/ABS: Mindel and Polysulfone (PPSU)/SAN; Acrylon T, Ucardel (US), see Table 4.48; ABS/PVC: Abson, Cycovin KAB, Cycoloy EH, Kralastic FVM, Polyman 509 (US); Ryulex (JP); Volkaril (FR) Roufaloy (NL); ABS/PC: Cycolac, Cycoloy (US), Bayblend, Terblend B (DE), Moldex A (IT), see Table 4.14; ASA/PC: Terblends (DE); ABS/TPU and TPU/ABS Sec. 4.1.8.4; ABS based high temp.: Stapron (NL), ABS/PA 66: Triax (USA).

Modifier resins for PVC (Sec. 4.1.4.3)

ABS based: Blendex, Kraton (U.S.) Baymod, Novodur A 50/70, Vinuran (DE). MBS based: Blendex, Paraloid (U.S.); Kane ACE, Kureha BTA, Metablen (JP). SMA based: Cadou (US) ABS/NBR: Baymod (DE).

Coupling materials based on SB blends are used as tie layers for laminating sheets and coextruded films of styrene polymers with polyolefins, PVC, PMMA and PA and as "alloying" agents in recycled compounds made of mixed plastics scrap.

3. General properties

Standard polystyrenes are crystal clear, glossy and relatively brittle. Their strength is minimally dependent on time and temperature. According to grade, moldings can be used at 60 to 90°C for prolonged periods. PS is insensitive to moisture and resistant to salt solutions, alkaline liquors and non-oxidizing acids. Acetone, ether, esters, aromatic and chlorinated hydrocarbons are solvents that are also usable for bonding. Some fuels and synthetic oil, plasticizers, alcohols, most vegetable oils and fragrances can attack PS by swelling and/or environmental stress cracking.

SAN copolymers are somewhat more rigid and temperature resistant, not quite so corrosion resistant, but less brittle and not liable to stress cracking from contact with the above-mentioned substances. SAN is clear but becomes slightly yellow on warming. The related SMS (styrene-α-methylstyrene) copolymer has better optical qualities.

The wide variety of grades and types of *copolymers, terpolymers* and *blends* up to leatherlike toughness provides a range of impact strength. Graft polymers can give very favorable combinations of rigidity at higher temperatures and toughness at lower temperatures. The chemical behavior and tendency to stress cracking of SB resemble those of PS and ABS those of SAN.

PS, SB and ABS in glass-clear or light colors discolor on long exposure in the open. SAN copolymers and butadiene-free impact grades are more suitable. In

---

[1] see section 4.1.9.2.

4.1. Thermoplastics: Raw Materials and Molding Compounds

enclosed rooms with ordinary lighting and temperature, products made of all polystyrene plastics remain unchanged for long periods.

The basic range of styrene polymers available for many applications is supplemented for particular purposes by special grades such as:

- Easy-flow injection and extrusion grades for thin-walled packaging, etc.
- Highly stress-crack resistant, tough grades for refrigerator equipment (insensitive to fluorocarbon refrigerants) and for packing fat-containing materials
- Contact transparent and glass-clear SB and terpolymer types
- UV stabilized
- Compounded for antistatic behavior and for flame retardance
- For use in building (B.1) to DIN 4102
- For radio and television housings, according to UL specifications reinforced with glass fiber (SAN and ABS)
- Electroplated (ABS, Sec. 3.9.3.4)

Most styrene polymers contain only traces of the monomer ($<0.01\%$). This is less than is permitted as the statutory maximum for food.

4. Processing techniques

Styrene polymers are, for the most part, processed by *injection molding* (for processing conditions see Table 3.9). The low shrinkage makes accurately molded parts possible. Fast injection can lead to orientation of the melt and to considerable stress, in which case it is necessary to raise the mold temperature at the expense of output. For consumable mass moldings very easy-flowing, sometimes "lubricated" molding compounds are adequate. Rubber-modified SB compounds are advantageous for large moldings on account of their good flow, insofar as the stressing does not require SAN or terpolymer (ABS) compounds. Thermoformable SB and terpolymer sheeting (Sec. 4.2.3) is made by *extrusion*. ABS sheeting can also be cold drawn (Sec. 3.8.2). PS sheets are used in electrical engineering. Extruded profiles have many applications. ABS and ASA pipes are used for standardized piping material (Table 4.51). Transparent or pigmented PS, SAN or ABS films are blown or chill-roll extruded. ABS, including glass-fiber reinforced material, is well suited for extrusion blowing (Sec. 3.3.6.7).

5. Fields of application

Polystyrene plastics offer unique combinations of outstanding electrical properties (Table 4.15) with almost unlimited moldability into precision parts at moderate cost. Their application in insulation and other construction parts in *electronic and communications technology* is ubiquitous (telephone apparatus consists, apart from the conducting elements, of ABS for the housing and SB and SAN for the internal parts). Many polystyrene plastics are also used as

4.1.3.2. Styrene polymers for foam processes

Table 4.15. Electrical Properties of Styrene Polymers

Properties[1]	Unit	PS	SAN	SB	SB[2] "Y"	ABS	ASA
Volume resistivity	ohm cm	10^{18}	10^{16}	10^{16}	10^{12}	10^{15}	10^{14}
Surface resistivity	ohm	10^{15}	10^{14}	10^{14}	10^{9}	10^{13}	10^{13}
Dissipation fact. (10^6 Hz)	10^{-4}	≤ 1	80	4	120	200	250
Dielectric const. (10^6 Hz)	–	2.5	2.9	2.6	2.7	3.2	3.4
Dielectric strength	kV/cm	160	170	200	–	150	80
Dielectric strength[3]	V/mil		500–700		–	350–400	
Tracking resist. methods	KA	1	2	1	–	3a	3a
	KB	160	200	200	–	300	600
	KC	500	450	550	–	>600	>600

[1] ISO/VDE 0303 methods.
[2] SBY indicates antistatic-modified SB.
[3] ASTM D 149.

protective electrical insulation of equipment, so long as the heat resistance remains adequate.

In *refrigeration*, internal housings, shaped-door interiors and external coverings of cold impact resistant selected SB and ABS types are injection molded and thermoformed. Injection-molded mechanical parts and housings, scales, keyboards etc. are used in the precision engineering, photographic and office machinery industry.

The *automotive industry* employs SB and terpolymers for linings, covers, instrument boards and radiator grills. The bodies of small sports cars are deep drawn in ABS. ABS and other terpolymers are used for the hulls of sports boats. Laminates with ABS and PMMA covering layers are also used. Glass-fiber reinforced SAN finds application as *camera and projector housings* and functional parts. ABS-GF is used for self-supporting instrument panels. Starting-battery housings are injection molded in SAN and SB. Mining and industrial *protective helmets* made of cold impact resistant terpolymers fulfil the standardized safety requirements. *Seating furniture* and furniture framework are injection or blow molded or thermoformed in ABS or ASA. Good quality *household articles* are made mainly from SAN; domestic appliances require high-impact grades. Combs and decorative goods are additional applications for styrene polymers.

Packaging is a significant use for styrene polymers, particularly in the food industry. Injection molded and thermoformed parts are used.

4.1.3.2. Styrene polymers for foam processes

1. Structural foam moldings, mostly of (medium) impact-resistant compounds, in the density range of 0.7 to 0.9 g/cm³ with a wall thickness of at least 5 mm

and up to 30 kg in weight are injection molded with chemical blowing agent (Sec. 3.1.5.2.3). In small plants they are also made by foam molding. Because of the roughness of the outer skin most of the products must be polished and lacquered. Construction and furniture profiles are extruded (Sec. 3.1.5.1).

2. With the *extrusion of light EPS semifinished goods* (packaging films, $d_R \geq 0.1 \, \text{g/cm}^3$; insulating material, $\geq 0.025 \, \text{g/cm}^3$) PS is used with physical blowing agent incorporated or fed directly into the extruder. For details see Sec. 3.6.5.3.

3. The raw materials for the "Styropor" process for EPS particle foam are pearl polymer beads (Sec. 3.1.5.1, 2 mm dia for thin-walled, up to 3 mm dia for block foam), in which a low boiling point hydrocarbon blowing agent, usually pentane, is incorporated during polymerisation.
Trade names: Styropor, Vestypor (DE); Extir (IT); Styrocel (NL); Rigipore (GB); Jackodur (NO); Koplen (CS); Okirol (YU); Dylite, Fyrid, Montopore, Pelaspan (US); Snowpearl (JP); high temperature resistant cop. Dytherm (US).

In the first step of the three stage foam process, the beads are prefoamed, usually discontinuously, in cylindrical mixers linked to downstream fluidized bed driers. The material is heated to between 80 and 110°C by passing steam and is expanded to densities of 10 to 20 kg m^{-3} for block and solid-borne sound insulation material, 15 to 30 kg m^{-3} for moldings.

The second stage, ageing of the closed cell prefoamed particles for 1 to 2 days in silos permeable to air, is required for stabilisation. The under-pressure caused by condensation of the blowing gas during cooling is equalised by air diffusing in.

Particle foam blocks and moldings are then produced mainly in molds with fixed perforated walls on automated equipment. The degassed loose particle material is heated for 5 to 50 sec with water vapor from steam chambers to softening point and the foam particles sinter together at the contact point. Finally the water vapor is evacuated and the product cooled with water.

Large block molds, 6 to 8 m long, cross-section $1.25 \times 1.00 \, \text{m}^2$, with downstream hotwire cutters for slicing into panels are more efficient than web plant. Cycle time and energy consumption can be reduced by operating automatic molding machines (for moldings up to 1.2/1.6/0.7 m external dimensions) with an additional cooling mold to which the molding formed in the hot mold is transferred.

Hydrocarbon blowing agents in particle foams can form explosive mixtures with air. Quite dry EPS can spontaneously generate an electrostatic charge by vibration. This property demands structural (dividing walls) and operation

measures (aeration, smoking ban, relative humidity > 50% and earthing of metal parts) for fire protection in processing plants.
Particle foam made of SM-Cop (Dytherm) remains serviceable up to 121°C.

4.1.4. Vinyl Chloride Polymers

Vinylchloride ($CH_2 = CHCl$) today is made mainly by the single or two stage addition of chlorine to ethylene (Table 1.3). Homopolymer PVC contains 56.7 w/w % Cl. Chlorine is amply available from NaCl electrolysis. Its high content in VC polymers explains their special economic importance and their properties. Homo- and co-polymer raw materials and blends (p. 000, 000) and pastes (p. 000) of the same, suitably supplemented for special products are available.
Trade names: Dural, Geon, Kohinor, Vygen (US); Nipeon, Nipolit, Vinka, Vinychlon (JP); Vestolit, Vinnolit, Vinidur, Vinoflex, Scovinyl, Solvic, Trosiplast (DE); Ravinil (IT); Lucalor, Lucovyl (FR); Corvic, Welvic (GB); Benvic, Ultryl, Varlan (NL); Rosevil (RO); Ongrovil (Hungary); Bovil, Hiplex, Juvinil, Zadrovil (YU).

4.1.4.1. to 4.1.4.4. VC-polymer raw materials

4.1.4.1. Syntheses and delivery forms

Vinyl chloride homopolymers are manufactured by emulsion (E), suspension (S) or bulk or mass (M) polymerization (Sec. 3.1.1).
Copolymers are made by the E and S processes and also by graft polymerization. PVC-U molding compounds (Sec. 4.1.4.3.3) with improved impact resistance are obtained by the latter method and by blending PVC with flexible materials. The polymers contain < 1 ppm of vinyl chloride monomer.
Whereas other thermoplastics are commercially available as ready-for-use molding materials, vinyl chloride polymers are prepared in many processing plants by mixing the raw powdered polymer with the additives necessary for processing and application. If it is not technically necessary to work the raw material up into pellets through an extruder agglomerates or dry blends (Sec. 3.1.4) are used. This technique is economical and protects the polymer, which is thermally sensitive. Dry blends for processing by extruder must be free flowing and therefore should not be too fine grained. However, very finely divided powder is necessary for the incorporation of plasticizer for casting and brushable PVC pastes. Figure 4.4 shows particle shapes and sizes of various PVC homopolymers.

Emulsion, suspension and mass polymers
E-PVC emulsion polymers are formed as primary particles, 0.1 to 0.2 μm dia. The final size and shape of the powder particles are determined by the

4.1. Thermoplastics: Raw Materials and Molding Compounds

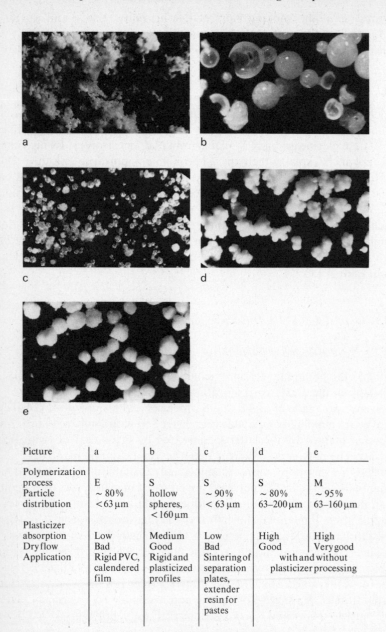

Picture	a	b	c	d	e
Polymerization process	E	S	S	S	M
Particle distribution	~80% <63 μm	hollow spheres, <160 μm	~90% <63 μm	~80% 63–200 μm	~95% 63–160 μm
Plasticizer absorption	Low	Medium	Low	High	High
Dry flow	Bad	Good	Bad	Good	Very good
Application	Rigid PVC, calendered film	Rigid and plasticized profiles	Sintering of separation plates, extender resin for pastes	with and without plasticizer processing	

Fig. 4.4. Particle structure of PVC powder (Hoechst).

processing conditions (drying) enabling products to be produced for specific applications as follows: very finely dispersed for plastisol processing; finely dispersed, easily broken down for calendering; coarse particled, free flowing with high bulk density for extrusion (PVC-U); coarse particled, porous for plasticized applications.

E-PVC contains up to 2.5% emulsifier and, sometimes, inorganic additives (e.g. alkaline pre-stabilisers). Depending on the type and level of such additives, transparency, water absorption, electrical insulation properties are in general inferior to those of S- and M-PVC. E-PVC possesses better processability leading to finished products distinguished by high toughness, smooth, closed surfaces, low electrostatic chargeability and thus lower tendency to attract dust.

Suspension polymers (S-PVC, sulfate ash < 0.1%) with a particle size distribution between 0.06 and 0.25 mm and the particularly uniform *bulk polymers* (M-PVC, sulfate ash 0.01%) with particles around 0.15 mm are, as a result of their manufacturing process, very pure.

Opaque and transparent grades of PVC are available appropriately stabilized to provide the required corrosion and weathering resistance for mechanical and electrical applications.

Types consisting of porous particles (bulk density 0.4 to 0.5 g/cm^3) are good for plasticizing, whereas the denser particles (bulk density 0.5 to 0.65 g/cm^3) are more suited for processing rigid materials. Free-flowing compounds are mixed in high-speed mixers at temperatures of 100°C. Paste-making Micro-pearl suspension polymer with a particle size of less than 10 μm is a special product.

E-PVC and S-PVC with fine, dense particles are used as sintering powders for porous accumulator separators.

For PVC-C and PVDC see Sec. 4.1.4.3.1.

4.1.4.2. Molecular weight and field of application

The function of the degree of polymerization of PVC is given by ISO/R 1628, the *viscosity number* (Sec. 3.1.2.1), which is obtained from a diluted solution of PVC in cyclohexanone. The inherent viscosity of PVC according to ASTM D 1234 is for practical purposes $\frac{1}{100}$ of the viscosity number. The K value, according to DIN 53726, parallels viscosity measurements using a different method of calculation. K values and viscosity numbers bear a definite ratio to one another, as shown in Table 4.16.

The higher the molecular weight of the PVC, the better are the mechanical and electrical properties of the product. On the other hand, processing, particularly of rigid PVC, becomes more difficult as the degree of polymerization increases. Table 4.17 shows the useful K values for individual processes. Details of

4.1. Thermoplastics: Raw Materials and Molding Compounds

Table 4.16. K values vs. Viscosity Numbers (DIN 7746, Part 1, 1986)

K values	Viscosity numbers (cm^3/g)	K values	Viscosity numbers (cm^3/g)	K values	Viscosity numbers (cm^3/g)	K values	Viscosity numbers (cm^3/g)
45	49.5	59	85.5	73	136	87	203
46	51.6	60	88.5	74	140	88	214
47	53.6	61	91.7	75	145	89	220
48	56.1	62	95.0	76	149	90	227*
49	58.4	63	98.3	77	154	91	233
50	60.9	64	101	78	169	92	240
51	63.3	65	105	79	164	93	247
52	65.8	68	103	80	168	94	254
53	68.4	67	112	81	174	95	261
54	71.1	68	116	82	179	96	269
55	73.6	69	120	83	185	97	276
56	76.6	70	124	84	190	98	284
57	79.5	71	128	85	196	99	292
58	82.4	72	132	86	202	100	301

Table 4.17. Areas of Application of PVC Types

PVC Types	Rigid PVC			Plasticized PVC		
	E	S	M	E	S	M
Process	K values			K values		
Calender (melt-roll machines, platen press)	(60–65)	57–65	57–65	70–80	65–70	70
Thermally compensated films	78	–	–	–	–	–
Flooring	–	–	–	65–80	65–80	
Extruder processing of PVC-U						
Pressure pipes	–	67–68	67–68			
Profiles	65–70	68–70	57–68			
Sheets and flat films	60–65	60	60			
Blown films	60	57–60	57–60			
Extruder processing of PVC-P						
General				65–70	65–70	65–70
Cable compounds,				–	65–80	65–70
preferable:				–	70	70
Blow molding	–	57–60	57–60		65–80	60–65
Injection molding	–	55–60	56–60	–	65–70	55–60
Paste technology	–	–	–	65–80	(70–80)	

4.1.4.3. Homopolymers, copolymers and blends

processing VC homo- and co-polymers of lower values (40–50) in solution are given in Sec. 4.1.1.4; for dispersions in Sec. 4.1.1.10.

4.1.4.3. Homopolymers, copolymers and blends

1. For ISO designations and ASTM specifications for VC homo- and co-polymers see Tables 4.18 and 4.19.
   Poly(vinyl chloride acetate), PVCA (ASTM) or VC/VAC (ISO), with 5 to 20% VAC content, is the most widely used copolymer. It is somewhat less resistant to distortion by heat, but it is easier to process than PVC at lower temperatures into injection-molded or deep-drawn (glass-clear) articles for office and drawing equipment and for thermoforming sheet and compression molded gramophone records. Copolymers with propylene (US) or vinyl ethers (JP) have also been considered for this purpose. For a graft polymer with 50% EVA see Sec. 4.1.4.6.
   PVC-C, the chlorine content of which is raised to 60% by post-chlorination (TempRite, Rhenoflex), is more difficult to process than PVC. It can, however, stand temperatures up to 100°C (PVC-HT) and is, for example, even more resistant to chlorine. It is used in hot-water piping and chemical equipment.
   The homopolymers *PVDC*, made from vinylidene chloride ($CH_2 = CCl_2$), decompose below their melting point. VDC-containing copolymers (Diofan, Ixan, Saran, Vilit, Viclan) are widely used as a barrier film or coating against water-vapor and aromas in food packaging. The coatings are made from dispersions (Sec. 4.1.1.10) or lacquer solutions (for cellophane). Weather-proof and corrosion- and tear-resistant filaments and tapes are additional applications of VC/VDC copolymers. For ABS/PVC alloys see Sec. 4.1.3.1.2 and Table 4.14.

2. *Processing aids*: Paraloid, Supercryl (US); Degalan, Irgamod, Plexigum, Vinuran (DE); Kane Ace, Metablen (JP); Diakon, Vinapol (GB) and Modarez (FR), are mainly high molecular weight methacrylate (multi-)polymers, miscible with PVC and therefore available for glass-clear products (e.g. Acry-alloy V, JP). They improve the flow of the melts and thermoformability, so that weather-resistant shock-proof articles can be made under favorable conditions. SAN modifiers raise the heat resistance of the product up to VST > 90°C, S/MA-based modifiers (Cadon, 30 to 50 m/m %) up to 120°C. For styrenic multipolymer modifiers see also Sec. 4.1.3.1.2.

3. *Increased and high-impact PVC grades* (see note 3 to Table 4.23) with 5 to 12% impact modifiers are two-phase systems. Typical impact modifiers are

Table 4.18a. ISO 1060/1 – 82   Designation of Vinyl Polymer, Data Blocks (p. 423) corresponding to DIN 7746–86

Data block 1: Symbols, for copolymers PVC content, after a hyphen polymerization process S, E, M or X for processes other than above.
Symbols for copolymers e.g.
VCE Vinyl chloride-ethylene polymer
VCEVA Vinyl chloride-ethylene-vinyl-acetate polymer
VCOA Vinyl chloride-octylacrylate polymer
VCVA Vinyl chloride-vinyl acetate polymer
VCVDC Vinyl chloride-vinylidene-chloride polymer

Data Block 2: G general purpose, P paste resins

Data block 3: Viscosity number (see table 39), three figures,
– for G resins, the apparent bulk density (ISO 60), two figures
– for P resins, the amount of DOP required for a paste with a defined viscosity, indicated by

Classes	4	5	6	7	8	9
DOP phr	≤45	>45 to 55	>55 to 65	>65 to 75	>75 to 85	>85

Data block 4: No entry, indicated by two commas

Data block 5: Optional, for G resins:

X	1	2	3	4	5	6	7	8

Sieve analysis

(a) Retention on sieve of 250 μm mesh, % (ISO 1624)

ND[1]	≤0.5	>0.5-5	>5-20	>20-40	>40-60	>60-80	>80 to 99.5	>99.5

(b) Retention on sieve 63 μm mesh, % (ISO 1624)

ND[1]	≤0.5	>0.5-5	>5-20	>20-40	>40-60	>60-80	>80 to 99.5	>99.5

Plasticizer absorption at room temperature, p.h.r. (ISO 4608)

ND[1]	<10	10-20	>20-25	>25-30	>30-35	>35-40	>40	–

[1] ND = not designated.

For P resins: – Apparent viscosity of a paste (ISO 4612), expressed by single digit class numbers 0 to 9
– Rheological characteristics:
D = Dilatant (viscosity increasing with shear rate)
N = Newtonian (nearly horizontal curve)
P = Pseudoplastic (decreasing viscosity)
X = Unspecified

4.1.4.3. Homopolymers, copolymers and blends

Table 4.18b. ISO 1060/1–82 Designation of Vinyl Polymers – Examples

	PVC, produced by mass polymerization, general use		PVC for pastes, produced in micro-suspension		VCVA-cop., prepared in suspension, general use	
	Value	Indication	Value	Indication	Value	Indication
Viscosity number (ml/g)	116	116	125	125	83	083
Apparent bulk density (g/ml)	0.56	56	–	–	0.79	79
Quantity of DOP (phr) to prepare a paste of viscosity 25 Pa s	–	–	60	6 (class)	–	–
Retention on sieve 250 μm mesh (%)	0	1 (class)	–	–	0	1 (class)
Retention on sieve 63 μm mesh (%)	2	2 (class)	–	–	96.5	7 (class)
Plasticizer absorption at room temperature (phr)	38	6 (class)	–	–	ND[1]	X
Apparent viscosity, Severs	–	–	18	3 (class)	–	–
Rheological characteristics	–	–	1.5	N	–	–
Designation	ISO 1060–PVC–M, G, 116.56,, 126		ISO 1060–PVC–X, P, 125–6,, 3N		ISO 1060–VCVA 90–S, G, 083–79,, 17X	

[1] ND = not designated.

polyacrylate-elastomers (PAE), chlorinated polyethylene (PE-C) and ethylenevinyl acetate (EVAC) as soft elastic components.

PAE are used as graft copolymers with vinyl chloride with 6 to 50% acrylate content (e.g. Hostalit H, Vestolit Bau, Vinnol K, Vinuran SZ, Solvic) or as copolymers with 60 to 90% acrylate content (e.g. Paraloid, Bärodur, Durastrength, Kane ACE).

PAE is blended with S-, M- or E-PVC to typical impact modifier concentrations of 5 to 7% or, in certain cases, above.

After blending with additives such as stabilizers, lubricants, fillers, etc., PVC-HI grades are processed mainly on single or twin screw extruders, calenders or injection molding machines. Depending on the type of plasticizer, the elastomers form island (PAE) or cross-linked structures (PE-C, EVAC) in the matrix during processing. They act as buffers and absorb part of the energy on mechanical stress resulting in increased impact strength.

4.1. Thermoplastics: Raw Materials and Molding Compounds

Table 4.19. ASTM Standard Specifications for Vinyl Resins

D1755 Polyvinyl chloride resins

Types: General purpose (GP), dispersion resins (D).

Grades: Designation order and extreme limits of 9 cells[1]:
(1) Inherent viscosity[2], GP $<0.7->1.4$, D[3] $<0.9-1.6$; (2) sieve (75 μm) analysis 0-100%; (3) apparent (bulk) density $<144->809$ g/1000 cm^3; (4) plasticizer absorption $<50->225$ phr DOP; (5) dry flow (only cells 2–8) $<2.0->10$ s/400 cm^3; (6) conductivity of water extract (only cells 3, 4) $<6->6$ εs/cm g.

D2474 Vinyl chloride copolymer resins

Types designated by a C and the cell numbers.
(1) 1: vinyl acetate, 2: vinylidene chloride, 3: maleic ester, 4: acrylonitrile, 9: polycomponent systems.
(2) Concentration of comonomer (9 cells) $<5-45\%$ by weight of resin followed after a dash (-) by

Grades (9 cells[1]): (3) inherent viscosity^2 $<0.4->1.1$, (4) bulk density <144 to >800 g/l, (5) spec. grav. $<1.3-1.429$, (6) spec. grav. $1.430->1.508$.

---

[1] The cell limits may be extended by half into the next higher or lower cell, indicated by a dash above or below the cell number.
[2] 10^2-viscosity number.
[3] Further only: (2) Brookfield viscosity $<24->190$ poise, (3) Severs viscosity $<49->1499$ poise of plastisols containing 500 g PVC and 300 g DOP.

In contrast to PAE, PE-C and EVC impact modifiers are somewhat shear sensitive, i.e. the achievable strength depends on processing conditions.

Styrene-modified polyacrylates can also be used to obtain transparent high impact moldings (e.g. Metablen, Vestolit).

All the PVC-HI mentioned above have long-term weather-resistance in outdoor use M/ABS are used as impact modifiers for crystal clear PVC-HI packaging and MBS for blow moldings. Their butadiene content makes them less suitable for outdoor use. PVC compounds with increasing levels of certain ABS grades provide translucent or opaque products with optimum combinations of flame retardance, impact strength and heat resistance in terms of cost and performance.

(PVC + MMA) (Acryalloy V, JP) is crystal clear. (PVC + NBR) (Elaster, JP) is a TPE.

4.1.4.4. Stabilizer systems and fillers

VC polymers cannot be processed in the absence of stabilizers. The discoloration and progressive deterioration as a result of oxidation and elimination of HCl during processing at high temperature, as well as by the action of heat and light in use, make stabilizers necessary.

Trade names of stabilizers reveal very little about their composition (e.g., Advastab, Cyastab, Estebex, Hostastab, Interstab, Irgastab, Mark, Meister, Naftovin, Nuostab, Polyfix, Sicostab, Synpron, Therm-Chek, Thermolite). To heat-stabilize E-PVC, small amounts of decomposable organic bases (e.g., diphenyl thiourea, phenol-indole) are adequate. Aminocrotonic acid ester and epoxidized soya-bean or castor oil in combination with calcium or zinc soaps are moderately effective in all PVC polymers. Certain di-n-octyl tin compounds provide good light stability. These stabilizers are physiologically inert and are permitted for use together with specific lubricants in the food industry.

The most widely used basic lead compounds (e.g., Dyphos, Tribase) in rigid and plasticized products are valuable in electrical applications and for potable water piping although they are not permitted for the latter in all countries. They are not suitable, however, for transparent articles. Lead phosphite and barium-cadmium stabilizers are used to meet the most demanding requirements in heat, light and weathering resistance as well as good retention of color, e.g. for window profiles and other outdoor applications.

Where the toxicity of cadmium compounds precludes their use, particularly in transparent articles, they have been replaced by barium-zinc and (e.g. for mineral water bottles) calcium-zinc stabilizer systems with diketone co-stabilizer. The stabilizer effect is synergistically enhanced through co-stabilizers such as the chelator complex-forming organic phosphites, epoxy compounds, antioxidant polynuclear phenols (e.g., bisphenol A) and benzo-triazole UV absorber (e.g., Cyasorb, Tinuvin).

Ready-to-use compounds containing the lubricants necessary particularly for the processing of rigid PVC are available as single component additive systems for the majority of processes and applications of rigid and plasticized PVC. They often also contain titanium dioxide, which must be used in the highly purified rutile form in combination with the stabilizing system. This is also required on dielectric grounds in insulation for communication cables.

Fillers such as certain types of finely ground calcium carbonate facilitate extrusion and can in proportions of up to 50%, increase the impact strength of the product by a factor of 2 by "sliding" the PVC molecule on the surface of the filler.

4.1.4.5. PVC-U compounds

Table 4.20 provides an overview of rigid PVC processing. Tables 4.21 and 4.22 give ISO and ASTM standards for PVC molding and extrusion compounds. Production of rigid PVC film on *calenders* (Sec. 3.7.3) is a large-scale operation. PVC powder stored in silos is pneumatically fed to largely automated weighing and premixing stations. The mixtures are then continuously

Table 4.20 Processing of Rigid PVC

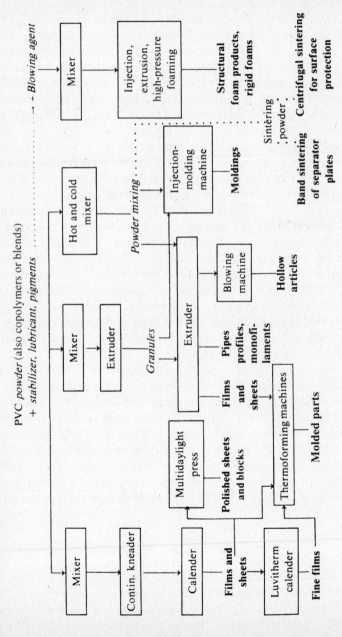

4.1.4.5. PVC-U compounds

Table 4.21. ISO/DIN Designations for Compounds of Vinyl Chloride Homo- and Co-polymers

For intended applications, method of processing, additives, and other information in data block 2 see Sec. 6.2.3

Data
PVC-U   Data block 3: Designatory properties

Vicat soft. temp. (VST/B)		Charpy impact notched			E-Modulus in tension			Density[1]	
Code	°C	Codes		kJ/m²	Codes		N/mm²	Code	g/cm³
060	<61	1[1]	03[2]	≤ 5	1[1]	18[2]	1500 to 2000	5	>1.30 to 1.35
062	>61 to 63	2	10	>5 to 20	2	23	>2000 to 25000	6	>1.35 to 1.40
:	:	3	30	>20	3	28	>2500 to 3000	7	>1.40 to 1.50
118	>117 to 119				4	33	>3000	8	>1.50 to 1.60
120	>119							9	>1.60

[1] ISO 1163/1 – 1980
[2] ISO 1163/1, DIN 7748/1 1985

Table 4.21. ISO/DIN Designations for Compounds of Vinyl Chloride Homo- and Co-polymers (continued)

PVC-P   Data block 3: Designatory properties

Shore hardness A or D[1]		Density		Torsional stiffness temperature at 309 MPa (TST 309)[3]		Tensile stress at 100% elongation	
Code	Values	Codes	g/cm³	Code	°C	Codes	N/mm²
A or D 40	≤ 41	1[2]	≤ 1.17	1[2]	< −5 to −15	1[2]	≤ 7
A or D 42	>41 to 43	20	>1.7 to 1.22	2	< −15 to −25	2	>7 to 12
⋮	⋮	⋮	⋮	3	< −25 to −35	⋮	⋮
				4	< −35 to −45		
A or D 88	>87 to 89	8	>1.52 to 1.57	5	< −45 to −55	6	>22 to 27
A or D 90	>89 to 91	9	>1.57	6	< −55	30	>27

[1] The shore D scale is used when the shore A values exceed 90.
[2] ISO 2898/1 – 1980, TST in the second, density in the fourth order.
[3] ISO 2898/1 – 1986, DIN 7749/1 – 1985.

4.1.4.5. PVC-U compounds

Table 4.22. ASTM Standard Specification for Vinyl Molding and Extrusion Compounds

D1784 Rigid vinyl compounds[1]		
Classification by designation order numbers. Number of cells in parentheses, maximum values of lowest and highest cells.		
1. (5) 1 PVC 2 PVC–C 3 E/VC 4 P/VC 5 VA/VC 6 alkylvinylether/VC		
(3) 1 PVC 2 PVC–C 3 copolymers		
2. Impact resistance	(6) 34.7-800 J/m	0.65-15 ft-lb/in
	(4) 40-267 J/m	*0.65-5.00 ft-lb/in*
3. Tensile strength	(5) 34.5-55.7 MPa	5-8 (10^3 psi)
	(5) 34.5-55.7 MPa	*5-8 (10^3 psi)*
4. Tensile modulus	(6) 1930-3034 MPa	250-440 (10^3 psi)
	(7) 1930-3860 Mpa	*280-560 (10^3 psi)*
5. Defl. temp. (1.82 MPa)	(8) 55-110°C	131-230°F
	(5) 80-110°C	*140-230°F*
6. *Hydrostatic design basis* (100,000 h)	*(5) 13.8-34.5 MPa*	*2000-5000 psi*

Suffix designation to 1784: A to D for (declining) chemical resistance.
[1] *Italics refer to D 3915 PVC and related plastics pipe and fittings compounds*

D2287 Non-rigid vinyl compounds
Types: PVC, PVCAC followed by the respective cell number: 1 durometer hardness, A scale (9) 45–94; 2 spec. grav. (9) 1.20–1.59; 3 tensile strength (9) 6.9–24.1 MPa (1000–3500 psi); 4 volatile loss 100°C (8) 1.0–7.0%; 5 brittleness temp. (9) −5 to −60°C; for electrical grades: E volume resistivity (3) 10^{11}–10^{15} ohm cm; X burning rate (1) extent of burning < 75 mm, time of burning < 10 s according to D 635.

plasticized in screw compounders (Sec. 3.1.4.1. Rigid PVC-film calenders, in F or L configuration, are particularly heavily built. Crystal clear and opaque films made of specially stabilized S-PVC or M-PVC, 0.02 to about 1 mm thick, but mainly 0.1 to 0.2 mm thick, are produced by processing with the calender rolls at 160 to 210°C.

In the Luvitherm process E-PVC films with a high K value are calendered at temperatures falling from 175 to 145°C. The product is taken over a roll at 240°C and then stretched in two directions to a thickness of 0.02 to 0.2 mm. Calendered films, about 0.5 mm thick, are intermediate products for stress-free blocks and sheets, which are laminated by pressing several layers between metal sheets in a multi-daylight press.

Dry blends or granules may provide the feed for *extruder* processing of rigid PVC. Single screw extruders used for dry blends require a screw length of at least 20 D and compression of 1:2.5 to 1:4. For granules, 15 D and 1:1.5 to 1:2.5 are adequate. For the planetary roller extruder see Fig. 3.48.

PVC pipes and deep-draw sheets up to 50 mm thick and profiles of high-impact PVC are made from powder feed with high efficiency twin screw extruders (Sec. 3.6.4). PVC melts are viscous, shear and temperature-sensitive and adhere to the processing contact surfaces less than other plastics.

Compounds containing S-PVC with low K values (Table 4.17, p. 214) and appropriate processing aids (Sec. 4.1.4.3.2) are suitable for *blow-molding* of bottles and for *injection molding* (Table 4.39), for example of tubular fittings. Special LF (low flow), mostly high impact LMW (low molecular weight) grades and alloys, e.g. with ABS, have been developed for injection molding of large general purpose and engineering components. Antistatic and V-0 grades are available.

In general, processing of rigid PVC is slower than that of most other thermoplastics. The configuration of the screw, the head and the die to avoid "dead spots" is extremely important, and to cope with the possible evolution of hydrochloric acid, corrosion-resistant provision is necessary. Burned material acts as a catalyst for further decomposition, so that when it occurs, the machine must generally be taken apart for cleaning.

Properties

Data for the properties of plasticizer-free PVC products are assembled in Table 4.23 and isochronous stress-strain curves are shown in Fig. 7.10.

PVC is physiologically inert. The utility of rigid articles for the food sector and their resistance to light and weathering depends on their stabilization (Sec. 4.1.4.4). Most unplasticized PVC products are flame resistant even without flame retardants (B1 of DIN 4102 to UL 94V-0). PVC is resistant to salt solutions, dilute or concentrated alkali solutions, and most dilute or concentrated acids including oleum, sulfuric acid, mixed acids and concentrated nitric acid. Gaseous chlorine forms a protective layer of chlorinated material, but liquid halogens are aggressive. Several of the other vinyl chloride polymers are attacked by concentrated acids including glacial acetic acid. They are all resistant to alcohols, aliphatic hydrocarbons, mineral oils, fatty oils and fats. Esters, ketones, chlorinated hydrocarbons and aromatic hydrocarbons partially swell or dissolve. Tetrahydrofuran and cyclohexanone are solvents for PVC.

Applications

Many applications of PVC-U are extruded pipes, profiles, sheets and films. PVC is widely used for pipework and chemical equipment (there are also explosion-proof types with volume resistivity of $< 10^9$ ohm cm, e.g. Trovidur E, DE); general services and sewage installations; window frames or translucent glazing profiles and panels with improved impact strength; office equip-

4.1.4.5. PVC-U compounds

Table 4.23. General Properties of Polyvinylchloride Molding and Extrusion Compounds according to ISO/DIN

Properties[1]	Unit SI	Unit U.S.	PVC-U[2] E, S and M types	PVC-C[4]	Modified PVC-U[3] Elevated impact a_k 5-10	Modified PVC-U[3] High impact a_k 30-50	DOP plasticized PVC-P 80/20	DOP plasticized PVC-P 60/40
Mechanical								
Specific gravity	g/cm³	lb/ft³	1.39	1.55	1.38	1.37	1.28	1.19
Tensile strength	N/mm²	10³lb/in²	50-65 86.9 7.1-9.2	75 96.9 10.7	45-50 86.3 6.4-7.1	35-48 85.6 5.0-6.8	25-28 80.0 3.6-4.0	16-18 74.4 1.1-1.3
Elongation, break	%		20-50	10-15	20-70	30-100	170-200	370-400
Tensile modulus	kN/mm²	10³lb/in²	~3000	3500	~2400 ~3.4	~2100 ~3.0		
Flexural stress, yield	N/mm²	10³lb/in²	70-110 10-14 ~4.3	125 17.8	75-80 ~11	60-80 8.5-11	22 3.1	3 0.43
Ball indentation hardness (10s)	N/mm²	10³lb/in²	110-130 15.6-18.5	155 22	~100 ~14	75-100 11-14		
Shore Durometer hardness (test)			D 83-D 84	>20	D 81	D 75-D 80	D 60/A 95	D 30/A 78
Charpy impact, unnotched	kJ/m²		n.b.	~2	n.b.	n.b.	n.b.	n.b.
Charpy impact, notched, +20 °C	kJ/m²		2.5		5-10	30-50	3-4	n.b.
Charpy impact, notched, −20 °C	kJ/m²		2-3		3-7	4-10	2-3	n.b.
Izod, notched, ASTM D 256	J/m	ft·lb/in	34-85		107-160 2-3	534-1282 10-24		
Thermal								
Vicat softening temp. B (50 N)	°C	°F	70-90 158-194	110 230	~80 ~176	~75 ~167	~40 ~104	
Linear thermal expansion	10⁻⁵ K⁻¹	10⁻⁵ °F⁻¹	7-8 4.4-5	6 3.3	8 4.5	8 4.5	15 8.3	21 12
Thermal conductivity	W/mK	BTU·in/h·ft²·°F	0.16 1.11	0.14 0.97	0.14 0.97	0.14 0.97	0.13 0.90	0.13 0.90
Electrical								
Volume resistivity	ohm cm		10¹⁵−>10¹⁶	>10¹⁵	10¹⁵	10¹⁵	≤10¹⁵	~10¹¹
Surface resistance	ohm		~10¹³				~10¹²	10¹¹
Dielectric strength (1 mm)	kV/mm		20-50				32-34⁵	24-26
Dielectric constant (50 to 10⁶ Hz)			3.2-3.7 to 2.9-3.2				4.2 to 3.2	8.0 to 4.0
Dissipation factor (50 to 10⁶ Hz)			0.011 to 0.015, pigmented 0.02 to 0.03				0.06 to 0.03	0.08 to 0.12
Tracking resistance (method)			KA 2-3b, KB, KC 300->600					
Water absorption	mg/4 d		E: 14-18; S,M: 3-4	2	10-20		E: 25-30; S: ~5	E: 40-60; S: 5-10

[1] Unfilled compression molded test specimens. Fillers and other additives can raise the specific gravity to 1.5 g/cm³ and influence other properties.
[2] Extrusion compounds, e.g. for pipes. Generally, the mechanical and electrical property values of S-PVC and M-PVC are slightly better than those of E-PVC.
[3] PVC-U compounds have been classified on the basis of Charpy impact strength, a_k, into the following grades: a_k <5 kJ/m² normal impact; a_k 5-20 kJ/m², elevated impact, and a_k >20 kJ/m², high impact.
[4] Chlorine content about 64 %.
[5] Electrical grades 40 to 50 kV/mm.

ment; packaging. For details of such products s. Sec. 4.2.1 and 4.2.4, for blown bottles for oil, wine, weakly carbonated beverages Sec. 4.3.6.7. The newer LF-LMW injection grades have found applications in the sanitary and computer sectors, for casings of all kinds, outdoor furniture and facade claddings.

4.1.4.6. PVC primary and secondary plasticizers

Table 4.24 gives a survey of common PVC *primary plasticizers*, their symbols and ranges of application. For information on other plasticizers and their symbols see the standards named in the first remark.

1:1 PVC/EVA graft polymer (Vinnolit K 550) enables soft films (Shore A ca. 80) to be produced without plasticizer. Similar rubbery elastifiers (Sec. 3.1.3.6 and 4.1.4.3) in PVC-P compounds diminish plasticizer demand and volatility. *Extenders or secondary plasticizers* are relatively cheap materials that gel with PVC without themselves providing adequate plasticization. They include fatty acid esters, alkylated aromatic hydrocarbons or naphthenes, which simultaneously serve to improve the flow properties of pastes. Chlorinated paraffins are flame retarding. Monomeric glycol methacrylate is used as an additive to pastes. It reduces their viscosity but polymerizes in the presence of an added catalyst on gelation to the end product and increases its hardness.

Most plasticizers are non-toxic, but only selected ones are permitted for use in food or medical applications.

4.1.4.7. PVC-P compounds

Table 4.25 summarizes the processing of VC polymers with plasticizers and additives; Tables 4.21 and 4.22 give ISO and ASTM standards for PVC-P molding and extrusion compounds. The trade names are for the most part those used for PVC in general (Sec. 4.1.4).

Delivery

Pellets of about 3 mm dia are generally used for injection molding, blow molding and extrusion. Ready-for-processing free-flowing agglomerates with stabilizers and lubricants are also used. Powder mixes are available for fluidized bed and electrostatic coating.

Composition and properties

PVC-P compounds are commercially available with different plasticizers in ratios of 80:20 to 50:50. Small amounts of plasticizer insufficient for homogeneous gelation, produce brittleness. Obversely, with increasing plasticizer

Table 4.24 PVC Primary Plasticizers

Group No.	Name	Abbreviation[1]	Characterization
1.	Phthalate plasticizers		
	Diisoheptyl phthalate	DIHP	Special plasticizers for plastisols
	Di-2-ethylhexyl phthalate[2]	DOP	Standard plasticizers for PVC, high gelling capacities, low volatile, balanced heat, cold, water resistance and electrical properties.
	Diisooctyl phthalate	DIOP	
	Diisononyl phthalate	DINP	The plasticizing effect; volatility, low-temperature resistance decrease from DINP to DITDP; heat resistance, electrical values increase. High processing temperature (Bisphenol-A additive), special plasticizers for cable compounds.
	Diisodecyl phthalate	DIDP	
	Diisotridecyl phthalate	DITDP	
	$C_7 - C_9$ phthalates of	no standard symbols exist for mixtures	Mixed alcohol esters, compared with DOP: lower viscosity (for pastes), better cold resistance and water resistance, less volatile (important for synthetic leather, floor coverings)
	$C_9 - C_{11}$ predominantly linear alcohols		
	$C_6 - C_{10}$ n-alkylphthalates		
	$C_8 - C_{10}$ alkyl phthalates		
	Dicyclohexyl phthalate	DCHP	Limited application, resistant to fuel extraction, good gelling, for foam pastes, spread floor coverings
	Benzylbutyl phthalate	BBP	
2.	Adipic, azelaic and sebacic esters		
	Di-2-ethylhexyl azelate	DIOA, DOA	DOA outstanding low-temperature resistant, plasticizing, lightfast, more volatile and more water sensitive that DOP.
	Diisononyl adipate	DINA	DINA-DIDA: less low-temperature resistant and less volatile than DOA
	Diisodecyl adipate	DIDA	
	Di-2-ethylhexyl azelate	DOZ	Less water sensitive than adipates, similar to DOS.
	Di-2-ethylhexyl sebacate	DOS	Best low-temperature resistance, less volatile

Table 4.24 PVC Primary Plasticizers (cont.)

Group No.	Name	Abbreviation[1]	Characterization
3.	Phosphate plasticizers		
	Tricresyl phosphate	TCF[3]	Flame resistant, heat resistant, for heavy duty mechanical and electrical articles, not suitable for low temperature and food applications
	Tri-2-ethylhexyl phosphate	TOF (DPOF)	Flame retardant, light resistant, less heat resistant than TCF,
	Diphenyl alkyl mixed phosphate		Similar to TOF, fuel resistant
4.	Alkyl sulfonic acid phenyl ester	ASE	Similar to DOP, less volatile than phthalates, tendency to discoloration, weather resistant
5.	Acetyl tributyl citrate	–	Similar to DOP, for food applications
6.	Tri-2-ethyl hexyl trimellitate	TOTM	Low volatile, thermally highly resistant,
	Triiso-octyl trimellitate	TIOTM	high price (cable compounds)
7.	Epoxidized fatty acids esters		Butyl-, octyl-, epoxy stearate low-temperature resistant, very low volatility, synergetically stabilizing with Ca-Zn stabilizers
	Epoxidized linseed oil	ELO	ELO and ESO primarily for improving heat stability, resistant to extraction
	Epoxidized soya oil	ESO	
8.	Polyester plasticizers	–	Polyester (from propane, butane, pentane and hexane) diols with dicarboxylic acids of groups 1 and 2. Non-volatile, less temperature dependent, highly resistant to extraction and migration.
	Oligomer plasticizers		Viscosity < 1000 mPa s, also mixed with monomer plasticizers, for pastes
	Polymer plasticizers		Viscosity up to 300 000 mPa s, suitable for extruding and calendering

[1] ISO/DIS 1043/3 · 1985, DIN E 7723 – 1987, corresponding to ASTM D 1600 – 83
[2] For 2-ethylhexyl 0=octyl is generally used
[3] P may be used in place of F for phosphate, however TCP is a registered trade mark in the U.K.

4.1.4.7. PVC-P compounds

Table 4.25 Processing Plasticized PVC

80-50 parts PVC powder
20-50 parts plasticizer
(extender, fillers) stabilizers, pigments

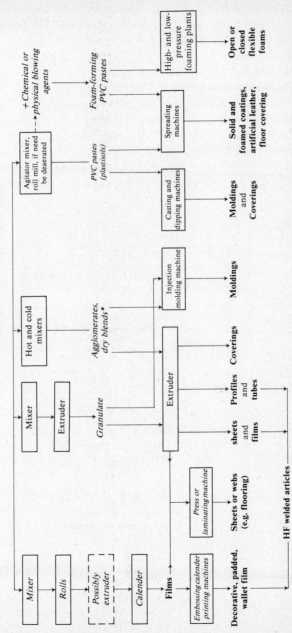

* Special types for powder processing (p. 61).

4.1. Thermoplastics: Raw Materials and Molding Compounds

content, at room temperature, relatively high tensile strength is obtained compared to corresponding rubber grades while extension at break of the products rises to values of up to 400% (Fig. 4.5, Table 4.23).

The Shore-Durometer hardness to ASTM D 2240/ISO 8608 (Table 4.26) is a significant value for PVC-P. Shore A values and those for the harder varieties, normally Shore D values, are not exactly mutually convertible. Shore hardnesses of A 96 to A 60 are used for extrusion compounds. A 85 to A 65 are blow molding compounds, whereas injection molding compounds are used down to A 50. Heavily filled PVC-P compounds, e.g., for flooring and cable sheathing with Shore hardnesses of A 85 to A 70, have reduced elongation at break and are less flexible at low temperatures than those that are unfilled. The temperature characteristics of PVC-P depend to a large extent on the choice of

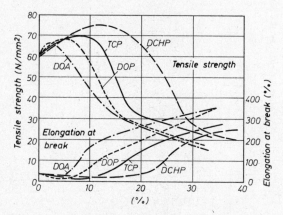

Fig. 4.5. Effect of concentration of plasticizer. Tensile strength and elongation at break at 23°C are given in relation to the type and quantity of plasticizers. DOP = Dioctyl phthalate, DOA = Dioctyl adipate, DCHP = Dicyclohexyl phthalate, TCP = Tricresyl phosphate.

Table 4.26. Shore Durometer Hardness A and D for PVC-P

Shore Durometer		General characteristics	Low temp. brittleness range*	
A	D		°C	°F
98–91	Approx. 60–40	Semi-rigid	0 to −20	32 to −4
90–81	39–31	Sole leather-like	−10 to −30	14 to −22
80–71	Not used	Taut rubbery	−10 to −45	14 to −26
70–61	Not used	Medium rubbery	−30 to −50	−22 to −58
60	Not used	Very soft injection-molded parts	−40 to −50	−40 to −58

* Measured by bend brittle point drop hammer tester, similar to ASTM D 746.

plasticizer. For flexibility at low temperatures (Table 4.26) special plasticizers are required. Increasing stiffness becomes more important, practically, about 20 K (36°F) above brittle temperature. Plastic deformation of PVC-P under load becomes critical between 40 and 60°C (104–140°F) and with no load 80°C may be regarded as an extreme service temperature for most types of PVC-P in regard to mechanical behavior and plasticizer loss. For higher temperature resistant compounds (e.g. cable compounds), which are less suited to low temperatures, trimellitic polyester plasticizers should be considered. Compounds adequate at extremely low and elevated temperatures are "internally plasticized" PVC/EVA, PVC/TPU (Sec. 4.1.8.4) and PVC/NBR (Sec. 5.1.8) alloys. Such compounds have good electrical properties; the TPU- and NBR-grades are resistant to oils, fats and fuels.

Monomeric plasticizers are soluble in many organic solvents and can do damage at the interface with other organic substances by migration. The hazard of migration should be considered when compounding for products to be used in combination with thermoplastics, rubbers or lacquers, nitrolacquers in particular.

There are regulations in all countries limiting the use of PVC-P in the food industry. When used for long periods, some plasticized PVC-P grades can harden owing to loss of plasticizer and "fogging" can be a problem for car interior applications. With exceptions, PVC-P compounds do not withstand prolonged contact with concentrated acids and alkalis. They are quite resistant to aqueous solutions of neutral salts and moderate concentrations of inorganic acids provided the plasticizer is not susceptible to saponification (e.g., many fatty acids). PVC-P plasticized with tricresyl phosphate (Table 4.24) and/or chlorine-containing extenders is flame retardant. PVC plasticized with purely organic materials continues to burn after ignition.

Processing and applications

PVC-P flows under relatively low pressure, but it must be processed at the highest temperature the compound will tolerate. With a lower temperature the products do not achieve maximum mechanical and electrical properties, shrink considerably and irregularly and have non-uniform mat surfaces. Fast cooling has the same effect. Machines and tools must be corrosion resistant.

Shrinkage in *injection molding* (see Table 3.6 and Table 3.9) depends largely on the configuration of the mold and the conditions of injection. The normal values of 2 to 4% in the machine direction and 1 to 2% across it can be considerably exceeded depending on the circumstances. Gaskets and protective sleeves, service knobs, suction pads, frames for protective and diving glasses, flexible toys and handle-bar grips are typical examples of plasticized PVC injection moldings. Electrical plugs are injected around movable conductors.

The footwear industry makes boots and sandals out of PVC, where necessary with inserts, on multistation injection-molding machines and injects in flexible soles and shoes. Oil resistant compounds are used for automotive parts.

Armrests for the vehicle industry are made by *blow molding*. Tubes and bottles are also blown from plasticized PVC. Dolls, toys and balls are blown from non-toxic compounds. *Paste casting* (Sec. 4.1.4.9) is preferred for the manufacture of larger hollow toys.

When extruding on machines with a single three-zoned screw with a length of $20\,D$ and compression of 1:2.6 to 1:2.8, compounds with a Shore hardness of A 60 to A 80 are processed at temperatures of 120 to 165°C. Harder mixes require 150 to 190°C.

Bowden cable, washing lines, fence wires and wood and steel window profiles are covered with plasticized PVC by extrusion.

Further details on PVC semi-finished goods, particularly extruded and calendered films, sheets, coated fabrics, floor coverings and foamed PVC-P are given in Sec. 4.1.4.10.

4.1.4.8. Special grades for insulation and coverings

The use of PVC-P insulating and cable sheathing compounds in accordance with relevant electrotechnical standards is a highly developed specialist area. The material is resistant to abrasion, moisture, oil and ozone. The general use of PVC-P in telecommunication is limited only by its characteristics at high frequencies. For heavy current circuits and cables PVC is used up to 10 kV. (For PE and PE-X in cable technology see Sec. 4.1.2.1.5.2 and 4.1.2.2).

Distributing frame PVC-P insulated wires, cross-linked by electron-beam irradiation, are prescribed by British Telecom. They withstand temperatures up to 150°C; they exhibit only a slightly reduced tensile strength and deformation stability under load. Other applications include vehicle wiring and flexible cords for domestic equipment.

The specific volume resistivity of plasticized PVC for cable coverings and antistatic special measures (as are, for example, required for mines) can be brought down to 5×10^4 ohm cm by the appropriate choice of plasticizer and the addition of carbon black.

4.1.4.9. PVC pastes (plastisols)

Starting materials, manufacture and properties

PVC pastes are, in the main, free-flowing, low viscosity dispersions of paste-making PVC powders (K values between 56 to 80, Tables 4.16/4.17) in plasticizers (p. 226). Up to 50% fine grain extenders (extender resins, mean particle size approx. 50 μm) are added to reduce viscosity and control the rheology. VC-VAC copolymers are used to obtain low gelling temperatures (> 160°C). The viscosity can be reduced with liquid extenders (p. 226) or viscosity depressants. Precipitated chalks or colloidal silicas are used to increase viscosity. The high processing temperatures necessitate the use of stabilizers. In chemical foaming, Zn-containing products act simultaneously as kickers (see p. 58).

PVC pastes contain 40 to 100 phr plasticizer and can be filled with up to 300 phr (fine chalk). Extremely high levels of colloidal silica or metal soaps result in kneadable plastigels. Pastes with high levels of thinners such as aliphatic hydrocarbons and glycols are called organosols.

The pastes are manufactured in high-speed mixers followed by filtration and degassing, particularly for dipping and casting pastes. Slow mixers followed by passes through water-cooled single or twin roll mills are used only in special cases. Heating above 35°C should be avoided as PVC pastes are temperature-sensitive and their shelf-life is reduced. Depending on the formulation, pastes gel between 150 and 220°C to melts with Shore hardnesses of A 50 to > A 90. The properties of these products resemble those of conventional plasticized PVC (see Sec. 4.1.4.7).

Processing and applications

PVC pastes can be processed by spreading, dipping, casting, injection molding or rotation screen printing. Spreading is used to apply foamed or solid coatings to diverse substrates including textiles, paper, glass and mineral non-woven and sheet metal to obtain synthetic leather, tarpaulins, floor and wall coverings and facades. In general coating is carried out in several linked stages.

Dipping is used to manufacture gloves and apply protective coatings to metal substrates.

Hollow parts, such as balls, dolls, etc., are usually manufactured by rotational casting (see Sec. 3.2.3). Spray plastisols are usually applied by the airless method to coat metals, e.g. vehicle underseals, etc. In rotary screen printing, PVC pastes are used to coat partially or fully paper, textiles, etc. to manufacture tarpaulins, synthetic leather and wall and floor coverings.

4.1.4.10. Foamable rigid and plasticized PVC

PVC-U molding compounds containing blowing agent are commercially available and used for the extrusion of structurally foamed pipes and profiles (Sec. 3.6.5.3) and structural foam molding (Sec. 3.3.6.4). They are worked up into door frames, roller shutter boxes, window boards and furniture parts with densities of 0.7 to 0.9 g/cm^3.

4.1. Thermoplastics: Raw Materials and Molding Compounds

Plasticized PVC compounds, containing 0.5 part blowing agent, e.g., 100/70 PVC/DOP are blended with 20 to 30 parts nitrile rubber or EVA to make shoe soles and sealing profiles, buffers and stops ($d_R \sim 0.65\,g/m^3$, Shore A 35–40). For discussion of the *high-pressure process* for hard PVC and plasticized PVC foams see Sec. 3.1.5.1; for products, Sec. 4.2.4.3. High-pressure PVC foams have a closed cell system. Cross-linking with diisocyanate raises the heat resistance.

For foaming on carriers, pastes are designed to gel and foam in one process. This may be performed in the heating channel of the coating machine at 165 to 175°C.

Single and multi-layer foamed synthetic leather and self-supporting expanded films are manufactured by this method. In chemically embossed cushioned vinyl floor and wall coverings, the pre-gelled foam layer is imprinted at low temperatures (approx. 140°C) with an inhibitor (TMSA, benzotriazol) which locally inhibits expansion during subsequent foaming (190 to 220°C) (see Sec. 4.2.4.2.4). Structurally foamed vinyl wallpapers are rotary screen printed.

4.1.5. Fluoropolymers

The basic chemical structure of the linear molecules is shown in Table 4.28.

Products, delivery, trade names and specifications

1. *Polytetrafluoroethylene (PTFE)*

    Suspension polymer powder, free flowing and non-free flowing grades for compression molding, sintering and ram extrusion; some grades anti-static. Emulsion polymer powder for paste extrusion. Micro-powder for use as anti-friction additive.
    Compounds with inorganic (e.g. glass fiber, graphite, carbon, bronze, mica, metal oxides) and/or organic fillers (polyimide, PPS, PEEK, aromatic polyesters).
    Emulsion polymer dispersions for coating and impregnation.
    Halon, Teflon (US); Polyflon (JP); Hostaflon (DE); Algoflon (IT); Fluon (GB); Ftorlon (SU); Tarflen (PO).

2. *Tetrafluoroethylene-perfluoropropylene copolymer (FEP)*

    Coarse granules in grades of increasing melt viscosity for injection molding, extrusion, lining, fluidized bed powder.
    Teflon (US), Neoflon (JP); Ftorlon 4 MB (SU); Hostaflon FEP (DE).

3. *Perfluoro-alkoxy (PFA) and propyl vinyl ether (TFA) copolymer, ethylene propylene vinyl ether (EPE) and vinylidene fluoride (TFB) terpolymers*

Molding compounds, dispersions and coating powders.
Teflon PFA, EPE (US); Neoflon PFA (JP); Hostaflon TFA (DE).
Hostaflon TFB: m.p. 160 to 180°C.

4. *Ethylene-tetrafluoroethylene copolymer (ETFE)*

Granules for injection molding and extrusion, dispersions.
Granules with 25% GF reinforcement for injection molding.
Tefzel (US); Neoflon (JP); Hostaflon ET (DE); Ftorlon 40 (SU).

5. *Polyvinylidene fluoride (PVDF)*

Powder and granules for melt processing, particularly injection molding, pipe and shrink-tube extrusion and coating powder. Dispersion in dimethyl phthalate-diisobutyl ketone for stoving enamels and cast film.
Kynar (US); Kureha KF (JP); Vidar (DE); Foraflon (FR); Solef (BE); Dispersion grades: Dalvor (US); Vidar L (BE); Ftorlon 2 (SU).
Hexafluoroisobutylene-VDF copolymer (US), higher temperature and corrosion resistance than PVDF. Can be injection molded.

6. *Polyvinylfluoride (PVF)* $(CH_2-CHF)_n$

Raw material for cast films; Tedlar (US, Sec. 4.2.5).

7. *Polychlorotrifluoroethylene (PCTFE), plasticized and plasticizer-free injection molding and extrusion compounds*: Kel-F (US), Edifren (IT), Voltalef (FR). PCTFE-telomer plasticizer: Faifoil, Fluorolube, Kel-F oils and waxes.

8. *Ethylene-chlorotrifluoroethylene copolymer (E/CTFE)*

Granules and powders for injection molding, extrusion, rotational molding and fluidized-bed sintering: Halar (US).

9. *Cross-linkable fluoroelastomers* (see Sec. 5.1.15):

CTFE + VDF: Kel-F elastomer (US)
Hexafluoropropylene + VDF + TFE: Viton, Fluorel (US); DAI-el (JP).
TFE + perfluoromethylvinylether: Kalrez (US).
TFE + PP: Aflar (JP).
Highly fluorinated hydrocarbons: Tecnoflon (IT).
Fluorinated polyacrylate (Poly-FAB, US) and siloxanylene (Fasil, US).
Polyfluor oxyphosphacene: Eypel (US).

ISO designations for fluoroplastics do not yet exist. *For ASTM specifications* see Table 4.27.

4.1. Thermoplastics: Raw Materials and Molding Compounds

Table 4.27. ASTM Standards for Fluoroplastics

ASTM D	Fluoro-plastics	Types	Properties*
1457	PTFE	I General Purpose granular, III Ram extrusion with volatile processing aid (4 grades, 2 classes), IV, V Powder resins < 100 μm Ø, VI Pre-sintered resins	*resins*: bulk density particle size, water content (0,0.4%) 1:327°C±10, III, extrusion pressure *molded*: 2, 3, 4, thermal instability
2116	FEP	I Extrusion, II Injection, III, IV Low MF, improved crack-resistance I, III, IV: MF 372°C/5000 g: 12 to 2 II: MF 372°C/2160 g: 5 to 12	1, 2, 3, 4, 5, 6
3159	ETFE	I, II, III Low to high MF MF 297°C/49 N: 2 to 28	1, 2, 3, 4, 5, 6
3222	PVDF	I Emulsion polymer, m.p. 160°C II Suspension polymer, m.p. 168°C	1, 2, 3, 4 Shore D79
1430	PCTFE	I Homopolymer, m.p. 210 to 220°C II Modified, m.p. 200 to 210°C III Copolymer, m.p. 180 to 205°C	1, 2, 5, 6 zero strength at 250°C Class A to D: 175 to 2500 s
3275	E/CTFE	I, II, III Low to high MF MF 271.5°C/2160 g: 0.05 to 4.1	1, 2, 3, 4, 5, 6

*The standard values of: 1, melting point; 2, specific gravity; 3, tensile strength; 4, elongation at break; 5, dielectric constant; 6, dissipation factor, correspond to the respective values in Table 4.28.

General properties

Polymers of perfluorinated monomers are non-flammable, outstandingly weatherproof and physiologically inert in use. The upper and lower limits of their service temperatures are superior to those of practically all other plastics. They also exhibit extreme corrosion and solvent resistance. Their excellent electrical properties depend to some extent on their individual chemical structures.
PTFE has an optimal combination of technically applicable properties. It is barely wettable or adhesive. Of all solids PTFE has the lowest static and dynamic coefficients of friction. For cost reasons it is used more widely than the perfluorinated polymers FEP and PFA even though its extremely high melt viscosity (melting temperature of 327°C) makes it unsuitable for conventional

4.1.5. Fluoro Polymers

melt processing. Its compression strength at elevated service temperatures is poor. For bearing and sealing components PTFE is therefore compounded with 5 to 40 % v/v of ground E-glass fibers, graphite or bronze together with MoS_2. (Fluorocomp + RT/Duroid, Tetraloy US). These reduce thermal expansion and improve the thermal conductivity and abrasion resistance.
Extrusion and injection molding of FEP, PFA/TFA and EPE is difficult because of their high melting and processing temperatures (Table 4.28, p. 240). FEP tends to melt fracture. The main fields of application are extruded wire insulation, shrink tubes and shrunk-on roll coverings. PFA (processing temperature, 380–450°C) is processed into moldings and linings for chemical equipment that have a better mechanical stability up to 260°C than PTFE. It is rather like EPE in behavior. The processability of E/FTE copolymer with about 25% ethylene within a more workable temperature range is achieved at the expense of a reduction of about 100 K in the maximum service temperature. The stability under load of E/FTE, even at elevated temperatures, is very significantly improved by glass-fiber reinforcement. TFB is used mainly for non-flammable weather-resistant textile coatings.

Films cast from PTFE and PVDF and extruded from E/TFE are discussed on p. 311. FEP, PFA and TFA are used as welding agents for PTFE and PFA and also as melt adhesives for metals, glass and ceramics.
As a result of their polar molecular structure (similar to that of PVDC and PVC) PVDF and PVF do not have the extremely low dielectric loss factor of the non-polar polyfluorocarbons, which is so important in high-frequency applications. PVDF is a highly crystalline, stiff material with good dimensional stability up to 150°C used for rigid pipes and in plant construction. The material meets the most demanding specifications for use in high purity water lines and moldings for the semi-conductor industry or packaging for high purity chemicals. PVDF + PMMA compounds are relatively inexpensive and are highly suitable for outdoor use. Compounds for structural foams with up to 0.97 g/cm³ density provide another application that reduces material costs.
PVDF pressure tubes for a service range of −25 to 130°C still have a creep rupture strength of $\sigma_B/10,000 \text{ h} = 10 \text{ N/mm}^2$ at 125°C. They are welded to one another and to injection molded fittings of the same material. Equipment parts are deep drawn from PVDF sheets (laminated with stretched fabric or mats) at 180°C. Practically unbreakable bottles impermeable to gas and aromas are blown from PVDF. Metal assemblies are coated with PVDF powder electrostatically or by a fluidized bed. Blown or cast films are piezoelectric. PVDF, like CTFE, is highly resistant to irradiation. Chlorinated processible fluoropolymers, also with fluorocarbon-polymer plasticizers, are used in radiation-resistant equipment parts or highly impermeable membranes.

4.1. Thermoplastics: Raw Materials and Molding Compounds

Safety measures in processing

Above about 350°C (660°F) fluoropolymers decompose, even with short-period stressing, with the formation of aggressive and poisonous hydrogen fluoride. Machines and molds must be made of corrosion-resistant alloys, the working area must be well ventilated and smoking must be prohibited where the atmosphere may be polluted with polymer dust. The machine fittings must under no circumstances be cleaned by burning.

Compounds filled with metals must not be used in connection with foodstuffs because of the possibility of producing soluble fluorides. Light metals, their oxides and TiO_2 and SiO_2 in finely divided form react explosively with PVDF at processing temperatures. Boron and boron trioxide, components of alloys of tools or in glass-fiber reinforcement reduce the melt stability of PVDF.

Processing of PTFE

At 19°C PTFE undergoes a phase transition with a 1.2% increase in volume, which must be considered in view of tolerance requirements and machining of shapes at 23°C. The volume increase of 30% on heating through 20 K in the crystalline melting point range of PTFE at 327°C is accompanied by conversion to a clear gel-like mass and leads to a corresponding process-dependent, anisotropic shrinkage on cooling the shaped material.

PTFE preforms are pressed from suspension polymer powder at 20 to 30°C and sintered above 327°C in one of the following processes:

(a) Mold-free sintering: Simple preforms are compressed with 20 to 200 N/mm^2 at 20 to 30°C. Those with undercuts or voids are subject to all-around pressure (isostatic) with flexible molds. Finally they are heated, free-standing in an oven, according to a predetermined temperature program, sintered and slowly cooled. Such "free" sintered moldings ($d_R \sim 2.1$ g/cm^3) are not pore-free.

(b) Pressure sintering or sintering with after-pressure: sintering the shaped piece in the tool under pressure, or applying pressure subsequently on the hot molding in the tool that has been sintered without pressure and then cooling under pressure, or shaping by impact pressing on a pre-sintered preform. The processes give accurate pore-free moldings of the highest density and strength. Parts that have been formed somewhat below the melting point have a tendency to revert, and that tendency is used in designing lip seals that adhere on warming.

(c) Ram extrusion (powder extrusion) from rods and thick-walled tubes: the inflowing molding material is intermittently pressed into tablets by the back and forth motion of the piston in the initial section of the long,

4.1.5. Fluoro Polymers

cylindrical mold. In the next section of the mold it is heated to 380°C under counter-pressure generated by thermal expansion and wall friction and then emerges continuously as a fully sintered rod or tube.
(d) Films are peeled from a sintered cylindrical block. They can be improved by passage through mill rolls.

Thin-walled products of great length, in the main tubes up to 250 mm in diameter with wall thicknesses of 0.1 to 4 mm and cable coverings, are made by paste extrusion of emulsion polymer using a ram extruder. The product is milled to a doughy mass with 18 to 25% white spirit. In a tunnel oven following the extruder, the lubricant first volatilizes and the product is then sintered at 380°C. Threadsealing tapes, which must remain porous, are not sintered after extrusion but simply rolled and dried. Films are cast from PTFE dispersions. The latter are also used to impregnate asbestos and glass-fiber articles and moldings of graphite or porous metal, which are finally sintered or pressed at 380°C. Anti-friction coatings made from PTFE dispersions on metallic or ceramic surfaces, are not impermeable even when coupling agents and stoving are employed, and are therefore not suitable as anti-corrosion coatings.

PTFE is welded at 380 to 390°C at 0.2 to 0.3 N/mm^2 pressure. Welding is used for edge-welded bellows and for lining, etc. In welding thicker films or sheets an unsintered welding tape or a fluorocarbon film is inserted in between. PTFE is capable of adhesion after etching with alkali metal solutions and there is a specific adhesive for temperatures up to 130°C. Machining the materials requires sharp tools.

For semi-finished fluoropolymers see Sec. 4.2.5.

Applications of PTFE and related fluorocarbons

Chemical industry, laboratories: Equipment, apparatus, transit bottles (up to 20 l), piping, film linings and jackets, seals, mobile bellows, internal heat exchangers, Raschig-rings, porous filter media, packings made of fibrillated fiber.

Textile, paper and food industry and domestic applications: Thread guides, conveyor belts, anti-adhesive coverings and coating of rolls, welding equipment, baking tins, frying pans, clothes irons, dispersion sprays as dry lubricants.

Electrical engineering and power electronics: Wire and cable insulation particularly E/TE telephone wiring and non-metallic cable ducts within buildings, foamed FEP and PFA insulation for computer coaxial feeders, shrink insulation tube, flexible printed circuits, components for the semi-conductor industry.

Aerospace industry: Fuel- and lubricant-resistant seals, pressure tubes, gearing parts, anti-icing protection.

4.1. Thermoplastics: Raw Materials and Molding Compounds

Construction: Highly stressed compression and slide bearings with minimal friction to absorb building movement and the displacement of buildings. Solar energy equipment, coated glass fabric covers for tents and pneumatic structures, E/TFE films for greenhouses and recreation centers.

4.1.6. Methylmethacrylate Polymers

DIN 7745 (1986) Part 1: PMMA – Molding compounds – Classification and designation [11.6]
DIN 7745 (1989) Part 2: PMMA – Molding compounds – Manufacture of test specimens and determination of properties
ISO 8257-1 (1987); ISO 8257-2 (1990); ISO 1628-6 (1990)

4.1.6.1. MMA polymers

Methylmethacrylate (MMA), $(CH_2 = C \begin{smallmatrix} CH_3 \\ COOCH_3 \end{smallmatrix})$

a liquid which boils at 100°C, is the main property-determining monomer of reactive resins (p. 165), homopolymer (PMMA) and copolymer molding compounds (and blends) and cast or extruded semi-finisheds (acrylic sheet, p. 314). It is mainly the MMA content in amorphous, brittle and easily machined (by sawing, drilling, milling, grinding and polishing) polymer molding compounds which results in very high weathering resistance without stabilisation unless impaired by other components. Typical optical properties: crystal clear, colorless, high transparency, practically no absorption in visible spectrum; can be colored to a high degree of saturation, colorfast and weather-resistant, either transparent using soluble colorants or translucent to opaque with pigments; opacity to UV can be regulated by incorporating suitable UV absorbers.

The glass transition temperature T_g of pure PMMA is around 115°C.

It is chemically resistant to inorganic acids and alkalis up to moderate concentrations, aliphatic hydrocarbons, ligroin (aromatics-free), other mineral oils and turpentine; it has limited resistance to 30% ethyl and methyl alcohol, fuel mixtures, fats and is not resistant to many organic solvents.

Resistance to materials with which it is in contact such as seals, paints, rubber, etc., should be checked with the manufacturer as stress cracking can occur.

4.1.6.2. PMMA standard molding compounds

1. *Trade names:* Degalan, Lucryl, Plexiglas (DE); Altulite (FR); Diakon (GB); Vedril (IT); Acrypet, Delpet, Shinkolite, Sumipex (JP); Acrylite, Lucite, Oroglas (US).

2. Marking, synthesis, forms of delivery and processing

Processability and properties of standard molding compounds to DIN 7745/ ISO 8275 are varied either by adjusting the molecular weight M_w (see p. 44) between 1×10^5 and 2×10^5 g/mol or by copolymerization with up to 20% acrylate ($CH_2=C{<}^H_{COOR}$ R = acrylic residue) which provides internal plasticization.

Table 4.29 summarises the standard description data block (see p. 423) for PMMA standard molding compounds. Low VST and VN values and high MFI values characterize easy-flow injection molding compounds; high VST and low MFI values characterize extrusion compounds. The molding compounds are synthesized by discontinuous bulk or suspension polymerization and by a continuous method. They are supplied as uniform size granules or beads and in an easily demoldable grade.

All machines with standard screws are suitable for melt processing at cylinder temperatures between 200 to 250°C. Degassing screws remove the need for pre-drying. The filling hopper should be of stainless steel to prevent contamination. The duration and level of holding pressure affect the properties of the injection molding. High quality moldings require the tool not to be excessively cooled.

3. Properties

Table 4.30 shows the indicative values for a range of products covering the range of applications of PMMA standard molding compounds. Their optical properties and resistance are those of pure PMMA (see p. 240).

4. Applications

Automotive industry: colored vehicle rear lights, indicator lights, warning triangles, light conductor systems, tachometer covers, reflectors for vehicles and signs.

Table 4.29. Characteristic properties of PMMA molding compounds to ISO 8257-1

Vicat softening temperature (VST)		Melt index (MFI)		Viscosity number (VN) (to choice)	
Code	VST range	Code	MFI range	Code	VN range
076	≤ 80	005	≤ 1	43	≤ 48
084	>80 to 88	015	>1 to 2	53	>48 to 58
092	>88 to 96	030	>2 to 4	63	>58 to 68
100	>96 to 104	060	>4 to 8	73	>68 to 78
108	>104 to 112	120	>8 to 16	83	>78 to 88
116	>112	240	>16	93	>88

4.1. Thermoplastics: Raw Materials and Molding Compounds

Table 4.30. Indicative values for a product range of 7 standard (S1 to S7) and a heat resistant (hr) PMMA molding compound – values correspond to the Campus database

Properties	Units	S1	S2	S3	S4	S5	S6	S7	hr
* VST/B 50	°C	88	96	95	103	104	108	106	119
* MVI (230/3.8)	ml/10 min	25	13	2.3	5.6	1.3	3.0	0.8	3.0
* VN	ml/g	53	53	72	53	72	53	72	60
Density	g/cm^3	1.19	1.19	1.19	1.19	1.19	1.19	1.19	1.19
Mechanical properties									
Tensile strength (5 m/min)	MPa	49	55	74	61	76	70	78	72
Elongation at break (5 mm/min)	%	1.5	2	8	2.5	7	3	10	2.7
Tensile modulus (secant; 1 mm/min)	MPa	3 200	3 200	3 200	3 200	3 200	3 200	3 200	3 600
Izod impact strength									
+23°	kJ/m^2	14	14	16	14	16	14	16	12
−30°C	kJ/m^2	14	14	16	14	16	14	16	11
Izod notched impact strength									
+23°C	kJ/m^2	2	2	2	2	2.2	2	2.2	2
−30°C	kJ/m^2	2	2	2	2	2	2	2	1.8
Thermal properties									
Heat distortion temperature HDT/B	°C	90	95	95	100	100	98	98	106
Coeff. of thermal expansion	10^6/K	75	75	75	75	75	75	75	70
Electrical properties									
Dielectric constant (50 Hz)	–	3.8	3.7	3.7	3.7	3.7	3.6	3.6	3.5
Dissipation factor (50 Hz)	–	0.05	0.05	0.05	0.05	0.05	0.05	0.04	–
Volume resistivity	ohm cm	> 10^{15}	> 10^{15}	> 10^{15}	> 10^{15}	> 10^{15}	> 10^{15}	> 10^{15}	> 10^{15}
Tracking resistance	Ω	> 10^{13}	> 10^{13}	> 10^{13}	> 10^{13}	> 10^{13}	> 10^{13}	> 10^{13}	> 10^{13}
Optical properties									
Light transmission	%	92	92	92	92	92	92	92	92
Refractive index	–	1.491	1.491	1.491	1.491	1.491	1.491	1.491	1.509
Water absorption									
23°C (saturation)	%	1.7	1.8	1.9	1.9	2	2	2	2.2
23°C/50% rel. hum.	%	0.6	0.6	0.6	0.6	0.6	0.6	0.6	0.6

* Characteristic property to ISO 8257-1

Lighting industry: covers and lamp bodies.
Precision engineering: scales, measuring, drawing and writing instruments; plant engineering; lenses and spectacles. Injection molded video disc blanks with and without embossed information; optical data carriers. The purity and low double refraction of injection molded parts is advantageous in data storage technology.
Building: light domes, glazing, illuminated signs, profiles, handles.
Special purpose lacquers can be used to render acrylic components scratch-resistant (e.g. Acriplex), antistatic, water-spreading, reflecting (e.g. IR) and non-reflecting (matt finishes).

4.1.6.3. Copolymers and blends

1. *Heat resistant PMMA molding compounds:* Copolymerization of MMA with chain stiffening comonomers (e.g. maleic anhydride) enables the heat resistance (Vicat softening temperature approx. 120°C) to be raised above that of PMMA homopolymer. Stress crack resistance to alcohol (components of de-icing fluids for vehicles) is also improved. Weathering and light resistance and optical properties correspond to those of unmodified PMMA. Properties of Plexiglas hr S5 see Table 4.30. Typical applications include lenses for illuminated signals in the transport sector (vehicles, shipping, aviation and rail) and rear fog lights and police flashing units, covers for street and operating theater lighting and monitor lights which get very hot.

2. *Impact resistant PMMA molding compounds*
Trade names: Lucryl, Plexiglas (DE); Diakon (GB); Oroglas (USA).
The impact resistant molding compounds are two-phase systems (see p. 5) manufactured by suspension or emulsion polymerization. The toughening disperse phase is formed in the PMMA matrix by, for example, styrene-modified acrylic elastomers ($<30\%$, with matched refractive index). This composition results in impact resistant molding compounds which can be blended as required with PMMA and which are as weather resistant and crystal clear as pure PMMA. Their stress sensitivity is less, their cyclic resistance to hot water better. Other properties are listed in Table 4.31.
The compounds are extruded at cylinder temperatures of 210 to 230°C and injection molded at 190 to 250°C. High gloss moldings require high processing and die temperatures (60 to 80°C). Degassing cylinders eliminate, as in the case of PMMA (p. 241), predrying.
Typical applications: anti-weathering layer for construction profiles, e.g. for PVC window profiles (see p. 307) applied by coextrusion; potting or encapsulation of metal inserts; household appliances, drawing and writing instruments, sanitary components, special purpose light covers.

Table 4.31. Indicative values for a range of 4 impact modified PMMA molding compounds (Z1 to Z4) (arranged in order of increasing impact or notched impact strength) – values according to the Campus database

Properties	Unit	Z1	Z2	Z3	Z4
Density	g/cm^3	1.19	1.18	1.17	1.16
Mechanical properties					
Yield stress (50 mm/min)	MPa	78	75	65	45
Elongation at yield (50 mm/min)	%	–	5	5	5
Tensile strength (5 mm/min)	MPa	60	57	51	38
Elongation at break (5 mm/min)	%	8	17	25	70
Tensile modulus of elasticity (secant; 1 mm/min)	MPa	2900	2800	2400	1800
Izod impact strength					
+23°C	kJ/m^2	17	19	26	52
−30°C	kJ/m^2	14	15	16	19
Izod notched impact strength					
+23°C	kJ/m^2	2.2	2.6	3.8	6
−30°C	kJ/m^2	1.5	1.6	1.6	1.6
Ball indentation hardness H 961/30					
Thermal properties					
Vicat (VST/B 50)	°C	103	102	100	95
Heat distortion temperature HDT/B	°C	102	99	98	97
Coefficient of linear expansion	10^6/K	70	75	85	110
Electrical properties					
Dielectric constant (50 Hz)	–	3.7	3.7	2.9	2.9
Dissipation factor (50 Hz)	–	0.05	0.05	0.05	0.05
Volume resistivity	ohm cm	>10^{15}	>10^{15}	>10^{15}	>10^{15}
Surface resistance	Ω	10^{13}	10^{13}	10^{13}	10^{13}
Optical properties					
Transmittance	%	92	92	92	92
Refractive index	–	1.49	1.49	1.49	1.49
Water absorption 23°C (saturation)	%	2.2	2.1	2.0	1.9
23°C/50% rh	%	0.6	0.6	0.6	0.5

3. PMMA/ABS Blends
Trade names: Plexalloy (DE).
This type of blend is used increasingly in automotive components (housings, reflectors), plant engineering, and in the electrical industry for technical components. The material can be metallized, has good welding properties and superior weathering resistance and stiffness to ABS. Its Vicat softening temperature is 104°C.

4. MBS molding compounds
Trade names: Cryolite (DE).
MBS molding compounds are polymers of MMA, butadiene and styrene. They are impact resistant, transparent two-phase systems even at low temperatures. Because of the butadiene component, they are not weather resistant but are resistant to oils, fats and fuels. They are suitable for sterilisation by gamma rays.
They can be processed on conventional injection molding, extrusion and blow molding machines. Heat sealing, bonding, ultrasonic, heated mirror and rotational welding, hot embossing and screen printing are possible.
Typical applications: bottles for cosmetics, spray and cleaning products and fine chemicals; technical blow moldings, hose connectors and equipment components, packaging (particularly for medical applications), medical disposables and fittings.

5. Acrylonitrile copolymers
Homopolymeric polyacrylonitrile (PAN) decomposes above its softening temperature, $> 200°C$, and is therefore not melt processable. The acrylonitrile component of copolymers and terpolymers (p. 203) raises their temperature resistance and stiffness. The limited gas permeability of PAN accounts for the designation of the material known as "barrier plastic". This is a 70% nitrile-containing (graft) copolymer with methacrylate or styrene and elastifying components based on butadiene.

Trade names: Barex (US, CH), Cycle-Safe, Cycopac, Lopac (US); Soltan (BE).

Properties

The strength of the highly transparent rigid plastic (Shore D 90, VST/B 78–81°C) is doubled to $\sigma_B = 164 \, \text{N/mm}^2$ (23×10^3 psi) by biaxial stretching during processing. The impact strength is increased many-fold and the permeability to CO_2 and oxygen is further reduced.
PAN copolymers are resistant at room temperatures to moderately concentrated acids and alkalis and to most solvents. Methanol and ketones soften them, and dimethylformamide acts as a solvent. Insofar as PAN copolymers are permitted in the food trade, their applications have to conform to the

4.1. Thermoplastics: Raw Materials and Molding Compounds

requirement of < 10 mg/kg (10 ppm) of monomer in the manufactured product. This is to ensure that the acrylonitrile content of the food itself does not exceed 0.05 ppm.

Processing

Extrusion is performed at 160 to 190°C; injection molding at up to 210°C, 1500 bar; shrinkage is 0.5%. In the thermoelastic heat-forming range around 120°C the deep draw can extend to 30:1. The material must be kept dry. The melt is highly viscous. The design of the processing machines is as for PVC.

Applications

PAN copolymers are used to make unbreakable, transparent and aroma-impermeable packaging blow moldings for UV- and oxidation-sensitive goods and for packaging of carbonated soft drinks and beer, which can be stored for several months.

4.1.6.4. Polymethacrylimide (PMI) molding compounds

Trade names: Pleximid (DE); Kamax (US); PMMI-Resin (JP).
PMI is formally a copolymer of MMA and glutaramide but is manufactured by reacting PMMA with methylamine (MA) at high temperature and pressure.

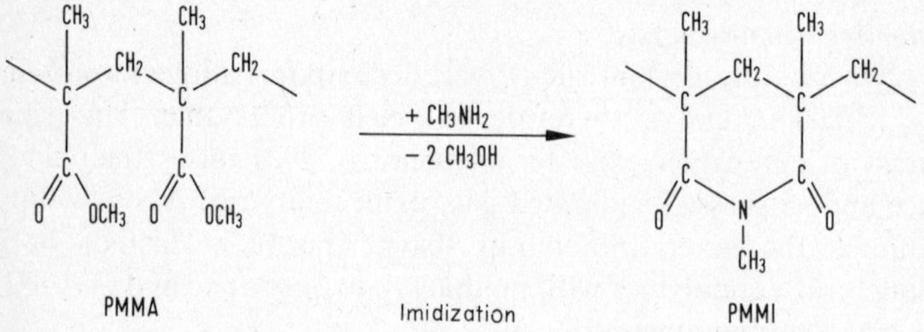

Fig. 4.6. Manufacture of PMMI from PMMA by reaction with methylamine.

Colorless and with high transparency, crystal clear and free of cloudiness, its properties closely resemble those of heat resistant PMMA. The ring termination provides increased chain stiffness and dimensional thermal stability.

Table 4.32 shows indicative properties; depending on the degree of imidization, it is possible to obtain all intermediate values of Vicat softening temperature up to that of PMMA at 175°C as well as high modulus of elasticity; for easily machinable carbon fiber reinforced products it can reach 25 000 N/mm^2. PMI exhibits barrier properties to oxygen; it is less susceptible to stress cracking caused by ethanol, ethanol/water and isoctane/toluene mixtures than PMMA or PMMA kw 55 (see Table 4.30).

Table 4.32. Indicative properties of a PMMI with high Vicat softening temperature (VST = 170°C)

Properties	Unit	Test standard	Value
Density	g/cm^3	ISO 1183	1.22
Fire performance	–	UL 94	HB
Water absorption 1d/23°C	%	DIN 53 495	0.45
Shrinkage	%	ISO 2577	0.3–0.5
Mechanical properties			
Tensile strength	MPa	ISO 527	90
Elongation at break	%	ISO 527	3
Modulus of elasticity	MPa	ISO 527	4350
Impact strength	kJ/m^2	ISO 180 1C	17
Notched impact strength	kJ/m^2	ISO 180 1A	2
Ball indentation hardness H 961/30	MPa	ISO 2039	230
Thermal properties			
Vicat VST/B 50	°C	ISO 306	170
Melt index MFI (260/21.6)	g/10 min	ISO 1133	2
Coefficient of linear expansion	K^{-1}	DIN 53 752 A	45×10^{-6}
Optical properties			
Refractive index	–	DIN 53 491	1.534
Transmittance τ_{D65}	%	DIN 5036 T3	88

Injection molding of PMI with VST around 170°C requires melt temperatures in the region of 200 to 310°C, mold temperatures of 120 to 150°C and predrying at 140°C.

Interest in this recently commercialized plastic has been shown by the automotive industry, e.g., for headlamp diffusers and by the lighting industry for street lighting covers. PMMI could also be used as a blend component and in fiber reinforced engineering materials.

For high dimensional stability PMI foam see Section 4.2.6.6. For linear polyimides see Section 4.1.11.

4.1.7. Polyacetals or Polyoxymethylenes (POM)

Acetal homopolymers and copolymers

Acetal homopolymers (POM 1, m.p. 170 to 180°C) are formed during the polymerization of formaldehyde:

$$x \;\; \begin{array}{c} H \\ | \\ C=O \\ | \\ H \end{array} \longrightarrow \left[\begin{array}{ccc} H & H & H \\ | & | & | \\ -C-O-C-O-C-O- \\ | & | & | \\ H & H & H \end{array} \right]_{x/3}$$

$x \sim 10\,000 \text{ bis } 30\,000$

4.1. Thermoplastics: Raw Materials and Molding Compounds

Because of the dense clustering of alternating molecular chains built of oxygen and methylene groups, they are highly crystalline and are amongst the stiffest and strongest unreinforced thermoplastics. The initial unstable hemi-acetyl terminal groups, –O–C–OH, are stabilized by esterification with acetic anhydride. Chemical attack by alkalis or water on these ester bonds leads through hydrolysis of the latter to progressive decomposition of the polymer chain. This decomposition can be slightly retarded by stabilization.

Acetal copolymers (POM 2, m.p. 160 to 170°C) are resistant to alkalis and even more resistant to hot water. This is achieved by a modified polymerization process, in which the main monomer is the cyclic trimer of formaldehyde (trioxan) and the "acetal" structure is interrupted by the stable –C–C– carbon bonds and the chain terminated by $HO-CH_2-CH_2-$ end groups. There is a slight reduction in the degree of crystallization and the concomitant mechanical strength and hardness (Table 4.32).

Homopolymers and copolymers are attacked by strong acids (pH < 4) and oxidizing agents. Neither are soluble in commonly used organic solvents, fuels or mineral oils and barely swell in them.

There are many grades, differentiated by their flow rates. ISO, in the draft proposal statement, proposes as a designatory property 7 cells for MF 190/2.16 from ≤ 4 to > 64, ASTM D 2133-81 gives for extrusion grades MF 190/2.16 < 4.1, for injection grades > 7.5. Other detailed requirements in this standard correspond to general values provided in Table 4.33.

The classification of glass-reinforced and filled POM molding and extrusion materials given by ASTM 2948 comprises cell numbers for

(1) Density, min.   9 cells: 1.45 to 1.70 kg/m^3
(1) Tensile strength, min.   9 cells: 48 to 103 MPa
                                 $(7 - 15 \times 10^3$ psi)
(3) Elongation at break, min.   4 cells: 1.0 to 8.0%
(4) Tensile modulus, min.   9 cells: 3400 to 900 MPa
                                 $(0.5 - 1.3 \times 10^6$ psi)
(5) Deflection temp., 1.82   9 cells: 116 to 160°C (240–320°F)
    MPa (264 psi)

ASTM D 4181-83 is a "calling out" specification according to D 4000 (Sec. 6.2.3) to facilitate the incorporation of future or other unspecified special material giving an enlarged cells, symbols and suffixes format.

Trade names:
Homopolymers: Delrin (US); Tenac (JP).
Copolymers: Celcon, Kematal (US); Duracon, Iupital, Tenac (JP); Kepital (Korea); Hostaform, Ultraform (DE); Sniatal (IT).

4.1.7. Polyacetals or Polyoxymethylenes (POM)

Table 4.33. Properties of POM Homopolymers and Copolymers

Properties	Unit		Homopolymers		Copolymers		
	SI	U.S.	Non reinforced		Non-reinforced		Reinforced[1]
Density	g/cm^3	lb/ft^3	1.42	88.8	1.41	88.1	1.56	97.6
Mechanical								
Tensile str., yield	N/mm^2	10^{3}lb/in^2	67–85	9.7–12	62–71	8.8–10	–	–
Elongation at yield	%		8–12		8–10		–	
Tensile str., break	N/mm^2	10^{3}lb/in^2	–	–	–	–	130	18.5
Elongation at break	%		15–70		20–75		2–3	
Tensile modulus	kN/mm^2	10^{5}lb/in^2	2.9–3.6	4.2–5.2	2.8–3.1	4.1–4.4	10	14
Charpy impact strength:								
unnotched, 23°C	kJ/m^2		N.B.		N.B.		30	
−40°C	kJ/m^2	–	–		80–N.B.		–	
notched, 23°C	kJ/m^2		3.5		4–9		6	
−40°C	kJ/m^2		3.0		6–7		–	
Izod impact, notched								
74° F(23°C)	J/m	ft-lb/in	70–125	1.3–2.3	54–87	1.0–1.6	60	1.1
−40°F (−40°C)	J/m	ft-lb/in	54–98	1.0–1.8	44–67	0.8–1.2	–	–
Ball indentation hardness, 358/30	N/mm^2	10^{3}lb/in^2	–	–	150–160	21–23	185^2	26
Rockwell hardness	Scale		M 90-M 94		M 78-M 80		(R 120)	
Thermal								
Melting temperature	°C	°F	175–181	347–357	165–175	329–347	166	331
Deflection temperature								
A 1.85 MPa (264 psi)	°C	°F	124	255	110	230	161	322
B 0.46 MPa (66 psi)	°C	°F	170	338	160	320	166	331
Linear thermal expan.	10^{-5}·K^{-1}	10^{-5}·°F^{-1}	9	5	11	6	3	2
Thermal conductivity	W/mK	$\frac{BTU \times in}{hr \times ft^2 \times °F}$	0.23	1.6	0.31	2.2	0.41	2.9
Electrical, all: Volume resistivity > 10^{15} ohm cm; dielectric strength, 50–60 kV/mm resp. 400–500 V/mil; dielectric constant, 3.8; dissipation factor, 0.002 to 0.005 at 50 to 10^6 Hz; 50% RH; track resistance, method KC 600.								

[1] Approximately 26% GF, chemically coupled. [2] H 961/30.

High impact POM elastomer alloy: Delrin T (US); Ultraform N2640 X (DE). Reinforced compounds: see pp. 290, 408.

Forms of Delivery
Natural opaque or pigmented granules. Basic grades for extrusion and extrusion blowing (MFI 190/2: 2.5-1), for injection molding (MFI approx. 9), these also modified with MoS$_2$, other mineral additives, PTFE, PE, special chalk or silicone oil to improve dry sliding and wear characteristics, reinforced with 10 to 40% glass fiber (MFI: 4-3), anisotropically reinforced with minerals or glass spheres (MFI: approx. 10). Easy-flow grades for precision thin-wall injection moldings (MFI: 13-50), high impact types modified by tailored micro-multiphase alloying with elastomers containing 50% TPU (p. 262) to specific stiffness/toughness ratios for moldings with high extensibility and high energy absorption. Also antistatic, electrically conducting and UV-protected formulations.

4.1. Thermoplastics: Raw Materials and Molding Compounds

Processing

For injection molding, see Table 3.9. For the formation of good crystalline and surface structure the molds (or polishing stack in the case of extrusion) should be heated to between 60 and 130°C. Process shrinkage decreases with mold temperature from more than 3% to about 1% while after-shrinkage decreases. The latter can be obviated by annealing at 110 to 140°C; solid profiles require treatment at the higher temperature. Shrinkage of glass fiber reinforced material is direction-dependent and between 0.2 to 0.8%.

Parts can be bent and edged in the crystalline temperature range. They can be welded (by a heating element, by friction or ultrasound), but they cannot be high-strength bonded with adhesives. Nails or rivets can be used. To finish an article by lacquering or metallizing in vacuo the surface must be treated with acid etching agents. There are special grades for electroplating (Sec. 3.9.3.4). For coloring in thermodiffusion printing see Sec. 3.9.3.3.

Properties

POMs are engineering materials of considerable strength and dynamic load-bearing capacity over an extended temperature range. These characteristics are improved by reinforcement with glass fiber (Table 4.33). With a glass transition temperature of about $-60°C$, they retain impact resistance down to $-40°C$. Characteristic properties to DIN 16781/88 are summarised in Table 4.34.

Because of the surface hardness and low coefficients of friction (against steel, static: 0.3 to 0.2, dynamic: 0.25 to 0.15), POMs are extraordinarily wear resistant. They are not prone to stress cracking. The upper temperature limit under

Table 4.34 Characteristic Properties of POM Molding Compounds to DIN 16 781/88

Melt Volume Index to DIN 53 735[1]		Yield Stress/Tensile Strength[2] to DIN 53 455		Modulus of Elasticity to DIN 53 457	
Designation	Melt Volume Index MVI 190/2.16 cm³/ (10 min)	Designation	Yield Stress/ Tensile Strength MPa	Designation	Elastic Moduls GPa
00	up to 3.5	05	up to 60	02	up to 2.5
02	over 3.5 to 6	07	over 60 to 75	03	over 2.5 to 4
06	over 6 to 9	08	over 75 to 90	05	over 4 to 6
12	over 9 to 14	10	over 90 to 110	07	over 6 to 9
24	over 14 to 30	12	over 110 to 130	11	over 9 to 12
48	over 30 to 50	14	over 130 to 150	14	over 12 to 16
96	over 50	16	over 150	20	over 16

[1] ISO/DIS 9988 states, similarly grouped, 17 to 20% higher ranges of MFI 120/2.15.
[2] If the yield stress σ_S is not measurable, the tensile strength σ_B is quoted.

load in air or hot water of homopolymers is 80 to 85°C and of copolymers above 100°C. Permeability to gases, vapors and organics is low.
UV and high energy radiation damage POM. They burn with a weak blue flame and drip. They are non-toxic and certain grades are permitted for use with foodstuffs. The good insulation and dielectric properties are only slightly temperature and frequency dependent. Because of their low dissipation factor, POMs cannot be HF welded.

Applications
POM injection moldings have largely substituted metallic precision parts. Applications are in close-tolerance, dimensionally stable components for wristwatches, keyboards, controls and counting mechanisms in office and measuring equipment, electronics and precision engineering. In the "outsert" injection molding process, preholed blanks up to $500 \times 250\,\text{mm}^2$ are filled simultaneously with up to 120 POM functional parts positioned to within 0.05 mm. The POM outsert technique saves up to 75% in production costs in, for example, time-switches, dictating machines, car radio-cassettes and video-recorders. The same aim lies behind efforts to integrate several functions in a single component. The springy POM copolymer is very suitable for snap closures and clips used for pipe assemblies, exterior and interior automotive claddings, etc. Friction bearings can be run without lubrication up to high pv values. The small difference between the static and dynamic friction coefficients of POM results in low starting torque. Gear wheels and other transmission components, program control elements, tank gauge and carburettor components, pump components in contact with hot water or fuel, kettles with steam shut-offs, mixing faucets, shower heads, valves and fittings are typical appliances in the general, automotive, domestic appliance and sanitary sectors. Other applications include hooks, screws, hinges, lock parts, aerosol containers, gas lighter tanks, fruit machine mechanisms, sports equipment, office equipment, etc.

Elastomer alloys with up to tenfold increased impact resistance and higher abrasion resistance are used, for example, for chain wheels subject to impact, housing parts with spring snap connections, and film hinges, fasteners in motor vehicles, ski bindings, and heavy-duty zip fasteners.

4.1.8. Polyamides (PA, Nylons) and Thermoplastic Polyurethanes (TPU)

The carboxylic acid amide group, $-\overset{\overset{\text{O}}{\|}}{\text{C}}-\underset{\underset{\text{H}}{|}}{\text{N}}-$ which occurs at regular intervals in the basic linear molecule of all PA homopolymers, determines the fundamental characteristics of the widely applicable PA engineering plastics. The carbamic acid ester unit, $-O-\overset{\overset{O}{\|}}{\underset{\underset{H}{|}}{C}}-N-$ of polyurethane (PU, see Table 4.68) is similar in its structural influence on linear (co-) polymer chain molecues (TPU, Sec. 4.1.8.4). The strongly polar character of these groups results in hydrogen bonding between neighboring molecules (Fig. 4.7), of which the toughness, rigidity and heat resistance of nylons is a consequence.

The presence of smooth aliphatic hydrocarbon segments between OCNH groups (see Sec. 4.1.8.1) makes them highly crystallizable. The OCNH groups determine the amount of water absorption of solid PA, increasing with the shortening of the hydrocarbon sequences between them. For the influence of "conditioning" on PA 6 and PA 66, see Table 4.37. Homopolymeric PA is resistant to almost all organic solvents, fuels, fats and mineral oils, dilute inorganic acids and alkalis up to concentrations of 20% NaOH and KOH, liquid ammonia and sulfur dioxide. Concentrated sulfuric acid, 90% formic acid, *m*-cresol and phenol are solvents.

4.1.8.1. Crystallizing PA molding compounds

These compounds are made

(A) From *one* monomer by ring-opening polymerization of a lactam. Caprolactam PA 6

$$n \begin{array}{c} CH_2-CH_2-CO \\ | \qquad \qquad \qquad \\ CH_2-CH_2-CH_2 \end{array} \!\!\!\! >\!\! NH \longrightarrow \{NH-(CH_2)_5-CO\}_n$$

Correspondingly $\{NH-(CH_2)_{11}-CO\}_n$, PA 12, is made from laurin-lactam $\{NH-(CH_2)_{10}-CO\}_n \rightarrow$ PA 11 from aminoundecanoic acid by poly-condensation with the elimination of water.

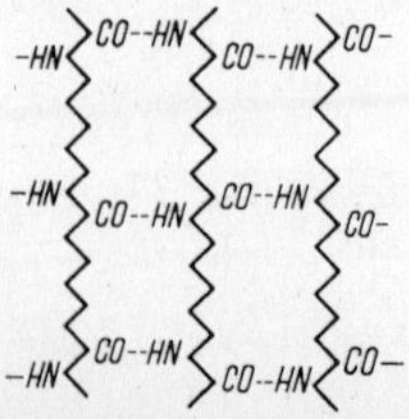

Fig. 4.7. Hydrogen bonding in PA 6.

4.1.8.1. Crystallizing PA molding compounds

(B) From diamines and dicarboxylic acids by polycondensation, e.g., PA 66 is made from hexamethylenediamine and adipic acid:

$$\{ NH-(CH_2)_6-NH-CO-(CH_2)_4-CO \}_{n/2}$$

PA 610 is made from hexamethylenediamine and sebacic acid:

$$\{ NH-(CH_2)_6-NH-CO(CH_2)_8-CO \}_{n/2}$$

There are uniform international symbols stating all the carbon atoms (including the CO) between the NH groups; in case B, according to ISO without, but sometimes written with a separation between the number of carbon atoms of the diamine and dicarboxylic acid (e.g. 6:6, 6/6 or 6–6 instead of 66). There is a potential conflict between this and ISO in which, for example PA 6/12 refers to a copolymer of PA 6 and PA 12. Table 4.35 reviews the basic commercial PAs and their symbols according to DIN/ISO and ASTM.

The designatory properties according to *ISO 1874/1, DIN 16773 Part 1* in Data-Block 3 (Sec. 6.2.3) are

1. Viscosity numbers for PA 11 and PA 12: ≤ 110 to > 240 (Code 11 to 24), measured in *m*-cresol solution, for all other PA: ≤ 90 to > 340 (Code 09 to 34), measured in sulfuric acid solution.

The values correspond to ranges of average molecular weight between about 10^4 and $> 5 \times 10^5$. For their significance in practice see Table 4.36.

2. Tensile modulus in the ranges

$\leq 150 \text{ N/mm}^2$	mean value code 001
> 150 to 250 N/mm^2	mean value code 002
$> 20,000$ to $23,000 \text{ N/mm}^2$	mean value code 220
$> 23,000 \text{ N/mm}^2$	mean value code 250

The extraordinarily wide overall range of the elastic moduli encompasses values for plasticized materials or elastomers and highly reinforced compounds.

Data-Block 3 can be supplemented by N (nucleated) for rapid setting.

Type *designations* according to *ASTM D 789* or of group and classes according to *ASTM D 4066* are supplemented with a "grade" number, primarily indicating rising relative viscosity. For classes in D 4066 see Table 4.35:

1 general purpose – 2 heat stabilized – 3 (except groups 5 to 7) nucleated – followed by (in between the groups' different sequences) impact modified, weather stabilized, flexural modified or plasticized or combinations of such modifications.

As an example, PA 123 means PA 66, heat stabilized and (grade 3) with a minimum relative viscosity of 100 and the general properties required for that grade. Using the codes for reinforcements, special properties tables for reinforced and unreinforced compounds, suffixes for e.g. electrical and

4.1. Thermoplastics: Raw Materials and Molding Compounds

Table 4.35. Survey of polyamides according to DIN/ISO and ASTM

Symbol DIN/ISO [1]	Chemical Structure	ASTM D 789	ASTM D 4066	
	A, B, C: (semi-)crystalline	Type	Group	Classes
A)	$(-NH-(CH_2)_x-CO-)_n$			
PA 6	homopolymer based on ε-caprolactam	II	2	8
PA 11	homopolymer based on 11-amino undecanoic acid	–	3	6
PA 12	homopolymer based on ω-amino-dodecanoic acid or laurinlactam	–	4	4
B)	$(-NH-(CH_2)_x-NH-CO-(CH_2)_y-CO-)_{n/2}$			
PA 46	homopolycondensate based on butendiamine and adipic acid	–	–	–
PA 66	homopolycondensate based on hexamethylenediamine and adipic acid	I	1	9
PA 69	homopolycondensate based on hexamethylenediamine and azelaic acid	IX	5	2
PA 610	homopolycondensate base on hexamethylenediamine and sebacic acid	III	7	0[2]
PA 612	homopolycondensate base on hexamethylenediamine and 1,10-decane-dicarboxylic acid	VIII	6	3
C)	Semi-aromatic polyamide			
PA MXD 6	polyarylamide $\cdots[-\underset{H}{\overset{H}{N}}-\underset{H}{\overset{H}{C}}-\phenyl-\underset{H}{\overset{H}{C}}-\underset{H}{\overset{H}{N}}-\overset{O}{\overset{\|}{C}}-(CH_2)_4-\overset{O}{\overset{\|}{C}}-]_{\frac{n}{2}}\cdots$ homopolycondensate based on m-xylylenediamine and adipic acid	–	–	–
D)	Amorphous Polyamides (Examples)			
PA 6-3 T	homopolycondensate based on trimethylhexamethylene and terephthalic acid $(-NH-CH_2-\underset{\underset{CH_3}{\|}}{\overset{\overset{CH_3}{\|}}{C}}-CH_2-\underset{\underset{CH_3}{\|}}{CH}-CH_2-CH_2-NH-CO-\phenyl-CO-)_{\frac{n}{2}}$	–	–	–
PA 6I	homopolycondensate based on hexymethylenediamine and isophthalic acid $[-NH-(CH_2)_6-NH-CO-\phenyl-CO-]_{\frac{n}{2}}$	–	–	–

[1] The symbols can be supplemented by – P for plasticized, by – G for cast PAs; however, DIN/ISO designations are not applicable for PA-G.
[2] Group 7 nylons are presently used commercially only as reinforced materials.

4.1.8.1. Crystallizing PA molding compounds

flammability requirements, D 4066 renders possible full specification to ASTM D 4000 (Sec. 6.2.3).
The general properties of standardized nylons (Table 53) are of the same order for DIN/ISO and ASTM.

Trade names

Mainly for PA 6 and PA 66: Adell, Ashlene, Capron, Comco, Ertalon, Firestone 200/210, Fosta, Fosta Nylon, Interpact, Minlon, Nyocoa, Nypel, Schulamid, Texalon, Vecana, Vekton, Vydyne, Wellamid, Xylon, Zytel (US); Amilan, Leona, Novamid (JP); Durethan, Ultramid (DE); Nivionplast, Sniamid (IT); Orgamid, Technyl (FR); Akulon (NL); Fabenyl (BE); Beetle, Maranyl (GB); Silon (CS); Tarnamid (PL); Grilon (CH).

PA 46: Stanyl (NL), a semi-crystalline PA, m.p. 295°C, HDT A > 150°C, reinforced (30 to 46% GF) up to HDT 285°C. PA 11 and/or PA 12; Vestamid (DE); Rilsan (FR); Grilamid (CH), PAMXD 6: IXEF (BE).

Reinforced and filled compounds: "preconditioned" alloys: Bergamid, Polyloy, Resistan (DE); Beetle ACI (GB); Igopas (CH).
Delivery is as dry granules in airtight packing. PA 11 and PA 12 are also ground for powder technology.

Table 4.36. Areas of Application and Viscosity Numbers of Nylons according to DIN 16773.

Viscosity number designations		Applications	
PA 6 PA 66	All others		
11 to 18	11 to 16	Injection molding: Extrusion:	moldings with large flow path/wall thickness ratio monofilaments, thin-walled wire insulation
14 to 22	16 to 18	Injection molding: Extrusion:	standard moldings monofilaments, wire, bristles, cable sheating, flat films, extrusion coating, semi-finisheds
18 to 22	18 to 34	Injection molding: Extrusion:	thick-walled moldings flat and tubular film, cable and hose sheathing, tube, semi-finisheds
22 to 24	20 to 34	Extrusion: Blow molding:	tubular film, tube, hose sheathing, semi-finisheds hollow articles

4.1. Thermoplastics: Raw Materials and Molding Compounds

Special types

PA 6 and PA 66 absorb 9 to 10% of water at full saturation in water (compared with only 1.5% for PA 11 and PA 12). Dry PA 6 and PA 66 moldings are brittle and liable to stress cracking. To improve their impact strength for use in standard conditions 23/50, they must be conditioned to more than 2.5% water content (Table 4.37). The moisture equilibrium is reached only very slowly in air, therefore fluctuations in atmospheric humidity are of little consequence.

Dry impact resistant PA molding compounds contain 10–15% PE, which is coupled using ionomers (Sec. 4.1.2.4.4) or chemically grafting with maleic anhydride or acrylic acid.

High impact PAs are alloys with, usually finely dispersed, EP(D)M, SBR, or similar synthetic rubbers up to (notched) impact resistances "without fracture" at $-40°C$ through to flexible compounds (4.1.8.3). For interpenetrating networks with silicone (Rimplast, DE) see Sec. 5.2.

High-flow injection-molding compounds have been "nucleated" to accelerate the formation of fine crystallites in the mold.

Reinforced nylon compounds are worked into more or less anisotropic products according to the adjusted proportion of fibers and glass sphere or mineral reinforcements. They can exhibit very favorable combinations of high stiffness, shock resistance and dimensional stability up to about 150°C (Fig. 4.8 and 4.9). 5 to 10 mm long fiber reinforced compounds have been cut from pultruded (Sec. 3.4.3 and 4.8.8) strands (Verton, GB). The moldings show very high ductility as well as E-moduli up to 60% higher than those reinforced with

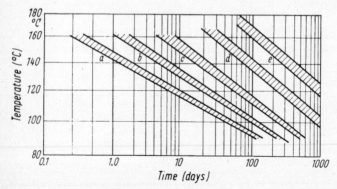

Fig. 4.8. Time-temperature limits for shock stressing of PA 66 moldings. For other stresses, e.g., electrical values, the temperature limits are in part significantly higher. (a) PA 66, (b) PA 66 stabilized, (c) PA 66 heat stabilized, (d) PA 66 reinforced with 25–30% glass fiber, (e) glass fiber reinforced PA 66, special heat stabilized. Temperature-time limits see also Table 7.5.

4.1.8.1. Crystallizing PA molding compounds

short fiber. 15% CF gives about the same reinforcement as 45% GF. For MXD 6 (Ixef, BE), supplied only in reinforced grades, see Tables 4.35, 4.37. The most widely used engineering plastics are PA 66 because of its very high heat resistance and stiffness and the more shock resistant (when dry and at low temperatures) PA 6. Water absorption decreases and toughness increases with increasing length of the (CH_2) segments up to PA 12. These characteristics together with good flow length/wall thickness ratio make such homo- or copolymer-based molding compounds suitable as engineering materials for impact resistant, elastic moldings which retain their dimensional stability even when subjected to moisture, heat and repeated flexural stress.

Semi-aromatic, partially crystalline or terephthalate-based copolymers (Ultramid T, m.p. 298°C, E 3.5 GPa) or poly-m-xylene adipamide products (Ixef, Nyref, BE; Reny, JP; Table 4.35) with 30 to 60% GF reinforcement are also available in high impact and flame retardant grades for electrotechnical applications. They are rigid (E_z 12 to 20 kN/mm^2), possess good heat resistance (F_{iso} A 220 to 230°C) and low creep even at elevated temperatures. Other characteristics which make them suitable as replacements for metals include water absorption after saturation of 3.5 to 1.9%, coefficient of thermal expansion 1.1 to 2×10^{-5} K^{-1} and good resistance to fuel and lubricants up to 120°C.

Additives include stabilizers against UV, thermal oxidation or hot-water hydrolysis and flame retardants not based on halogens or phosphorus up to UL 94VO/5V. Unreinforced PA 6 is low flammability without flame retardant. Molybdenum disulfide, graphite, coarse PTFE or PE-HD reduce friction in bearings and gears. Carbon black is used for semiconducting antistatic applications. High proportions of barium ferrite are used in magnetic materials. Apart from compounds with special additives most nylons are permitted for (boiling water resistant) sterilizable medical articles and for use with food and drugs. Some natural colorants can discolor nylon.

Processing

Crystallizable nylons have very low melt viscosities and sharp melting and setting temperatures. The solidification of the melt occurs suddenly with a volume contraction of from 4 to 7% for unfilled compounds. Stiffness, hardness and wear resistance increase with growing degree of crystallization. Slow cooling raises the crystalline proportion. Shock cooling reduces it and produces more transparent and more flexible moldings. Nylons should only be processed moisture-free after vacuum predrying.

In *injection molding*, difficulties are created by the sudden freezing of the melt with decreasing temperature, the tendency to form voids in thicker sections because of the marked contraction in volume and the oxidation sensitivity of the melt in contact with air. Heating the nozzle and enlarging the sprue and runners provide a remedy. A locking (anti-drip) nozzle is necessary because of

4.1. Thermoplastics: Raw Materials and Molding Compounds

the thinness of the melt. The highest possible degree of crystallization or stress-free molding of uniform structure and high surface hardness is achieved at the expense of efficiency, using high mold temperatures (> 100°C). Annealing the molding at 140 to 170°C works in the same way. Finished parts can be subsequently colored with special dyes in an aqueous bath at 70 to 95°C.

For extrusion a machine with a 20 to 24 D screw length is required with 1:25 to 1:4 compression (a three-zoned screw) and finely adjustable heating for temperatures of 240 to 300°C. The low viscosity of the melt makes high take-off speeds possible but requires effective cooling in the "Technoform" process (Sec. 3.6.5.2) for solid profiles with cooled pipe-shaping under pressure. Transparent cast films are made by the chill roll process (Sec. 3.6.5.7). For polyolefin-PA combined blown films and blow moldings see 4.1.8.2.

Ground PA 11 and PA 12 can be worked up by all the powder-processing techniques. The Nyrim process is used to manufacture high-strength reinforced cast PA *block copolymer* moldings made of caprolactam (Sec. 4.1.1.1.4). There are also special techniques for laurinlactam casting for PA 12, see Table 4.37. Polyamides can be welded, and they can be bonded together with special adhesives based on resorcinol, cresol, cyanoacrylate or two-component EP resins. They can be machined easily. Making parts from semi-finished material on automatic machines can be economical with longer runs. Certain grades can be metallized by galvanizing (Sec. 3.9.3.4).

Applications

Nylons, about 50% as reinforced grades, are the most widely used engineering plastics. The main fields of applications of moldings are those requiring impact strength and shock resistance, also dynamic loading, noise and vibration absorption, abrasion and wear resistance. The fuel- and mineral-oil-resistant industrial moldings possess good sliding properties without lubrication. Friction bearings, bearing cages, driving elements, rollers, transporting-chain slide guides, ships screws, housings, (hot-water) fittings, safety dowels for building are made of PA; C-fiber reinforced grades are used for rod connectors. Polyamides are used in electrotechnical applications as insulators in precision components, subject to high mechanical and temperature stresses, such as multipoint plugs and for wear-resistant wire and cable covering. In the automotive industry they are used in fuel-resistant moldings under the hood; the Nyrim process is of interest for body parts. Hollow articles are blown from PA 6 and PA 11/12 and rotation molded from lactam melts and PA 11/12 powders. Garden furniture, boat davits, motor vehicle parts, parts for piping system installations and food plants are coated with PA 11/12 using powder techniques. For pressure tubes and semifinished products see Sec. 4.2.7.

Alloys of PA with ABS (Elemid, Triax, US) and ASA (Ultramid/Luran S, DE), PPO (Noryl, US; Ultranyl, DE) and other (Bexloy, US), mostly reinforced, have been used for computer housings and for car body parts.

4.1.8.2. Amorphous polyamides

Trade names:

Trogamid, Durethan T40 (DE); Grilamid TR55, Grivory (CH); Zytel 330 (US). For Selar PA and RB (US).
Incorporating bulky segments such as aromatic dicarboxylic acids, branched aliphatic or alicyclic diamines in PA molecules instead of the smooth $(CH_2)_{x/y}$ segments results in semi-crystalline or amorphous hard and tough, crystal clear PA. Predominant two-sided linking of the CONH-groups to aromatic rings results in resistance to extreme temperatures (see aramides).
Amorphous PAs are supplied in unmodified, stabilised, reinforced and processing aid modified form. Blends with partially crystalline PA and impact modified amorphous PA (Grivory (CH); Bexloy C (US)) are also available.

Properties

Amorphous PAs exhibit the typical good characteristics of PA. Characteristic of these materials is the advantageous combination of hardness, rigidity and toughness. Water absorption is generally significantly lower than that of the partially crystalline PA 6 and PA 66 and they are consequently not discolored by beverages or ink. The mechanical properties are largely retained when water is absorbed; toughness is considerably improved. Amorphous PAs are dimensionally stable and low-warpage since they undergo hardly any after-shrinkage. They are heat resistant, impact resistant down to $-40°C$ and exhibit good creep characteristics practically up to their high glass transition temperature (Fig. 4.9, Table 4.38).

Processing

The wide softening range and high melt viscosity facilitate processing on conventional machines into injection moldings, sheet, profiles and film. Shut-off nozzles and non-return valves can be dispensed with in injection molding; processing temperatures are between 250 and 320°C. Total shrinkage is approx. 0.5%. Because of the high melt viscosity, extrusion and blow molding pose no problems though the material must be dry to avoid blisters.

Applications

Typical applications are filter catch pots for compressed air lines, water and diesel oil filters, mechanically stressed components in machinery and plant,

4.1. Thermoplastics: Raw Materials and Molding Compounds

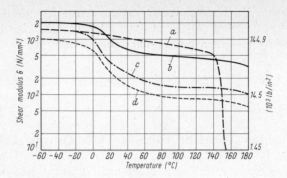

Fig. 4.9. Shear modulus-temperature curves for
(a) Amorphous polyamide: Trogamid T,
(b) PA 6, reinforced with glass fiber with 1.5% water,
(c) PA 6, injection molding grade with 2.5% water,
(d) PA 6, extrusion grade with 2.5% water.

low-warpage and impact resistant components for electrical engineering and electronics, transparent parts in the medical, laboratory and dairy sectors, flowmeters, spectacle frames, etc.

4.1.8.3. Flexible and thermoplastic elastomeric PA copolymers and PEBA

For copolymeric and copolymer-modified PA resins for melt adhesives and lacquers see Sec. 4.1.1.8.
The basic polymers described in Sec. 4.1.8.1. can be flexibilized to a limited degree by plasticizers of similar polarity. Some PA copolymers exhibit extensive elastomeric behavior, unaffected by low temperature, up to the limits of the Shore hardness range of soft rubber (Fig. 4.10).

1. *PA 12 copolymers*

Trade names: Vestamid PAE (DE), Rilsan F 25/15 (FR), Grilamid L (CH).
Properties see Table 4.39.

Applications: Pneumatic tubing, especially for air-brake lines in the automotive industry, straps previously made of leather, ski boots and bindings, cycle saddles, bearing runners and sealing gaskets.

2. *Polyether-block amides (PEBA)*

Trade names: Dynyl, Pebax (FR); Keltaflex (NL).
Polycondensation of a melt of polyetherdiols (Sec. 4.6.1.8.1) and PA intermediate products with carboxylic end groups or RIM of lactam and diol (Sec. 4.1.1.1.4) leads to block copolymers of

4.1.8.3. PA copolymers and PEBA

Table 4.38 Properties of amorphous polyamides

Properties	Standard	Units	Trogamid T5000	Trogamid T-GF35	Grilamid TR55	Durethan T40	Grivory G355NZ
Density	DIN 53479	g/cm^3	1.12	1.40	1.06	1.18	1.08
Mechanical properties							
Yield stress		N/mm^2	90	140	75	110	55
Elongation at yield		%	8	2,0	9	6	6
Elongation at track		%	>50		>50		35
Tensile modulus of elasticity	DIN 53457	kJ/m^2	2800	10000	2200	2800	1800
Impact strength 23°C	ISO 180/1A	kJ/m^2		20	55		
−30°C				16	45		
Notched impact strength 23°C	ISO 180/1C	kJ/m^2	12.0	5.4	2.0	10.0	20.0
−30°C			8.0	5.4	2.0	8.0	
Shore hardness D	DIN 53505	–	86	89	83		
Thermal properties							
Heat distortion temperature	DIN 53461						
Method A		°C	120	140	126	110	108
Method B		°C	140	150	148	118	124
Vicat softening temperature	DIN 53460						
ISO 75 VST A/50		°C	148	158	155		135
VST B/250		°C	142	151	150	125	129
Glass transition temperature	DSC	°C	150		159		135
Coefficient of linear expansion	DIN 53752	10^{-4} × m^{-1}	0.54	0.31	0.70	0.65	0.80
Electrical properties							
Volume resistivity	DIN/VDE 0303 T.2	Ohm cm	>10^{15}	10^{13}	10^{13}	10^{15}	10^{14}
Surface resistivity	DIN 53482	Ohm	>10^{15}	>10^{15}	10^{11}	10^{15}	10^{12}
Dielectric constant (50 Hz)	DIN 53483		4.2	4.5	3.0	4.8	3.0
Dissipation factor (50 Hz)	DIN 53483		0.021	0.016	0.008	0.040	0.005
Water absorption, saturation	DIN 53495	%w/w	7.7	5.1	3.5	8.0	7.0

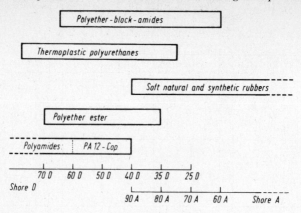

Fig. 4.10. Polymers at the Shore D/Shore A boundary (refer to Tables 46/108 Shore durometer hardness.

the general formula $\mathrm{HO}\left[\underset{\underset{O}{\|}}{C}-PA-\underset{\underset{O}{\|}}{C}-O-PE-O\right]_n \mathrm{OH}$

where PA = polyamide block and PE = polyether block.

PA:PE ratios of 80:20 to 20:80 in these polymers make possible wide variation of their elastic behavior which remains relatively constant between −40 and 80°C (Figs. 4.10, 4.11). The incorporation of different PA blocks (as shown in Table 4.35) makes it possible to vary melting points between 120 and 210°C. The use of various polydiols can give products with minimal water absorption or, for sweat-absorbing textiles, hydrophilic materials absorbing 110% water.

Applications

Bladders for sports balls, skiboots and sports shoes, bellows for drive shafts, windshield-wiper blades, pump membranes, catheters, noiseless flexible drives in precision engineering and coating of optical fibers.

4.1.8.4. Thermoplastic polyurethane elastomers (TPU)

Polyaddition of mixtures of diisocyanates, short-chain diamines or diols and long-chain polyesters or polyether diols results in block polymers consisting of flexible segments and rigid crystalline segments. Industrially this is performed by web casting or on reaction-screw equipment. Besides "pure" TPU, blends with other polymers (ABS, PC) and glass fiber reinforced compounds (R-TPU) are available.

This class of materials ranges from rubber-elastic to hard polyamide.

4.1.8.4. Thermoplastic polyurethane elastomers (TPU)

Table 4.39. Special Nylon Homopolymers and Copolymers

Properties	Unit		Amorphous polyamide			Flexibilized PA 12 copolymer				Polyether-block-amides			
	SI	U.S.	Trogamid T (PA 6-3-T)		Trogamid T + 35 % GF		Rilsan F 15 general purpose		Rilsan F 25 general purpose		Pebax 5533 SN general purpose		Pebax 2533 SN general purpose	
Examples:														
Density	g/cm³	lb/ft³	1.12	70	1.38	86	1.06	67	1.06	67	1.01	63	1.01	63
Water absorption, 20°C/65 % RH	%		2-3		~1		1		1		0.5		0.5	
Mechanical														
Tensile str., yield	N/mm²	10³ lb/in²	85	12	–	–	11	1.6	7	1.0	22	3.1	–	–
Elongation, yield	%		9.5		–	–	17		17		28		–	
Tensile str., break	N/mm²	10³ lb/in²	60	8.5	130	19	40	5.7	30	4.3	33	4.7	29	4.1
Elongation, break	%		70		3		350		280		510		680	
Flexural modulus	N/mm²	10³ lb/in²	2900	412	9700	1380	150	21	120	17	200	28	20	2.8
Torsion rigidity (Clash-Berg modulus) 23°C	N/mm²	10³ lb/in²	–	–	–	–	47	6.7	37	5.3	85	12	<15	<2.1
−25°C	N/mm²	10³ lb/in²	–	–	–	–	160	23	130	19	–	–	–	–
−40°C	N/mm²	10³ lb/in²	–	–	–	–	–	–	–	–	160	23	45	6.4
Charpy unnotched 23°C	kJ/m²		N.B.		–		N.B.		N.B.		N.B.		N.B.	
unnotched −40°C	kJ/m²		>60		30		N.B.		N.B.					
impact: notched 23°C	kJ/m²		13		25		N.B.		N.B.					
notched −40°C	kJ/m²		3-5		~7		76		80					
Hardness:														
Ball indentation	N/mm²	10³ lb/in²	125	18	150	21	–	–	–	–	–	–	–	–
Shore	D-scale		–		–		52		43		55		25 (70 A)	
Thermal														
Vicat soft. point: A, 9.81 N	°C	°F	–	–	–	–	145	293	125	257	156	312	63	145
B, 49.06 N	°C	°F	145	293	145	293	95	203	78	172	–	–	–	–
Defl. temp. 0.46 N/mm²	°C	°F	140	284	145	293	105	221	62	144	–	–	–	–
Defl. temp. under load 1.85 N/mm²	°C	°F	130	266	130	266	45	113	–	–	–	–	–	–
Linear thermal expan.	10⁻⁵ K⁻¹	10⁻⁵ °F⁻¹	8	4.5	7	4	12	7	12	7	23	13	21	17

4.1. Thermoplastics: Raw Materials and Molding Compounds

Trade names: Desmopan, Elastollan, Irogran, Luvoflex, (DE); Uceflex (BE); Avalon, Davathane, Europolymer (GB); Apilon, Laripur (IT); Pandex, Paraprene (JP); Estane, Estaloc, Pellethane, Ornaflex, Texin (US).

Delivery: Granules of various particle shapes, generally natural. Coloring is easily carried out with pigment masterbatches or pigment pastes during plasticizing (injection molding, extrusion).

Processing: Injection molding, extrusion of tube, cable sheath, blown and flat film. Special grades are used for calender coating of substrates. Predrying is sometimes necessary. Processing temperatures are between 180 and 240°C depending on grade and processing technique. Cooled or temperature-controlled molds (20 to 60°C) are necessary. Optimum material properties, in particular compression set (Rdv), require conditioning for 15 to 20 hours at 80 to 120°C.

Properties:

Density	1.10 to 1.26 g/cm^3 (R-TPU: 1.30 to 1.40 g/cm^3)
Shore hardness A	70 to 98
Shore hardness B	25 to 75
Ultimate tensile strength	25 to 60 N/mm^2
Elongation at break	300 to 700% (R-TPU: 10 to 50%)

Service temperatures up to 90°C for long-term exposure, 110°C short-term.
TPU exhibit good resistance to weathering, many solvents, alcohol-free fuels and lubricants. They are decomposed hydrolytically by acids, alkalis and hot water. Hydrolysis-resistant and translucent grades are also available. Hard grades possess good elastic strength, toughness, dynamic load bearing capacity and abrasion resistance.

Applications: Harder grades are used for silent-running drive components for transmitting small forces, bearing bushes, thrust rings, rollers and roller wear surfaces, automotive control and textile machinery components and impact-resistant car-body parts. Softer types are used for robust cable sheathing, sealing rings and sleeves, membranes and bellows, dust caps, fire-fighting hose linings and solid profiles for drive belts and seals.
The impact strength is exploited in hammers; weathering and wear resistance in sports shoes. Dropping containers are manufactured from film; bad weather and sports clothing and roofing materials are coated.
For TPU powders for plasticized PVC films (Baymod PU).
ABS + TPU or TPU + ABS alloys (Cycoloy, Desmopan, Estane, Pellethane) are produced by extrusion compounding the components. 2 to 10% of soft TPU improves the impact strength and abrasion resistance of ABS. 10 to 50% ABS blended with a soft TPU increases the latter's flexural modulus and load-carrying capacity and lowers the cost.

4.1.9. Linear (Semi-)Aromatic Polyesters

Polyesters with –C–O– ester links are, in most cases, produced as condensation polymers from diacids and diols or their derivatives. For unsaturated polyester cross-linking to give thermosets see Sec. 4.6.1.5. Polyesters containing only alkyl groups, $(-CH_2-)_x$, between the ester links are widely used intermediates, for example in polyurethanes (Sec. 4.6.1.8.1) or for lacquer resins. They are generally of low molecular weight. High molecular weight *polycaprolactones* $(-O-(CH_2)_5-CO-)_{\sim 300}$, made by ring-opening polymerization, are soft, partially crystalline materials. They are used as flexibilizers, blending agents or components for different plastics. They are biodegradable (Sec. 4.1.15.2). Polyesters containing benzene rings either in the diol (e.g., bisphenol A) or in the diacid (e.g., terephthalic acid) are basic polymers for engineering thermoplastics.

Polycarbonate (PC)

Polyterephtalate (PTP)

x: ethylene $(-CH_2-CH_2-)$;
butylene $(-CH_2-CH_2-CH_2-CH_2-)$ etc.

Bisphenol A | Terephtalic acid | Polyarylate
(components see Table 4.1, p.168)   (aromatic polyesters)

The aromatic rings stiffen the molecular chains thereby raising the deflection and melting temperatures. The more frequently they occur in the chain, the higher are these temperatures. Polyarylates have outstanding thermal stability (see Sec. 4.1.9.5). *Polyester carbonates*, mixed esters of bisphenols, carbonyl chloride and terephthalic acid, have glass transition temperatures of $> 195°C$. The semi-aromatic polyesters are not attacked or swelled by aliphatic hydrocarbons or ethanol. They absorb only minimal amounts of water and are physiologically inert. Because they have saponifiable ester groups in the chain they are attacked by alkalis and their resistance to oxidizing acids and continuous exposure to water or steam at temperatures above 70°C (160°F) is limited. PC is more resistant to strong acids but is less stable than PTP to benzene and methanol.

4.1.9.1. Polycarbonates (PC)

General trade names: Calibre, Lexan, (US), Makrolon (DE, US), Jupilon, Novarex, Panlite (JP), Sinvet (IT), Orgalan (FR), Xantar (NL)

Polyestercarbonates: Lexan PPC (US), Apec (DE)

General

Polycarbonate is made by the melt condensation process from bisphenol A and diphenyl carbonate, with the vacuum removal of the eliminated phenol or by passing phosgene (carbonyl chloride) through an aqueous alkaline solution of a bisphenol. The polymer product is taken up by the solvent or by passing phosgene into the pyridine solution of the bisphenol. PCs have molecular weights of 20,000 to 30,000 for injection molding compounds and up to >60,000 for extrusion.

Polymers for blow-molding compounds are given a high pseudoplasticity by incorporating chain branching. Polymers for cast electrofilms are made strongly flame retardant by the incorporation of brominated components. Thiodiphenol (Bisphenol S) increases the n_D at 1.62 by Abbe number 27; tetramethyl bisphenol A raises the heat resistance to almost 200°C and the hydrolysis stability in boiling water.

Designatory properties to *ISO 7391/DIN 7744*, Part 1, Block 3 (Sec. 6.2.3) are viscosity number (0.5 PC in CH_2Cl_2), MFI 300/1.2 (see Table 4.40) and impact strength, for reinforced materials Charpy unnotched (A 0 to A 9 $\sim$ < 10 to >90), for others Charpy notched (B 0 to B 9 $\sim$ < 8 to > 40) or Izod notched (ISO 180 C 0 to C 9 $\sim$ < 10 to >90), all in kJ/m^2.

ASTM D 3925 uses the three digit designation (ASTM D 4000, Sec. 6.2.3) of basic polymers Group 1 Polycarbonates, Classes 1 General, 2 Special Purpose – Group 2 Copolymers, Class 1 Special Purpose, each followed by a Grade description number whose first cell gives the Flow Rates (Table 4.40). Cells 2 to 6 give properties with dimensions as those as in Table 4.41. For products that cannot be specified in this way, there is the possibility of characterization by 8 cell limiting values in the designation arrangement:

PC materials:	unreinforced	reinforced
1. Tensile strength, min. MPa	40–110	40–180
2. Flexural modulus, min. MPa	1200–3300	2100–12,000
3. Izod impact strength, min. J/m	55–750	58–200
4. Deflection temp., 1.82 MPa, min., °C	120–155	125–160
5. Flow rate, g/10 min	3.0->24	—

Izod values in kJ/m^2 (ISO) and in J/m (ASTM) are not comparable (Sec. 7.6.4). For the significance of other designatory properties see Table 4.40.

4.1.9.1. Polycarbonates (PC)

Table 4.40. Approximate Classification of PC-Materials

Viscosity number ISO[1]	Melt flow codes ISO[2]	Rates grades ASTM	Applications
46	24	1.2	Easy flow injection
49	18	3.4	Injection molding
55	09	5.6	
61	05	7	Injection, Extrusion
67	03	8	Pipes, Profiles, Sheets, Blow moldings
70			

[1] Codes are mean cells values from ≤ 46 to > 70.
[2] Codes are mean cells values MFI > 24 to ≤ 3 g/(10 min.).

Forms of delivery

Glass-clear, transparent, translucent, or opaque-colored PC compounds are delivered as pellets, usually packed in PE sacks. Unfilled PC compounds, including those that have been UV and thermally stabilized, those made more resistant to hydrolysis and those of food (FDA) quality and easily ejected thermostabilized injection molding materials all have similar properties (Table 4.41). The requirements of UL 94 V-1 are mostly met by the inherently low flammability PC injection-molding compounds reinforced with 20 to 40% glass fiber. Flame retardant modified filled and unfilled grades 0.8 mm thick can meet the requirements of UL 94 V-0. The flame retardant, *polyphosphonate*, has a similar structure to PC.

Processing

Refer to Table 3.9, for injection molding. For material stored in non-airtight packaging, prewarming or predrying at 120°C to a moisture content of $\leq 0.2\%$ (this can be detected by not blistering at 250°C) is also necessary for extrusion. Injection molds should be maintained at 80 to 120°C. Shrinkage of unreinforced PC is uniform in all directions at about 0.7 to 0.8%; shrinkage of glass reinforced material almost independent on the direction of flow is 0.25 to 0.45%. On account of the stiffness of the material, undercuts in the mold result in difficulties in ejection. PC moldings can be easily polished and lacquered or printed with special coatings. They can be bonded with solution or reaction-resin adhesives and can be welded by HF or ultrasonically.

Properties

Largely amorphous PC is a transparent, hard, elastic and, above −90°C (−130°F), tough plastic (Table 4.41). It is also creep resistant at high tempera-

Table 4.41 General Properties of PC and PC-ABS Alloys

Properties	Unit SI	U.S.	Polycarbonate Standard types		35-40% GF		PC-ABS alloys	
Density	g/cm³	lb/ft³	1.20	75.0	1.52	95.0	1.12-1.16	70.0-72.5
Mechanical								
Tens. str., yield	N/mm²	10³lb/in²	≥60	≥8.5	–	–	45-55	6.4-7.8
Elongation, yield	%		6-8		–	–	4-5	
Tens. str., break	N/mm²	10³lb/in²	≥65	≥9.24	120	17	35-45	5.0-6.4
Elongation, break	%		≥100		2-4		40-50	
Tensile modulus	kN/mm²	10⁵lb/in²	2.3	3.3	~10	14	2.0-2.2	2.8-3.1
Flexural stress, yield	N/mm²	10³lb/in²	100	14	170*	24	75-85	11-12
Charpy impact:								
unnotched 23°C	kJ/m²		N.B.		25		N.B.	
−40°C	kJ/m²		N.B.		25		N.B.	
notched 23°C	kJ/m²		20-60		6-10		25-30	
−40°C	kJ/m²		–		–		10-15	
Izod impact, notched, RT	J/m	ft-lb/in	700-900	13-17	100	1.9	100-500	1.9-9.4
Ball indentation, hardness, 358/60	N/mm²	10³lb/in²	90-110	13-16	140	20	80-90	11-13
Rockwell hardness	Scale		M 70/R 120		>M 90		–	–
Thermal								
Vicat soft. temp. B (50 N)								
50 K/h	°C	°F	160-170	320-338	165-175	329-347	–	–
120 K/h	°C	°F	145-150	283-302	150-160	302-320	110-129	230-264
Defl. temp. A: 1.82 MPa (264 psi)	°C	°F	135-140	275-284	150	302	95-105	203-221
46 MPa (66 psi)	°C	°F	140-145	284-302	155	311	100-125	212-257
Linear therm. expansion	10⁻⁵ K⁻¹	10⁻⁵ °F⁻¹	7	4	2	1	7-8	4-4.5
Thermal conductivity	W/mK	BTU in / h ft² °F	0.21	1.5	0.25	1.7	0.19-0.20	1.3-1.4
Electrical								
Volume resist.	ohm cm		≥10¹⁶		10¹⁶		10¹⁶ -> 10¹⁶	
Dielectric const.	≤1 kHz/1MHz		3.0/2.9		3.8/3.6		2.9-3.0	
Dissipation fact.	≤1 kHz/1MHz		0.0009/0.01		0.0009/0.009		0.004/0.008	
Tracking resist.	Method KC		100-300		150-175		300->550	
Equilibrium water cont.								
23°C/50%RH	%		0.15		0.13		0.2	
23°C in water	%		0.36		0.27		0.6-0.7	

* Flexural strength at break.

Table 4.42 Indicative values for (PC + ABS) and (PC + PBT) blends

Properties	Unit	Standard grades (PC + ABS)	(PC + ABS) GF grades 10–30%	(PC + ABS) FR grades	(PC + ABS) FR-GF 10–20% GF	(PC + PBT)
Density	g/cm³	1.12–1.16	1.2	1.27–1.35	1.52	1.21–1.24
Mechanical properties						
Yield stress	N/mm²	40–60	60–80	50–60	–	43–45
Elongation at yield	%	3–5	2–3	3–5	3–4	4
Tensile strength	N/mm²	35–55	55–75	40–50	70–80	55–57
Elongation at break	%	>50	–	>50	–	120
Tensile modulus of elasticity	N/mm²	1600–2200	4000–8000	2200–3000	4000–6000	1.9–2.2
Flexural strength	N/mm²	60–90	100–140	90–110	110–130	85
Impact strength a_n +23°C (Charpy)	kJ/m²	n.g.	20–30	n.g.	25–30	n.b.
−40°C (Charpy)	kJ/m²	n.g.	5–10	n.g.	–	n.b.
Notched impact strength a_K +23°C (Charpy)	kJ/m²	25–40	–	10–30	–	40–45
−40°C (Charpy)	kJ/m²	7–20	–	–	–	6–8
Ball indentation hardness H 30	N/mm²	80–90	110–130	100–130	120–130	100
Thermal properties						
Vicat softening temperature (VST B/120)	°C	100–135	130–140	85–135	130–140	120
Heat distortion temperature HDT						
A (1.82 N/mm²)	°C	90–120	110–120	75–115	115–125	85–90
B (0.45 N/mm²)	°C	95–130	120–130	85–130	120–130	108
Coefficient of linear expansion	$10^{-4} \cdot K^{-1}$	0.7–1.0	0.3–0.4	0.7–0.9	0.3–0.5	0.7–0.72
Thermal conductivity	W/mK	0.19–0.2	0.20–0.22	0.19–0.20	0.20–0.22	–
Oxygen index	%	20–25	20–25	28–30	30–33	–
Flammability	Flammability class	H B	H B	V-0	V-0	–
Electrical properties						
Volume resistivity, dry	ohm cm	$>10^{14}–>10^{16}$	$>10^{14}$	$>10^{15}–>10^{16}$	$>10^{14}–>10^{16}$	–
Dielectric constant, dry (50 Hz)		2–3	3–4	3–4	3–4	–
Dissipation factor, dry (50 Hz)	10^{-4}	20–40	20–30	25–50	20–30	–
Tracking CTI	Step	250–500	200–300	300–600	150–250	–
Water absorption 23°C (saturation) 23°C/50% rh	%	0.2–0.3	0.15–0.25	0.2–0.3	0.15–0.25	–

tures (Fig. 4.13). Maximum long-term service temperatures of reinforced and unreinforced solid and structurally foamed PC molding material lie only marginally below the stability under load (Table 7.5). When subjected to abrasion and long-term dynamic stress (machine components) PC is only of limited applicability. Transmission of visible light is 88% for a thickness of 3 mm. The refractive index is 1.586. Scratch resistance of PC is increased by siloxane-containing stoving lacquers. All types of PC, including structural foams, are low flammability, to DIN 4102. The electrical characteristics are practically independent of the moisture content of the surroundings or their temperature.

Applications

Applications of moldings of unreinforced PC include impact-resistant streetlights, traffic signals and motor vehicle signal lights, protective helmets, transparent cover boxes and plates for switch and measuring equipment, protective glasses, high-grade domestic utensils, extrusion or injection molded baby bottles, returnable milk and beverage bottles, canteens, canisters, rotomolded large containers, temperature- and impact-stressed moldings and housings for electrical, electronic and consumer equipment. Glass reinforced grades are used for load-bearing parts and housings for measuring instruments, cameras, projection equipment, telescopes, chronometers, temperature-stressed switch gear, insulators, spools, plugs with injected contacts and soldering tags for soldering directly in proximity to the plastic. The mechanical and thermal stability under load and the fire safety of reinforced and unreinforced PC structural foams are properties that are used in large moldings such as cable distributor cabinets, protective covering, assembling sheets, housings for electrical plants, sports equipment and lamp masts. Special easy flowing grades are used for compact discs and laser data carriers.

Rationalized mass production of large injection moldings and PC's lacquerability have led to growing use in motor vehicle manufacture, because of its almost unchanged properties from -30 to $120°C$ and because of improved adjustable stiffness and high impact strength. It is used in the internal and external equipment of cars and for car body parts such as side doors, front elements, bumper bars and spoilers.

PC is swelled by benzene and therefore liable to stress cracking in contact with super-fuels unless it has been lacquered. Low temperature high impact PC/PBTP alloys are fully resistant to these as well as to methanol-containing fuels. The Vicat softening temperature of copolymers of polycarbonate with bisphenol TMC (trade name Apec HT) increases from 160 to $205°C$ as the level of bisphenol TMC is raised (the theoretical limit of $238°C$ has not yet been achieved). Like PC they are amorphous thermoplastics with high transparency, low inherent colour, good UV resistance and good processability. Toughness is somewhat less than that of PC. Applications include diffuser lenses for lamps and other transparent, highly thermally stressed moldings.

4.1.9.2. Polycarbonate + styrene component blends

Trade names
PC+ABS: Bayblend T, Bayblend FR, Terblend B (DE); Koblend, Moldex, Mablex (IT); Cycoloy, Pulse, Triax (US); Stapron (NL); Exelloy (JP)
PC+ASA: Bayblend A, Terblend S (DE)
PC+SMA: Arloy (US)
PC+AES: Koblend (IT), Exelloy (JP)

General
Most PC+ABS blends (see Section 4.1.3.1) are based on bisphenol A polycarbonate; emulsion and bulk ABS are used as the other component. PC+ABS blends are amorphous and distinguished by good processability and low temperature strength. Corresponding blends with ASA or AES have better weathering resistance and reduced toughness at low temperatures. Heat resistance can be increased by the use of, e.g., SMA, α-methylstyrene based ABS or special-purpose polycarbonates. Reinforced flame-retardant grades with 10 to 30% glass fiber are also available.
Flame-retardant PC+ABS blends (UL 94 V-O from 0.8 mm) generally incorporate organic bromine compounds, phosphoric acid esters and an anti-drip agent. Blends with chlorine- and bromine-free flame-retardant systems are increasingly important. Foamed grades also achieve UL classification V-0 and V-1.

General properties
Unreinforced standard grades cover the Vicat range VST/B/120 from 100 to 135°C with increasing PC content and possess notched impact strengths (DIN 53 453, R.T.) of 25 to 40 kJ/m^2. Reinforced grades incorporating up to 30% glass fiber exhibit increased modulus of elasticity (approx. 2000 mPa per 10% GF) but lower toughness than that of unreinforced grades.
FR grades lie in the Vicat VST/B/120 range 85 to 135°C and exhibit notched impact strengths (DIN 53 453, R.T.) of 10 to 30 kJ/m^2.
PC+ABS are easily lacquered and electroplated and can be processed by conventional means (machining, thermoforming, welding, bonding). Table 4.36 provides a summary of indicative values.

Forms of delivery:
PC+ABS blends are supplied as opaque granules (die-face, diced or cylindrical) packed in 25 kg sacks or large containers.

Processing:

Before injection molding, drying of the granules (dependent on grade, approx. 10°C below the Vicat temperature) to residual moisture <0.05% is recommended. The melt temperature should not exceed 280°C. Mold temperatures of 60 to 100°C have proved suitable.

For extrusion blow molding, melt temperature is approx. 245°C and mold temperature 60 to 80°C.

Applications:

The main area of application for standard PC+ABS grades is the automotive sector. Of particular interest are applications in vehicle interiors (control panel supports, instrument surrounds, column cladding, ventilation nozzles, etc.) and external parts (spoilers, decorative strips, wheel covers, etc.)

Flame retardant PC+ABS blends are used in the information technology sector (housings for computers, printers, etc.), decisive characteristics being the UL 94 V-0 (1.6 mm) and 5 V-B (2.3 mm), good processability, light resistance, stiffness and toughness. Additional advantages include ecological aspects and recyclability which have resulted in strong growth in the use of the new FR generation with chlorine- and bromine-free flame retardant systems.

PC+ABS is also used in the electrical engineering sector (switchboxes, connector strips).

4.1.9.3. Polyalkylene terephthalates (PTP)

At room temperature the very hard and stiff homopolymer *polyethylene terephthalate (PET or former PETP)* and the somewhat softer *polybutylene terephthalate (PBT or former PETP)* (formulae, Sec. 4.1.9) are partially crystalline, with crystalline melt temperature ranges of 255 to 258°C (491–496°F) and 224 to 228°C (435–442°F), respectively. Both polymers contain amorphous components with glass transition temperatures of about 50 to 70°C (140–158°F). This brings about a marked reduction of the E modulus of the unfilled products. Their deflection temperatures, as measured by Method A with a load of 264 psi (182 MPa), lie within this range. At smaller loads using Method B (66 psi or 0.44 MPa), they are about 170°C (338°F). Tailor-made reinforced PTP compounds, stiffened by specially coated reinforcements such as milled glass fibers and/or glass microspheres, mica, wollastonite, or talc reach deflection temperatures using Methods A or B above 250°C (392°F). ASTM D 3221/3220 show the effects of reinforcement on the upper limits of the ranges (Table 4.43).

ISO DIS 7792, DIN 16779 Part 1 propose viscosity numbers ≤ 60 to > 140 (signs 06 to 15) for PET and ≤ 90 to ≥ 170 (08 to 18) for PBT as the desig-

4.1.9.3. Polyalkylene terephthalates (PTP)

Table 4.43. ASTM D 3221/3220 Specifications for Different Cell Ranges of PTP

PTP	Not reinforced	Reinforced and filled
1. Specific gravity, 23°C min.	4 cells: 1.18–1.40	7 cells: 1.30–1.70
2. Tensile strength, min.	4 cells: 6.5–9.0×10^3 psi, 45–62 MPa	6 cells: 6.5–21×10^3 psi, 45–145 MPa
3. Flexural modulus, min.	6 cells: 2.5–4.0×10^5 psi, 1.7–2.8 GPa	7 cells: 3.3–18.0×10^5 psi, 2.3–12.4 GPa
4. Izod impact strength, min.	4 cells: 0.4–1.0 ft;lb/in, 21–53 J/m	6 cells: 0.4–1.5 ft-lb/in, 21–80 J/m
5. Deflection temp., min.	3 cells: 54–60°C, 113–140°F	6 cells: 50–210°C, 120–410°F
6. Flow rate, 250°C/2160 g*	5 cells <3–35	–

* Not applicable for PET

natory property. Codes for intended application and other important properties (data block 2) and for fillers or reinforcements (data block 4) are given in Chapter 6, Standardization.

PET crystallizes slowly and therefore mold temperatures of 125 to 150°C (257–302°F) may be used in injection molding. Most molding compounds have, in consequence, been developed on a PBT base, which permits faster cycles and lower mold temperatures. Quenched to 20 to 50°C (68 to 122°F) PET is amorphous and becomes crystal-clear. Crystal-clear stretched film (Sec. 3.6.5.8) and bottles are the main PET products. However, many PET-PBT modifications have been produced that enhance processability and extend the range of properties and thus the performance of PTP. These modifications, especially those with improved notched sensitivity for special applications, can only be hinted at in the following list.

Trade names

PET-based

Tenite, Carodel, Cleartuf (US), Hostadur E, Ultradur (DE), Rhodester, Techster (FR), Arnite A (NL), unfilled specially for bottles, and reinforced compounds.

Kodar (US) modified with CHDM (HOCH$_2$–⬡–CH$_2$OH), clear, amorphous and tough for extrusion of films, sheets, tube, profiles and blow molding, T_g 81°C (178°F).

Petlon, Petra, Rynite (US), Arena FR-PET (JP), impact-modified and highly filled for optimal processability, thermal and mechanical performance (Table 4.44), mold temperature 90 to 100°C (149–212°F).
Ektar PCTG (US), modified by CHDM, reinforced.
Ropet (US), PET-PMMA blends; Mindel (US) PET + PSU blends.
Ekkcel (US) PTP + polyparabenzoylester, see Sec. 4.8.4.
Xenoy (US), Makroblend, Ultrablend (DE), PC/PTP alloys see Sec. 4.1.9.1.
Selar PT (US) polyester blend barrier resin.
Illen (DE), PTP high impact with PE.
Rimplast (DE) SI-PTP interpenetrating network (IPN), see. Sec. 5.2.

PBT-based molding and extrusion compounds series comprising HI and PET (modified) products

Celanex, Gaftuf (Gafite), Valox (US); Hostadur B Pocan B, Ultradur, Vestodur B (DE); Novadur, Shinko-Lac, Toray, PBT, Tufpet PBT (JP); Pibiter (IT); Deroton (GB); Arnite T (NL); Orgater, Techster (FR); Crastin (CH).

Forms of delivery

PTP is delivered in airtight packages of pelletized unfilled or (see above) reinforced injection molding compounds in all colors, as high-flow, tough, stiff and temperature-resistant grades, and as flame-retarding compounds meeting requirements up to UL-94-V0, 0.8 mm. Incorporation of rubber improves impact strength. Blending PBT/rubber with PC increases this property further. Most unfilled and reinforced grades are permitted for use with drinks and foodstuffs.

Processing

Air-damp material must be dried for about 5 h at 120°C to avoid ester saponification at processing temperatures up to 30 K above the melt temperatures (Table 3.9, p. 96). Heating above the processing range may cause decomposition. For mold temperatures, see above. In spite of high process shrinkage (1.5–2.5%; reinforced 0.4–0.8%) the encapsulation of metal parts, even with unfilled compounds, is feasible if wall thickness is adequate. GF-reinforced materials with blowing agents are processed into structural foam injection moldings. Ultrasonic and friction welding can be generally used; hot plate and hot gas welding are suitable for unfilled PTP. Cyanoacrylate, EP or PUR adhesives are used for bonding. In stretch blowing of PET bottles for drinks with grades 9 through 12, temperatures a little above the melting point (265°C) are optimal and avoid the elimination of acetaldehyde (Sec. 3.3.6.7).

Properties

PTPs are not prone to stress cracking. They are abrasion resistant and have good slip properties. In air the long-term service temperature range is −40 to 140°C. For short periods, unreinforced PBT are heat resistant up to 185°C, reinforced grades up to 125°C. These have good stability under load and little tendency to creep. Mechanical data are summarized in Tables 4.43 and 4.44, temperature time limits in Table 7.5, creep-moduli diagrams are shown in Fig. 7.13. PBT is resistant to fuel, oil and fat, most chemicals and solvents (except alkalis, ketones and phenols) even at elevated temperatures, but it is hydrolyzed by boiling water. The density of unmodified amorphous PET is 1.33, of PBT 1.31 g/cm^3. The coefficient of linear expansion is 3 to 8×10^{-5} K^{-1} ($2-4 \times 10^{5\circ}$F^{-1}); water absorption at saturation (0.3–0.5%) is low. Electrical properties are moisture and temperature dependent to only a small extent: volume resistance $> 10^{15}$ ohm cm; dielectric constant 4 and 3.5 at 60 and 10^6 Hz respectively; dissipation factor $2-4/10-20 \times 10^3$ for filled products.

Applications

Rapid hardening, rigid PBT parts, injection molded in short cycle times, substitute metals in, for example, components for headlamps, radiator grilles and bumpers, fuel tank caps, ignition plugs, pump housings, door handles, wing mirror parts in automotive engineering, electrical and lighting components that are subject to thermal loading, housings for domestic irons, food friers, toasters and other domestic appliances, furniture hinges, joints, slideways and drive and bearing elements in precision engineering.

Crystal-clear PET is used for optical parts and other impact-resistant, transparent moldings. There is a mass market for blown bottles of 0.25 to 3 l with good barrier properties for household chemicals as well as for beverages. Beer bottles need an oxygen barrier coating. Returns are recycled for molding compounds or filling fibers. Multilayer ovenable dishes, Kraft paper coated pastry moulds, are considered on p. 329. Further applications of sheets and films are given in Sec. 4.3.2.

4.1.9.4. Thermoplastic poly(ether)ester elastomers

Trade names: Arnite (NL), Ecdel Elastuf, Gaflex, Hytrel, Lomod (US), Pelprene (JP), Pibiflex (IT), Illenoy (DE).

Structure and properties: block copolymers of soft segments of poly-alkylene ether diols and/or long chain aliphatic diacid esters with semi-crystalline PBT comprise the border area of stiff rubbery through highly flexibilized engineering polymers (Fig. 4.11, Table 4.44). In a variety of compounds they combine

4.1. Thermoplastics: Raw Materials and Molding Compounds

Table 4.44. General Properties of Some Special Polyesters

Properties	Unit SI	Unit U.S.	PET, mod., +45% GF		Polyester ethers		Polyarylates, amorphous	
Density	g/cm^3	lb/ft^3	1.69	105	1.15–1.23	71.9–76.9	1.21	75.6
Mechanical								
Tens. str., yield	N/mm^2	10^{3}lb/in^2	–	–	14–23	2.0–3.3	71	10
Elongation yield	%		–	–	50–100		6–9	
Tensile str., break	N/mm^2	10^{3}lb/in^2	193	27	15–40	2.1–5.7	62	8.8
Elongation, break	%		2		100–800		20–60	
Tensile modulus	kN/mm^2	10^{5}lb/in^2	14.6	21	0.17–0.7	0.24–1.0	2.1	3.0
Flexural strength	N/mm^2	10^{3}lb/in^2	283	40	6–20[1]	0.9–2.9	81	11.5
Izod notched: RT	J/m	ft-lb/in	128	2.4	N.B.	N.B.	220–280	4.1–5.2
–40°C/°F	J/m	ft-lb/in	123	2.3	N.B.–17	N.B.–0.3	ca. 210	ca. 4
Hardness:								
Rockwell	Scale		M100/R120		–		M95/R125	
Shore	Scale		–		D40–D63		–	
Thermal								
Melting temp.	°C	°F	245	473	190–213	374–415	–	–
Vicat soft. point, Rate A, method:								
A, 10 N	°C	°F	–	–	178–194	352–381	–	–
B, 50 N	°C	°F	–	–	82–120	81–248	~ 190	~ 374
Defl. temp.:								
A, 1.82 MPa (264 psi)	°C	°F	226	438	n.a.[2]–57	n.a.–135	165–175	329–347
B, 0.46 MPa (66 psi)	°C	°F	–	–	n.a.[2]–147	n.a.–300	–	–
Linear therm. expansion	10^{-5}× K^{-1}	10^{-5}× °F^{-1}	2.7	1.5	15	8.3	6.2	3.4
Electrical								
Volume resist.	ohm cm		10^{13}		10^{10}–10^{13}		> 10^{16}	
Dielectric const.	1 kHz/1 MHz		4.0/3.9		4.6/4.0		3.4/3.2	
Dissipation fact.	1 kHz/1 MHz		0.005/0.011		0.01–0.03		0.002/0.017	
Water absorption 24 h	%		0.04		0.49–0.33		0.26	
In water at 23°C 30 days	%		–		0.63–0.58[3]		0.71	

[1] At yield. [2] Not applicable. [3] Saturated.

good aging resistance with fatigue- and hysteresis-free rubber elasticity from –40°C to 100°C. They can be made resistant to fuels, lubricants, hydrolysis, UV and weather.

Processing

Film and tube extrusion, blow molding and injection molding at 250°C.

4.1.9.5. Amorphous polyarylates

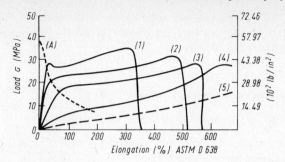

Fig. 4.11. Strain-stress diagram (see Fig. 7.6) of polyether-block amides.
(1) Polyamide, (2) 30% Polyether, (3) 50% Polyether, (4) 80% Polyether, (5) Rubber, (A) Yield stress.

Applications

Membranes, tubing, coverings, bellows, cover caps, bumpers, dual-joint clutch and drive elements, seals, ski and running shoe soles, bodywork components.

4.1.9.5. Amorphous polyarylates

The basic formula for polyarylates and copolymers is given in Sec. 4.1.9.
Trade names: Ardel, Arylon (Bexloy M); Durel (US), U-Polymer (JP), Arylef (BE).
Amorphous, yellowish, transparent, tough thermoplastics with softening (glass transition) temperatures above 180°C (356°F). They are inherently difficult to burn, and either unreinforced or GF reinforced. They can be formulated to meet the requirements of UL 90-V0. With good weather resistance, stability under load and electrical properties they can be used long-term at temperatures up to 150°C. For information on chemical resistance see Sec. 4.1.9. They are affected by aromatic hydrocarbons, ketones, esters, cyclic ethers and chlorinated hydrocarbons (except carbon tetrachloride). Other properties are given in Table 4.44.

Processing

Polyarylates are predried at 120 to 140°C (248–284°F), injection molded and extruded at 310 to 380°C (610–716°F) with the mold temperature at 40 to 100°C (100–212°F). Mold shrinkage is 0.2% in the direction of flow and 0.7 to 0.9% at right angles to it.

Applications

Amorphous polyarylates are used as transparent sheets for special glazing (solar panels) and as functional moldings for electronic and electrotechnical applications, instruments, medical applications demanding superior fire safety and dimensional stability at high temperatures, parts with snap fasteners as well as for microwave cookware.

4.1.9.6. Self-reinforced crystalline polyarylates (LCP)

Trade names: Ekkcel, Vectra, Xydar (US), Ultrax (DE)

Long chain macromolecules of arylate copolymers with reinforcing groups, e.g. poly p-hydroxybenzoate (Sec. 4.8.4) or 2.6 naphthalene diacids and diols (NTP), in the chain form liquid crystalline phases when molten. Decreasing their extremely high melting points ($> 400°C$) by 100 to 150 K, renders "thermotropic" LCP more susceptible to conventional processing. From the anisotropic, readily flowing melt, highly crystalline moldings can be injection molded at approx 200 K lower mold temperature from very low temperatures to near the melting point. They are dimensionally stable with high (aligned) creep rupture strengths (Table 4.45), tough, with high quality electrical properties (Fig. 4.12), chemically inert, resistant to exposure and hydrolysis, flame resistant without evolving gases or fumes.

4.1.10. Linear Polyarylene Ethers (Oxides), Ether ketones, Sulfides and Sulfones*

The linear binding of aromatic ring structures by oxygen or sulfur bridges resulting from oxidative coupling (Sec. 3.1.1.2) in the case of polyaryl ethers (or polyaryloxides) or by condensation reactions in the case of polyether ketones, polyaryl thioethers (or polyaryl sulfides) or through the SO_2- bridges in the sulfones, leads to high melting polymers with stiff molecular segments (Table 4.46). Such engineering plastics made of amorphous polymers have useful mechanical properties from low to almost glass transition temperatures (Table 4.47). Partially crystalline polymers, at least with reinforcement, have such properties up to temperatures below the crystalline melting point. They have particularly high creep strengths with a low tendency to creep (Fig. 4.13). These polymers are very resistant to acids and alkalis, they are not hydrolyzed by hot water or steam and they are not oxidation-sensitive. Unmodified polymers are flame retardant (see Table 4.47, Limiting Oxygen Index). They are suitable for applications in electrical engineering and electronics where

---

* Phenoxy resins, high molecular weight linear epoxide resins (Table 4.47), are mixed aromatic-aliphatic polyethers with T_m of 70 to 250°C. They are used as impregnating, lacquer and adhesive resins. Similar aliphatic polysulfones are used as masking lacquer in the production of integrated control circuits.

4.1.10. Linear Polyarylene Ethers (Oxides), Ether ketones

Table 4.45. Typical properties of LCP

	Units SI	Units U.S.	Extrusion Grades[1]	
Mechanical Properties				
Tensile strength	MPa	kpsi	ca. 600	ca. 85
Elongation, break	%		< 2	
Tensile modulus	GPa	Mpsi	ca. 50	ca. 7
			Injection Grades	
Tensile strength	MPa	kpsi	100–240	14–35
Elongation, break	%		1–7	
Tensile modulus	GPa	Mpsi	8–40	1.1–5.5
Flexural strength	MPa	kpsi	130–300	19–43
Charpy notched	kg/m^2	–	20–60	–
Izod notched	J/m	ft-lb/in	53–530	1–10
Physical Properties				
Density	g/cm^3	lb/ft^3	1.35–1.9[2])	84–119
Water absorption at equlibrium	%		0.02–0.04	
Thermal Properties			**Injection Grades**	
Melting Point	°C	°F	275–>400	530–>750
HDTA (1.82 MPa)	°C	°F	120–>250	350–>480
Lin. therm. expansion	10^{-5} K^{-1}	10^{-5} F^{-1}	F 0.1–2.5 T 2.5–>5	0.1–1.5 1.5–>3
Flammability				
Lim. oxygen Index	% O$_2$		35–50	
Flammability 0.8 mm	UL 94		V-0	
Electrical				
Volume resist.	ohm cm		10^{15}–10^{16}	
Arc resistance	sec		63–180	
Tracking resistance	KC, Volts		150–200	
Dielectric constant			2.6–3.2	
Dissipation factor			0.003–0.004	

[1] 0.76 mm (30 mil) rod properties
[2] wear resistance reinforced

temperature resistance and low flammability are of major importance. The UL 94 V0/5V classification is achieved without the need for halogenated flame retardants (Fig. 4.12). Chlorinated hydrocarbons attack all amorphous products. Amorphous grades are more liable to solution or stress cracking in contact with ketones, esters and aromatics, but better than semi-crystalline ones with respect to molding shrinkage and warpage. Copolymers combining good qualities of both (e.g. Ultrason KR, Vestoblend (DE), Table 4.47) are available.

4.1. Thermoplastics: Raw Materials and Molding Compounds

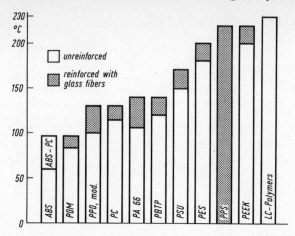

Fig. 4.12. Maximum long-term service temperatures for electrical and mechanical properties of engineering thermoplastics moldings according to the grading of Underwriters' Laboratories.

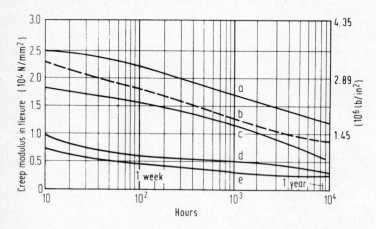

Fig. 4.13. Creep modulus values at higher temperatures for thermoplastics with benzene rings in the polymer chain. Creep moduli in flexure for
(a) Polyether sulfone   at 150°C and 6.9 N/mm² bending stress,
(b) Polyether sulfone   at 150°C and 21.0 N/mm² bending stress,
(c) Polysulfone   at 150°C and 6.9 N/mm² bending stress,
(d) Modified polyphenylene oxide at 100°C and 6.9 N/mm² bending stress,
(e) Polycarbonate   at 120°C and 3.5 N/mm² bending stress.

Resistance to β and γ rays is high. The polymers are physiologically inert. They are permitted for use with food in many countries. The high softening points of the unmodified polymers require processing temperatures of 300 to 420°C for injection molding, the extrusion of profiles and film processing. Regrind can be

4.1.10. Linear Polyarylene Ethers (Oxides), Ether ketones

Table 4.46. Polyarylene Ethers, Ketones, Sulfides and Sulfones

Generic Name	Symbol	Structural elements	$T_m\ °C^1$	$T_s\ °C$	Trade names	Remarks
Polyarylene Ether or Oxide:	PPE "PPO"	not modified	265	120	Arilex (SU), Biapen (PL)	not commercial
from 2,6-dimethyl-1,4-phenylene ether homogeneously modified and/or graft polymers	(Cop.: PEC) PPO mod. PPOS		—	<80–200	Noryl. (US, NL), Prevex, (US), Xyron (JP), Luranyl, Vestoran (DE)	many grades for various purposes
Alloys	PPO/PA	PPO + PS, HIPS micromultiphase, compatibility agent	—	≥ 210	Noryl GTX, Prevex (US), Ultranyl, Vestoblend (DE)	
Polyarylene Ether Ketone	PEEK	e.g.	340–375	167*	Victrex (GB), Hostatec, Ultrapek (DE)	*T_g not reinforced, reinforced up to ≥300°C
Polyarylene Ether Ether Ketone	PAEK			145*	Kadel (US), Victrex (GB)	
Polyphenylene Sulfide	PPS		280–290	85**	Ryton (US), Tedur (DE), Craston (CH), PTFE alloy: Alton (US), Supec (US)³; Primet (NL)	**T_g not reinforced, reinforced annealed (usually) up to ~250°C
Polyarylene Sulfone Polysulfone	PSU		—	170–185	Ultrason S (OG), Udel (US) ABS alloy Mindel (US)⁴	T_g ca 185°C Transparent, amorphous
Polyarylene Ether Sulfone	PES		—	180–215	Ultrason E (DE) Radel (US),	T_g 225°C transparent
Polyaryl(ene) sulfone (also polyphenylsulfone)	PAS PPSU		—	190–215		T_g 296°C T_g 288°C

4.1. Thermoplastics: Raw Materials and Molding Compounds

reprocessed. Welding, bonding with solvent or reactive-resin adhesives, thermoforming of semi-finisheds and powder coating are all possible.

Product groups

1. The *polyarylene ether* (PPE or "PPO") plastics are unique among linear arylene polymers because of the possibility of unlimited modification with varying amounts of low-cost styrenics. Many grades of great technical and commercial significance with continuous service temperatures between 90°C and 150°C can be molded easily and precisely at melt temperatures between 260°C and 360°C. They include unreinforced and reinforced varieties, electrical and antistatic grades up to V-0 and 5V and food grades that can provide Izod notched impact strengths of 100 to 400 J/m, and Charpy a_K 5 to 25 kJ/m^2 at 23°C, 4 to 25 kJ/m^2 at -40°C. The amorphous, low shrinkage and low deformation compounds have broad fields of application in precision engineering, electrical and automotive parts with good stiffness at elevated temperatures, especially for large-area and structural foamed components, computer cabinets and special formulations for cable and wire insulation. Laminated or coextruded composites with sulfur-vulcanizable rubbers (Vestoran, DE) have greater strength after vulcanization than the tear strength of the cured rubber.

    PPE/PA alloys containing mineral fillers have been used for dimensionally accurate on-line paintable car body parts, GF reinforced compounds for fuel-, oil-, brake fluid- and anti-freeze resistant components in the engine compartment.

2. The *polyarylene polyether ketones* (PEK and PAEK) are inherently flame resistant (V-O), highly resistant to chemicals, especially solvents, radiation and weathering, tough and abrasion resistant from cryogenic temperatures (-250°C) up to (reinforced) 250°C to 300°C. They meet the most stringent specifications of the aerospace, aircraft, nuclear and petroleum engineering industries for injection molded or machined parts, films (Stabar, Sec. 4.2.7), pipes, wire insulations and monofilaments. PAEK or PEK impregnated CF-rovings or -webs are highly advanced composite materials. PEKK, PAEK are development products.

3. *Polyphenylene sulfide* (PPS) has, on account of its low glass transition temperature and low toughness, been used only in highly filled and reinforced compounds that are relatively low-cost, (Sec. 4.8.8).

    Injection molded parts recrystallized by annealing at 200°C have continuous service temperatures ≥ 200°C. ASTM 4067 classifies eight grades by limiting cell values for
    1 Tensile strength 60 to 300 MPa
    2 Flexural modulus 7000 to 26,000 MPa

4.1.10. Linear Polyarylene Ethers (Oxides), Ether ketones

3 Izod impact notched 75 to 110 J/m
4 Flexural strength 85 to 285 MPa
5 Density 1.40 to 2.10 g/cm^3.

There are many applications for injection moldings.

4. The hard, amorphous, slightly yellow *polyarylene sulfones* (PSU, PES) are available in unfilled and glass fiber reinforced grades (up to 40% GF) for, in the main, injection molding. The dimensionally stable moldings are resistant to aqueous media, steam, lubricants and transmission oils but not to polar and aromatic solvents. PES moldings exhibit outstanding creep resistance from -100 to $+200°C$, PSU to $+170°C$. The long-term limit for GF reinforced PES is 180°C. PES film (0.5 mm thick) and 1.5 mm thick PSU 30 GF are classified V-0 without added flame retardants. Because of its minimal water absorption PSU is suitable for electrotechnical applications. Transparent insulating films are also of importance because their dielectric properties are almost independent of frequency and temperature. The good resistance to hydrolysis renders PSU suitable for steam-sterilisable transparent medical apparatus. Other applications include transparent parts for domestic appliances involving high temperatures such as cookers, water heaters, etc. PSU and PES are permitted materials in many countries for contact with foodstuffs.
Polysulfones are not UV-resistant.
Thermoformable grades are widely used in aircraft interiors because of their flame resistance, low smoke emission and liberation of heat.
Unidirectional composites with 60% CF-rovings in PES matrix achieve 1800 to 2200 MPa tensile strength and 120 to 130 MPa tensile modulus.
Because of their characteristics and the ability to mass-produce cost effectively even complicated, dimensionally stable moldings without subsequent machining or waste, these high-temperature resistant engineering thermoplastics are increasingly replacing metal components which required several stages of manufacture and were used in combination with glass, thermosets or ceramics.
Ultratect polyethersulfone foam is produced by block foaming ketone-swelled gel under vacuum at high temperature. The thermoplastic, hard and tough foam has a compressive strength of 0.6 N/mm^2 (DIN 53421) and heat distortion temperature of 210°C (DIN 53424) at bulk density of 0.05 g/cm^3. It meets the flammability classification for use in aircraft interiors. Light, rigid sandwich structures are manufactured by laminating with PF or PES prepregs.

4.1.11. Linear Modified Polyimides

Polyimides with the characteristic imide group $O=\underset{|}{C}-\underset{|}{N}-\underset{|}{C}=O$ or related groups (see Sec. 4.8.6), e.g., benzimidazols (PBI) (Table 4.48) in ladder or semi-ladder polymers (Fig. 4.23 b,c) with aromatic or heterocyclic nuclei are thermosetting, highly heat-resistant specialty plastics. The basic structure and trade names of some modified, linear thermoplastic polyimides are summarised in Table 4.48. General properties are tabulated in Table 4.49.

1. *PMMI*, modified polymethacrylimide (PMI) is a new development which complements the related PMMA molding compounds for outdoor use by providing superior heat resistance. Unmodified PMI is a high temperature resistant core material for structural components.
   Airtex R 82.90 polyetherimide foam is manufactured from Ultem by a process similar to that described for the product from polyethersulfone. Both foams have very similar properties. Chlorinated hydrocarbons are used as foaming agents.

2. *PAI* is a speciality material for refrigeration as well as for the aerospace industry. It has a tensile strength of 220 N/mm^2 (31 × 10^3 psi) and 6% extension at break at −196°C and 53 N/mm^2 (7 · 5 × 10^3 psi) tensile strength (84 N/mm^2 (12 × 10^3 psi) if reinforced) at 260°C (500°F) and a high creep resistance. It is resistant to aliphatic, aromatic, chlorinated and fluorinated hydrocarbons, ketones, esters, ethers, weak acids and bases and high-energy radiation. PAI is attacked by water vapor and alkalis at elevated temperatures. Grades modified with PTFE or graphite have been used at temperatures up to 250°C for unlubricated bearings with minimal frictional values. For highest service values the moldings must be tempered for 5 days at 245°C to 260°C. A pilot gasoline engine with all movable parts made of PAI is an example of its strength at extreme working stress.

3. *PEI* is, thanks to its excellent flow, processible by sprueless injection molding, injection blow molding, extrusion and foaming. It has good creep resistance under load up to high temperature and corresponding electrical values. As measured by the Limiting Oxygen Index, it has the lowest flammability of all thermoplastics, apart from polyfluoro-olefins. With its moderate price it is suitable for a wide range of applications. The amorphous, unpigmented, amber-yellow transparent plastic is soluble in methylene chloride and trichloroethylene. It is not attacked by alcohols, vehicle and aircraft fuels, lubricants and cleaning materials even under stress. It is resistant to acids and weak alkalis (pH < 9), to hydrolysis by hot water and steam and to UV and high energy radiation. Applications are high-tempera-

Table 4.48. Linear Modified Polyimides

General type	Abbr.	Basic configuration	Components	Trade name
1. Modified polymethacrylimide	PMI	See Fig. 4.8	Methacrylonitrile, methacrylic acid; see also Sec. 4.1.6	Kamax (US) (Oroglas XHTA, US)
2 Poly(amide-imide)	PAI		Trimellitic anhydride (Table 4.81.) Diisocyanates (Table 4.67) and/or diamines, R unknown	Torlon (US)
3 Poly(ether-imide)	PEI		m-phenylenediamine, phthalic anhydride + Bisphenol A (Table 1.4)	Ultem (US)
4 Poly(imide-sulfone)	PISO		Trimellitic anhydride, diisocyanates, aromatic intermediates	Vectra (DE)
5 Maleimide copolymers		Maleic imide base reactions see Table 4.81	Bismaleic imide and comonomers	Malecca (JP)

4.1. Thermoplastics: Raw Materials and Molding Compounds

Table 4.49. Properties of Thermoplastic Polyimide Compounds

Properties	Unit SI	Unit U.S.	Polyamideimide General purpose		Polyetherimide Unreinforced/reinf. 30% GF	
Injection molding						
Predrying temp.	°C	°F	150–180	200–360	150	300
Processing temp.	°C	°F	330–380	626–716	340–425	640–800
Mold temp.	°C	°F	230	110	65–175	150–350
Linear mold. shrink.	%		–		0.4–0.7/0.2	
Density	g/cm^3	lb/ft^3	1.40	87.5	1.27/1.51	79.4/94.4
Mechanical						
Tensile strength	N/mm^2	10^{3}lb/in^2	186	27	105^1/160	15^1/25
Elongation, yield	%		–		7–8/–	
Elongation, break	%		12		60/3	
Tensile modulus	kN/mm^2	10^{5}lb/in^2	4.6^2	6.7	3.0/9.0	4.3/13
Flexural strength	N/mm^2	10^{3}lb/in^2	212	31	145/230	21/33
Izod, notched	J/m	ft-lb/in	136	2.5	50/100	1.0/2.0
Rockwell hardness	Scale		M 119		M 109/M 125	
Thermal						
Vicat soft. point (50 N)	°C	°F	–	–	219/228	426/442
Defl. temp.						
A: 1.82 MPa (264 psi)	°C	°F	274	525	200/210	392/410
B: 0.44 MPa (66 psi)	°C	°F	–	–	210/212	410/414
Cont. service temp.	°C	°F	220^3	428^3	170^3	338^3
Linear therm. expans.	10^{-5} K^{-1}	10^{-5} F$^{°-1}$	3.6	2.0	6.2/2.0	3.4/1.1
Lim. oxygen index	%		43		47–50	
Electrical						
Volume resistance	ohm cm		10^{17}		$7 \times 10^{17}/3 \times 10^{16}$	
Dielectric constant	1 kHz/1 MHz		4.0/–		3.2–3.5/–	
Dissipation factor	1 kHz 1 MHz		0.001/0.009		0.0014/0.0025	
Water absorption:	%		–		0.25/0.18	
24h saturation	%		5 (2000h, 90°C)		1.25/0.9	

[1] At yield.
[2] Flexural modulus.
[3] UL relative thermal index.
[4] also compression molded or sintered and autoclave tempered.

ture automotive, electronic and other appliances, aircraft equipment such as seats, window frames and safety belt clasps.

4. *PI-PPS alloy* is generally filled with PTFE or graphite and can be injection molded at melt temperatures of 370°C (698°F). The moldings are annealed at 270°C (518°F). They are used in self-lubricating bearings, etc. Without

admixture of PPS, linear PI of this type no longer flows above T_g 310°C but only assumes a rubber-like softness. Films can be cast from dimethyl formamide solution.

5. *Malecca*, with a variety of general purpose, high gloss, impact and reinforced grades easy flowing at 280°C, T_g 140 to 170°C, HDT (meth. A) 130 to 160°C, has a good balance of heat resistance and moldability. With low density (1.06 to 1.08 g/cm^3), thermally stable mechanical properties ($\sigma_B > 40$ kg/mm^2, Izod notched strength 100 to 200 J/m^2), dielectric constant 2.5 to 3.0, tan δ 0.001 to 0.002, tracking resistance C 600, 0.3% water absorption, it provides high performance at relatively low cost. It is resistant to fuels and chemicals.
Applications are in automotive components, domestic appliances, medical instruments and aviation components.

4.1.12. Cellulosics

4.1.12.1 Cellulose esters (CA, CAB, CP)

Esterification of pure cellulose (cotton linters) with acetic acid and sulfuric acid as catalyst in solvents of medium polarity produces readily soluble resins (Sec. 4.1.1.9.2) with varying acetate content depending on the process. The most highly acetylated are used for triacetate fibers and cast films. Secondary acetates with lower acetate content within certain limits and related esters are flexibilized by plasticizers and granulated into molding compounds. After predrying, if necessary, they are processed by injection molding (Table 3.9), extrusion (for semi-finished forms see Sec. 4.2.9.1) and blow molding. The machines are corrosion resistant to cope with possible evolution of acid.

Cellulose acetate (CA)
Trade names: Bergacell, Saxetat (DE); Setilithe (BE); Dexel (GB); Cellolux, Sicalit (IT); Acetyloid (JP); Tenite Acetate (US).

Standards: ASTM D 706, DIN 7742.

Cellulose acetobutyrate (CAB)
Trade names: Tenite Butyrate (US); Cellidor B(DE).
Standards: ASTM D 707, DIN 7742.

Cellulose propionate (CP)
Trade names: Tenite Propionate (US); Cellidor (CP (DE).
Standards: ASTM D 1562, DIN 7742.

Many grades of cellulose ester compounds are tailor-made by varying the ester and plasticizer. The flow temperature, measured by ASTM D 569 Method A correlates with the hardness of the materials. They are divided into high impact and very soft (S2) to medium hard (MS), flow 130 to 150°C, and high strength and hard (H) to very hard (H6), flow 155 to 180°C. DIN 7742 gives for CA-H (acetate content $\geq 55\%$), CA-N ($\geq 52\%$), CAB ($\geq 40\%$ butyrate) and CP ($\geq 55\%$ propionate) the designatory properties VST/B/50: <0.55 to $>115°C$ (10 cells) and mass lost at 80°C: <0.25 to $>6.5\%$ (10 cells, code 01 to 70).

General properties of the common cellulosics are given in Table 4.50. These are quite hard and scratch resistant and sufficiently tough for encapsulating metal parts while being glass-clear, brilliant and intensely colorable. As a result of their relatively high water absorption the products are antistatic (they do not attract dust), but because of this and the hazard of plasticizer migration they are not suitable for precision parts. Approval for food, and flammability depend on the type of plasticizer. Cellulosics are not attacked by fatty oils, mineral oils and aliphatic hydrocarbons. Their resistance to fuel mixtures, benzene, chlorinated hydrocarbons and ether depends on their composition. Their resistance to alkalis and acids is low.

Applications

CA: Electrically insulating encapsulated tool grips, writing equipment, combs, buckles, buttons and other personal requisites, H-types (can be intensely colored, sound absorbing) are used for telephones and sound pickups, etc.
CAB: Car components such as sweat resistant steering wheel covering, service knobs, etc.; vacuum cleaner parts; suitcase grips; weather-resistant materials for external lights, light domes and heraldic shields.
CP: Like CAB is used in high-quality sunglasses, optical spectacle frames, infra-red-absorbing material used for welders' protective goggles and screens.

4.1.12.2. Ethyl cellulose

Trade names: Ampec (US).

Alkali cellulose is reacted with ethyl chloride to form ethyl ether molding compounds containing 44.5 to 48% ethoxy groups.
General properties: The densities of Type I, general purpose, and Type II, improved impact resistance, according to ASTM D 787, are $d = 1.11$ to 1.13, respectively. Rockwell hardness is R110 to R50 and minimum tensile strengths are 26 to 45.8 MPa (2.9–6.5 × 10^3 psi). Izod notched impact strength at 23°C for Type I is 91 to 117 J/m (1.7–2.8 ft-lb/in), for Type II it is 186 to 320 J/m (3.5–6.0 ft-lb/in). At −40°C tensile strength for Type I is 27 MPa (0.5 ft-lb/in) and for Type II, 54 to 80 MPa (1.0–1.3 ft-lb/in); the deflection

4.1.12.2. Ethyl cellulose

Table 4.50. General Properties of Cellulosics to ASTM D569

Properties	Methods		CA S-H2 135–160°C		CAB S-H 135–155°C		CP MH-H4 150–170°C	
	Flow							
	Flow temps.							
Density	SI g/cm^3	U.S. lb/ft^3	1.26–1.30	78.8–81.3	1.15–1.21	71.9–75.6	1.18–1.23	73.8–76.9
Mechanical								
Tensile strength minimum acc. to ASTM	N/mm^2	10^3 lb/in^2	18–36	2.6–5.2	19–34	2.8–4.9	19–34	2.8–4.9
Practical values (yield)	N/mm^2	10^3 lb/in^2	30–50	4.3–7.1	26–45	3.7–6.4	30–47	4.3–6.7
Elongation, yield	%		2.5–3		4–4.5		3.5–4	
Limiting flex. stress[1]	N/mm^2	10^3 lb/in^2	35–62	5.0–8.8	38–56	5.4–8.0		
Flex. mod. min[2]	kN/mm^2	10^5 lb/in^2	0.9–1.6	1.4–2.3	0.7–1.5	1.0–2.1	0.8–1.5	1.2–2.5
Charpy impact: unnotched	kJ/m^2		N.B.–0.5		N.B.		N.B.	
notched	kJ/m^2		2–1		3.5–0.5		2.6–0.6	
Izod, notched[2]	J/m	ft-lb/in	200–90	3.7–1.5	285–90	5.3–1.7	400–80	7.5–1.5
Rockwell hardness	Scale		R77–R107		R72–R106		R80–R107	
Thermal								
Vicat soft. Temp. (50 N)	°C	°F	45–70	113–158	75–105	167–221	68–100	154–212
Heat defl. temp. A: 1.81 MPa	°C	°F	48–65	118–149	62–70	144–158	61–73	142–163
B: 0.45 Mpa	°C	°F	52–82	126–180	67–90	153–194	72–98	162–208
Linear therm. expansion	10^{-5} K^{-1}	10^{-5} F^{-1}	16–10	9–6	13–10	7–6	14–11	8–6
Therm. conduct.	W/K m	BTU in / h ft^2 °F	0.22	1.52	0.20	1.39	0.21	1.46
Electrical								
Volume resist.	ohm cm		10^{13}–10^{15}		10^{14}–10^{15}		10^{15}–10^{16}	
Dielectric str.	kV/mm	V/mil	30–35	290–600	35–38	250–400	34–36	300–450
Dielectric const.	50 Hz	1 MHz	5.8/4.6		3.7/3.5		4.2/3.7	
Dissipation fact.	50 Hz	1 MHz	0.02/0.03		0.006/0.02		0.01/0.03	
Water absorption, max. including soluble loss[2]	%		4.7–3.0		1.9–1.6		2.3–1.8	
	%		1.4–0.4		0.6–0.2		0.2–0.1	

[1] Limiting flexural stress according to DIN 53 452 (1952). [2] According to ASTM.

temperatures are A/B, 49 to 82°C/77 to 94°C; water absorption (24 h), max., is 1.0 to 1.7%; weight loss at 72 h at 82°C, max., is 0.3 to 2.0%.
Applications: Flashlight cases, fire extinguisher components, electrical appliance parts, alkali-proof sheets and films for special food formulations.

4.1.13. Thermoplastic Composites

4.1.13.1. Molding materials

Compounding polymers with fillers, enhancers and reinforcements such as those described in Sec. 3.1.3.8, considerably improves the stiffness, mechanical strength and creep resistance at elevated temperatures, dimensional stability and the friction and abrasion characteristics, etc., of thermoplastics. Low cost additives, faster cycling and reduced wall thickness can sometimes reduce the price of the finished product in spite of special compounding and machine costs. Examples of the general properties of filled and reinforced compounds compared with those that have not been reinforced have been given in

Table	page	for
4.5	182	PP
4.6	183	PP
4.13	209	SAN
4.33	249	POM
4.37	256	PA
4.41	268	PC
4.44	276	PETP
4.45	279	LCP
4.46, 4.47	280, 281	PPO, PPS, PES
4.49	286	PEI

Besides plastics producers who supply reinforced compounds and other compounds under their own brand names, there are many compounding specialists supplying thermoplastic composites under trade names such as
Ferro, Fiberfil, Thermocomp (US), Arpylene (GB), Verafil (CH).

Rovings, tapes or webs of GF-, CF- or Aramide-fibers are combined with high temperature thermoplastics (4.1.10/11) to form high performance *Advanced Thermoplastics Composites (ATPC)*.
These can be processed without curing by automated precision winding (Sec. 3.2.4), pultrusion (Sec. 3.4.3) and autoclave prepreg thermoforming (Sec. 3.2.4).

4.1.13.2. Glass mat reinforced thermoplastic prepregs (GMT)

Trade names: Azdel, STX (US), Baydur GMV Elastogran (DE), Symalit (CH)

2 to 4 mm thick GMT contain 30 to 50% continuous strands and long glass fiber mats in a thermoplastic matrix. The fibers are usually randomly distributed, but there are also GMT-UD with predominantly unidirectional mats. Cut-to size blanks, preheated freely to flow temperature, are transferred in cold or slightly preheated molds fixed in conventional fast-cycling presses for molding under 50 to 100 bar pressure. Press times are 20 to 30 s, cycle times 60 to 80 s. According to the complexity of shape, flow pressing (flash is reusable for injection molding) or positive mold pressing without flash can be used. GMT parts are stiffer and more impact resistant than reinforced injection moldings, GMT-UD parts are isotropic. Typical properties are PP-GMT E_t 4500 N/mm^2, Charpy impact 40 kJ/m^2 at 23°C, 50 kJ/m^2 at −40°C, PP-GMT-UD E_t to 16,000/1300 N/mm^2, impact 80 kJ/m^2, 90 kJ/m^2 at −40°C.

Continuous service temperatures with high mechanical strength are 100 to 150°C, dependent on the matrix material (TPU, PE-HD, PP, PA, PBT or PET).

Applications: Large area machine covers, underbody and other cladding in vehicles, GMT-bumper carriers.

4.1.14 Reclaiming, Reprocessing, Recycling

Much effort is currently devoted by raw materials manufacturers and processors and their trade associations to reclaiming scrap from plastics processing for further use. Recycling of used plastics products is showing dramatic increases although there are still some not inconsiderable difficulties to be overcome.

Processing scrap such as edge trim, sprues from injection molding, parison waste, out of tolerance moldings, etc. is reground and mixed in certain proportions with virgin material where possible. Such "internal recycling" is on the increase to feed valuable, separated, clean material back into the production process. Such regrind is often termed reclaim to distinguish it from recyclates which contain regrind from used plastics products and can be expected to be inferior in some respects as a result of degradation, swelling, contamination, etc. Some 13.5 million t of plastics residues occur every year in Western Europe (1990). Approximately 1 million t are recycled, 2.1 million t are incinerated and 10.5 million t end up in landfill or other dumping sites. The waste consists of (w/w):

60% polyolefins (PE, PP),
15% styrenics (PS, SB, ABS),
13% polyvinylchloride (PVC),
6% polyurethanes (PU),
4% polyethylene terephthalate (PET),
ca. 2% other plastics.

4.1. Thermoplastics: Raw Materials and Molding Compounds

Plastics represent 5 to 6% w/w of communal waste, i.e., 1.3 million t/y although they take up a disproportionate amount of space because of their low density.

Separated or sortable plastics scrap, i.e., rejects, used packaging materials, disposable tableware, disposable syringes, agricultural film, car batteries, palettes, dustbins and leftover granulated molding compounds have been recycled by specialist companies for many years. Reprocessors offer secondary injection molding and extrusion compounds with defined properties.

In many countries, plastics parts are marked in accordance with standards to ensure efficient recycling. In Germany, DIN 54840 E (Oct 91) specifies that the marking indicates the type of plastic, the type of filler or reinforcement and its % by weight. The characteristic letters and numbers are enclosed by > and < signs, e.g.

>PA 66-GF30<

Abbreviation for basic polymer
Filler and reinforcement
Filler content by % w/w

In the case of copolymers, the two main components are marked, e.g. >PA 6/12< and polymer blends are shown by, for example, >PC + PBT<. If a take-back system for plastics is in operation, a special logo is also shown:

Further processing of mixed plastics waste is more problematical (see Section 4.1.14.3) and the different types of plastics must first be separated.

4.1.14.1. Mechanical separation

1. The Flotation Process

The flotation process enables plastics mixtures to be separated into fractions with over 98% purity provided there is a difference in density of $0.05\,g/cm^3$. For example 98% pure polyolefin was separated from a granulated plastics fraction similar to that from segregated domestic refuse and free from other impurities using water as a separating medium. Impurities in the floating fraction are caused mainly by fine particles in the feedstock or foam materials. These could be separated by wind classification before flotation.

Successful segregation requires a separation zone free of turbulence which could cause the heavy fraction to float thus limiting output from the flotation process. Discontinuous flotation separators are less economic than continuous processes such as separation with hydrocyclones.

2. AKW Process and Andritz Process

Reprocessing is intended for recovering an almost completely segregated polyethylene fraction with a maximum 5% polypropylene content as regrind from presorted domestic refuse. The purity of the polyolefin constituent is 99.7% (AKW = Amberger Kaolin Werke).

After washing, the material is fed into a hydrocyclone using pressures of 0.5 to 1.5 bar. The material is separated into a heavy fraction (PS, PVC, etc.) and a light fraction (PE plus approx. 5% PP). Separation is sharper than in the flotation process. By feeding in containers and films separately, PE-HD rich and PE-LD rich fractions are obtained.

3. Continuous Melt Filtration

A continuous self-cleaning melt filter has been specially developed for reclaiming segregated polymers. To use the screen several times, part of the purified melt is taken off the main stream via a narrow side channel. The branch channel is located in the filter block after the screen disc and the secondary melt flow is thus fed to the rear of the screen. The cross-section of the melt channel is slit-shaped so that a good degree of screen cleaning is achieved with relatively little loss of material.

The processes described above are already in use and have proved cost-effective providing the price of molding compound is adequate.

4.1.14.2. Chemical and thermal processes

1. Hydrolysis and Alcoholisis

In contrast to the commodity plastics polyethylene, polystyrene and polyvinylchloride, macromolecular materials manufactured by polycondensation such as polyesters, polyamides, polyurethanes and polycarbonates possess certain linkages which are easily attacked chemically thus enabling tailored decomposition to monomers.

This can be exploited to recover starting materials (monomers) under relatively mild conditions. Using certain reactants the molecule is split in a reverse reaction at the position where polycondensation has resulted in linking. Suitable reactants are water, alcohols, acids or amines leading to various decomposition products from saponification depending on the process. This process is undergoing pilot testing.

2. Pyrolysis

Pyrolysis is the thermal decomposition (carbonization) of organic material in the absence of a gaseous medium (oxygen, air, carbon dioxide, etc.). Volatile fission products are produced at temperatures between 450 and 1200°C.

4.1. Thermoplastics: Raw Materials and Molding Compounds

Recycling of plastics waste by pyrolytic decomposition involves the splitting of the macromolecules into smaller molecule units which occur as pyrolysis carbon, oil and gas. The products of pyrolysis can be used either as chemical raw materials or as fuel thus providing the energy for heating the plant.

Smouldering and fluidised bed pyrolysis are used on a small scale at present. Another way of using mixed plastics waste has been suggested by Menges et al. of the IKV. Using a cascade extruder, the melt is degraded into low molecular components and chlorine is liberated. The oily liquid solidifies at room temperature into a brittle wax. It is suggested that this product is fed in powder or pellet form into other recycling plant (pyrolysis or hydrogenation plant). This process is at the prototype stage.

3. Hydrogenation

In principle, plastics are cracked during hydrogenation resulting in molecular fragments with reactive terminal groups. The reaction takes place in a hydrogen atmosphere so that hydrogen reacts with the radical terminal groups forming standard hydrocarbons. Two companies are operating pilot plants.

4. Incineration

Incineration of plastics waste is also termed energetic recycling. Up to 60% of the energy used to manufacture the plastic can be recovered. Commodity plastics, in particular, can be burnt in suitably equipped refuse incinerators without endangering the environment. Numerous tests have demonstrated that dioxin emissions are not measurably increased by the presence of PVC in the process. It is predicted that even with the most intensive efforts, a considerable proportion of plastics waste will not be able to be recycled via the materials route. Incineration of the residues is thus ecologically the most sensible path. Because of the high energy content of plastics, it would make sense to develop in the near future special-purpose high temperature incineration plant to process a certain proportion of plastics waste.

4.1.14.3. Processing of Mixed Plastics

Besides pyrolysis of mixed plastics, it is possible to melt-process mixed used plastics directly into thick-walled moldings using a planetary gear extruder which melts and homogenizes the material. At least 60 to 80% must be of the same type of plastic (e.g. polyolefins as found in coarsely sorted domestic waste) and the remaining 40 to 20% of other plastics or impurities. Simultaneous processing of different types of plastic does not result in a homogeneous structure since they have different melting temperatures. Thus at a particular melting temperature, one plastic can be liquid while another is not yet molten or is even decomposed. Thus the proportion of one type of plastic

4.1.15. Degradable Plastics

Degradable plastics are generally divided into photodegradable and biodegradable polymers. Water-soluble polymers, such as polyvinyl alcohol and polysaccharides, which biodegrade in water can be considered a sub-group.
Photo- and bio-degradability is normally undesirable because of the accompanying deterioration in physical properties and plastics are thus optimized for long life by incorporating stabilizers and preservatives. The objective of making a proportion of plastics degradable is to reduce the volume of plastics refuse through decomposition of the polymer chain into carbon dioxide and water, if possible, and additionally, in the case of biodegradation, into non-toxic metabolites of microorganisms.

4.1.15.1. Photodegradable plastics

Selective incorporation of UV sensitive molecular structures such as keto groups (e.g. E/CO-Cop) and addition of photosensitizers (e.g. iron dialkylthiocarbamate and other organometallics) enables photodegradation of polymers to be controlled relatively precisely. The main applications of photodegradable polymers are agricultural films, carrier bags and refuse sacks (Ecolyte-P and S for PE and PS; Ecolon; Eslen-PS; Ercoten; Plastor, Plastopil).

4.1.15.2. Biodegradable plastics

Biodegradation can occur aerobically or anaerobically but always only in the presence of humidity. The surface of biodegradable plastics must therefore be hydrophilic. Absorption of moisture impairs the mechanical properties. These products are manufactured as follows:

1. Special strains of bacteria form hydroxycarboxylic acid polyesters, e.g., polyhydroxybutyrate and valerate (Biopol, ICI), polylactic acid and polycaprolactone.

2. Compounds of biodegradable polymers (polysaccharides, starches) with conventional, non-biodegradable polymers (PE) with approx. 94% PE. Higher starch levels cause processing problems. Normally a starch-PE-colorant masterbatch is prepared and then extruded with further PE (St. Lawrence Starch Comp., Archer Daniels Midland Co., Epron Ind. Ltd., Amylum). Packaging is the main area of application.

The PE-encapsulated starch particles are degraded after diffusion of moisture. The PE matrix is however not biodegraded.

Of particular interest is a product consisting of a specialty starch (up to 80%) and a non-olefinic thermoplastic polymer matrix (mater-Bi, Montedison). In contrast to other biodegradable polymers it can be melt-processed and is almost completely degradable. The high starch content provides the material with good oxygen and grease barrier properties and it is thus used for films and coatings in hygiene applications and for thermoformed and injection molded articles.

3. Several chemically modified "thermoplastic" starch products are at the development stage. These include hydroxypropylated starches (Ems Chemie/Batelle, Warner Lambert, Fluntera AG, National Starch and Chemical Co., American Excelsior Corp.)

4. Potato starches, maize and rice starches compressed under high pressure and humidity serve as replacements for foam PS chips. Similar compressed products contain plant fiber.

5. Polyvinyl alcohol (PVAL) water-soluble packaging is produced by casting or the cheaper extrusion blow molding process. Both cold water soluble and hot water soluble grades are available (Aquafilm Ltd, Aicello), the former for packaging toxic powders, e.g. plant protection products, the latter for bacteria-impermeable packaging for hospital laundry.

4.2. Semi finished Thermoplastic Material

4.2.1. Standardized and Hallmarked Piping Material

Comprehensive international basic and applications-related standards continue to be developed for thermoplastic pipes, tubes, fittings and jointing materials (Table 4.51)[1]. The last include welding sleeves (PE-HD), solvent cements (ABS, PVC, CAB), metallic parts and seals and gaskets for detachable joints (Table 4.52). General standards are complemented by commercial and military standards. To these must be added building codes or rules of good practice developed by the gas, water and sanitary authorities. These regulations differ in detail in individual countries. However, they are all based on the structural behavior that is common to thermoplastics. The basic standards encompass

---

[1] ISO 161/1978: Thermoplastic pipes for the transport of fluids, normal outside diameters and nominal pressures. Part I: metric series; Part II: inch series. The standards of the relevant ISO/TC 138 supplement the general basic standard with technical application dimensional standards for fittings and valves and with testing standards for piping materials. For these see ISO Standards Handbook 28, Volume 2.

4.2.1. Standardized and Hallmarked Piping Material

Table 4.51. Thermoplastic Piping Systems: Material, Basic Standards, Applications

Materials	PE-LD PE-HD	PE-X	PP PP-R	PB	ABS (ASA)	PVC PVC-HI	PVC-C
Molding and extrusion compounds, page	176	190	196	201	202	219	215
Details semi-finished products, page	300	300	300	300	305	306	306
ASTM[1] cf. Annual book of ASTM Standards, vol. 08.04	D 2239 D 3035 F714	F 876 F 877	–	D 2662 D 3000 F 809	D 2282 D 2661 D 2751	D 2241 D 2665 D 2729	D 2846 F 442
DIN[2] cf. DIN Taschenbuch 52 for basic, 190 for application standards[3]	8072/3 8074/5	16892 16893	8077 8078	16968 16969	16890 16891	8061 8062	8079 8080
Gas supply Water supply Hot and cold water installations (Floor) heating	+ +	+ +	+ +	+ +	+	+ +	+
Pneumatic post Air conditioning						+ +[4]	
Guttering Drain, waste, vent (DVW) Accessible tube and tubular fittings Buried drain and sewers[4] Soil draining (perforated, corrugated)	+ + +		+ +	+	+ +	+ + +	+
(Buried) electric conduit	+				+	+	

[1] SDR-PR series, some main standards for drain, waste, vent and sewer pipes and fittings. Basic (D 2446) and application standards, also for CAB (Sec. 4.2.9.2). For reinforced thermoset pipes see Sec. 4.6.3.7.
[2] Dimensions and requirements, applying also to industrial piping. Basic dimension standards also for PA (Sec. 4.2.7), PC (Sec. 4.2.8.1), and PTP (Sec. 4.2.8.2). Similar dimensional series, but not (yet) standardized. PVDF (Sec. 4.2.5), PMMA (Sec. 4.2.6.1), PPO (Sec. 4.2.7). For reinforced thermoset pipes see Sec. 4.6.3.7.
[3] Also containing an ISO directory and lists of related regulations.
[4] Also large diameters ($\leq$ 3000 mm): wound pipes with profiled walls (DIN 16961).

pipe series with nominal diameters between about 10 mm ($\frac{1}{2}$ in) and, according to type of pipe, 160 to 1000 mm (ASTM 6 to 12 in). Wall thicknesses generally follow the ISO equation[1]

$$S = P\frac{OD - t}{2t} \text{ or } S = P\frac{ID + t}{2t}$$

where S = hoop stress, P = pressure, OD = average outside diameter (ISO: DN), ID = average internal diameter and t = minimum wall thickness. In the ISO/DIN and in the American SDR (Standard Dimension Ratio) series the ratio OD to t is maintained constant for all dimensions of a particular series. The nominal pressure PN, called PR (pressure rating) in the US, is constant for all pipes and fittings belonging to the same series. The individual

4.2. Semi finished Thermoplastic Material

Table 4.52. Plastic Pipe Joints

Material	Use	Detachable joints		Permanent joints	
		Sliding sockets	(supported) flanged or screwed joints	Adhesive-bonded bell and spigot joints	Butt or socket heat fusion
PE	All uses		+		++
PE-X	Floor heating		+	(+)	
PP	Industrial, floor heating		+		++
	DVW*	+			
PB	Hot water installations		+		+
ABS	Most applications		(+)	++	
	DVW*	+			
PVC	Industrial		+	++	+
	Gas supply, pneumatic post		(+)	+	
	Water supply, sewers	++	+	++	
	DVW*	++		(+)	
	Soil draining	+			
	Electric conduit	+		+	
PVC-C	Industrial		+	+	+
	Hot water installations		+	+	
	DVW*	+			

*DVW: Drain, vent and waste pipes

series of standards for given types of pipes are, for example, based on PN = 2.5, 4, 10 or 16 bar or PR = 50 to 250, max. 315 psi (0.34 to 1.72, max. 2.17 MPa)[2].

PN and PR relate to expected continuous service for at least 50 years at PN or PR internal pressure with water or non-hazardous transported fluids at room temperature. This is on the basis of results of the long-term hydrostatic strength (LTHS), i.e., the circumferentially oriented (hoop) stress measured on the pipe itself and its reduction, depending on the type of pipe, by safety factors of between 1.4 and 2.8 on the hydrostatic design stress (HDS) that can be continuously applied.

According to ASTM D 2837-76 the mean values of time-to-failure measurements (to D 1598) at various pressures and temperatures in the prescribed distribution between 10 and 10 000 h are made linear by applying the least squares method to a log stress-log time curve for the lower confidence limit (LCL). From these curves the LTHS at 100 000 h and the hydrostatic strength

---

[2] Alongside SDR-PR standard specifications there are ASTM specifications, Schedules 40, 80 and 120, for PE, PVC, PVC-C, ABS and CAB pipes with their assimilated Iron Pipes Standards series of dimensions. With increasing diameter the pipes decrease within SDR and PR.

4.2.1. Standardized and Hallmarked Piping Material 299

at 50 years are extrapolated. By comparison of these values, if necessary taking into account a limiting circumferential expansion, the values are categorized as the hydrostatic design base (HDB) for further practical calculations. In ASTM D 2837, such values are tabulated from 190 to 5000 psi (1.31–34.47 MPa). In European pipe standardization the relationships evident from the logarithmic minimum time curves for various temperatures (e.g., Fig. 4.14), which have been experimentally confirmed over several decades, are used to extrapolate the LTHS to 50 years and to predict the temperature resistance of plastic pipes (Fig. 4.15). Both processes lead to agreement to a first-order approximation

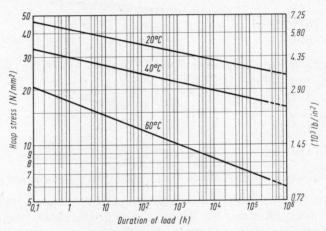

Fig. 4.14. Long-term behavior (minimum values) of PVC pipes at various temperatures.

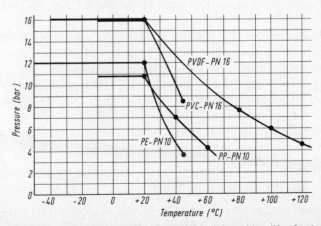

Fig. 4.15. Pressure and temperature loading of industrial piping with a life of at least 20 years. Media influences must be taken into account, when appropriate, with "resistance factors".

4.2. Semi finished Thermoplastic Material

but not to identical results. Unified ISO processes for the determination of long-term stability of pipes under load are currently under development. According to ISO method No. 1, 10 000 h curves, measured at three elevated temperatures differing by at least 10 K, can be extrapolated to 50 years at temperatures at least 40 K lower than the highest test temperature.

ISO methods No. 2 and No. 3 resemble the ASTM methods. Creep curves also give materials data for calculation of the design of plastic piping under permanent external pressure, e.g., for underground or submarine pipelines and cable conduits. For free-flowing systems, apart from the deformation, the buckling strength under external pressure must be assessed.

4.2.2. Polyolefins

For materials and general properties see PE, Sec. 4.1.2.1, PP, Sec. 4.1.2.6, PB, Sec. 4.1.2.7, and PIB, Sec. 4.1.1.2.3.

4.2.2.1. Pipes

For general applications see Table 4.51.

Forms of delivery, nominal diameters (DN) and special applications

PE: Generally UV stabilized with carbon black, DN 10 to approx. 160 mm ($\frac{1}{2}$ to 6 in) in 300 m lengths in ring bundles or on cable drums, above these diameters in fixed lengths, e.g., 6 m or 12 m (20 or 40 ft).

PE-LD: DN 15 to approx. 160 mm ($\frac{1}{2}$ to 6 in) in 300 m lengths in ring bundles or on cable drums, above these diameters in fixed lengths, e.g., 6 m or 12 m (20 or 40 ft).

PE-HD: Pressure pipes DN $\leq$ 450 mm (18 in); extruded pipes feasible to DN 1600 mm (63 in). Long-distance supply lines, buried or underwater (e.g., for islands), sewage disposal systems or outlets and underwater cable conduits can all be produced without joins by extruding PE-HD pipes continuously on-site with mobile extruding equipment. Hot-wound pipes (Sec. 3.6.5.1) stiffened by hollow profiled walls with integrated bell ends, DN 500 to 3600 mm (20–140 in) are made for sewage and ventilating systems. Extruded gas supply pipes (DN $\leq$ 200 mm, 8 in) are often yellow.

PE-X: For cross-linked PE-pipes, used for hot water installations and floor heating, see Sec. 4.1.2.2.

PP: Off-white pressure pipes in fixed lengths up to approx. DN 450 mm (18 in), hot-wound pipes for industrial purposes, flame-retardant hot water resistant drain and waste installations inside buildings, DN = 40 to 150 mm (Europe).

Floor-heating copolymer pipes, DN = 16 to 20 mm, in continuous lengths of 120 to 240 m.

PB: Hot and cold water distribution systems up to 82°C (180°F), 0.25 to 2 in (6.4 to 50.8 mm) nominal size. Also floor heating pipes as described.

4.2.2.2. Sheets and sheeting

PE and PP sheets 0.5 mm to 100 mm (0.02 to 4 in) thick and 1000 by 2000 mm to 2500 by 5000 mm (40 to 200 in) in size are used for chemical plant and equipment. Laminating with GF gives the base for an outer reinforcing layer with GF mat or fabric. Flexible PIB sheeting, 1.5 to 3 mm thick, filled with carbon black or graphite, can be bonded by adhesives onto concrete or metals and seam welded. It is used to provide corrosion-resistant linings. For highly abrasion-resistant linings of PE-UHMW see Sec. 4.1.2.1.8.

Roof covering and membrane liners are as a rule 1 to 2.5 mm thick, 1 to 1.5 m wide, filled but flexible sheeting without plasticizers and with or without felt or fabric carriers. They are made of bitumen-modified ethylene copolymers (ECB, Sec. 4.1.1.2.4), chlorinated PE (PE-C, Sec. 4.1.2.3) or sulfochlorinated PE (CSM, p. 414), VAE (Sec. 4.1.2.4.2), APP (Sec. 4.1.1.2.1) or PIB (Sec. 4.1.1.2.3): Various standardized types are used for all kinds of flat roofing as well as for sealing membranes in applications such as garbage landfills, tunnels, reservoirs and shipping or irrigation canals, which must be permanently resistant to hydrostatic pressure and chemical and biological attack. They are compatible with bituminous materials but can be solvent or hot-seam welded and can cover large areas. Extruded PE-HD sheets, 10 m wide, about 3 mm thick and up to 150 m long joined by only a few extrusion welds can be used to meet the heaviest demands. Thinner PE sheeting and films are used as damp-proof membranes in construction.

For thermoformable sheets used for, e.g., car door paneling and made of chalk- or wood-filled PP see Sec. 4.1.2.6.3. Special 0.3 to 2.0 mm (12–80 mil) thick 50% chalk-filled PP sheeting is solid phase pressure formed (SPPF process, Sec. 3.8.2) with a heated plug but a cold cavity to produce packaging that can be filled with hot contents (< 130°C).

4.2.2.3. Films

For general properties of packaging and electrofilms see Tables 4.54 and 4.55.

1. *Blown PE-LD film* (Sec. 3.6.5.6). Service temperature range −60 to 60°C (−76 to 140°F), supplied as tubular film or cut into webs.

4.2. Semi finished Thermoplastic Material

Packaging film for machine processing in working widths of < 2000 mm (< 80 in), $\sigma_B = 18$ to 30 N/mm^2, also pretreated on one side by corona discharge for color printing, slip or nonslip (for stacking), cold-needle perforated for oxygen exchange or with a coating against exudation of the packed contents. PE-LLD or EVA modified forms enhance the toughness and impact strength. The films are glossy, mat transparent or translucent.
Gauges:
- < 10 to approx. 50 μm (0.4 to 2 mil): sealed and wrapped packages for (frozen) food and non-foods (e.g., textiles), laminating films.
- 50 to 100 μm (2 to 4 mil): carrier bags.
- ≤ 150 μm (≤ 6 mil): drum and Kraft paper sack liners.
- 150 to 250 μm (6 to 10 mil): rubbish and fertilizer sacks and other heavy-goods sacks.

Shrink films, also radiation cross-linked, shrink temperatures > 110°C (230°F): Biaxial 20–50 μm, for the shrink wrapping (and freezing) of poultry, fresh meat, cucumbers and non-food items (e.g., books and writing materials); thicker and uniaxial shrink films are used for packing of bundles and pallets.

Stretch films: Mostly 25 μm (1.0 mil) thick, highly stretchable, with adhesive coating, to fix the packaged goods on pallets by wrapping under tension at room temperature.

Wide films: Double layered, 2000 to 6000 mm (80 to 240 in) wide: Webs for membranes under pitched roofs (50 μm) also perforated or reinforced with latticed fabric; covering films in building (100 to 200 μm) and for greenhouses; black mulch film (80 μm) and silage film (200 μm) in agriculture; clothes protection bags, rain capes, and the like.

2. *Cast (chill roll extruded) PE-LLD cling film*, 10 to 30 μm (0.4 to 1.2 mil) for commercial and domestic food wrapping made of VLD grades (Sec. 4.1.2.1.2) is highly transparent and puncture resistant. EVA copolymers are also used in this field.

3. *PE-MD to PE-HD film*: Premium grades for mechanical strength and for temperature resistance (100°C boil-in bags), for more highly stressed light and heavy packaging, for hot-filling packaging, also in PB films. The lower cost high-strength, thin PE-LLD film (Sec. 4.1.2.1.2) is also used in this market.

4. *PE-HMWHD "paper" film*: Strongly stretched tubular film, not fully transparent, approx. 10 μm (0.4 mil) thick, tough with tissue-paper-like handle in the packaging field. For supermarket carriers 100 μm thick, highly filled like printing paper.

ASTM D 2103 provides a *classification system* for all types of PE film and sheeting up to 0.3 mm (0.012 in) thickness. It is based on cell numbers for (1) density (0.910 to 0.965), (2) dart impact strength (< 40 to > 50 g/mil), (3) coefficient of friction (< 0.2 to > 0.7), (4) haze (< 5 to > 9%), and (5) light transmittance (< 1 to 100%).

5. *PP "cast" film*: Chill roll extruded, (see p. 137) 10 to 90 μm (0.4 to 3.6 mil), up to 1750 mm (69 in) wide, $\sigma_B = 30$ to 60 N/mm^2, higher clarity and gloss, stiffer than PE films, cellophane-like touch, sealable for packaging textiles, stationery, flowers and food. It is also available perforated, antistatic, treated for print or laminating and microporous, bacteria-proof for ultra-filtration and sterilizable for medical use. For classification by haze, gloss, coefficient of friction, secant modulus and elongation at break (> 600 to < 400%) see *ASTM D 2530*.

6. *OPP film*: Biaxially stretched; 10 to 50 μm (0.4 to 2 mil); up to 3000 mm (120 in) wide; elongation at break, MD 150%/TD 50%; crystal clear; σ_B "unbalanced" MD 120 to 180 N/mm^2; TD 300 to 400 N/mm^2; for "balanced" BOPP film by annealing (according to ASTM D 2673) > 103 N/mm^2 in both directions, differing by no more than 55 N/mm^2 or 60% absolutely (cf. Fig. 3.64, p 140). It is used for high quality see-through packaging of moisture-sensitive articles. It is coated for heat sealing with
   - EVA (slight water vapor, high gas permeability) for baked wares, confectionery, pharmaceuticals and tobacco items
   - PVDC (slight water vapor and gas permeability) for high quality foods, fat-containing products, spices; special grades: electro-films (condensers).

7. *TQ film*: The downward extrusion of PP through an annular die followed by rapid cooling (quenching) using a water ring is a simple, but effective process for making nearly clear, heat sealable carrier bags. This process is widely used in Latin America and South East Asia.

For multi-layer films PE-LD/PE-HD, PE-HD/PE-HMWHD, PE/PP/PE, PE/ionomer-PA etc. see Sec. 4.3.2, for PMP film Sec. 4.2.7.

4.2.2.4. Fleeces and flat filaments

For PE and PP "fibrides" see Sec. 4.1.1.1.10 and 4.1.2.7. Spin-bound PP fleece serves to consolidate the soil in road and water system construction. Filaments and tapes cut from film and strengthened by stretching when processed to (raschel-knit) fabrics, nets or ropes have many applications e.g. big bags for heavy goods, construction site and other protective webs and geotextiles. Woven fabric sacks are coated with PE in the extrusion process.

4.2. Semi finished Thermoplastic Material

4.2.2.5. Foams

Types and trade names (examples)
PE Particle foam: Arpak, Volara (US): Neopolen (DE). PE-HD: Novawood, Furukawa structural foam (JP). Extruded foam cross-linked chemically or by radiation: Frelen, Trocellen (DE); Softlon (JP); Plastazote (GB); Alveolit (CH). Extruded foam, non-cross-linked: Ethafoam (US); Alkozell, d 35 kg/m³, Synthen (DE); Lightlon, Eftlon (JP); Alveocel, d 30 to 40 kg/m³, Alveolan, d 500 to 600 kg/m³ (CH).

PE-LD particle foam (generally cross-linked) and closed-cell cross-linked extruded foam are tough-rigid to semi-rigid. Non-cross-linked extruder foam is soft. The semi-rigid cross-linked foams ($d = 30 - 200$ kg/m³, $\tau_B = 0.3 - 2.0$ N/mm², $\varepsilon_B = 90 - 200\%$, $\lambda = 0.04$ to 0.05 W/mK) with their high compressibility and good capacity for recovery are serviceable from about $-70°$C to between 80 and 100°C. At 80°C they shrink by 1 to 2%. Their water absorption is minimal. PE foam can be supplied in flame retardant grades.

Forms of delivery: Rolls of 1 to 10 mm thickness in webs or sheets, mostly 1000 mm wide. The foams are thermoformable and can be welded. Applications include shock damping and insulating for vehicles, packings, sports equipment, protective clothing, prosthetics and directly molded orthopedic supports, packaging, buoyancy elements in life jackets and water sports equipment, membranes in building and in sports center construction. Low density extruded non-cross-linked foam is used for filling and heat insulating profiles and pipe covering. Higher density material is used as films or webs of 0.15 to 2 mm thickness for carrier bags, coverings, mats and inserts.

Irradiation cross-linkable PP extruder foam is used for car interior fittings.
Cross-linked EVA foam (Evazote, GB, d 40 to 135 kg/m³) is rubber-like and is used for garments that protect against cold. Closed cell copolymer foam, d 150 to 265 kg/m³, with a tensile strength of approx. 1 N/mm² and an extension at break of 60 to 80% (Herex, JP) is supplied in rolls with a thickness of 0.5 to 5 mm. Conductine (GB), closed cell EVA/PE foams more than 3 mm thick are available for aircraft and electronics.

For light wood-like structural foams with a board profile and density of about 200 kg/m³ made of PE-HD (Woodlite, JP) see Sec. 3.6.5.3, for PE foam cable insulation see Sec. 4.1.2.1.5.4.

4.2.3. Styrene Polymers

For properties and general considerations regarding behavior and applications of styrene polymers see Sec. 4.1.3.

4.2.3. Styrene Polymers

1. *Standard polystyrene* sheet is used in high-frequency applications and, UV-S stabilized, in lights and lamps. Special grades of cross-linked polystyrene can be used at temperatures up to 100°C in HF applications. Undrawn PS films are brittle. Biaxially oriented they are supplied as electrical insulation, brilliant crystal clear laminating films, printing films and packaging and thermoforming films. Thicknesses range from 4 to 100 μm, (0.2–4 mil). Properties are summarized in Tables 4.58 and 4.59, pp. 327, 328. For biaxially oriented polystyrene sheet (<1.02 mm, 0.04 in thick) ASTM D 1463 specifies an orientation ratio of at least 2:1 in each direction and a minimum tensile strength for PS of 50 MPa (8000 psi); for SAN, 72.5 MPa (10 500 psi).

2. *SAN, SB and ABS sheet*
   Forms of supply: Rolled sheeting or sheets, 0.5 to >10 mm (0.02 to 0.4 in) thick and in large dimensions. Natural, white and in many colors. Surface machine smooth and glossy on one side (also high-gloss laminated), with leather-grained and other decorative effects. For general properties of HT, HI and very high HI in leather-like types see Tables 4.13 and 4.14.
   Processing by mechanical thermoforming is described in Sec. 3.8.4
   Applications include refrigerator and refrigerator door linings, office machine housings and coverings, basins, transport containers, cases, displays, toys and car-body and aircraft interior fittings (flame retardant). Traffic shields subject to external weathering, seating furniture, protective headgear, bodies of sports vehicles and sports boats are made from stabilized ABS or ASA. Multi-ply ABS sheets with a foamed core (Royalite, US) are used for boats and sanitary fittings as are coextruded laminates of ABS and PMMA with coupling agents (p. 315). Films >0.1 mm (4 mil) and sheeting are raw materials for the mass production of thermoformed packaging containers and throwaway beakers. Multilayered coextruded composite films are used for multi-colored products. The films consist, for example, of 0.3 mm colored impact resistant PS and 0.05 mm standard crystal clear PS as a glossy outer layer.

3. *For ABS pipe* see Table 4.51.

4. *EPS boards and sheeting* (see Sec. 4.1.3.2)

Flame-retarded *building materials* include:
– Heat insulation panels (up to 100 mm thick): Particle foamed in the 15 to 30 kg/m^3 (1 to 2 lb/ft^3) density range, cut from thick blocks, continuously produced with smooth skin on both sides, molded in steam-heated closed molds with formed edges and surface patterns or foamed by extrusion, d 30 to 40 kg/m^3.

4.2. Semi finished Thermoplastic Material

- Solid-borne sound-insulation panels, elastified by milling or pressing to dynamic stiffnesses of $<30\,MN/m^3$.

In the *packaging sector* blown film, d 60 to $200\,kg/m^3$, is foamed from a thickness of 0.1 to 3.5 mm and freely after-foamed to 6.6 mm. It can be thermoformed up to a ratio of 1:1 into fast-food and egg packaging, menu covers and disposables. Boxes are made from two-sided paper-coated continuous blown film.

Extruded PS structural foam (d 400 to $500\,kg/m^3$) resembles wood in appearance and behavior; it is used for interior building profiles by after-embossing surface-patterned panels (Denkalite wood, Woodlac, Woodlite, JP).

4.2.4. Vinyl Chloride Polymers

For materials and general properties see Sec. 4.1.4.1/4.

4.2.4.1. Rigid PVC, unplasticized (PVC-U)

1. *Pipes*

   For standards and general applications see Sec. 4.2.1. Typical standard lengths are 6 and 12 m (20 and 40 ft), with or without bell ends.
   Main applications: Pressure pipes for industrial uses (PN up to 10 bar), DN 10 to 315 or 110 mm; for water mains (bell and spigot gasket or adhesive-bonded joints) to DN 400 mm; and for gas mains (adhesive bonded only) to DN 225 mm. DVW (drain, vent and waste) pipes (rubber ring socket), DN 40–150 mm and standard sewer dimensions (with socket for elastic sealing ring joints), DN 100 to 500 mm. Glass clear PVC pipes, DN 6 to 160 mm (e.g. Simona, DE).

2. *Chemical engineering*

   Blocks (up to 100 mm thick), sheets 2000 by 4000 mm and up to 30 mm thick, continuous lining sheeting (up to 1 mm) and welding profiles are used for equipment, pumps and ventilators in the chemical and related industries. They are suitable for temperatures up to 60°C. PVC-C, e.g., in chlorine plants, can be used up to 90°C. PVCHI is used in air-conditioning equipment. Partially conducting PVC is explosion-proof. Apparatus can be armored by laminating with GRP. PVC pipes armored with standard GRP are standard products. Sintered porous filters and accumulator separator plates are available as special products. For general properties see Table 4.23.
   ASTM D 1927 for rigid PVC sheets (see also DIN 16927) differentiates between three types. Type I is for the highest chemical and mechanical

requirements, Type II is for high impact resistance and Type III is for medium impact with moderate or low chemical resistance. Standard D 2123 for VAC-cop sheets refers to Type I for the highest general requirements (including clarity) and Type II for higher impact strength. The ASTM standards give only general definitions and not specific requirements.

3. *Structural engineering*

   Raised impact strength PVC, suitably stabilized and in light colors, has proved itself as a material for outdoor use capable of withstanding long-term weathering and service in different climates. Applications include

   – Transparent, translucent or opaque, flat, lengthwise or transverse corrugated sheeting and sheets for roofing and façades. Biaxially stretched sheets are hailstorm-proof;
   – Translucent double-wall extruded hollow panels connected lengthwise by tongue and groove joints to heat insulating walls and roofings;
   – 2 to 3 mm (0.08 to 0.12 in) thick, mainly off-white pigmented sheets for thermoforming tailor-made cold façade cladding or mass produced (also wood grained) wall cladding, louvres and interior decoration;
   – Single or multicavity sections for manufacturing window and door frames by machine mitering and butt welding.

Hollow profiles for *roller shutter blades*, a widespread application in Europe, and profiles for interior decoration and furniture are made largely of standard PVC-U grades.

PVC foils, 0.08 to 1 mm thick, are used principally for packaging such as pots, lids, blisters, folding cartons, etc. and for industrial applications in the building, printing and computer sectors.

4. *For consumables, office, drawing, measuring and graphics equipment* the dimensional stability and corrosion resistance of high-gloss or mat-surfaced products (for lettering), made commonly from PVC-PVAC, are of major importance. Typical applications include card index and punched cards, slide rules, set squares, stencils, rulers and measuring scales, engravings from multicolored laminated materials, repro- and cartographic drawing materials, wallboard coverings and engraving block material for various printing processes.

5. *Other applications of thermoformable sheet materials* include refrigerator parts, machine coverings, street signs and car number plates, clothing inserts, displays and lighting parts, raised type for the blind and relief maps.

6. *Packaging sheeting and film* is mainly made by high-temperature calendering (see Table 3.14) of PVC-S or the more glass-clear, flexible, low temperature-

resistant copolymers or blends. There are as many types as there are requirements, also FDA approved for food. Sheeting 0.3 to 0.6 mm (12 to 24 mil) thick is stable enough for (partly welded) cans, boxes and thermoformed cups; 0.1 mm thick sheet for rigid nesting packaging of small objects.

7. *Thermally treated PVC-E film* (see Table 3.14) when unstretched is somewhat cloudy, but rather tough. Longitudinally stretched, it is used for adhesive tapes, cable winding and sound-carrier tapes.
For properties of rigid PVC-S and PVC-E packaging films and of *vinylidene chloride (VDC) copolymer* films see Table 4.54. VDC-copolymers, because of thermal instability mainly used as barrier interlayer or coating (Sec. 4.3.2.1), are extruded, with 4 to 9 phr plasticizer, to 15 to 70 μm thick (shrink) wrapping films.

8. Furniture foils are supplied usually in widths of 1200 to 2000 mm either as type I, rigid (0.1 to 0.4 mm thick) or type II, semi-rigid (0.15 to 0.3 mm thick) multicolor printed with wood grain, embossed or smooth, frequently with transparent PU gel coats, sometimes plain colored, veloured or metallized. They are suitable for machining in combination with chipboard, plywood, steel or aluminum sheet. They can be thermoformed and folded for mitred corners on furniture carcasses. Type I foils are used for abrasion and chemical resistant door leaves and furniture fronts and visible surfaces on ceiling panels and room dividers. Type II foils are used mainly for furniture interiors.

9. Very thin (down to $< 50\,\mu$m, < 2 mil) *blown film* (Sec. 3.6.5.6) biaxially oriented by blowing is used as wrapping or shrink film for overpacks, for impact-proof windows in letter envelopes and for garlands and similar decorations.

4.2.4.2. Plasticized PVC (PVC-P)

For structure and material properties see Sec. 4.1.4.7.

Main groups of PVC-P products:

Plasticizer content below 25%, Shore hardness approx. A 95: semi-rigid for thermoformable products such as shrink tubes (Sec. 3.8.1.2), 20 μm thick transparent cling films for food wrapping and other films for flexible molded packaging.
Plasticizer content about 30%; Shore hardness A 85 to A 80: sole leather-like profiles and sealing sheet gaskets, waterproof liners and roof-covering membranes, clothing and decorative films, floor covering scuff-resistant top layers.
Plasticizer content, 35 to 40%; Shore hardness, A 75 to A 60; bag and padding film, flexible gasket profiles, tubes, filled PVC flooring.

1. Tubes and profiles

Tubes with internal diameters of 1 to 80 mm (0.04 to 3 in) and wall thickness of 0.3 to 10 mm (0.012 to 0.4 in) are commercially available as machine tubing, for carbon dioxide, propane gas, water and acids. Special grades are suitable for garden sprays and for drink and beer piping, for blood transfusion and other medical purposes and for electrical insulating tubes and abrasion protective covering.

Portion packagings of lighter fuel, lubricants, cleaners and detergents are made by cross-welding tubing with wall thicknesses of 0.1 to 0.35 mm (4 to 14 mil) after filling. Tubes with a screw cap or similar closures are also made of such tubing.

Flexible profiles are used as welting or beading in footwear and leatherware and upholstery and in the automotive and other industries, as truck tarpaulin ropes, belt drives and, e.g., clothing-lines coated by extrusion.

Construction profiles for interior use such as stair edging, skirting, bannister handrails, table and door edging, window seals and glazing profiles are made in many forms and sizes. In construction and structural civil engineering, specially designed high-duty profiles serve as expansion joint sealing anchored on both sides.

2. Sheets, webs and band for technical and packaging applications

Sheets are used for seals exposed to corrosive attack and for movable parts of equipment, for the lining of galvanizing baths, for wear and tear protection, as punch supports and so on. Pressed glass-clear material serves for the most demanding transparency requirements, e.g., for factory swing doors or windows for convertible cars.

Water-proof membranes, 0.7 to 3 mm thick, also reinforced by woven or non-woven fabrics, are on the market in many types. Generally the sheets are lap jointed by heat or solution welding in the workshop and on-site to form large impermeable tarpaulins. There are special methods of fixing these mechanically for tunnel linings and other construction engineering purposes and for light-weight flat roofs. They are laid as floating membranes on other flat roofs. Waterproofing by such one-ply techniques does not necessitate sealing by several bituminous layers, and the sheets used for them are mostly non-bitumen-resistant grades. Bitumen and mineral oil resistant types are used, e.g., under asphalt road bases, in protecting dams surrounding oil tank fields and, pore-free laminated, for leak-proof blisters in heating oil tanks. Translucent sheets reinforced with synthetic fiber fabrics are used for air-inflated structures and tents.

Heavy work *clothes* are made of reinforced material; light work clothes and rainwear are made from unreinforced sheeting of 0.2 to 0.4 mm thickness. For

4.2. Semi finished Thermoplastic Material

sacks, tubular sheeting or film 0.15 to 0.5 mm (6 to 30 mil) thick, 40 to 75 cm (16 to 30 in) double-wide material is used. Transparent to crystal-clear material 0.08 to 0.5 mm thick is converted into removable hoods and clear-view covers. Medical equipment, self-adhesive and insulation tapes and self-adhering corrosion-resistant wrappings are made with PVC grades containing non-migrating polymeric plasticizers.

3. *Flooring, wall and table coverings*

Homogeneous single layer PVC floorings up to 2 mm (80 mil) thickness are produced either as webs up to 160 cm (4 ft 8 in) wide by extruders with special arrangements for color mixing and polishing or – the most expensive design – by pressing single sheets between polishing plates (Sec. 3.5.3). Webs and sheets are also cut in square tiles. They are supplied in one color and many oriented or non-oriented patterns. Such homogeneous floorings are extremely wear and corrosion resistant, but rather expensive. They can be considered for ranges of applications that are subject to aggressive wear as in hotels or in industrial situations such as laboratories. By welding mostly color-contrasting plasticized PVC filler rods into the joints with welding machines running on rollers, it is possible to produce seamless, easy-clean floor-coverings in any desirable size.
Heterogeneous PVC floor coverings are made continuously as webs up to 4 m (12 ft) wide on complex production lines. Such lines combine the facilities for spreading (foamable) PVC pastes on flexible bases, laminating several layers and multicolored printing and, if necessary, mechanical embossing.
The printed picture, e.g., of glazed tiling, simulated natural products or modern designs, is protected by a hardwearing covering layer of glass-clear PVC and in many cases by a PU lacquer. Mechanically embossed relief surface flooring 1.2 to 2 mm (50 to 80 mil) thick is raised on a base of, e.g., mineral fiber fleece, strengthened by a PVC-spread. *Cushioned vinyl* (CV) coverings are chemically embossed by the local addition of foam inhibitors to the printing inks. Four-to five-layered CV coverings up to 4 mm (0.16 in) thick with the embedded intermediate stabilizing glass-fabric carrier combine the highest foot comfort with considerable heat and sound insulation.
Wall coverings about 1 mm thick are made by both processes. PVC *table coverings* are single layered and homogeneous.

4. *Artificial leather and films for welded articles*

Heavy deeply embossed artificial leather or leather cloth for public transport and commercial vehicle seats is produced by calender laminating or coating fabric with PVC (foam) plastisols (Sec. 4.1.4.9 and 4.6.1.8.2.4.2). Calendered sheeting is used as
– leather-like embossed sheeting for flat upholstery and car furnishing 0.4 to 0.6 mm (16 to 24 mil);

- high gloss or embossed plastic "leather" for bag making 0.2 to 0.5 mm (8 to 20 mil);
- film for welded rainwear and other garments, bathing bags, laminated nonporous inflatable articles, etc., as well as for curtains, table cloths, etc ≤ 0.2 mm.

4.2.4.3. PVC foams

For foaming processes for PVC-U and PVC-P see p. 233. PVC-U structural foam panels, d 0.7 g/cm^3, are thermoformable lightweight materials for indoor and outdoor construction.

PVC-U foam, d 0.04 to 0.13 g/cm^3, closed celled in sheets and blocks, 3 to 160 mm thick, is used as sandwich core material for boats, fishing net buoys and rescue floats and for cryogenic containers of liquid gases.

Divinylcell (SE) is an aramide cross-linked rigid PVC foam with high compression strength (0.7 N/mm^2 at 0.05 g/cm^2 to DIN 53421) and heat resistance (103° to DIN 53424) Densities are between 0.03 and 0.3 g/cm^3. It is used as a core material for sandwich components. Applications are in the marine, aviation, automotive and off-shore industries.

Plasticized PVC foam, d 0.05–0.15 g/cm^3, closed celled in sheets and blocks, serves as highly resilient material in gymnasium mats and for helmet shock and machine vibration damping.

Plasticized PVC foam, d 0.07–0.33 g/cm^3, open-celled, is a material used for sound absorption and for "breathing" upholstery.

4.2.4.4. PVC-coated metal sheets

A deep-draw material for walls and roofs, in metalworking and vehicle building, for refrigerator furnishings, housings and cassettes is mass produced by applying an adhesive layer to one or both sides of a zinc-coated metal sheet 0.25 to 1.5 mm (10–60 mil) thick. Onto this is a semi-rigid PVC is rolled or applied as a paste by reverse roll coating. Laminated metal sheets with plastic plies, which are vibration damping, can also be deep drawn.

Sheet metal is protected against corrosion by PVC stoving lacquer layers about 0.025 mm (10 mil) thick on one side or both sides.

4.2.5. Fluoropolymers

Polyfluorocarbons and copolymers

An overview of the group, processing, general properties and fields of application is given in Sec. 4.1.5.

4.2. Semi finished Thermoplastic Material

Trade names for semi-finished groups of various fluoroplastics include: Chemfluor, Chemloy, Fluororay, Rulon (US); Fluroflex, Klingerflon, Polyfluron (DE); Lubriflon, Polyflon (IT); Gaflon (FR); Fluolion, Permaflon (GB).

ASTM specifications

D 3294 PTFE: basic shapes, dimensions and tolerances not specified
D 1710 PTFE: rods, $\frac{1}{8}$ to $1\frac{1}{2}$ in (3.2 to 38 mm) diameter
D 3293 PTFE: molded sheets, $\frac{1}{32}$ to $\frac{1}{2}$ in (0.8 to 12.7 mm) thick
D 3308 PTFE: skived tapes, 0.002 to 0.125 in (0.05 to 3.18 mm) thick

These specifications differentiate between Type I, Premium; Type II, General Purpose; Type III, Commercial, i.e., non-critical use in electrical, mechanical or chemical applications, and Type IV, Utility, having no electrical requirements. The mechanical requirements decrease from I to IV from 4000 to 1300 psi (27.6 to 9.0 MPa), for rods from 2200 to 1200 psi minimum tensile strengths, and 300 to 75%, for rods with 200% to 40% elongation at break. The types may be subdivided into grade 1, made from virgin resin only, and grade 2, using reprocessed resin. They may be further subdivided into classes A to C in terms of their thermal dimensional change ($\leq 0.5\%$ to $\geq 5\%$).

D 3369 for TFE resin cast film stipulates 0.001 to 0.005 in (0.025 to 0.127 mm) thickness, 4300 to 4500 psi (29.67 to 31.05 MPa) minimum tensile strength and 400 to 370% elongation at break, dependent on thickness. All the above listed ASTM standards give further electrical and dimensional requirements.

D 3295 PTFE, tubing, and D 3296, for FEP tubes, specify dimensional series in several classes from thin walled to heavy wall (chemical) tubing.

Type I, based on American wire gauge, DI 0.01 to 0.325 in (0.25 to 8.26 mm).
Type II, based on fractional inch sizes from $\frac{1}{32}$ to 2 in (0.79 to 50.8 mm) with minimum dielectric breakdown values.

D 3368 FEP, sheet and films, 0.005 to 0.095 in (0.013 to 2.4 mm) thick differentiates between Type I, General Purpose, Type II, cementable films (with specific dielectric requirements) and Type III films for applications requiring unusual flex or extreme thermal or chemical service (2500 psi tensile strength, 250% elongation at break, 0.005 to 0.095 in thick).

Type IV film is for mold release applications.

The quality rules laid down by trade organizations in several countries are similarly based, but there are no ISO specifications for these engineering polymers.

PTFE semi-finished products (apart from the materials mentioned above) are available as rods up to 300 mm in diameter; sheets and blocks up to 120 mm thickness; composites with graphite, molybdenum disulfide, bronze for bearings, PTFE-impregnated GF fabrics for industrial purposes and tent roofing membranes, and PTFE fibers for gland packing. Paste or ram-extruded pipes up to DN 300 and wound pipes up to DN 600 for the internal lining of steel pipes are supplemented by isostatically pressed bends and fittings.

Extruded sheets and films of FEP, ETFE, PFA or TFA, are used for flexible printed circuits and flat cables (heat shrinkable), tubing and tapes of all fluorocarbons for low (cryogenic) temperatures as well as heat-resistant, practically non-combustible cables and wire insulation for electronic, aerospace and security equipment.

ETFE films 150 μm (Hostalen ET) are crystal clear, with about 90% transmission of visible and near UV light, but not of infrared radiation. The flame retardant film is weather and puncture resistant, even to hail and sand storms. They are used for skylights in large recreation centers (allowing natural suntanning) and for greenhouses, in thermal insulating multi-layered inflatable constructions, as well as for solar collector technology. For the last mentioned, PFA and ECTFE (Halar, US) films are also available. PCTFE Aclar (US) films (0.5 to 20 mil, 0.013 to 0.51 mm) are used for vapor barrier pharmaceutical packaging. ASTM D 3595 classifies them into three types, differentiated by impact strength and (very) low water vapor transmission rate, minimum tensile strengths 2800 to 3100 psi (19.32–21.40 MPa), and minimum elongation of 50%.

PVDF pressure pipes and other semi-finished products supplement PVC in chemical equipment and piping systems by extending the temperature range of application from -40 to $140°C$ (-40 to $284°F$) and for increased chemical stress.

Trade names: Kynar (US); Dekatemp. Trovidur (DE); Solef (B); Sygef, Symalit (CH).

Laminated sheets with fiber fleece as adhesion promotor are used for bonded linings in steel equipment. Equipment made of such sheets is also reinforced by laminating GRP onto the outside.

(Bi-) oriented PVDF films and sheets (Kynar, US; Kureha, JP and DE) are piezo- and pyro-electrical. Metallized, they are used for high-tech electronic equipment (ultrasonic sensors, infrared detectors, sonars, medical devices).

Polyvinyl fluoride PVF is commercially available only in the form of semi-finished: as glass-clear laminating film (Tedlar, US), 0.5 to 4 mil (12.5 to 100 μm); as packaging film (Vac-Pac, US), 1 to 2 mil (25 to 50 μm), and as UL-classified panels (Resolite Fire Snuf), 45 to 90 mil (1.8 to 3.6 mm).

The main application of the material which is resistant to UV, IR, weather, corrosion and fire is long-term surface protection for external and internal construction, aircraft interiors and aerospace equipment. PVF adhesive-coated films are laminated onto vinyls, GRP, plywood, steel or other metals and are used to glaze solar energy collectors.

4.2. Semi finished Thermoplastic Material

The following are some significant properties of the material:

Specific gravity: 1.38–1.6⁶ g/cm³ (86.3–100 lb/ft³)
Tensile strength: 45–125 N/mm² (6.5–18 × 10³ psi)
Elongation at break: 115–200%
Useful temperature range: −100 to 150°C (−148 to 302°F)

4.2.6. Methylmethacrylate polymers, PMI foams

Semi-finished PMMA materials are standardized in
DIN 16957 (1985) Cast PMMA sheet
ISO 7823-1 (1987)
DIN 16958 (1981) Extruded PMMA sheet
ISO 7823-2 (1988); ISO 8257-1 (1987)

4.2.6.1. Manufacture of acrylic sheet

Trade names: Oroglas (US); Paraglas, Plexiglas, Resartglas (DE); Altuglas (FR); Perspex (GB); Shinkolite, Sumipex (JP)

PMMA semi-finished products are cast from MMA reactive resins (see 4.1.1.1/3) or extruded from PMMA molding compounds (see 4.1.6). These have similar structures and can be combined.

Cast PMMA sheet
PMMA sheet is made by casting MMA with polymerization aids, such as initiators, catalysts, plasticizers, UV absorbers, colorants, etc., between optically flat silica glass plates separated by elastic sealing strips. This assembly is clamped together at the edges.
Approximately 25% volume shrinkage occurs during polymerization with a consequent reduction in the distance between the plates. The sealing strips must be designed to accommodate this movement. Heat is liberated during polymerization and reaction temperature and heat dissipation (i.e. reaction rate) must be optimized to prevent overheating and bubbles formed by evaporating MMA. Polymerization is carried out in a water bath for good heat removal, air autoclave at increased pressure with high air speeds or in ovens at atmospheric pressure.
Water bath polymerization is followed by tempering in an oven to ensure complete polymerization (residual monomer content, 1%). The process parameters depend mainly on the sheet thickness. The average molecular weight M_W of PMMA is at least 3×10^6 g/mol. Easily formable cast sheet is often produced from syrup resulting in approximate M_W of 6×10^5 g/mol. A

continuous casting process with a double web plant and steel belts is sometimes used in the USA and Japan.

Extruded acrylic sheet

This is usually produced from standard PMMA molding compounds (4.1.6.2) with M_W approximately 2×10^5 g/mol and Vicat softening temperature (VST/B) between 103°C and 108°C.

Double and triple-walled ribbed extruded sheet and other complex profiles can now be produced with outstanding optical quality. Corrugated sheet, usually 3 mm thick, is also produced by extrusion.

4.2.6.2. Properties of PMMA semi-finished products

The typical property profile of PMMA plastics is described in 4.1.6.1. Specific properties of cast (CS) and extruded (XT) semi-finisheds are summarized in Table 4.53.

High molecular weight cast acrylic sheet can be formed in the thermoelastic range between 140°C and 210°C but is not melt-processable (see Fig. 3.70). Cast sheet requires higher forming pressures and temperatures than the extruded product. Because of the higher molecular weight, creep and stress cracking behavior of cast acrylic sheet is somewhat better, (Table 7.8, p. 454). Extruded acrylic sheet is, however, easier to solvent bond. Polymerizing adhesives based on MMA are preferred for cast sheet.

Special optical grade cast PMMA is utilized for applications requiring freedom from optical distortion such as aircraft canopies and passenger windows.

Cross sections of typical double- and triple-walled sheet are shown in Fig. 4.16. Their design provides low weight, high rigidity and relatively good thermal insulation (Table 4.54), hence their use in insulating glazing. There are no major differences in the machining of cast and extruded acrylic sheet.

A special purpose PMMA product with an outstanding property profile is oriented Plexiglas GS 215 (4 mm) which was used for roofing the sports stadia in the 1972 Munich Olympics and Berlin (1973/74). It is a flame retarded, transparent, very weather-resistant, 70% biaxially oriented PMMA. It is the only acrylic sheet meeting the low flammability B1 classification of DIN 4102 and does not give rise to burning droplets. The material on the roofs mentioned above can be walked on and must perform well against wind and snow loads. High temperature resistant copolymers of PMMA and acrylonitrile are no longer produced in Europe.

4.2. Semi finished Thermoplastic Material

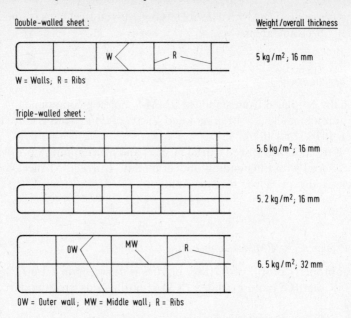

Fig. 4.16. Common profiles of walled sheet.

4.2.6.3. Forms of delivery and applications of acrylic sheet

Tables 4.55 and 4.56 provide an overview of the common forms of delivery and specialist formulations of acrylic sheet. Table 4.57 summarizes areas of application for these semi-finished products.

The "no-drop" coating mentioned in Table 4.57 is used not only for greenhouses, where it spreads condensation on the inside thus avoiding dripping onto plants and light loss, but also for roofing over terraces and conservatories. Because of its exceptional resistance to weathering, the film can be applied to the outside. This results in improved self-cleaning and appearance during damp weather or condensation.

4.2.6.4. Acrylic foil, manufacture, forms of delivery, applications

Trade names: Plexiglas, Shinkolite

These 0.05 to 1 mm thick foils are extruded from impact-resistant modified PMMA (see 4.1.6.3). They are supplied in sheets 1000×2050 mm or in rolls 1400 mm wide, both in colorless and colored grades with smooth or textured surfaces and sometimes with functional coatings (e.g. to provide scratch resistance or anti-reflectance).

4.2.6.4. Acrylic foil, manufacture, forms of delivery, applications

Table 4.53. Typical properties (23°C/50% r.h.) of cast PMMA: CS ($M_W > 3 \times 10^6$ g/mol) and extruded PMMA: XT ($M_W > 2 \times 10^6$ g/mol)
XT corresponds to the S5 molding compound from Table 4.30 in 4.1.6.1.
The values correspond to the Campus database.

Properties	Units	Test standard	CS	XT
Density	g/cm^3	ISO 1183	1.19	1.19
Water absorption				
23°C (saturated)	%	DIN 53495	2.1	2.1
23°C (50 r.h.)	%		0.6	0.6
Shrinkage (in direction of polishing)	%		1.6	–
Thickness 1.5 to 2.5 mm	%		–	< 10
Thickness 3.0 to 8.0 mm	%		–	< 5
Mechanical properties				
Tensile strength (5 mm/min)	MPa	ISO 527	80	72
Elongation at break (5 m/min)	%	ISO 527	5.5	4.5
Tensile modulus of elasticity (secant: 1 mm/min)	MPa	ISO 527	3300	3300
Charpy impact strength +23°C	kJ/m^2	ISO 179/1D	15	15
−30°C	kJ/m^2		14	14
Izod notched impact strength				
+23°C	kJ/m^2	ISO 180/1A	1.6	1.6
−30°C	kJ/m^2		1.4	1.4
Thermal properties				
Vicat softening temperature (VST B/50)	°C	ISO 306	115	102
Coefficient of linear expansion	K^{-1}	DIN 53752-A	70×10^{-6}	70×10^{-6}
Electrical properties				
Volume resistivity	ohm cm	DIN/VDE 0303 Part 3	$3 > 10^{15}$	$> 10^{15}$
Surface resistance	ohm		5×10^{13}	5×10^{13}
Optical properties				
Degree of light transmission	%	DIN 5036 Part 3	92	92
Refractive index n_D^{20}	–	DIN 53491	1.491	1.491

Applications: shades, scales, coverings, signs, light diffusers, optical filters, drawing stencils, etc.

A special purpose, weather-resistant Plexiglas foil, 0.05 and 0.1 mm thick, contains a high level of polymer-bound UV absorber. It absorbs more than 98% of solar UV radiation impinging on the earth.

Applications: weather protection for wood, metals and plastics such as PVC, ABS and PC.

4.2. Semi finished Thermoplastic Material

Table 4.54. Typical characteristics of ribbed multiwall sheet

Type	Dimensions mm	Weight kg/m^2	Thermal conductivity W/m^2 K
Double walled sheet		5.0	2.9
overall thickness	16		
rib height	approx. 1.8		
rib thickness	approx. 1.7		
distance between ribs	approx. 32		
Triple walled sheet		6.5	1.9
overall thickness	32		
outer wall thickness	approx. 1.9		
middle wall thickness	> 0.3		
rib thickness	approx. 1.5		
distance between ribs	approx. 31		

Table 4.55. Forms of delivery of PMMA semi-finished products

	Products	Common dimensions	
		Thickness or ϕ mm	Width × length or length, mm
Cast PMMA	Sheet[1]	1.5 to 25	2000 × 1200
			3000 × 2000
	Blocks[3]	30 to 80	2000 × 2000
		90 to 250	2000 × 1000
	Tube	3 to 10	up to 2600
	Rod	ϕ2 to 100	
	round section	10 to 50	up to 2000
	square section		up to 2000
Extruded PMMA	Sheet[2]	1.5 to 8	2000 × 1200
			3000 × 2000
			4000 × 2000
	Tube	1 to 5	up to 4000
		ϕ5 to 200	–
	Twin-walled sheet	16	W: 600, 980, 1200
			L: 2000 to 6000
	Triple-walled sheet	16	W: 1053, 1200
			L: 2000 to 7000
		32	W: 1230
			L: 2000 to 7000

[1] For 4 to 8 mm thickness, various widths and 2700 mm length available to special order.
[2] For 10 to 15 mm thickness, various widths and 2000 mm length available to special order.
[3] For 80 mm thickness, 3500 × 1800 mm available to special order.

Table 4.56. Grades of acrylic sheet

---
colorless, transparent
colorless, light diffusing (textured)
colorless, permitted for use with foodstuffs
anti-reflecting (weakly diffusing)

collrless and colored for acoustic insulation walls

white: translucent, opaque
colored: transparent, translucent, opaque

UV transparent $\lambda \geq 260$ nm)
UV protected against specified wavelength ranges, e.g. for solaria, transparent to UV-A and opaque to UV-B and UV-C

stabilized for arid areas
easily formed
weakly cross-linked/improved resistance to chemicals and solvents

low inflammability (B1) to DIN 4102,
oriented PMMA with flame retardent

with scratch-resistant coating
with anti-static coating
no-drop coating (water-spreading, e.g., multi-walled sheet)
improved impact resistance
with mirror coating

4.2.6.5. Polymethacrylimide (PMI) rigid foam

Trade name: Rohacell
Structural formula:

$$\left[-CH_2-\underset{\underset{O}{\parallel}}{\underset{C}{\overset{CH_3}{\overset{|}{C}}}}-CH_2-\underset{\underset{O}{\parallel}}{\underset{C}{\overset{CH_3}{\overset{|}{C}}}}- \right]$$
$$\diagdown\underset{H}{N}\diagup$$

Manufacture:
PMI foam is manufactured in two stages from monomeric methacrylonitrile and methacrylic acid and chemical (formamide) or physical (isopropanol) blowing agents. In the first, a cast polymer (Casting, see 4.2.6.1) is produced.

4.2. Semi finished Thermoplastic Material

Table 4.57. Main areas of application of semi-finished acrylics (CS = cast, XT = extruded)

Area of application	Type
Construction	
Thermally insulated roof glazing, conservatories, terrace glazing, greenhouses	XT multi-walled sheet (with and without no-drop treatment)
Parapets, single glazing, partitions	CS, XT
Light domes	CS, XT
Barrel vaults	CS, XT
Large roofs	CS, low flammability (B1) to DIN 4102
Interior fittings (e.g. stairs)	CS
Sanitary	
Bath tubs, shower trays, basins	CS (XT not suitable for water conduits)
Parts not in contact with water	CS, XT
Solaria	CS, transparent to UV
Signs and Displays	
Signs, displays, traffic signs, exhibition equipment	CS, XT, mirrors
Lighting	
Lamp covers, optical fibers, profiles	CS, XT
Aviation	
Windows, canopies, sandwich components for cockpit glazing	CS, oriented CS, scratch-resistant, coated CS
Plant construction/electronics	
Safety glass (e.g. food processing, chemical industry, medical)	CS, XT
Electronic instruments	CS, XT, anti-static grades
Transportation	
Acoustic, roadside barriers	CS (≥ 15 mm thick)

In the second stage, a suitable section of sheet is foamed at temperatures between 170 to 220°C to densities between 30 and 300 kg/m^3. Imidization occurs during the foaming process.

Forms of delivery: sheet, max. 2500 × 1250 mm, 1 to 65 mm thick in densities of 30 to 300 kg/m^3. (Maxima, do not apply to all types.)

Rohacell IG = industrial grade, P = pressed, A = aircraft, WF = heat resistant, HF = high frequency, s = self extinguishing (with flame retardants).

Properties and applications: closed cell, vibration-resistant rigid foam with high strength and high heat resistance.

Properties are dependent on grade and density.

Elastic modulus to DIN 53457	20 to 380 N/mm^2
Compression strength to DIN 53421	0.2 to 15.7 N/mm^2
Shear strength to DIN 53294	0.4 to 7.8 N/mm^2

Heat distortion temperature to DIN 53424 up to 215°C, continuous service temperature up to 180°C.

All grades are resistant to most solvents and fuel components but not to alkalis.

PMI foam has very good creep characteristics even at elevated temperatures and is thus suitable for manufacturing sandwich components in autoclaves up to 180°C and 0.7 N/mm^2. It is suitable for "co-curing", i.e., curing of the entire sandwich structure (prepreg outer layer + core) in a single operation. It is used as a thermoelastically formable sandwich core for IMP (In Mold Pressing), a process based on pressure generated in the mold by the foam core. In the case of standard Rohacell grades, this pressure is generated by after-foaming, with prepressed grades by expanding the core. Since expansion of the core in the closed mold is hindered, pressure builds up and the outer layer is pressed against the tool wall. In the expansion of prepressed, i.e. thermoelastic compressed, cores, use is made of the memory effect to produce pressure.

Rohacell can be bonded easily with conventional two component adhesives to itself or with outer layer laminates.

Applications: core material for mass produced structural components for aircraft (flaps, covers, stringers); for helicopter blades (Rohacell A, WF), in ship building as a core for the stern or in superstructures (Rohacell IG, S), in the automotive sector as a core for body parts; sandwich cores for high performance bicycle frames/rims; cross country and downhill ski cores, cores for tennis racquets (Rohacell IG, WF); antennae and radomes (Rohacell HF), satellites (Rohacell WF, HF), models, couches transparent to x-rays (Rohacell IG, WF).

4.2.7. Engineering Thermoplastics

For general properties, references to extrusion and applications of semi-finished products see:

   4.1.7. Acetal resins (POM)
   4.1.8. Polyamides (PA) and thermoplastic polyurethanes (TPU)
   4.1.9. Polyarylates (aromatic polyesters, APE)

4.2. Semi finished Thermoplastic Material

4.1.10. Modified polyphenylene oxide (PPOS), polyarylether ketone (PEEK), polyphenylene sulfide (PPS), polysulfones (PSU, PES, PPSU)
4.1.1.11. Poly- (methacrylic, amide, ether) imides (PMI, PAI, PEI)

Extruded pipes and hollow sections (e.g. for heat exchangers), thermoformable sheets and films of the above-named amorphous high-temperature plastics (also of PMP, Sec. 4.1.2.8) are commercially available (Europlex, DE). Clear optical PES films, 50 to 250 μm thick (Stabar, GB), with long-term service temperature of 180°C are amorphous (but crystallize on thermoforming). These and opaque crystalline PEEK films compare advantageously in aerospace applications with non-thermoplastic polyimide (Sec. 4.8.6) films. Extruded rod, sometimes compounded with additives, is machined into technical components. For electrical insulating films see also Table 4.60.

PA (mostly PA 6), POM and PPO semi-finished products are of general commercial importance. Trade names for rod, block and sheet include Allied STX, Cadco, Danko, Wgulon (US); Aeternamid, Igopas, Oilex, Lamigamid, Nylatron, Supramid and Supranyl (DE).

Pressure pipes and tubing made of PA (DN < 2.5 mm to about 40 mm/0.1 to 1.5 in) are used for lubricating, cooling and brake lines in vehicles and aeroplanes and for hydraulic control systems. Trade names include Mecanyl, Tecalan (US, DE) and Technoflex (TE). PA 612, 11 and 12 are used for standardized non-kinking spiral connecting lines for trucks and cars and for coating seamless aluminium tube. PA 66 up to about 800 mm/32 in dia abrasion-resistant pipes are used, for example, as feed lines in the brewing industry because they are non-toxic and resistant to disinfectants at temperatures from −30 to 110°C/−22 to 230°F.

PPE pipes (DN 12 to 225 mm/0.5 to 9 in) are suitable for use with water up to 100°C and can be sterilized at 115°C. Trade names include Dekaryl (DE) and Durapipe (GB).

PA cast or blown films are mainly PA 6. They are also combined with PE, EVA, ionomers and PVDC (Tables 4.59 and 4.60) as laminated films. Trade names include Capran, Cryovac, Polyseal, Upolam, Vac Pac (US); Alkoron, Platilon, Walomid (DE); Diamiron, Harden (JP), and Rilsan (FR).

Trade names of biaxially oriented products are Emblem (JP), Filmon BX (IT) and Orgamid (FR).

PA and TPU films can be used from −60 to 130°C for mineral oil and fat-resistant and odor-proof tube and flat-film packaging for lubricants and fuels, hot-filling bitumens, foods, sterilizable and disinfectable medical equipment and sharp-edged objects. Additional applications include heat-resistant release films, laminating films for surface-finishing, lining films, electrowinding film and membranes. Unstretched PA films are thermoformable and therefore suitable for blister packaging. Stretched films have extremely good tensile and bursting strength.

Rubber-elastic TPU sheeting (Platilon U, Walopur, DE; Duraflex, Hi-Tuff, Scotch, US) is suitable for vibration-stressed membranes and sleeves, etc. because of its high folding endurance and tear strength. It is extremely elastic and puncture resistant, permeable to water vapor and gases but water- and wind-proof. Applications include hot-melt adhesives, expanded film for car seats, weather-proof clothing, sanitary and medical purposes. Roofing membranes (Difutex, Delta-Purofol, DE) have good water vapor permeability and UV resistance.

Wires and filaments

PA 6 wires, 2 to 4 mm in diameter, 0.15 to 0.35 mm thick monofilaments, are used for zippers, fishing nets and lines, musical instrument strings, etc. Multifilaments are used for dolls' hair.

4.2.8. Thermoplastic Polyesters

4.2.8.1. Polycarbonate (PC)

Material properties, Sec. 4.1.9.1.

1. *Semi-finished technical material and glazing elements*

Trade names: Lexan, Shelfield, Tuffpak, Twinwall (US); Jupilon (JP); Makrolon, Riag, Thermodet (DE). Hollow-profile transparent glazing elements, e.g., Everlite, Rodeca (DE); Akroplast (IT).

Forms of supply: Sheets for industrial and glazing applications, some reinforced with short glass fibers, 1 to 10 mm (0.4 to 4 in) thick in sizes up to 2 by 3 m (79 by 118 in), clear (with light transmission 85% at 3 mm), milky translucent (30 to 45% transmission), one or both sides smooth or surfaced. Hollow double or triple wall sheets. Profiles for heat-insulating glazing 3 to 10, 16 and 40 in total thickness.

Processing: PC can be nailed; within certain limits it can be cold pressure formed. Dried and heated to 150 to 180°C, it can be thermoformed with tools from 115°C. It is weldable by all processes except HF.

Applications: Components in medical and electrical equipment, glass-fiber reinforced and flame-retarded solid and multi-wall components for aircraft interiors. Unbreakable door panels, telephone kiosks, glass houses, industrial glazing, sports stadiums and bridge balustrade panels, burglar-proof slot-machine and bullet-proof bank counter window glazing. There are also scratch-proof products (Margard, US) and flame retarded types.

4.2. Semi finished Thermoplastic Material

2. Sheeting and films

Trade names: Coburn, Lexan (US); Diafoil (JP); Europlex, Makrofol, Pokalon, (DE); Altuchoc PC (FR). PC-blend films: Bayfol (DE).

Forms of supply

Extruded film, up to 1200 mm wide, 0.1–0.5 mm thick. Continuous thermal resistance 130°C, heat moldable at 190 to 200°C, hot stampable up to 120°C, good punching properties, dimensionally stable. Glossy on both sides, >80% light permeable, structured or glass fiber filled, light scattering, available in flame resistant and antistatic formulations.
Permanently reverse printed (hot formed), light transmitting, operating and warning display panels for equipment and vehicle cockpits, membrane contact switches.
Thermoformed microphone membranes, chocolate casting molds etc. Low birefringence special film for compact disks, technical drawing and reprographic films. Carbon black filled, electrically conductive for typeless typewriters and blister packaging for electronic components. Solvent-cast electrical films (Table 4.58), e.g.
Makrofol N, 20 to 200 μm, 1200 mm wide, recognition color yellow;
Makrofol G, 10 to 60 μm, 1000 mm wide, recognition color yellow, stretched lengthwise at 150 to 160°C for 50% shrinkage;
Makrofol KG, 2 to 60 μm, 1000 mm wide, recognition color green, stretched lengthwise and crystallized, high heat and solvent resistance, also vacuum metallized;
Makrofol SN, SG (red), SKG (turquoise), heavily flame retardant with tetrabromobisphenol.

4.2.8.2. Polyethylene glycol terephthalate (PETP)

Material properties, Sec. 4.1.9.2.
Trade names for sheeting and films: Mylar, Petra, Scotchpak, Scotchpar, Sheldal (US); Hostaphan (DE); Espet (JP); Meliform, Melinex (GB); Terphane (FR).
General properties are given in Tables 4.58 and 4.59 pp. 327, 328. The films are glossy, crystal-clear and very strong and tough, dimensionally stable with service temperatures from −60 to 130°C (−76 to 226°F), flame retardant and nontoxic, with a low permeability to gases, aromas and water vapor. They cannot be heat-sealed without coating.

Supply forms, applications

PET films in the thickness range 1.5–350 μm are extruded in large scale plants of up to 8 m frame width in the following basic types:

		density g/cm^3	glass trans. temp. °C	melting temp. °C	Young's modulus N/mm^2
A	Amorphous	1.33	70–75	–	1600
B	Longitudinally stretched, semi-crystalline, uniaxially oriented	1.38–1.42	75–80	258	15 000
C	Multiaxially stretched on two sides, fully crystalline	1.50	–	310	140 000

Main fields of application of A-films are photographic films, reprographic and drawing films (50–180 μm), packaging (incl. shrink films), metallizing, embossing films, typewriter ribbons, adhesive tapes etc. (12 to 25 μm), capacitors and wound films (1.5 to 30 μm), other electrical films and printed circuits (20 to 350 μm). Other applications include multilayer laminates, e.g. for circuit boards, packaging, with bitumen for waterproofing felts in building.

The highest specifications for purity, dimensional stability and surface quality are required for magnetic storage media. Disks (floppy disk, 75 μm) and computer tapes (26 to 36 μm) are made of A-film. With increasing recording density in the high-tech. audio and video fields, B and C type film, currently 4.5 to 23 μm thick are increasingly important in the professional entertainment sector. 1.5 mm thick tapes are being developed.

Films made electrically conductive by coating with carbon black (Hostatherm), with integrated copper contact strips and breakdown resistant cover film are used as elements for ceiling and floor heating.

PET films can be vacuum metallized without pretreatment. Films 12 to 60 μm thick coated with 0.01 to 0.1 μm aluminum form better barrier layers than aluminum alone for packaging (e.g. "bag in box") and construction. Reflecting 90% of visible light and infrared, they are used for mirrors and as a base for decorative films.

4.2.8.3. Mixed terephthalate-isophthalate polyesters

Mixed polyesters (Kodar, Lustro, US) are extruded to 0.07 to >0.5 mm (0.028 to >0.2 in) thick crystal-clear films, which are satisfactorily thermoformed at 130 to 177°C and sealed at 177 to 200°C without additive. They can be laminated with almost all substrates. They are used as packaging, cake molds and abrasion-resistant light- and corrosion-resistant linings for building panels.

4.2. Semi finished Thermoplastic Material

4.2.9. Cellulose Esters

4.2.9.1. Celluloid and related semi-finished goods

Because of its fire hazard, celluloid has only special applications. Semi-finished products are made into "acetyl celloid" from CA and from CAB with effective imitation or tortoiseshell, horn, mother-of-pearl, agate and abstract patterns. Extruded sheets (see following paragraph) can be made only in limited multi-colored layered and patterned forms.

Fields of application

Celluloid is used for table-tennis balls, CA and CAB material, for combs and hair decorations, fitting out of musical instruments, spectacle frames, toiletries and door protectors.

4.2.9.2. Extruded products from cellulose ester compounds

For cellulose acetate (CA), cellulose acetate-butyrate (CAB) and cellulose acetate propionate (CP) and their properties see Sec. 4.1.12.1.

1. *Sheets and films*

CAB and CP sheets 0.5 to 8 mm (0.02 to 0.3 in) thick are tough thermoformable materials with good heat and weather stability (CP for moderate and cold climates, CAB also for the tropics).
Applications: Light domes, illuminated signs, impact-proof light coverings, protective shields, mudguards and tent windows. Glass-clear secondary acetate films 0.1 to 0.3 mm (0.04 to 0.12 in) and 1200 mm wide in rolls and thicker cut sheets 1000 by 1200 mm are used for packaging (bags and boxes) in glossy laminations and drawing films. Anti-mist optical sheet is obtained by surface saponification.
Properties: See Table 4.58.

2. *Tubes and profiles*

Decorative profiles from cellulose esters with metal bands embedded manifest long-lasting metallic sheen and are used for bicycles and vehicle construction.

4.2.9.3. Cast films for electrotechnical applications

Cellulose ester cast films, have been largely replaced in electrotechnical applications by newer thermoplastics such as PC and polyester. For properties of the electrically insulating films see Table 4.59.

4.2.9.3. Cast films for electrotechnical applications

Table 4.58 Typical Properties of Packaging Films

Description pages	Material	Density (g/cm³)	Tens. strength MD/TD (N/mm²)	Elongation at break MD/TD (%)	Service temp. range (°C to °C)	Permeability coefficients at 20 °C in SI units* Multiply the figures here for water vapor by 10^{-9}, all gases by 10^{-12}					
						Water vapor	Air	O_2	N_2	CO_2	H_2
300 f.	PE, low density	0.92	22/15	300/700	−60 to 80	1.5	9	19	6	75	55
	PE, high density	0.95	33/25	800/1000	−50 to 100	0.5	3.5	8	2	32	25
	PP, not oriented	0.90	50/40	430/540	−20 to 100	1.1	2.8	7	2	25	65
	PP, monoaxially oriented	0.90	250/40	10/700	−50 to 90	0.4	−	5	−	18	−
	PP, biaxially oriented	0.91	200/200	80/80							
304	PS, biaxially oriented	1.05	70/70	10/10	to 80	25	3.5	14	2.2	85	150
207 f	PVC-E, thermally treated and oriented	1.38	53/50	90/30	−15 to 80	6.5	0.12	0.4	0.05	0.9	10
		1.38	110/45	30/10							
	PVC-S, glass-clear	1.4	55/55	30/30	−10 to 70	5	0.1	0.4	0.03	0.8	8
208	PVDC copolymer	1.6	~80	~30	−20 to 100	<0.2	−	0.2	0.08	1.1	6
313	PVDF, extruded	1.78	/65	400	−30 to 135	5.2	−	0.7	0.2	<1	3
	CTFE, extruded	2.10	/40	150	−200 to 180	<0.1	−	~2	<0.1	1.5	15
322	PA 6, not oriented	1.13	80/60	400/	−30 to 120	35	0.25	0.6	0.1	3.5	5
	PA 6, biaxially oriented	1.13	300/300	70/70		~20	−	0.3	−	−	−
	PA 12, extruded	1.03	60/40	400/250		~10	−	~10	~0.1	~70	15
323 f	PC, cast film	1.20	85	110	−100 to 130	30	7.5	37	1.2	110	200
	PET-A**, biaxially oriented	1.40	200/220	130/110	−60 to 130	4	0.06	0.13	0.03	0.65	4
326	CA, 2½ acetate	1.3	90	20	−10 to 120	225	1.6	5	1	34	52
331	Cellophane, plain	1.45	155/55	20/55	Short time, 150	>300	−	1	1	−	10
	Cellophane, polymer coated	1.45				<4	−	1-2	~2	−	2

* For other units and transmissions rates see Sec. 7.5, p 456.
** A = amorphous, common designation not yet standardized.

4.2. Semi finished Thermoplastic Material

Table 4.59. Typical Properties of Electrical Insulation Films

Description p.	Material	Type	Mechanical properties		Electrical properties				Thermal properties	
			Tensile strength, MD (N/mm²)	Elongation, break (%)	Dielectric constant (50 Hz/ 20 °C)	Dissipation factor (50 Hz/ 20 °C)	Volume resistivity (ohm cm)	Dielectric strength (kV/mm)	Tensile heat distortion temp.(°C)	Continuous service temp.(°C)
303	Polypropylene	Biaxially oriented	150	75	2.3	0.7	10^{17}	300	150	105
304	Polystyrene	HT	70	4	2.5	<0.2	10^{17}	200	110	90
312	Polytetrafluoroethylene		17	350	2.1	<0.3	10^{17}	100	190	<180
324	Polyethylene terephthalate	Biaxially oriented	210	111	3.3	2	10^{17}	300	240	130
323	Polycarbonate	Normal	80*	100	3.0	1	10^{17}	350*	150	130
		Crystalline and oriented	220*	240	2.8	1	$2 \cdot 10^{17}$	350*	240	140
282	Polyphenylene oxide		65	25	2.7	1.5	10^{16}	300	170	110
283	Polysulfone		80	65	3.1	1.2	$5 \cdot 10^{14}$		185	165-170
405	Polyhydantoine		100	119	3.3	1.5	$4 \cdot 10^{16}$	250	260	160
257	Polyamide imide		180	45	4.2	9	$3.5 \cdot 10^{17}$	200	>250	
404	Polyimide		180	70	3.5	2	10^{17}	270	>350	>180
326	Cellulose triacetate	Normal	90	23	4.5	12	10^{14}	220	190	120
		Soft	80	27	4.3	21	10^{15}	200	170	120
326	Cellulose acetobutyrate	Normal	80	25	4.1	11	10^{15}	230	150	120

* film thickness 0.04mm (test at 50 Hz under potting compound, ball/plate)

4.3. Electrical Insulation Films and Packaging Materials

4.3.1. Characteristics and Properties

For mechanical, packaging and electrotechnical reasons, packaging and electrical films are required almost plasticizer-free, in thicknesses less than 0.1 mm (4 mil); packaging film for the most part is 0.02 to 0.05 mm thick, and electrofilms 0.002 mm. These, according to grade, "rigid" (shear modulus $\geq 5 \times 10^2$ N/mm^2) but flexible fine films are characterized by particular groups of property values, which are summarized in the following
   Table 4.58: Typical properties of packaging film;
   Table 4.59: Typical properties of electrical insulating films.
The tables contain the relevant mid-range values for comparative purposes in SI units.

4.3.2. Multilayer Packaging Film and Packages

4.3.2.1. Extruded multilayer composite materials

Plastic packaging for foodstuffs, beverages, chemicals, and medicaments must exhibit a complex combination of properties. They must be unaffected by the packaged goods and prevent permeation of air, CO_2, water vapour, aromas and other materials from both sides. In some cases, such as cold sterilization and CAP (controlled atmosphere packing) for fresh meat, they must permit defined gas permeation. They must also withstand mechanical stresses at various temperatures (hot filling at 80°C, conservation and sterilization at 120 to 125°C, boil bags, microwave oven portion packages, deep frozen foods) and heat sealability is also often required.

These complex demands are met by composite blown films for packaging bags, multilayer parisons for blow molding of bottles and tubs and composite flat sheets for thermoforming of stand packs manufactured on multilayer coextrusion plants (Sec. 3.6.5.6), in which up to five extruders work together.

Barrier core layers for gases are manufactured predominantly from EVOH (Sec. 4.1.2.4.2), PVDC compounds (Sec. 4.1.4.3.1) or PAN (Sec. 4.1.6.3.5). EVOH is easily processed in combination with polyolefins and can be recycled as mixed reclaim. It needs, however, to be covered with a sealing layer that is impermeable to water vapor. PVDC is more difficult to process, diminishes the reusability of scrap but is more heat resistant and is also impermeable to water vapor. Copolymers are used as coupling resins suitably tailored for the layer structure (Sec. 4.1.2.4.3). The variety is as great as that of the outer layer polymers. Special purpose amorphous PA and PTP barrier resins (Selar, US) are considered in Sec. 4.1.8.2 and 4.1.9.2.

Multilayer packaging films with outer layers of polyolefins on both sides used as bags or outer packages can make the packaged material storable for several years with no deterioration. PA and TPU films (Sec. 4.2.7) are resistant and impermeable to hydrocarbons, many solvents, oils and greases. An example from the industrial field is film packing for sharp edged metal objects with puncture resistant PA inner layer and water vapor impermeable PE outer layer bonded with an ionomer intermediate layer.

For carrier layers on *stand packs* made by multilayer extrusion, filled PP, PET, PC or thermoplastics with even higher temperature resistance are required (Sec. 4.1.9.4 and 4.1.10.4). Such plastics packaging is increasingly used in the food industry where temperature resistance and product storability previously necessitated glass or metal containers.

An example from the field of engineering is the PE-HD vehicle fuel tank, made by coextrusion blow molding, which is protected against fuel permeation by PA6, PA 12 or PE+PA blend (Selar RB technology). For fluorination and sulfonation of PE-HD tanks and packaging see Sec. 4.1.2.1.4. EVOH is used as a barrier material for fuel lines and underfloor heating pipes.

4.3.2.2. Coatings, laminates

Packaging films and packages (PET bottles) are coated on one side with dispersions (Sec. 4.1.1.10) or lacquers (Sec. 4.1.4.3.1), usually based on PVDC. PE-LD copolymers or ethylene copolymers can be extrusion coated as melts onto films or other packaging materials. Composite films are made by multiple lamination of different films, e.g. for thermoforming packaging materials for dairy products (Sec. 3.8.4.3). Decoration can be applied permanently under cover films. Lamination of aluminum films provides complete gas and vapor impermeability. Vacuum metallized films require a protective lacquer on the metal layer.

4.3.2.3. Special products

Composite films with incorporated PP or PA bonded fabrics and longitudinally stretched PE-HD sheets laminated diagonally to each other have high mechanical strength (Valeron, US).

Other packaging materials include air cushion composite films for shock-sensitive materials (e.g. Alkorthylen L.P., DE) and edge welded air cushions (Pneupak).

4.4. Regenerated Cellulose

4.4.1. Cellophane

Manufacture: A solution of cellulose, made by the viscose (or copper oxide-ammonia) process is spun from a slit die into a precipitation bath, and the resulting film is cleaned of chemicals and dried.
Forms of supply: Films 0.002 to 0.45 mm (1 to 10 mil) thick, up to 1300 mm (51 in) wide, crystal-clear and transparent colors, heat sealable lacquered or polymer coated.
Properties: Table 4.58.
Applications: Generally usable strong packaging film and sheeting, if extremely low water permeability is not required. This is considerably reduced in the "weatherproof" quality. Polymer lacquered cellulose film is also impermeable to fats and oils. In tube form it is used as sausage skin. Materials for artificial flowers and separating tapes in multicolored cables are cut from the film.

4.4.2. Foamed Products

With an open-celled structure from precipitated cellulose hydrate, foamed products are used as cleaning cloths and sponges.

4.4.3. Vulcanized Fiber (VF)

Vulcanized fiber is a laminate made by the "parchmentization" of unsized paper using a zinc chloride solution or sulfuric acid. To make sheets, the paper web is drawn slowly through the zinc chloride solution and tightly wound onto an iron cylinder to the desired thickness. After cutting the winding the sheet is freed of chemicals by washing. The washing period for a 2 mm thick sheet is about 3 weeks; for a 20 mm sheet, it is 6 months. After drying they are pressed flat and smooth. Tubes are made by winding around a mandrel. After satisfactory parchmentization the individual layers form an almost homogeneous mass. Processing is mainly machining. However, bending of thin material on hot bending machines and also pressing and drawing are possible, after softening in hot water as necessary.

Forms of supply

Sheets 0.1 to 50 mm thick, 100 to 130 cm wide and 140 to 200 cm long, single-layered rolls (fish paper) 0.1 to 12.5 mm thick and 105 to 300 mm wide; cylindrical rods, tubes, shaped pieces, e.g., suitcase corners; colors, brown, also grained leather-like, gray, red, white and black.

4.4. Regenerated Cellulose

Material property data: Table 4.60

Laminated vulcanized fiber with phenolic resin cured by hot pressing has a higher strength and swells only slightly in water.
Chemical resistance: VF is resistant to organic solvents and fuels.

Applications: Because of its tenacity, scuff resistance and smoothness under wear, VF has applications:

– *in mechanical engineering and the automotive industry*: ring and sheet gaskets for water and compressed gases, friction and pressure rings and discs, cylinder rings, rollers, sleeves, pneumatic post shells, vacuum cleaner nozzles, backing-sheet ring discs, gear wheels, grinding discs;
– *in the textile industry*: brake rings, press roll covers, thread rings and rollers, pickers, dobby cards, spinning cans, spinning rings, bobbin flanges and buttons, transport boxes and carts, shuttle coating, gear wheels for, e.g., ribbon looms;
– *in electrotechnology*: terminal nipple inserts, control rolls, short wedges, switch grips, intermediate track layers, slider contacts, plug discs.
Further uses are sheets, corners and beading for the suitcase industry, index cards, lacquer fiber for cap peaks, paper baskets.

Table 4.60. General Properties of Vulcanized Fiber and Artificial Horn

	Unit SI	Unit U.S.	Vulcanized fiber		Artificial horn	
Density	g/cm^3	lb/ft^3	1.2–1.3	75–81	1.3–1.4	81–88
Tensile strength (MD/TD)	N/mm^2	$10^3 \, lb/in^2$	100/60	14/8.5	–	–
Flexural strength (MD/TD)	N/mm^2	$10^3 \, lb/in^2$	120/100	17/14	100–180	14–26
Charpy impact:						
unnotched	kJ/m^2		120		20–40	
notched	kJ/m^2		30		–	
Martens soft. temp.	°C	°F	–	–	50–60	122–140
Moisture, conditioned	%		7–10		~10	
Water absorption, immersed	%		> 50		~30	

4.5. Artificial Horn, Casein Plastics (CS)

Trade name: Galalith
The effect of formaldehyde is to harden rennet casein to resemble natural horn. Hardening times, when immersed in a 5% formaldehyde solution may be up to several months depending on the thickness.
Forms of supply: Sheets; rods; tubes in all colors, especially light, bright ones; for machining as carving material.
Properties: Table 4.60.
Galalith is resistant to alcohols, ethers and most other organic solvents; it is not resistant to stronger acids and alkalis. It can be colored by aniline dyes dissolved in 50% acetic acid, bent at 100 to 120°C after pretreatment with hot water and oil of glycerol. The high water absorption affects the dimensional stability of the products.
Application: buttons, chips, fancy and stationery goods.

4.6. Thermosetting Plastics; Raw Materials, Molding Compounds and Semi-finished Products

4.6.1. Curable Engineering Resins

4.6.1.1. Phenol (PF), cresol (CF), xylenol and resorcinol formaldehyde resins

The structure of curable polycondensation resins from phenol and formaldehyde is based on the following combinations of fundamental reactions, which involve the formation of a large variety of bridges during curing.
Addition to produce phenol alcohols:

Phenol + 1-3 $H_2C=O$ ⟶ Methylol groups (on substituted phenol ring with $HOCH_2$, CH_2OH, CH_2OH)

Condensation with the elimination of water:

(1a) n (phenol-CH_2OH) + n (phenol) ⟶ (— methylene bridge —)$_n$ — nH_2O

4.6. Thermosetting Plastics

Condensation with the elimination of water and formaldehyde:

(1b)

[chemical scheme showing phenol derivatives with CH₂OH groups condensing, eliminating H_2CO and H_2O to form CH₂-bridged structure]

Preparation, resin stages and properties: The batch condensation of phenol, methyl and dimethyl phenols, cresol and xylenol or mixtures of these with 30% aqueous solution of formaldehyde at elevated temperatures leads to many different engineering resins. The product depends on the ratio of the reactants, the catalyst used and the way in which the water is removed from the intermediate products. Baekeland ("Bakelite") distinguished and named the main products in 1909 as follows:

a. *Novolak*: Molecular ratio of phenol-formaldehyde is 1: <1, acid catalyst, hard fusible, linear resin with CH_2 bridges linking phenolic groups (formula 1a). It is not self-curing and is used as an alcohol-soluble lacquer resin. By the addition of hexamethylene tetramine (abbreviated "hexa"), which is decomposed at elevated temperatures (from ca. 110°C) into formaldehyde and ammonia, novolaks become heat curing. These mixtures are binders for storage-stable "fast-curing compounds" (Sec. 4.6.2.2) and used in powder form for friction linings, abrasive wheels, textile non-wovens and fire resistant products, etc.

Hexa   + 2   Novolak

4.6.1.1. Phenol, cresol, xylenol and resorcinol formaldehyde resins

Phenol-aralykyl resins based on precondensates of the type

from phenol and p-xylyl-dialkylether (Xylok, GB, Table 4.82) can also be cured with hexa. With polyadditive cross-linking of epoxidized novolaks volatile substances are not eliminated.

b. *Resols*: Excess formaldehyde condensed in an alkaline medium gives rise to methylol group containing resins in the "A stage" (similar to formula 1b), which are soluble but not stable. Rather, they condense further spontaneously, slowly at room temperature and more quickly at higher temperatures. Resols condensed to a low degree (molecular ratio phenol-formaldehyde up to 1:3) are water soluble. They are supplied in aqueous solution as liquid resins. More highly condensed resols (phenol-formaldehyde below 1:1.8) are dissolved in alcohol to a thick viscous liquid or supplied

4.6. Thermosetting Plastics

as a solid resin. The use of resols as binders is described below. "Single-stage" (p. 337) resins for ammonia-free moldings are discussed on p. 371.

On curing resols and novolak-hexa mixes go through
- The *Resitol*, or "B", stage, which is rubbery, swells only in solvents and is not readily melted

to the
- *Resit*, or "C", stage, which is fully cured, cannot be melted, is insoluble in all normal solvents and, as a pure resin, is resistant to most chemicals other than strong acids and alkalis.

In the transition from the A to the B stage there is an increase in the molecular weight with the elimination of water, formaldehyde and, possibly, ammonia. In the additional curing to the C stage spatial cross-linking occurs.

With "overcure", cross-linking advances without further enlargement of the molecules until embrittlement and brown discoloration occur. Similar reactions and those with atmospheric oxygen cause darkening of the cured phenolic resins.

Hot and cold curing: Most of the products bound with PF resins are cured at 140 to 180°C. On hot compression or injection molding, the eliminated volatiles must be allowed to escape (Sec. 3.3.7) to prevent blister formation. Resols can be cold cured with strong acid catalysts (sulfuric acid, aryl sulfonic acid) above 25°C. The more reactive resorcinol, with two phenolic OH groups, results in resins which can be cured without pressure or acid catalysts at room temperature. Their main applications are weatherproof adhesives for wood, leather, rubber and plastics. *Low-pressure resins* (e.g., Urafen 76, US; Norsophen 1200, FR) are used in fire resistant glass fiber reinforced composites for aircraft and vehicle interior equipment.

Phenolic resin foams are produced by exothermic, acid catalysed curing of resols, usually in the presence of low-boiling (halogenated) hydrocarbons as blowing agents. Blocks and strips are the usual forms. Previously only open-celled products were known but for some time closed cell foam webs have been manufactured (Exeltherm xtra, Eurothane xtra, Fenomo) using special resols (e.g. Cellobond K).

The heavy gas blowing agent which remains in the cells confers very good heat insulation properties on the product. The foam shows minimal smoke development and, suitably treated, is flame resistant. Other properties are given in Table 4.61.

4.6.1.2. Urea (UF) and melamine (MF) resins

Urea (UF) and melamine (MF) resins are known collectively as aminoplastics because they contain amino groups. They are very suitable for a wide range of

4.6.1.2. Urea (UF) and melamine (MF) resins

Table 4.61. Applications and types of phenolic resins

Applications	Proportion %	Types of resin	Amount of binder in the material %
1. Wood materials and glues (plywood, resin-compressed wood, chipboard and hardboard)	26–30	Aqueous phenol-resol Resorcinol resin cold glues	ca. 10
2. Molding compounds	16–18	Phenol-novolaks and phenol-resols in powder form	30–50
3. Thermal and acoustic insulating materials a) inorganic (e.g. mineral wools) b1) organic (e.g. textile mats) b2) foam	 12–14 3–5 1–2	 Aqueous phenol-resols Phenol-novolak/hexa powder resins Liquid resols with blowing agent cured in acid	 2–4 3–40
4. Coatings	6–7	Resols and novolaks, solid and in solution; alkyl phenolic resins; etherified resols	up to 50
5. Industrial laminates (paper- and fabric-based)	6–7	Aqueous resols, alcoholic solutions of phenolic and alkyl phenolic resols (also modified)	up to 50
6. Foundry resins (core- and molding-sand-binders for metal casting)	4–5	Aqueous resols, solvent containing resols and epoxide resin formulations for cold curing and gas curing Novolak and novolak solutions for hot curing	0,6–2
7. Abrasive material a) Grinding discs b) Grinding medium on paper	3–5	Modified and unmodified powder resins based on phenol novolak/hexa, aqueous and solvent containing resols	ca. 15 up to 20
8. Friction linings (disc, drum, clutch and machine linings)	2–3	Modified and unmodified powder resins based on phenol novolak/hexa and aqueous phenolic resols for wetting abrasives Aqueous phenolic resols	5–25
9. Fire resistant products (molded and un-molded products)	3–5	Novolak/hexa powder resins novolak solutions, aqueous resols	3–7
10. Other applications (chemically resistant building components and cements, adhesives, industrial filters, rubber additives, carbon and graphite materials, lamp-socket cements, fiber composites (carbon fiber reinforced plastics, glass fiber reinforced plastics, flame resistant fibers, foam for flower displays and similar)	8–12	All types of resins, solid, liquid, dissolved, powder, modified and unmodified	up to 40

industrial applications because of the many possibilities of chemical modification of the intermediate products.

1. Urea Formaldehyde Resins (UF)

UF resins are produced by an addition reaction between urea ($H_2N-CO-NH_2$) and formaldehyde (HCHO), followed by polycondensation.

Areas of application for UF resins

a) Molding compounds: Colorless, light-resistant, molding compounds can be produced using UF- and, especially, MF-condensation products. Such molding compounds can be colored with permanent pigments. Molding compounds are produced by condensation of urea with formaldehyde in a mole ratio of 1:1.2–1.5 or of melamine with formaldehyde in a mole ratio of 1:1.5–4. Flow properties are improved by addition of benzoguanamine and/or toluene sulfonic acid amide. Inert fillers such as cellulose fiber, wood flour and inorganic stone flour are added to the resin solution. Inorganic or organic pigments (e.g. zinc oxide, titanium dioxide, ultramarine or phthalocyanine) are used for coloring. After addition of a lubricant (zinc stearate, glycerine monostearate) and a curer the molding compound mixtures are kneaded until homogeneous, dried and ground. Processing to molded items involves hot pressing or injection molding.

b) Coating resins: Manufacture of aminoplastic coating resins is by condensation of aliphatic alcohols with aldehydes (mainly formaldehyde), urea or melamine and in certain special cases with urethanes, benzoguanamine and sulfonic acid amides. As aminoplastic coatings resins are generally too brittle, they are combined with other binders. Such combinations allow production of physically drying paints, acid curing paints and stoving enamels. Physically drying paints are obtained by adding aminoplastic resins to nitrocellulose paints. Combinations with alkyd resins or polyester resins are used for acid curing coatings. Addition of acids causes curing and produces films with high hardness, scratch resistance and resistance to solvents and light which are therefore suitable for coating furniture and sealing parquet flooring. Stoving enamels are obtained by combination of aminoplastic resins with alkyd resins, unsaturated polyester resins, heat curable acrylic resins or epoxide resins.

c) Binders for the wood industry in the manufacture of plywood and chipboard.

d) Paper auxiliary materials, leather auxiliaries, textile finishing, mat binders, foundry auxiliaries and nitrogenous fertilizers.

Urea foam is produced by stirring and spraying a mixture of aqueous solutions of foaming agent plus acid catalyst and UF precondensate. Cross-linking and drying leads to mechanically stable open-celled *in situ* foam. For properties see Table 4.78, p. 398. Applications include sound and heat insulation of pipe shafts and double-skinned walls, safety foaming in mining, ground-covering layers, nutrient carriers for infertile soil and in nursery cultivation (Plastoponik) and in the medical field.

2. Melamine-Formaldehyde Resins (MF)

MF resins are produced by addition and condensation reactions involving 1,3,5-triamino-triazine (melamine) and formaldehyde.
Adducts of melamine and formaldehyde (methylolmelamine) can be acetylated (etherified) with short chained alcohols. Under acid conditions, reaction products of melamine, formaldehyde and sometimes short chained alcohols form stable aqueous solutions in which the aminoplastic part exists predominantly as a cation (see reaction scheme p. 339).
Products of reaction between melamine and formaldehyde are available commercially as aqueous solutions or, after spray drying, as water-soluble powders under the tradenames Cymel, Madurit, Melan, Melolam, Kauramin and Supraplast.
Etherified methylolmelamines are available dissolved in water or alcohol, or, after exhaustive conversion in material ratio of 1:6, in solvent free liquid form. Trade names are: Cymel, Maprenal and Luwipal.
Methylolmelamine and etherified products in acid formulation are available in aqueous solution. Trade names are: Madurit, Paramel and Urecoll.
Special characteristics of these product classes in cured form and in combination with suitable substrates are:

mechanical: high hardness, scratch resistance, flexural strength and relatively good impact strength;
thermal: high temperature resistance and resistance to yellowing;
chemical: good resistance to chemicals and hydrolysis;
electrical: high volume resistivity and surface resistance, low tracking;
optical: good light fastness, good pigmentability.

New developments are directed towards reducing the amount of formaldehyde in the manufacture of MF resins and to lowering the amount of formaldehyde released in processing and when in use.

Areas of application for melamine resins
a) Surface improvement
– impregnating resin for decorative papers (decorative wood coatings and furniture edges);
– impregnating resin for decorative papers and core layers for the manufacture of high pressure laminates (HPL) and continuous laminates (CPL);
– cross-linker for binders in industrial coatings, principally automotive finishes and for fillers.

4.6. Thermosetting Plastics

Reaction of Melamine and Formaldehyde

Melamine $\xrightarrow{\text{Methylolization}}$ $+ n\ CH_2O$ → Methylolmelamine I $\xrightarrow[\text{Condensation}]{-H_2O,\ -CH_2O}$ Water-soluble condensate III

Methylolmelamine I $\xrightarrow[\text{Etherification}]{+n\ ROH,\ -H_2O}$ Methylolmelamineether II

Water-soluble condensate III $\xrightarrow{HX}$ Water-soluble cation

$n = 1-6$
$R = -CH_3$, water-soluble
$\quad -C_4H_9$, soluble in organic solvent

I, II oder III
$R = -CH_3$
$\quad -H$

b) Binders
- adhesives for wood materials;
- binders for curable molding compounds, in conjunction with phenolic resins for clutch and brake linings, consumer goods, electric insulating parts;
- cationic in aqueous solution for giving wet strength to paper;
- binder for glass fiber mats and weaves, flame resistant to DIN 4102/A2;
- textile finishing, tanning of leather;
- liquefier for hydraulic binders;
- cross-linker of reinforcing resins in automobile tyres.

Melamine resin foam (Basotect) molding compound is produced from MF precondensate with an emulsified, low boiling blowing agent, and hardened with an acid curing agent. At 7.16 kg/m^3, the open pore foam (thermal conductivity 0.033 W/mK, degree of sound absorption 90–95%) is thin walled and flexible. It is heat resistant up to 220°C and highly flame resistant with minimum smoke evolution. Applications include insulating jackets for pipes subject to temperatures up to 150°C, and various kinds of acoustic protection. Thermal/mechanical treatment transforms the foam into fleecy materials (d_R approx. 100 kg/m^3).

4.6.1.3. Furane resins

Trade names: FurCarb, Hetron, QuaCorr (US).
Furane resin is an umbrella name for a group of linear curable resins whose main raw material, furfuryl alcohol,

$$\begin{array}{c} HC \!\!-\!\! CH \\ \| \quad \| \\ HC \diagdown_{O}\!\! C - CH_2OH \end{array}$$

is obtained by processing of corn waste (QO Chemicals, US) or other agricultural residues.

Furfuryl alcohol can self polycondense stepwise in the presence of specific catalysts either by elimination of water and formation of methylene bridges or by double bond polymerization between the rings leading to cross-linking. Additionally, furfuryl alcohol reacts with formaldehyde and can also be reacted with UF or PF condensates and pre-condensates.

Cured furane resins (some modified with PF and UF) are highly chemically and thermally resistant.

4.6.1.4. Reactive resins and reactants for casting and low-pressure processing

Reactive resins are liquid or liquefiable plastics preproducts that may be monomers, low-molecular prepolymers or monomer-polymer mixtures. By combining with specific reactants – which actually may be polymerization initiators

4.6. Thermosetting Plastics

Table 4.62. Applications and properties of furane resins

Areas of application	Binders	Property profile
1. Binders for the foundry industry – no-bake method – hot-box method – warm-box method – Hardox method	Preferably furfuryl alcohol condensate unmodified or modified with formaldehyde, UF and PF condensates	Good reactivity with acid or when heated in processing of core and molding sand; high strength and dimensional stability of the molding, good decomposition on casting
2. Binders for builidng materials – cement – mortar – concrete	Furfuryl alcohol and furfurol condensate and co-condensate as well as furfurol-acetone condensation resins	Good mixing and curing behaviour; high resistance to chemicals, acids and alkalis
3. Laminating resins for fiber reinforced plastics – tanks (containers) – tubes – structural components	Furfuryl alcohol condensation resins and diverse co-condensates	Good weathering resistance, resistant to solvents, bases and acids, good strength at higher temperatures, fire resistant, low smoke development
4. Impregnating and carbonizing material – carbon materials – fireproof materials	Different binders based on furfuryl alcohol and furfuryl aldehyde and co-condensates with PF resins	Good mixing behaviour and compatibility with fillers; high yield on carbonising
5. Other applications – curable molding compounds – foams – chipboard, wood glues – wood impregnation – grinding materials – polyurethane	Different condensation resins, also furfuryl aldehyde/phenol resins or furfuryl aldehyde and/or furfuryl alcohol as reactive thinner for PF and UF resins	Different property characteristics

(hardeners) and accelerators or components to be chemically bonded into the macromolecules – they are cured in exothermic addition polymerization reactions without the elimination of volatile substances. These reactions can be started *in situ* at room or elevated temperatures. Hence reactive resins are suitable for the no-pressure and low-pressure processing methods described in Sec. 3.2). The main groups are methacrylates (Sec. 4.1.1.3), unsaturated polyesters (Sec. 4.6.1.5), epoxides (Sec. 4.6.1.6) and isocyanate resins (Sec. 4.6.1.8), but there are also, for example, PF, furan resins (Sec. 4.6.1.3) and "hybrids" (Sec. 4.6.1.5, special granules). The DIN standards for reactive resins, e.g., casting, impregnating and laminating reactive resins, are

DIN 16945: Reactive resins, reactants, reactive molding materials, test methods

4.6.1.4. Reactive resins and reactants

DIN 16946: Casting resin molding materials in the molded state; Part 1, test methods; Part 2, types

DIN 16943/44/48: Glass-reinforced reactive resin materials in the molded state: 43, preparations of test specimens; 44, test methods; 48/1, classification and designations, 48/2, properties of special molding materials

Fig. 4.17 shows the cold curing process of reactive resins. The starting point is the moment of mixing all the components; for hot setting the elevated starting temperature is adjusted by external heating. Curing time of hot-set reaction resins is determined by the time required to reach maximum heat distortion temperatures measured to ISO/R75. With cold curing, the curing time is that after which the acetone-soluble portion or (with UP resins) the styrene monomer content no longer changes significantly.

Applications of reactive resins

1. Reactive resins, with mineral powders, are used as binders.
   Reactive-resin concrete (5–15% binder) is used for applications requiring very high mechanical strength (Table 4.63) and corrosion resistance such as sewage equipment parts, basement windows, window sills, decorative filled wall panels and sanitary (marble or onyx) items in mass production (MMA, UP); machine foundations and frames, surveyor's tables (MMA, UP, EP); engineering components, bridge or runway surfacing (EP), and lightweight structural elements made of foamed reactive resin (UP, PU). Concrete composites are a field of significant development, and include

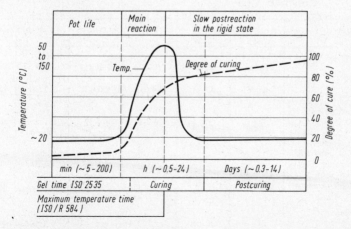

Fig. 4.17. The cold curing of reactive resins.
The time and temperature parameters are indicative. The maximum curing temperature and the main reaction time depend not only on the resin but also on the filler content and the thickness of the product.

4.6. Thermosetting Plastics

polymer concrete (PC, Table 4.63), polymer cement concrete (PCC) and polymer impregnated concrete (PIC).

Reactive-resin adhesive mortar (10–30%) is used for high strength bonds between large prefabricated concrete parts and for building reconstruction by injection (EP resin), as well as for other steel concrete and steel construction (UP resins).

Reactive-resin flooring and knifing compounds (10–40%) are used for jointless industrial flooring and for car body and other repairs, for elastic coverings and PUR sealants.

Reactive resin coatings (30–50%) are used for the protection of industrial buildings and containers against mechanical and corrosion stresses.

Sealing and impregnating compounds (> 50%) possibly containing solvent are available in many combinations.

2. *In electrical engineering* cold flow or melt flow processed cast resins (property values, Table 4.60) are used as construction and insulation materials in the dipping or impregnation of spools and windings, for the embedding of electronic parts and for high-voltage open-air line insulators mainly in EP resin. MMA (Sec. 4.1.1.1.3) and hydrocarbon resins (Sec. 4.6.1.7) have the best dielectric values in the HF range.

3. *For fiber-reinforced articles* UP resins and glass fibers predominate. EP resins are used where the highest mechanical, chemical and thermal stresses

Table 4.63. Some Important Values for Polymer Concrete (PC) and Cement Concrete (CC)

Property	Unit		Polymer Concrete		Cement Concrete	
	SI	U.S.				
Binder content:						
per vol.	kg/m^3	lb/yd^3	150–270	254–457	350	626
per weight	%		6–12		14–15	
Density	g/cm^3	lb/ft^3	2.4–2.2	150–138	2.5–2.3	156–144
At 20°C (68°F):						
Compressive strength:						
after 1 day	N/mm^2	10^{3}lb/in^2	70–100	10–14	2–5	0.3–0.7
after 28 days			95–150	14–21	40–50	5.7–7.1
Flexural strength:						
after 1 day	N/mm^2	10^{3}lb/in^2	23–30	3.3–4.3	0.5–3	0.07–0.4
after 28 days			30–40	4.3–5.7	6–8	0.9–1.1
Modulus of elasticity	kN/mm^2	10^{5}lb/in^2	15–30	21–43	30–60	43–85
Linear therm. expan.	10^{-5} K^{-1}	10^{-5} °F^{-1}	1.5–2.0	0.8–1.1	1.0–1.4	0.6–0.8
Resistance to acids			Fair to high*		Poor	
oils, gasoline			High		Poor	
freezing			High		Fair to poor	

* Depends on type of resin.

4.6.1.4. Reactive resins and reactants

Table 4.64. Properties of Cured Reactive-Resin Castings

Property[1]	Unit		Unsaturated polyesters*					Bisphenol-A Epoxies[2]						Cycloaliphatic epoxy[3]	
			Normal		High			RT cured	Fluid		Cured (and cast) at elevated temperatures		Solid		
	SI	U.S.	Flexural strength and HDT												
Density g/cm³		lb/ft³	1.2	75	1.2	75		1.2/1.6 75/100	1.2/1.8 75/113	1.2/1.8 75/113	1.2/1.8 75/113	1.2/1.8 75/113	1.25	78.1
Mechanical															
Tensile strength	N/mm²	10³lb/in²	30[4]	4.3	55[4]	7.8		50/35 7.1/5.0	Unfilled 3.5 to 4.0 kN/mm² | 5.0 to 5.7 · 10⁵ lb/in²	55/45 7.1/6.4	60/40 8.5/5.7		130	18.5
Tensile modulus	kN/mm²	10⁵lb/in²	3.5	5.0	3.5	5.0			filled > 10 kN/mm² | 14·10⁵ lb/in² depending on fillers				6	8.5
Flexural strength[1]	N/mm²	10³lb/in²	65	9.2	110	16		80/50 11/7.1	100/85 14/12	130/110 19/16			220	31
Compressive strength	N/mm²	10³lb/in²	160	2.3	150	21		85/100 12/14	110/170 16/12	120/200 16/28			200	28
Charpy impact: unnotched	kJ/m²		10–20		15–18			10/4	10/15	15/8			25	
notched	kJ/m²		—		—			1.2	1.4	1.8			—	
Thermal															
Defl. temp. A 1.82 MPa (264 psi)[1]	°C	°F	55[5]	131[5]	90[5]	194[5]		40/60 104/140	100–140 212–284	100/110 212/230			170	338
Linear therm. expansion	10⁻⁵K⁻¹	10⁻⁵F⁻¹	6–8	3.3–4.4	6–8	3.3–4.4		9/7 5/3.9	7/4 3.9/2.2	7.5/4 2.2			—	—
Loss of weight (7d, 140°C/284°F)	%		—		—			0.5/0.3[6]	0.5/0.2	0.5/0.2			—	—
Electrical															
Surface resistivity[7]	ohm		10¹²–10¹⁵					10¹³	10¹³	10¹³			—	
Volume resistivity	ohm · cm		10¹³–>10¹⁶					10¹⁴	10¹⁴	10¹⁴			10¹⁴	
Dielectric constant[1]	50 Hz–1 MHz		4.5–4		4			4	4	4			3–4	
Dissipation factor[1]	50 Hz–1 MHz		0.02–0.03		0.01–0.02			0.02–0.05[8]	0.01/0.02[8]	0.01/0.02[8]			0.01	
Tracking resistance[1]	Method KC		500		500			500/250	500	300/200			600	
Water absorption after 30 min boiling water (ISO/R 117)	mg		—		—			70/40	30/25	20/15			—	

[1] Minimal properties according to DIN 16946, Part 2. [2] Values for dian EP: unfilled / 50 to 65% granular inorganic filler. [3] Not standardized. [4] Elongation at break: normal <2%; HDT types, >2%. [5] Glass transition temperature 70–120°C. [6] At 100°C (212°F). [7] After 24 h in distilled water. [8] 23°C/50% r.h., at 10 K below HDT: 0.25. * ISO 3672: Tensile strength <40 to >70 N/mm² (3 cells), Defl. temp. <40 to >180°C (9 cells).

4.6. Thermosetting Plastics

occur. Figure 4.18 gives achievable glass-fiber contents and densities of products manufactured by various processes and incorporating various reinforcements. For GRP laminates see Table 4.61. Their specific strengths (absolute values/density), excluding the E-modulus, exceed those of metals. The elongation at break of GRP is 1 to 2%. Parts can be designed to allow elongation of 0.3 to 0.8% before the first separation of resin and reinforcement. This is done assuming about 50% of the short-period values for 10,000 hours of load, see Fig. 7.12. Figure 4.19, shows the varying influence of temperature on strength, depending on the resin.

Applications of UP-GF and some hybrid materials with C- or A-fibers, are leisure boats and working boats (minesweepers up to 50 m long), gliders and aircraft, railway and utility vehicles, racing-car bodies, highly stressed cantilever roof structures, swimming pools, transport containers, silos and large storage tanks (cf. Table 3.5), containers for sea transport, pallets, molded sheels, façade cladding, furniture and sports equipment. EP-GF products are used in high-pressure containers and parts for space vehicles and tools. Advanced composites are also considered in Sec. 1.3.2.

4. *Syntactic foams* are used to reduce weight in some of the products mentioned above; see Sec. 3.1.5.1.

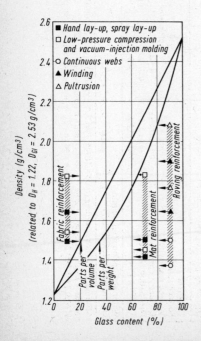

Fig. 4.18. Glass content and density of various GRP products.

4.6.1.4. Reactive resins and reactants

Table 4.65 Indicative Values for GRP-Laminates

Properties			UP-GF laminates									
Glass content by weight	%		25		45		50		65		65	
Glass filament structures			Mats, quasi-isotropic				Woven or non-woven fabric laminated crosswise		Woven or non-woven fabric laminated crosswise ≤450 g/m²		Woven or non-woven fabric 90% unidirectional	
	SI	U.S.										
Density	g/cm^3	lb/ft^3	1.35	84.3	1.45	90.6	1.60	100	1.80	113	1.80	113
Tensile strength	N/mm^2	$10^3\,lb/in^2$	70	10	140	20	200	28	300	43	500	71
Tensile modulus	kN/mm^2	$10^5\,lb/in^2$	5	7.1	9	13	10	14	19	27	28	40
Flexural strength	N/mm^2	$10^3\,lb/in^2$	120	17	180	26	220	31	350	50	550	78
Compression strength	N/mm^2	$10^3\,lb/in^2$	120	17	160	23	160	23	280	40	400	57
Linear therm. expan.	$10^{-6}\,K^{-1}$	$10^{-6}\,°F^{-1}$	35.	19	25	14	18	10	15	8.3	12	6.7
Thermal conductivity	W/mK	BTU in / h ft² °F	0.15	1.0	0.23	1.6	0.24	1.7	0.26	1.8	0.26	1.8

Properties			EP-GF laminates								For comparison: Cr-Ni-sheet steel	
Glass content by weight	%		50		65		65		67-78			
Glass filament structures			Woven or non-woven fabric laminated crosswise				Non-woven, almost completely unidirectional		S-glass 92% unidirectional special resin			
Density	g/cm^3	lb/ft^3	1.60	100	1.80	113	1.80	113	1.8-2.0	113-125	8.0	500
Tensile strength	N/mm^2	$10^3\,lb/in^2$	220	31	350	50	700	100	1300-1700	185-242	~500	71
Tensile modulus	kN/mm^2	$10^5\,lb/in^2$	10	14	18	26	30	43	~60	85	195	277
Flexural strength	N/mm^2	$10^3\,lb/in^2$	280	40	400	57	800	114	1200-1600	171-228	220	31
Compression strength	N/mm^2	$10^3\,lb/in^2$	220	31	300	43	600	85				
Linear therm. expan.	$10^{-6}\,K^{-1}$	$10^{-6}\,°F^{-1}$	18		15		12	6.7				
Thermal conductivity	W/mK	BTU in / h ft² °F	0.24		0.26		0.26	1.8				

4.6. Thermosetting Plastics

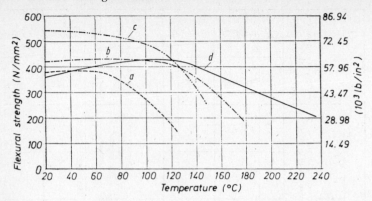

Fig. 4.19. Temperature dependence of the flexural strength of various glass cloth reinforced plastics.
(a) Standard polyester resin, (b) Heat-resistant polyester resin, (c) Normal epoxide resin, (d) Heat-resistant phenolic resin.

5. *Reactive-resin molding materials*, Sec. 4.6.2.4; *laminates*, Sec. 4.6.3.5; *building panels and pipes*, Sec. 4.6.3.6.

6. *Other applications* of UP and MMA include centrifugally-cast, decoratively-filled button sheet. The resins are also supplied pre-accelerated for "do-it-yourself" and for embedding biological specimens.

4.6.1.5. Unsaturated polyesters (UP)

Trade names:
Corezyn, Glastic, Polylite, Stypol (US); U-Pica (JP); Alpolit, Leguval, Palatal, Vestopal (DE); Norsodyne (FR), Crystic (GB); Synolite, Vicon (NL); Civic (FI); Norpol (NO); Viapal (AT); Chromoplast (YU).

UP resin chemistry

Precondensates for UP resins are relatively low molecular weight *linear polyesters* (mol. weight $< 10{,}000$) which contain unsaturated linkages. These are usually introduced using maleic acid anhydride, made by the oxidation of benzene, which in the reaction becomes spatially involved as a fumaric acid chain link.

$$\begin{array}{c} H-C-C \\ \parallel \parallel \\ H-C-C \end{array} O \;+\; HO-(CH_2)_y-OH \;\longrightarrow\; \begin{array}{c} H-C-C-O-(CH_2)_y- \\ \parallel \\ -(CH_2)_y-O-C-C-H \end{array}$$

Maleic anhydride   *Diol*   *Fumaric acid link in the UP chain*

4.6.1.5. Unsaturated polyesters (UP)

Precondensates for standard resins are made from polyglycols (HO–$(CH_2)_y$–OH, $y = 2 - 4$) and phthalic anhydride (for the formula see Table 1.4) as the normal saturated acid monomer.

Monomers for other resins are dealt with under the heading "Resin Types and Properties". The components are reacted together in approximately equimolecular ratio and with the elimination of water in a molten state at 170 to 200°C (338–392°F). The polyesters are subsequently dissolved while still warm in the stabilized comonomers required for the cross-linking copolymerization. When cooled and protected from light, the resulting liquid UP resins have a shelf life of about six months. Usually 30% styrene serves as the comonomer solvent, but vinyl toluene (p. 165), methyl methacrylate and, particularly for compression molding compounds, the relatively non-volatile derivatives of allyl alcohol ($CH_2 = CH–CH_2OH$) are used (Table 1.4).

Designatory properties

Many different types and grades are on the market with varying monomers and differences in the unsaturated element, varying degrees of polymerization of the precondensate and different comonomers in differing molecular proportions in relation to the precondensate. The significant designatory properties for the processor, i.e. viscosity, gel time at 25°C, reactivity at 80°C, maximum temperature (Fig. 4.17), etc., have to be included in the technical data sheets. ISO 3672/1-1979 gives a detailed designatory scheme comprising classes and cells, for the above-mentioned properties of unpolymerized liquid resins as well as tensile strength and deflection temperatures of polymerized resins (Tables 4.64, 4.65) and preferred processing methods.

Curing agents and curing

Exothermic cross-linking double-bond polymerization with, for example, short (2–3 styrene groups) PS bridges between the polyester chains is initiated by organic peroxides that have been retarded by making them into a paste with plasticizer or powder preparations. Hot curing (for example with benzoyl peroxide) occurs at 60 to 90°C. The following initiating systems are in use for cold curing:

(A) 1 to 4% methyl ketone peroxide or cyclohexanone peroxide paste with 0.25 to 2% cobalt accelerator (1% Co) or copper naphthenate-ketimine accelerator at $\geq 18°C$ yields light stable and, in large measure, tack-free products.

(B) 2 to 4% benzoyl peroxide paste, 1 to 2% amine accelerator in 10% solution which is active at temperatures as low as 5°C.

The catalyst and accelerator must not be mixed with each other (danger of explosion!). UP resins treated with accelerator can be kept for a long time; for small-scale fabrication they are delivered pre-accelerated. The curing agent is added directly before processing or, particularly with automatic metering, two separate portions, each containing accelerator or hardener, are mixed together. Liquid UP resins have a low flammability classification. Peroxide paste and amine accelerator are corrosive. Peroxides decompose with inappropriate handling.

Relevant industrial safety regulations should be observed, as should those for the limitation of styrene emissions. Laminating resins intended for open processing are designed for low styrene emission. UP dissolved in MMA to yield a lower vapor pressure are a premium price speciality which can be used underground. Gel coat resins for cold curing contain skin-forming materials (e.g., migrating paraffin additives) that protect the resin from hindering curing and access of air hindering curing and from styrene evaporation.

Cold curing with peroxide and accelerator is not completed on hardening of the resin (Fig. 4.17). GF polyester moldings from which good aging, chemical and heat resistance are required or that must satisfy food regulations must be post-cured in hot air (e.g., 4–5 h at 80 to > 100°C). With later post-curing there is the danger that the curing system will decompose in the interim. The minimum requirement is one to two weeks of dry storage of moldings at $\leq 20°C$ before use. UP resin screeds and coatings can be fully mechanically and chemically stressed only after several days curing. Water, in small amounts, slows down the rate of polymerization. Fillers and pigments can influence the course of polymerization.

Dielectric curing is a special kind of heat curing that is applied in the pultrusion process (Sec. 3.4.3).

Visible light curing of GF-UP laminates up to 20 mm thick, e.g. with normal fluorescent lamps, is possible with stable-in-the-dark one-pack "VLC" UP resins containing light sensitizers (Palapreg, DE; Synolite LNL). The process is used in the continuous production of laminates between covering sheets by the incorporation of light tunnels, in the vacuum-injection process with incorporation of light tunnels, in the vacuum-injection process with illumination through transparent GF-UP surface layers, in the winding process with lamp panels, in hand lamination for repair work and also for curing roof coatings by sunlight. Operating parameters can be so arranged that considerable energy savings result compared with hot curing, by dispensing with post-curing. Opaque products highly filled with reinforcing light-blocking carbon or Aramid fibers cannot be cured by light.

Resin types and properties

Standard resins, based on ortho-phthalic acid and simple diols, are resistant at room temperature to water, salt solutions, dilute acids and hydrocarbons but are attacked by chlorinated hydrocarbons, many solvents, alkali solutions and concentrated and oxidizing acids. The heat distortion temperature (HDT) is approximately 70°C (158°F). *Flame-retardant resins* contain brominated or highly chlorinated (HET acids) acid components and antimony trioxide. Low smoke, low-viscosity resins are filled with aluminium hydroxide, $Al(OH)_3$ up to a ratio of 1:1.8.

Resins with improved corrosion resistance are also suitable for gel coats, continuously exposed to warm water (bathing pools). They have good mechanical properties when based on iso- or tere-phthalic acid and neopentyl glycol ($HOCH_2C(CH_3)_2CH_2OH$) or similar diols and service temperatures of up to about 80°C.

Resins based on bisphenol A (Table 1.4,) are long-term resistant to hydrolysis and saponification by water and 20% hydrochloric acid up to 100°C and 70% sulfuric acid and 20% alkali solution up to 80°C. HDT is 110 to 125°C (230–257°F).

Flexible resins with aliphatic dibasic acid components, $HOOC(CH_2)_xCOOH$ ($x = 4$ adipic, $x = 7$ azelaic acid) are admixed or modified with isocyanate, e.g., for impact-resistant floor covering.

Highly transparent weather resistant resins involve methyl methacrylate as comonomer.

Vinylester resins (Corezyn, Corrolite, Derakane, US; Ripoxy, Spilac, JP; Atlac, GB; Diacryl, NL; Palatal, DE) from bisphenol A or novolak-epoxy resin (Table 4.66) and methacrylic acid dissolved in styrene are used in normal and flame retardant grades. After curing they exhibit 6 to 3.5% elongation at break and service temperatures of 100 to 150°C (212–302°F). Because of their high elongation at break they are particularly suitable for gel coats. They contain no ester groups in the polymer chain and resist 37% HCl and 50% NaOH, wet chlorine, chlorine dioxide and hypochlorite in all concentrations, as well as hydrocarbons and oxygen-containing organic media. Resins with increased temperature resistance are also stable to chlorinated and aromatic solvents.

UP resins suffer 6 to 8% volume shrinkage when unfilled. For hot compression molding there are low shrinkage (low profile) two-phase systems that are modified (Sec. 4.6.2.4.3. by inorganic and/or liquid polymer additives. Unfilled cured UP resins are transparent with visible glass fibers. The refractive indexes can be equalized by addition of MMA. Depolymerization begins above 140°C; beyond 400°C the gases ignite and non-flame-retarded materials then burn further. Highly filled mineral materials are practically non-flammable.

For properties and applications of products see Sec. 4.6.1.4 in particular Tables 4.64 and 4.65.

Special grades

"*Hybrid*" *resins* of low molecular weight UP with hydroxy end groups and diisocyanates as chain extenders dissolved in styrene, are soft elastomeric to

4.6. Thermosetting Plastics

hard impact-resistant casting resins. They are also used in RIM and SMC processes.

Water emulsifiable UP resins (Filabond, US; Wist, DE; WMC, JP) have been proposed as processible casting resins, with 40 to 70% water, for wood-like furniture parts.

Admixture of 3 to 4% Saran (Sec. 4.1.4.3.1) microspheres to laminating resins for boat building gives syntactic foam, similar to lightweight composites. Because Saran is soluble in styrene, the resins must be processed shortly after mixing. Structurally foamed mat laminates, d_R 0.4–0.8 g/cm³, are manufactured by a low pressure process with nitrogen generating Luperfoam (US) blowing agents in conjunction with special inhibitor/accelerator systems. The same system is used for fiber-resin spray lay-up with additional blowing agent for micro-cellular support layers of PMMA bath tubs. The binding agent in UP-resin (Legupren) lightweight concrete (Sec. 4.6.1.4) is foamed with a CO_2-generating blowing agent to d_R 0.05–0.2.

4.6.1.6. Epoxy resins (Epoxies, EP)

Trade names:
Conapoxy, Cyracure, Epiphen, Epocast, Epophen, Epolene, Isochemrez, Stycast (US); Epodite (JP); Beckopox, Biresin, Blendur Epoxin, Eurepox, Lekutherm, Rütapox (DE); Lopox (FR); Eponac Exatron, Ravepox (IT); Epicote, Epon (GB); Aicarpox (ES); Epidian (PL); Araldit, Grilonit (CH).

Epoxide chemistry

The richness of epoxide resin chemistry (Table 4.66) relates to the capacity of the epoxy group to combine with "active" hydrogen (alcohols, acids, amides, amines) under suitable reaction conditions and with suitable catalysts. The hydrogen is displaced relative to the epoxy oxygen in such a way that it gives rise to an active HO group in the addition product. This can be used for further epoxide but also for other addition reactions, e.g., with isocyanates (Tables 4.67 and 4.68) to form hybrid casting resins showing excellent impact strength at low as well as high temperature (e.g. Blendur E, DE).

With epichlorohydrin and bi- or multi-functional components, "resins" are produced that are mono- or oligo-molecular *diglycidyl-compounds* (G in Table 4.66) with epoxide terminal groups. With other reactants preproducts can be formed, e.g., according to formula 6.

Reactants for the synthesis of highly cross-linked EP resin products include bi- or multi-functional low molecular weight products containing active hydrogen. They do not act simply as catalysts in initiating cross-linking of pre-polymers, as is the case of curing agents and accelerators for UP resins. Rather they react chemically and it is therefore necessary to mix-in exactly measured quantities

Table 4.66. EP Chemistry

EP chemistry basic addition reaction

$$-CH_2-CH-CH_2 + H-R \longrightarrow -CH_2-CH-CH_2-R$$
$$\underset{O}{\smile}\underset{OH}{|}$$

Glycidyl [G] { Epoxy } Group

Examples of di-epoxy prepolymers

(1) Bisphenol-glycidyl-ether type (see Table 1.4, p. 17, fields 8, 12, 13)

$$CH_2-CH-CH_2Cl + (n+1)\ OH-\phi-C(CH_3)_2-\phi-OH + (n+1)\ Cl-CH_2-CH-CH_2 \xrightarrow[(n+2)\ NaCl]{NaOH\ powder,\ catalyst}$$

Epichlorohydrin   Bisphenol A   Epichlorohydrin

$$G\!-\!\!\left[O\!-\!\phi\!-\!C(CH_3)_2\!-\!\phi\!-\!O\!-\!CH_2\!-\!CH(OH)\!-\!CH_2\right]_n\!\!-\!O\!-\!\phi\!-\!C(CH_3)_2\!-\!\phi\!-\!O\!-\!G$$

Dian-resin:

$n = 0$: Bisphenol A diglycidyl ether (BADGE) $0 < n < 10$: fluid to rigid resins

Variations: flame retarded by halogenated dian, other bisphenols

(2) Epoxy (cresol) novolac type

[structure: three cresol rings bearing O-G and CH₃ substituents, linked by CH₂ bridges, middle unit repeated n times]

(3) Aliphatic diglycidyl ether type

$G-O-(CH_2)_n-O-G$ Flexible resins with, e.g., $n = 4$

(4) Diglycidyl amine derivatives

[structure: phenyl-N(G)₂]   [structure: G₂N-C₆H₄-CH₂-C₆H₄-NG₂]

Diglycidyl aniline   *Multifunctional,*
bifunctional diluent   HT-coating resin base

4.6. Thermosetting Plastics

(5) Cycloaliphatic diglycidyl ester type

Hexahydrophthalic acid diglycidyl ester. HT arc and tracking-resistant resin base

(6) Cycloaliphatic types with directly bound EP groups

3,4-Epoxycyclohexylmethyl-
3,4-Epoxycyclohexane carboxylate

Vinyl cyclohexene dioxide

Materials for electrotechnical encapsulating casting resins

of the cross-linking components. An additional possible variation is the use of "reactive diluents" with only one epoxy group to reduce the viscosity of liquid resins and to increase the flexibility of molded materials.

Liquid aliphatic polyamines and polyamidoamines are used mainly for cold curing liquid epoxy resins. Tertiary amines serve as effective curing accelerators. For hot curing at $>80°C$ either aromatic amines or their derivatives are used or alternatively phthalic or carboxylic acid anhydrides (HET acid anhydride for flame-resistant laminates), sometimes in combination with accelerators. The most chemically resistant materials are based on aliphatic systems, the solvent resistant on aromatic amines, and the best weathering and acid resistant on anhydride-cured systems.

Many reactants, particularly amines, are corrosive or otherwise dangerous. Attention should be paid to official health hazard regulations and to the instructions of the manufacturers.

Formulated epoxy resins in general manifest an extraordinary high adhesion to almost all materials, small shrinkage on cross-linking and good corrosion resistance, temperature stressability and electrical properties. About 50% of products are supplied as solvent-free liquid or powder coatings for surface protection or for high strength adhesives, about 20% for reactive resin concrete and adhesive mortar (Sec. 4.6.1.4) and another 20% in the electrical-electronic industries.

Resin types and properties

ASTM D 1763 Specification for epoxy resins (containing no hardeners) defines Types I to VI in terms of their chemical nature (similar to the arrangement in

4.6.1.6. Epoxy resins (Epoxies, EP)

Table 4.66). These types may be further subdivided into Grades 1 and 2 without and with reactive diluent respectively.
They are further differentiated by Classes A, B, C, etc., according to the weight per epoxy equivalent (WPE) and the viscosity of fluid resins or solutions of solid resins. *ISO 3673/1: Epoxide Resins; Part 1: Designations* with up to 10 cell numbers for chemical base, viscosity, epoxide equivalence, modifiers, density and additives, and special indications permit the characterization of a resin by a sequence of digits. *ISO 4597/1, Hardeners and Accelerators for Epoxy Resins, Part I: Designations*, is organized similarly.
Fluid-to-solid diglycidyl ethers (1) or epoxy moldable types (2) of Table 4.65, are the generally used standard EP resins; aliphatic diglycidyl ethers (3) are used as flexibilizers.[1] The cycloaliphatics (5,6) and multifunctional amine derivative (4) resin bases are used for high-temperature electronic components, adhesives, coatings and advanced composites. Cycloaliphatic-based coating systems for rapid UV curing (Degacure, DE) are also available.
EP resin/hardener systems are classified as

- Multifunctional resins
- Casting, impregnating and drip impregnation resins for electrotechnology (see p. 379 and Table 4.64)
- Laminating resins for highly stressed composites (p. 380)
- Tool resins
- Surface protection resins

Hot cured epoxy moldings can withstand continuous service temperatures of up to 250°C. Almost zero curing shrinkage, high dimensional and temperature stability under load and marked hardness and abrasion resistance are all properties of filled epoxy resins. These have widespread application for measuring, testing and master gauges and models and even for tools for shaping metals and plastics.
Because of the adhesion of EP resins to nearly all surfaces, it is always necessary to use release agents in molds. The adhesion and curing of liquid EP resins with aliphatic polyamines are impaired by moisture at the bonding surfaces. Special hardeners, particularly polyaminoamide adducts (Versamide, US, Eurelon, DE; Sec. 4.1.1.8), are insensitive to water, so that materials can bond to a moist base, sometimes underwater or as an aqueous dispersion. For application in building, low-temperature initiated "zero-grade" and water-activated "masked" hardeners (Ketimide) are important. The reactive epoxide groups can cross-link with many other materials, e.g., tar products or

---

[1] Long-chain linear epoxies without terminal epoxy group (phenoxy resins Bakelite, US) are used as solvent-free hot melt adhesives at 70 to 250°C and as lacquers dissolved in esters or ketones.

4.6. Thermosetting Plastics

liquid polysulfide rubber (Sec. 5.1.10). The generally good chemical resistance of EP resins can be modified by variation of the cross-linking agent. Their resistance to alkalis is particularly good.

EP foams (with d_R 0.03–0.3 g/cm^3) are made by vigorous agitation of powdered resins with chemical blowing agents and liquid resins with physical blowing agents. Electronic components are encapsulated in EP syntactic foams; these are continuously usable up to 200°C and have compression strengths of up to 60 bar for deep-sea applications.

Further information on properties and applications is given for epoxy molding compounds, Sec. 4.6.2.5; technical laminates, Sec. 4.6.3.5, and other semi-finished goods and pipes, Sec. 4.6.3.1 and 4.6.3.7.

4.6.1.7. Special casting resins

1. *Allyl ester*

Diethylene glycol-bis-allyl carbonate monomer (trade name CR39, US) cured by peroxide catalysts gives a highly transparent, impact-, scratch- and abrasion-resistant cast resin for lightweight spectacle lenses. For diallyl phthalate isophthalate based molding compounds see Sec. 4.6.2.6.

2. *Cross-linking hydrocarbon casting resins*

These resins exhibit favorable dielectric properties in the HF range and are therefore of interest for microwave technology (radar, 10^9 Hz). MMA casting resins behave in the same way, although their dielectric loss increases markedly in the two-day boiling test in salt water.

Trade names: Ricon (US) are liquid resins containing 1,2 vinyl and 1,2- and 1,4-butadiene structures. They are cross-linked with peroxide catalysts.

"Bisdiene" resins from cyclopentadienyl sodium and aliphatic dihalogens are hydrocarbon resins with exceptional dielectric properties up to 200°C. They serve as base materials for printed circuits. Dicyclopentadiene can be polymerized to large high impact parts by a "metathetic" RIM process (Metton, US).

4.6.1.8. Polyurethanes (PU): preproducts and products

Trade names of the main starting materials

Polyisocyanates: Hylene, Isonate (Papi), Multrathane, Nacconate, Nafil, (U.S.); Desmodur, Elastonat, Lupranat (DE); Sumidur (JP); Tedimon (IT); Scuranat (FR); Systanat (DD); Suprasec (GB); Elate (NL).
Polyols: Isonol, Multranol, Multron, Niax Polyol, Pluracol, Quadrol, Thanol, Voranol (US); Desmophen, Elastophen, Lupranol, Lupraphen (DE); Sumiphen (JP); Glendion (IT); Napiol, Scuranol (FR); Bermodol (SE);

4.6.1.8. Polyurethanes (PU): preproducts and products

Caradol, Daltolac, Daltorez, Diorez, Estolan, Propylan, Arcol, Arimax, Polyurax (GB).

1. Chemistry of the polyisocyanates

The variety of polyisocyanate chemistry arises from the fact that the characteristic isocyanate groups, which are common to all di- and poly-isocyanates, can take part in a series of different addition reactions (Table 4.67). Reactions with compounds which have active hydrogens (e.g. hydroxyl, primary and secondary amines) or carboxyl groups are particularly important (Table 4.68). Reaction with water also belongs to this class (Table 4.68, 3). These reactions lead to the formation of linear, branched and cross-linked polymer structures.

The products of such reactions can react further with excess isocyanate (Table 4.68, 5–7) leading to further chain branching. Finally, isocyanate groups also react with each other (Table 4.68, 8-10) or with other reactive double bonds or strained ring structures, e.g. epoxides.

The capacity of isocyanates to react stepwise is very advantageous in practice as more or less isocyanate groups can be pre-reacted with e.g. polyoles, amines

Table 4.67. Important di- and poly-isocyanates

1. TDI	3. NDI
Tolylene-2,4-diisocyanate + 2,6-diisocyanate 80% + 20% = TDI 80/20 65% + 35% = TDI 65/35	Naphthylene 1,5-diisocyanate
2. MDI	
MDI pure monomer MDI raw polymer Diphenylmethane-4,4′-diisocyanate	

4.6. Thermosetting Plastics

Table 4.67. Important di- and poly-isocyanates

4. IPDI	5. HDI
Isophoronediisocyanate (structure: cyclohexane ring with H$_3$C, H$_3$C, H$_3$C substituents and $-N=C=O$ and $-CH_2-N=C=O$ groups)	$O=C=N-(CH_2)_6-N=C=O$ Hexane-1,6-diisocyanate
6. Isocyanurate Triisocyanate (triazine ring with three C=O groups and three R-N substituents)	7. $R-N=C=N-R$ Carbodiimid-Diisocyanat R in 6 and 7: (benzene ring with $-CH_3$ and NCO substituents) or similar

1. Because of its poisonous nature (MAC value, 0.02 ppm) and high vapor pressure TDI is safely handled only in special industrial plant. 4-NCO in the 2,4 isomer reacts eight times faster than in the 2 position; the NCO groups in 2,6 TDI react with medium speed. TDI is thus suitable for the synthesis of prepolymer with residual NCO groups and corresponding adducts for lacquers and adhesives. Their main field of application is flexible foams.
2. 4,4-MDI with equally reactive NCO-groups forms regular networks in (cellular) elastomers and (integral) rigid and flexible foams. 4,4-MDI is crystalline at room temperature. To facilitate processing the melting point is lowered by MDI isomers, prepolymers and/or other modifications sufficiently to make these polyisocyanate mixtures liquid at the processing temperature. Liquid MDIs have minimal vapor pressures. They are safely handled.
3. Highly reactive NDI is used for cast elastomers of the Vulkollan type (Table 4.70).
4,5. Cycloaliphatic (IPDI) and aliphatic (DI) isocyanates and their derivatives are used for light- and weather-resistant products. They are not very toxic. Main applications: outdoor lacquers and coating compounds, e.g., colored sports coverings, IPDI electrocasting resins (Table 4.70).
6,7. These exist as dimerized and trimerized diisocyanates.

or other reaction partners mentioned above. In this way favorable conditions can be produced for further reaction to form high polymers, e.g. liquid consistency, stepwise reactivity or the preconditions necessary to obtain certain properties of the polymer. These pre-products are called pre-polymers (when the equivalence ratio isocyanate/reaction partner is 2:1) or semi-prepolymers (ratio $>2:1$).

Principal reactants for formation of polyurethanes are di- or poly-ols which can be grouped as follows.

4.6.1.8. Polyurethanes (PU): preproducts and products

(1) *Polyether polyols* in which chains of ethylene oxide (EO, Sec. 4.1.1.11) or propylene oxide (PO) are added to initiators (multifunctional alcohols or amines). To increase the reactivity of POX polyols (Table 4.68, reactions 1,3) which are liquid up to great chain lengths they can be grafted with EOX. According to required cross-linking density and reactivity the following are used, *inter alia*:

For elastomers, casting resins and coatings: long-chain polyols (mol. weight 2000–6000) based on propylene glycol and triols (bi- or tri-functional).
For rigid foam: very densely cross-linked polyols (mol. weight 300–1000) from triols, ethylene diamines, sorbitol (three- to six-functional).
For semi-rigid and "cold" cast flexible foam: long-chain, tri-functional combined with short-chain four-functional polyols.
For flexible block foam and "warm" cast foam: mainly long-chain di- and tri-functional polyols based on propylene glycol, glycerol or trimethylol propane.

(2) *Polyester polyols*, e.g., from adipic acid and glycols or glycerol, are used for certain flexible foams and solid or foamed elastomers. They are more expensive than polyether polyols and give products with higher oil but lower hydrolysis resistance; the flexible foams have lower resilience than those based on polyethers.

(3) *Other polyols*, e.g., polycaprolactones (Sec. 4.1.9), polybutadiene diol for sealants, pentaerythritol, glucosides, sucrose (the last three are four- or five-functional as polyols), castor oil (as flexibilizer) and other plant products.

Di- and poly-amines lead to the formation of polyurea structures.

(1) Long-chained polyether polyamides for polyurea polymer chains.
(2) Short-chained diamines (e.g. DETDA, diethyltolylenediamine) as chain extenders and cross-linkers.

Additives for polyurethane manufacture are:

(1) Catalysts for the polyaddition reaction: organometallic compounds such as dibutyltin laurate. Also for "autocatalytic" polyols.
(2) Catalysts for cross-linking and foaming: tertiary amines combined with (1) or alone (triethylene diamine DABCO).
(3) Water and/or physical blowing agents (see p. 362).
(4) Foam stabilizers: polysiloxane derivatives.

4.6. Thermosetting Plastics

Table 4.68. Addition reactions of isocyanates

---

1. Isocyanate + Hydroxyl → Urethane

$$R-N=C=O + HO-R' \longrightarrow R-N-C-O-R'$$
$$\overset{H}{|}\ \overset{O}{\|}$$

---

2. Amine + Isocyanate → substituted urea

$$R'-NH + O=C=N-R \longrightarrow R'-N-C-N-R$$
$$| | \ \| \ |$$
$$H(\text{or } R'') H(\text{or } R'')\ O\ H$$

---

3. Isocyanate + water-(carbamic acid) → Amine + CO_2

$$R-N=C=O + HOH \longrightarrow R-N-C-OH \longrightarrow R-N-H + CO_2 \nearrow$$
$$\overset{H}{|}\ \overset{O}{\|} \overset{H}{|}$$

---

4. Isocyanate + Carboxylic acid → Amide + CO_2

$$R-N=C=O + R'-COOH \longrightarrow R'-C-N-R + CO_2 \nearrow$$
$$\| \ |$$
$$O\ H$$

---

5. Urethane (from 1) + Isocyanate → Allophanate

$$R'-O-C-N-H + O=C=N-R \longrightarrow R'-O-C-N-C-N-R$$
$$\| \ | \| \ | \ \| \ |$$
$$O\ R O\ R\ O\ H$$

---

6. Substituted urea (from 2) + Isocyanate → Biuret

$$R'-NH-C-NH + O=C=N-R \longrightarrow R'-NH-C-N-C-N-R$$
$$\| \ | \| \ | \ \| \ |$$
$$O\ R O\ R\ O\ H$$

---

7. Amide (from 4) + Isocyanate → Acylurea

$$R'-C-N-H + R-N=C=O \longrightarrow R'-C-N-C-N-R$$
$$\| \ | \| \ | \ \| \ |$$
$$O\ R O\ R\ O\ H$$

Reactions are selectively accelerated by catalysts and reaction conditions

(5) Flame retardants: liquid organophosphorous and halogen compounds and halogenated polyether polyols (ixols with about 32% bromine content).

4.6.1.8. Polyurethanes (PU): preproducts and products

8. Isocyanates → Isocyanurate

$$3\ R-N=C=O \longrightarrow$$ (1,3,5-trisubstituted isocyanurate ring)

9. 2 Isocyantes → Uretidione

$$2\ R-N=C=O \longrightarrow$$ (uretidione ring structure with R–N and N–R)

10. 2 Isocyanates → Carbodiimide + CO_2

$$2\ R-N=C=O \longrightarrow R-N=C=N-R + CO_2$$

$$+ R-N=C=O$$

$$+ R-N=C=O \longrightarrow R-N=C \underset{R}{\overset{R}{\diagup\!\!\!\diagdown}} N \diagdown C=O$$

11. Isocyanate + Epoxide → Oxazolidone

$$R-N=C=O + R'-CH-CH_2 \longrightarrow$$ (oxazolidone ring with R–N, C=O, H_2C, CH–R')

(6) Other additives: UV absorbers, pigment pastes, fillers, moisture absorbers, antistatic agents, mold release agents.

Polyurethane forming additions (Table 4.68, reactions 1, 3) of mainly bifunctional active components lead to more or less long, linear and, in some cases, branched chain molecules. These primary reaction products contain NH groups with active hydrogen, which can contribute to further reactions with excess isocyanate (Table 4.68, reactions 4, 5) which contribute to the final cross-linking of the products. In the presence of water the reaction first produces carbamic acid, which decomposes spontaneously with the elimination of gaseous CO_2 (Table 4.68, reaction 2). For this reason solid PUR must be prepared in the complete absence of water. On the other hand, water serves as a (participating) blowing agent for cellular and foamed products.

Polyisocyanurates (PIR) from di- or poly-isocyanates alone are very stiff and brittle. If reaction mixtures of polyols and diisocyanates, with the latter in excess, are used, then less brittle PIR products are obtained which are heat resistant and contain greater amounts of isocyanurate units.

For the production of cross-linked copolymers with "interpenetrating networks" (IPNs, Sec. 1.3.1) other types of reacting monomer components are added to the cross-linking PU components, e.g., one component might be styrene or methacrylate, another might be a polymerization catalyst or preproduct such as unsaturated polyesters or epoxy resins. Simultaneous optimization of several technical properties is achieved, such as hardness, flexibility, impact strength or very high temperature resistance and strength under load at elevated temperature (see Table 4.70). Such product combinations are important for RIM as "simultaneous interpenetrating networks" (SIN).

General physicochemical properties of cross-linked PU products are as follows: Their individually very different mechanical behavior remains little changed over a wide temperature range and the upper service temperature limits lie between 80 and 140°C, (176 to 284°F). PU products can darken in the light; in other respects, such as resistance to surface water and salt water, biological pollution, disinfectants (apart from phenols), detergent solutions and fuels and lubricating oils, they range from adequate to good. The resistance of PUs to the hydrolytic effects of hot water, water vapor, alkalis and acids is limited. Thoroughly cross-linked PU is physiologically inert and, when in conformity with appropriate regulations, permitted for use in medical equipment and in the food industry. Flame-retarded pre-products are processed for the building and transport industries.

2. Processing and application, product-systems

In the manufacture of PU moldings, several liquid preproducts, mostly differing in viscosity and in part highly reactive, must be brought together for the polymer-forming reaction. In industrial mass production, extensively automated production lines are used with casting or spraying plants, in which up to nine feeding streams are conveyed as a reaction mixture batchwise or continuously (Sec. 3.2.2) with exact temperature control and metering by high- or low-pressure mixing heads. On the other hand, product systems in which

4.6.1.8. Polyurethanes (PU): preproducts and products

ancillary processing materials are admixed with diisocyanate and polyol components or in which prepolymers are used, permit PU liquid reactive resin preproducts with only two or three components to be cast or processed as binders after premixing by hand or mechanically. Highly differentiated frothing and foaming techniques extend the broad application ranges for rigid and flexible PU products. The following descriptions of PU systems are therefore subdivided to give only a general survey of the manifold PU applications. These include RIM articules (2.1–2.3), casting and binder resins (2.4, 2.5), and rigid and flexible slabstock and foams (2.6, 2.7). They are also used in lacquers, adhesives and systems for rubber processing. For TPU see Sec. 4.1.8.4.

2.1 Semi-rigid large RIM moldings

To reduce weight and to improve external and internal safety the automotive industry uses lightweight "soft face" front and rear bodywork sections, self-supporting bumpers, bumper coverings and spoilers, steering wheel padding, dashboard fascias, etc., the shock damping of which can allow local impact deformation to be withstood over the temperature range from -30 to $+65°C$ (-22 to $+149°F$). These demands are fulfilled by semi-rigid, large RIM moldings or even better by polyurea systems.

Depending on the formulation and geometry of the molding, the mold residence time may vary from 20 to 120 seconds. The efficiency of the process is increased by IMR-formulations containing release agent that eliminate the need to add release agent to the mold for each shot. With injection pressure 10 to 15 bar, large-part molds are lighter and less expensive than those used for corresponding thermoplastic injection moldings. Therefore PU-RIM confers greater design freedom, including the integration of functional elements, for cost-effective production. Body parts can be coated with light-stable elastic PU-based systems in any desired color after the release agent has been removed from the surface by vapor-phase degreasing. IMC (in-mold coating) is a similar process.

PU systems based on IPDI or HDI (Table 4.67, (4) and (5)) are light-fast and can be directly colored.

General properties of semi-rigid to semi-flexible foams in the density range about 1.0 to 1.3 g/cm^3 (62–77 lb/ft^3) are given in Table 4.69. To meet particular requirements, microporous moldings are filled or reinforced with chalk or aluminum trihydrate, barytes, mica, or specially cut or milled glass fibers. The RRIM (reinforced reaction injection molding) technique incorporates these additives, generally in the polyol component, without unduly raising the viscosity. The equipment must be highly resistant to abrasion.

2.2 Rigid integral foam RIM-systems

Rigid PU integral foam, density of about 0.4 to 0.7 g/cm^3 (25–44 lb/ft^3), with a solid external skin and density distribution within the foamed core correspond-

4.6. Thermosetting Plastics

Table 4.69. Properties of PU structural foams

Property	Unit		Flex. (microporous) systems for				Rigid integral skinned foam[3]	
	SI	U.S.	RIM[1]		RRIM[2]			
Density	g/cm³	lb/ft³	1.0-1.1	62.5-68.8	1.23	76.9	0.11-0.7	25-44
Hardness	Shore D		35-67		65		60-80	
Tensile strength	N/mm²	10³ lb/in²	14-30	2.0-4.3	25/24[4]	3.6/3.4	8-24	1.1-3.4
Elongation at break	%		250-140		65/160[4]		50-8	
Flexural modulus at								
−30 °C (−22 °F)	N/mm²	10³ lb/in²	290-1300	41-185	3270[5]	465	–	
20 °C (68 °F)	N/mm²	10³ lb/in²	55-700	8-100	1750[5]	249	500-1400	71-200
60/65 °C (140/149 °F)	N/mm²	10³ lb/in²	39-350	5.6-50	1210[5]	172	–	
$E_{-30°C}/E_{+65°C}$	–		7.4-4.0		2.7		–	
Charpy impact, unnotched:								
20 °C (68 °F)	kJ/m²		N.B		>75/>55[4]		7-60	
−30 °C (−22 °F)	kJ/m²		N.B-61		40/>60[4]		6-21[6]	
Deformation temp.	°C	°F	–		–		70-125[7]	158-257
Sag test[8] at								
120 °C (248 °F)	mm	in	10-2.5	0.4-0.1	–		–	–
160 °C (320 °F)	mm	in	–		1/3.5[4]	0.04/0.14	–	–
Linear therm. expan.	10⁻⁵ K⁻¹	10⁻⁵ °F⁻¹	~18	~10	4/15[4]	2/8	7.4-8.6	4.1-4.8

[1] Isonol RMA type systems. [2] Bayflex GR 110/50, 20% milled GF. [3] Baydur/Desmodur systems. [4] LD/TD. [5] LD dynamic modulus (DIN 53513). [6] At −20 °C (−4 °F). [7] Cantilever beam method, DIN 53432. [8] Cantilever beam, 100 mm (3.94 in).

ing to that in Fig. 3.5 for medium density 0.65, is rather like wood in its property profile. Individual properties can be tailored to meet application requirements (Table 4.69). PU rigid integral foam, processed by the RIM technique, is used to manufacture relatively large components such as acoustic housing for computer printers, cabinets for microfilm equipment, TV frames and parts of office furniture. In building, it is used for heat-insulating flat roof

4.6.1.8. Polyurethanes (PU): preproducts and products

gullies, light-dome installation rings and, reinforced with metal inserts, for window frames. Mass produced RIM parts have to be ground and lacquered. For interior trim in car door panels, switch consoles, etc., decorative surface finishes in the form of plastic sheets or textiles are back-filled with integral foam. Very lightweight elements ($d \leq 0.2 \text{ g/cm}^3$, 12.5 lb/ft^3) are used for imitation antique wood interior decoration.

2.3 Semi-flexible integral foam RIM systems

Semi-flexible integral foam parts, mostly in the density range < 0.2–0.6 g/cm^3 (12.5–37.5 lb/ft^3), have a tough, leather-like, abrasion-resistant skin. Typical examples of such RIM products are safety padding for ski boot linings, for working or military footwear, molded-on platform or other shoe soles, car interior trim and bannister handrails.

2.4 PU casting resins

PU casting resins must contain a water absorbent additive (generally a zeolite paste) to prevent undue porosity. Soft systems masked against the isocyanate/water reaction cross-link under water.

2.4.1. *Encapsulating resins* for heavy-duty electrotechnical equipment such as cable terminal boxes, current converters, transformers, ignition coils and high-voltage cascades are based on different polyether-polyols and MDI within the Shore D hardness range 19 to 86, corresponding to glass transition temperatures between -21 and $+130°C$ (-6 and $+266°F$). Cycloaliphatic isocyanate-based and/or IPN copolymer types (Sec. 4.6.1.8.1), which are used for external high-voltage insulation, exhibit the highest rigidity, electrical properties and temperature and weather resistance (Table 4.70). These casting resins need post-curing for about 16 h at temperatures up to $160°C$ ($320°F$).

2.4.2. *Elastomeric hot-casting resins*, developed on the "Vulkollan" principle, are three-pack systems. They consist of a long-chain adipic acid ester diol, which must be fully dehydrated and degassed by heat and vacuum in the fusion pot before casting. The very reactive NDI (m.p. $128°C$, $262°F$, Table 4.67) is then added in excess to build up a long chain, without the formation of stable intermediates, and this is followed by the addition of a small amount of a simple glycol or a similar chain-extending and cross-linking agent (reaction 4 or 5, Table 4.68). The cast products begin to cure immediately. However, they need afterheating at 80 to $140°C$ to obtain a full cure.

This type of "snappy" elastomer, with a wide service temperature range (Table 4.70), is distinguished by extremely high wear resistance and resistance to lubricants, many solvents and weathering. It is used for heavy-duty solid tires (e.g., on forklift trucks), guide rollers for elevator couplings and spring loaded elements.

4.6. Thermosetting Plastics

Table 4.70 Properties of Rigid and Elastomeric Products from PU Casting Resins

Property	Unit		Rigid cast resins				Elastomeric Vulkollan three-package system	
	SI	U.S.	Cycloaliphatic IPDI system		Mixed MDI-styrene-EP system[1,2]			
Density	g/cm³	lb/ft³			1.1/1.7	69/106	1.26	78.8
Tensile strength	N/mm²	10³ lb/in²	87	13	54/77	7.3/11	25-39	3.6-5.6
Elongation at break	%		4		–		600-300	
Flexural strength	N/mm²	10³ lb/in²	170	24	95/106	14/16	–	
Elastic modulus	kN/mm²	10⁵ lb/in²	–		3.2/9.4	4.6/13	≤0.6³	≤0.9³
Impact resilience[4]	%		–		–		50-35	
Charpy impact	kJ/m²		25		12.5/6.6		–	
Ball ind. hardness	N/mm²	10³ lb/in²	226	32	227/533	32/76	–	
Shore hardness	Grade		–		–		A65-99, corr. D70	
Glass transition temp.	°C	°F	150	302	300	572	~ –40³	
Defl. under load, Martens	°C	°F	133	271	208/250	409/482	–	
Thermal conductivity	W/mK	BTU·in / h·ft²·°F	0.20	1.4	0.17/0.76	1.2/5.3	0.29	
Linear therm. expan.	$10^{-5} \cdot K^{-1}$	$10^{-5} \cdot °F^{-1}$	7.9	4.4	6.7/3.4	3.7/1.9	–	

[1] Baymidur/Baygal pilot product. [2] Unfilled/2:1 by weight silica filled. [3] Deformation under load nearly constant to 100°C (212°F). [4] Pendulum test (DIN 53512).

Cellular elastomers of this sort with densities in the range 0.25 to 0.65 g/cm³ (12.5 to 41 lb/ft³) are produced by the measured addition of water. Because of their gas content they are compressible without lateral distortion and they have very favorable damping properties and resilience. They are used for shock breaking buffers as well as for vibration- and sound-damping elements in building and machine foundations and for railway bridges and underground railway lines.

Besides these types there are two-component stable polyether-MDI prepolymer types for hot casting. They can be processed with simpler equipment and, although they are not quite so good mechanically, they are more resistant to hydrolysis. They are used for wear-resistant linings, for example, for sieves and receptacles. Other grades are anti-abrasive topcoats for artificial leather (Sec. 4.2.4.2.4). They can be plasticized for very soft printing roller surfaces.

2.4.3. *Cold-castable two-component types* are used for flexible shuttering, bell and spigot joint gaskets, etc. Single component "blocked" PU-prepolymer compounds, which cure with humidity without detrimental elimination of CO_2 gas, are important as expansion joint sealing compounds in construction.

4.6.1.8. Polyurethanes (PU): preproducts and products

2.5. Prepolymer binder resins

Only a few of the many uses of PU prepolymers as binders can be mentioned here.

- Systems for large parts, similar to those mentioned in Sec. 2.1. and as binders for glass fiber mat prepregs for high-pressure molding (Sec. 3.3.4.2).
- Hot-curing MDI prepolymers for pressing formaldehyde-free heavy-duty plywood and chipboard.
- Cold-curing elastomeric single- and two-component systems for all types of indoor and outdoor sporting surfaces filled with scrap rubber and other aggregates. These are built up in several layers according to differing sports requirements, often with light-stable aliphatic isocyanates (Table 4.67, 5).
- Humidity-cured decorative flooring compounds, sealed by a second layer.
- Crack resistant elastic stoving enamels for cars and paints for the building industry, based on aliphatic diisocyanates.
- Solvent-free reactive adhesives for e.g. assembling car bodies instead of welding or riveting.

2.6. Rigid PU and PU-PIR foams

Heat insulating rigid foams (density 0.02–0.08 g/cm^3) were previously blown mainly with halogenated alkanes. Nowadays CFC-free rigid PU foams are commercially available. These rigid foams are based on MDI/polyether- and polyester polyols which are themselves based on TMP, glycerine, ethylene diamine or sucrose and/or phthalic acid anhydride esters or PET reclaim.

For a comparison of their general properties with those of other light-foamed materials see Table 4.78. These foams, with approximately uniform distribution of cells across the cross section at $\geq 95\%$ closed cell rigid foam, are distinguished by their heat-insulating capacity, which is better than that of most other foams, and because they can be used within the temperature range -200 to $> 100°C$ (-328 to $> 212°F$). To meet flammability requirements as prescribed in building regulations, PU rigid foam must be made with reactive (halogen- or phosphorus-containing) components or added flame retardants or processed with covering layers. Using preproducts and catalyst systems that result in the formation of increasing proportions of more highly condensed polyisocyanurate (PIR) structures (Table 4.67, 6) within the substance of the foam, PU-PIR and PIR rigid foams are made to satisfy normal and more demanding requirements for flame-retarded materials without flame retardants. They can be used for service temperatures up to $140°C$ ($284°F$), e.g. for district heating. Rigid PU foams of higher density (≥ 100 kg/m^3) are used as supporting cores in sandwich structures and are also used as binders in reactive resin-lightweight concrete (Sec. 3.1.5.1 and 4.6.1.4).

4.6. Thermosetting Plastics

Continuous slabstock-foam production plants are equipped with fixed or oscillating mixing and spraying heads arranged across conveyor belts, which work with moving guide strips on both sides and a cover belt. They produce rectangular slabstock at the rate of 3 to 4 m^3/min (106–141 ft^3/min), to a height of about 75 cm (30 in) and a density of 32 to 35 kg/m^3 (2–2.2 lb/ft^3). These are then cut up into flat or wedge-shaped sheets, pipe insulation, etc. Sandwich products with flexible or rigid covering layers which bond firmly to the foam are made on such plant. These products are used in many applications including civil engineering and cold stores. They can be laid with hot-bituminous adhesives (melt temperatures of up to 250°C). Sandwich sections with profiled steel or light metal outer layers and PU foamed cores with a density of about 60 kg/m^3 (3.7 lb/ft^3) are self-supporting elements for light walls and roofing. A reaction system based on polycarbodiimide (PCD) cross-linking (Table 4.67, 7) is used to manufacture light, elastically-flexible, open-celled foam 2 to 20 kg/m^3 (0.5–1.25 lb/ft^3) for sound insulation. For building applications PCD foam does not require added flame retardant.

Casting and spraying prepolymer mixes for in situ PU foam is used in building construction for the fabrication of wall blocks with integral heat insulation. Jointless, insulated and waterproof roofs are produced in a single operation with foaming machines installed on a specially adapted truck, from which the two components are pumped into a spray gun through long, heated hoses up to the flat or corrugated sheet roof. The foamed roof must be protected against UV by a PU-based coating. The same technique is used for external insulation of storage tanks and district heating lines. Refillable canisters for two-component foams and spray cans for moisture-curing single-component foams are used by installers for foam covering of joints and around door and window frames. Refrigerator cabinets are internally foamed with rigid PU of a density of about 28 kg/m^3. The "frothing" process (Sec. 3.1.5.1) is used for uniform filling of the hollow spaces. The foam mixture is physically prefoamed using an easily gasified blowing agent and the heat liberated by the chemical reaction. For deep hollow sections and for the insulation of long double-walled pipes, automated lances distribute the foaming material uniformly.

2.7. Flexible molded and block foams

Foamed flexible padding, with a density of 17 to 60 kg/m^3 (1 to 3.8 lb/ft^3), for car seats and furniture, is manufactured in single-shot processing plants at an output rate of 300 to 400 cushions per hour. Car seats combining two grades of foam hardness for seating comfort as well as safety have been made by double head machines in one shot. The "hot foam" technique used for the processing of polyether polyols with TDI or MDI, in which the products must be heated to 160 to 180°C (320 to 356°F) for the final cross-linking, is being increasingly replaced by the high resilient (HR) and cold-foam processes with their highly reactive systems. They require the same short times to achieve the final cross-linking. Fluoroalkane water blowing systems are being replaced by those based on water alone.

Water-activated filled dense foams made of polyether polyols and MDI to densities of 300 kg/m^3 (19 lb/ft^3), up to 1000 kg/m^3, depending on the amount and kind of filler, are doctor-knifed as back coating onto textile flooring webs up to 5 m wide or elastic flooring (p. 000). They are cross-linked on a rotating drum at 80°C (176°F). Open-celled foam permits improvements in sound insulation by more than 23 dB with good walking comfort. The harder water-activated open-celled flexible polyester foams are predominantly foamed as slabstock for pillow bases. Cylindrical stock is shaved for sheets and for heat laminating of textiles; with large holes it is used for sponges and for many other flexible foam domestic and consumer products. For properties of flexible foams see Table 4.79.

Special grades of reticulated foams with minimal densities act as carriers for filter media in air conditioning and environmental protection and for water purification and oil separation and at temperatures from −60 to 200°C (−76 to 392°F).

4.6.2. Thermosetting Molding Compounds

4.6.2.1. General characteristics

For manufacture from resin and granular, fibrous or web-based fillers see p. 00.
Forms of delivery are as far as possible made suitable for all processes: dust-free ground powder, pourable pellets and chips. For dough-molding (DMC, BMC) and sheet-molding compounds (SMC) with fluid reaction resins see Table 4.73, PF-SMC Table 4.72.

For processing by compression, transfer or injection (-compression) molding see Sec. 3.2.4. Processing conditions are indicated in Tables 3.11 and 3.12, processing shrinkage and tolerances in Table 3.6. Condensation resins (PF, UF, MF, Sec. 4.6.1.1) require venting and higher temperatures and pressures than the lighter and easier-flowing reactive resins (UP, EP, Sec. 4.6.1.5). Because the moldings can be ejected while still hot, shorter cycle times than for thermoplastics are possible.

In most of the applications of *thermoset moldings as engineering plastics* the decisive combination of properties is

- Suitable fillability for many requirements;
- Suitability for highly flame retardant (HB to V–0, UL 94) electrotechnical insulating moldings;
- Relatively small reduction of strength and stiffness with rising temperature and no change at low temperatures;

- Continuous elevated service temperatures from 100 to 200°C (212 to 392°F) combined with short-term overloading to several 100 K (i.e., a difference of >180°F) thus providing a high safety factor;
- Dimensionally stable to <0.1% especially if annealed; also limited processing shrinkage and aftershrinkage of the molding – according to material and processing.

Their usefulness for specific service requirements depends on the choice of an appropriate filler, enhancer or reinforcement (Sec. 3.1.3.8), which accounts for 40 to 60 weight percent of the total molding material.

Molding compounds with inorganic fillers have a higher heat resistance and long-term service temperature and are less affected by moisture than organically filled compounds (Fig. 4.20). With increasing linear or surface expansion of the filler there is an increase in notched impact strength of the compound. However, pourability, bulk factor and flow are reduced and the surface finish of the moldings is diminished.

The standardization of the many different molding compounds, based principally on phenolic and aminoplastic resins designed for different technical applications has led to an abundance of coded grades. This historically developed coding does not correspond to designatory properties of the types and grades. Within the ISO attempts are being made to classify, arrange and code types and grades of thermoset molding materials uniformly, according to application, properties and composition. This aim is to specify the grades by tables of general property requirements to be tested for classification (e.g. Table 4.71). These are "acceptable quality" requirements to be met by the standardized molding materials and moldings made from them. A control symbol similar to DIN 7702 (Fig. 4.21) will indicate compliance. However, the as yet unstandardized *special molding compounds* are gaining in importance, particularly those that can withstand very high temperatures and have even better electrical properties and to some extent also higher strengths.

4.6.2.2. Phenolic (PF) molding compounds

For marking commercial products producers generally use proprietary letter-digit symbols based on those used in standardized series (Tables 4.71, 4.72, pp. 371, 376) but giving more details on the performance of the many specially formulated compounds, as described below.

1. Compounds for technical moldings

Trade names: Durez, Genal, Nestorite, Plenco (US); Tecolite (JP); Trolitan (DE); Fenochem, Lerite, Moldesite (IT); Progilite (FR); Vyncolite (BE); Fenoform (YU).

4.6.2.2. Phenolic (PF) molding compounds

Phenolic molding compounds contain phenol-novolaks as binders and hexamethylene-tetramine as hardener or ammonia-free phenol-resols without this hardener. They are designated N and R respectively (2 or 1 in ISO); resol is also designated SS in ASTM (for single stage). R compounds have a limited shelf life. For technical reasons having to do with flow in compression, mixed resins with about 20% cresol are currently in use.

As the cured resins are yellowish-brown and darken subsequently in sunlight or on warming, PF compounds are supplied only in dark colors. The flow of the compound in processing (easy, medium, stiff) is determined by its resin content, its degree of condensation and its moisture content. Compounds with a small content of only lightly precondensed resin flow easily but give moldings with a longer curing time, greater shrinkage and lower quality than resin-richer compounds that have a medium or stiff flow. They are therefore used only for smaller moldings that are not much stressed. Compounds of similar grade can be prepared in different forms as powders, pellets, etc., to suit the molding process.

All PF moldings are resistant to organic solvents, fuels, fats and oils, even at elevated temperatures. They have only limited resistance to more concentrated acids and alkalis. Those with organic fillers, and consequently increased water absorption, are more strongly attacked on longer exposure than those filled with inorganic materials. The rate of water absorption (Fig. 4.20) is so small with all PF compounds that they are not damaged by short-period exposure to water. Adequately cured moldings should show no signs of damage after boiling for half an hour in water or in dyestuff solutions. Inorganic-filled compounds should be used for insulating parts exposed in the open or in damp spaces.

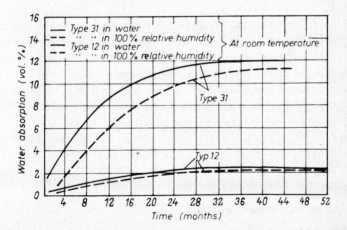

Fig. 4.20. Moisture absorption of organic (Type 31) and inorganic (Type 12) filled phenolic molding compounds, measured or standard rods 15 by 10 by 120mm. Linear change at saturation, Type 31, ~ +1.2%; Type 12, ~ +0.2%.

4.6. Thermosetting Plastics

Fig. 4.21. Control symbol according to DIN 7702.

The range of applications of individual standardized compounds (Tables 4.71, 4.72) stems from a combination of electrical, mechanical, temperature and moisture requirements. The cost of the molding compounds increases with more demanding requirement profiles corresponding approximately to the order of the group.

Special molding materials are mainly used to meet increased thermal, mechanical and/or precision needs, e.g., for rocket parts, brake and clutch linings serviceable up to 500°C, parts for domestic equipment (e.g., baking ovens) with service temperatures up to 280°C, heating armatures, slip rings for ships' generators, water-pump vanes, electrical ignition, liquid drive parts, carburetor heads, cylinder head covers in cars and metal-plated typewriter typing heads. Copper powder filled compounds are highly heat conducting. Those with iron powder or barium ferrite are magnetizable; with lead powder they are suitable for screening from X rays and with graphite filling they are semi-conducting.

Special compounds filled with wood flour and rubber have increased impact strength but considerably reduced heat stability. They are used for cable plugs and the like.

Mineral-filled "cold-pressing" compounds (those processed at room temperatures, Types 212 and 214 in DIN 7708.4), which are heat cured outside the mold, are used in tracking-resistant (KC > 400) and highly heat-resistant insulation moldings but are not important at present.

GF filled granulated compounds are used for high impact (Charpy notched 2.5 to 20 kJ/m^2) and heat resistant parts. Long fiber, high impact resistant reinforced compounds can be press or injection molded.

GF reinforced prepregs (Sec. 4.6.1.1) with special PF resins and curing agents are gaining in importance in building and transport sectors on account of their heat stability (T_g > 300°C, > 572°F) and low flammability (Durez, US; Norsophene, FR; Durostone (DE))

2. *Wood chip and veneered materials*

Wood-chip molding materials are made from wood chips or similar plant waste bound with special UF, MF and PF resin mixtures with added flame retardant and fungicide. Wood chip moldings are weather- and moisture-proof. They are

Table 4.71. Types and Grades of Phenolic and Aminoplastic Molding Compounds

Filler-enhancers	Resin base	Usual Types		
		ASTM D 700, 3881 (PF), 704 (MF), 705 (UF)	DIN 7708, part 2 (PF), part 3 (MPF, MF, UF)	ISO 800 (PF)[8], 4896 (MPF), 2112 (MF, UF)
I GP, mechanical and electrical				
Wood fiber flour	PF	2[1], 11, (14)[2]	30.5[1], 31, 31.5[1], 31.9[3], 32[3]	PF-N, WF 60, 55[1]/-R, WF 60[3], 55[1,3]
	MPF	–	180, 182[1,4]	MPF A 20
	MF	6[1]	150	MF B 20
	UF	–	–	UF A 20
α-cellulose. pulp	PF	3[1], (15)[2]	52[1], 52.9[1,3]	
	MPF	–	181[1], 181.5[1], 183[1,5]	MPF A 10[1], 11[1]
	MF	1[1,6], 7[1]	152[1], 152.7[6],	MF B 10, 11[6]
	UF	1, 2	131, 131.5[1]	
II High-impact, organic enhancers				
Cotton, cellulosic, synthetic fiber flock	PF	12, 24	51, 51.5[1], 51.9[3], 83, 85	PF-N, LF
	MF	–	153	–
Yarn or twine	PF	4-6	71, 75, 74, 84	PF-N. LY/L-T, PF-N-LC
Chopped fabric	MF	3	154	MF-D
III Heat-resistant, inorganic fillers and impact enhancers				
Stone powder, mica (asbestos-free)	PF	8[1], 9, 10, 13[1], 22[1]	11.5[1], 13[1], 13.5[1], 13.9[3]	PF-N. MP[1]/MS[1]
	MF	2[1,7]	155	MFC
Asbestos fibers or twine	PF	19, 20[1]	12, 15, 16	PF-N. AP[1]/AF
	MF	–	156, 157[1], 158	–
Glass fibers	PF	21, 25[1]	GF – types not yet specified	
	MF	8[1]		

The Roman numerals I, II, and III correspond to group numbers of DIN 7708 and to the ISO groups A, D, and C. The grades of the DIN groups IV, enhanced electrical properties, and V, other additional properties, have been attached to the main groups, marked by remarks 1 to 3. [1] Suitable for electrotechnical applications (general and special grades). [2] Rubber modified. [3] Free of ammonia (<0.2%). [4] Wood flour and inorganic filler. [5] Cellulosic and inorganic filler. [6] Also for foodstuff contact. [7] Mineral and other fillers. [8] acc. ISO 800 DP 1983. Types and grades acc. ed. 1977: 2 A1, 2 A2 ~ N, WF 60, 55; 1 A1, 1 A2 ~ R, WF 60, 55; 2 C1, 2 C2 ~ N, AP, AF; 2 C4 ~ N, MP; 2 D1, 2 D3, 2 D4 ~ N, LY, LT, LC; 2 E1 ~ N, MS

4.6. Thermosetting Plastics

used for profiles and panels for façades, inner walls and door linings, concrete shuttering elements, etc.

Veneered laminating compounds have similar applications and in addition are used for seat shells, benches for open-air stadiums, school furniture and components for effluent handling.

3. *Phenolic resins for corrosion protection*

Phenol-resins with inorganic fillers (previously asbestos but now other fiber materials) are used for corrosion protection, particularly when subject to higher temperatures ($> 100°C$), as well as for molding materials and as knifing compounds for liners. The knifing materials cure after the addition of suitable hardeners, without pressure, at room or moderate temperature.

4.6.2.3. Aminoplastic molding compounds

Types and properties see Tables 4.71, 4.72.

1. Urea resin (UF) molding compounds

Trade names: Bakelite, Pollopas (DE); Seritle, Urochem, Uroplas (IT); Beatle, Scarab (GB); Carbaicar (ES); Perstorp, Skanopal (SE).

Commercial types: Fine powders, mainly granules, also for automatic compression molding and injection, lightfast in white and all pastel colors.
The resin content is about 60% which must be this high to achieve adequate flow. The light-colored compounds with bleached cellulose give translucent moldings with intense colors.
The molding material is resistant to solvents and oils but has little resistance to acids and bases.
Applications: In white for threaded moldings in cosmetics packaging, for sanitary moldings, for domestic equipment and for electrical installation material. Small quantities of residual formaldehyde render UF compression moldings unsuitable for eating and drinking utensils.

2. Melamine resin (MF) molding compounds

Trade names: Cymel (US); Resart, Supraplast, Bakelite (DE); Melochem, Melsprea (IT); Beatle, Melmex (GB); Melaicar (ES); Melopas, Neonit (CH); Isomin (SE).
Melamine resin moldings filled with 60% α-cellulose are white and lightfast in all colors as well as resistant to boiling water and detergents. They are suitable for colored cooking utensil handles. Special grades are harder and more scratch resistant than MP (see under 3 below) or UF moldings.
The main field of application of the light-colored molding compounds is tracking- and moisture-resistant and mechanically highly stressable insulation parts.

Compounds filled with inorganic materials (including glass fibers) are arc and heat resistant. They are practically non-combustible and therefore are permitted for fire-endangered rooms and in shipbuilding. MF molding compounds are attacked by acids and strong alkalis but are resistant to fuels, oils, solvents, alcohol and many other chemicals.

MF materials can be injection or injection-compression molded only when specially formulated. Because of high after-shrinkage they can be liable to stress cracking.

3. *Melamine-phenolic (MP) resin molding compounds* are mixed compounds without the advantages of pure melamine resin compounds in relation to antitracking and color stability in light and heat but with adequate color stability for many purposes in their white and light-colored forms. The significantly reduced after-shrinkage of these mixed compounds is an important advantage compared with aminoplastics.

Applications: Light-colored moldings for the electrical industry, domestic and kitchen equipment, threaded moldings.

4. *Melamine-polyester molding compounds*

These compounds, filled with cellulose or mixed organic-inorganic fillers, combine the color brilliance and tracking and arc resistance of MF with minimal after-shrinkage, minimal tendency to crack and improved dimensional stability under thermal stress.

Applications: Electrotechnology, domestic and switch gear, lamp sockets and other electrical parts.

4.6.2.4. Polyester (UP) resin compounds

Trade names[1]: Aropol, Haysite, Hetron, Styrol (US); Vyloglas (JP); Durapol, Illandur, Keripol, Menzolit, Resatherm, Resipol (DE); Norsomix (FR); Beetle, Crystic Impel (GB), Uromix (NL); PMWP (SE); Esteform (YU).

For forms of supply and specifications see Table 4.73. As there are many different types that are better than the designatory properties of the standards given in Table 4.73, actual property ranges have been given in Table 4.74.

1. *General properties*

UP resins used in molding compounds flow readily with relatively low pressure at processing temperatures. They cure quickly and completely by addition

---

[1] Different sorts of dry, putty type bulk (BMC) and sheet molding (SMC) compounds, including (for US) "Alkyd" Compounds.

4.6. Thermosetting Plastics

Table 4.73. Types of Dry and Wet Molding Materials*

	Form of delivery	Main filler type	\multicolumn{2}{c}{Standardized UP compound types}	
			ASTM D 1201	DIN 16 911
Dry	Granules	Mineral + short glass fiber	1: General purpose	802: Pourable 804: Flame resistant, pourable
		Mineral + cellulose	2: Improved mechanical strength	
	Pellets 5–20 mm long	Long glass fiber	–	801: Normal 803: Flame resistant
Wet	Putty type, dough molding compounds (DMC) or bulk molding compounds (BMC)	Mineral filler	3: General purpose 4: Superior electrical qualities	
		Glass fiber	6: High impact	801, 803, not standardized strand type
	Sheet molding compounds (SMC), prepregs	Glass fiber mat	5: High impact	DIN 16913, part 3* 830, 832 normal, 831, 833, 834, flame resistant. Special (point 5) types for electrical use

* DIN 16913, part 1 and 2 standardize terms, types symbols for reinforced reactive resin compounds and the testing of prepregs in general. DIN 18820 E is a general standard for textile materials reinforced UP-resins for structural components.

polymerization without elimination of volatile products. Moldings undergo hardly any after-shrinkage and therefore remain dimensionally true and are not liable to cracking. The cured moldings are light stable with any light-colored pigment and are resistant to alcohol, ether, petrol, lubricants and fats. They are resistant under certain conditions to benzene, esters, weak acids and boiling water but not to alkalis and strong acids. (For higher corrosion resistant special binders and vinyl ester resins see Sec. 4.6.1.5). They manifest high glass transition temperatures, good electrical properties with high tracking resistance and low dielectric loss. Molding compounds with inorganic fillers are arc and heat resistant and through (additional) fillers such as aluminum oxide trihydrate can be made flame retardant to UL 94 HB V-0.

For indicative properties see Table 4.74.

4.6.2.4. Polyester (UP) resin compounds

2. Dry and putty-type compounds

Granulated and pelletized compounds are styrene-free. Most are cross-linked with diallyl phthalate (Sec. 4.6.1.7). They can be stored for up to one year, whereas the styrene-containing BMC has a shelf life of only several months. Dry and flock-like styrene wet compounds can be injection or compression-injection molded with short cycle times. The molds must be well vented to avoid burning by the "Diesel effect". Damp strands can be drawn in continuously from the roll. Articles with extremely high oriented strength are produced by appropriately positioning such strands or dry, long, fiber-reinforced, rod-form compounds into a compression mold. Reinforced rods can be made up to 1 m long and moldings produced by laying them horizontally in the mold are able to withstand extremely high stressing.

The main fields of application of these materials are highly stressed, complicated shapes for components in electrical engineering and electronics, car electrics and dishwasher and microwave resistant domestic equipment. Suitably formulated compounds satisfy the same electrotechnical safety requirements as steatite or porcelain. Light-weight moldings are produced by the incorporation of hollow glass spheres into the compound.

Large moldings and those with large surface areas in vehicle building are made from reinforced BMC by compression or transfer molding. Process shrinkage of 0.2 to 0.4% leads to rough surfaces requiring considerable finishing operations. Additions of up to 25% of thermoplastics such as PS, PMMA and CAB in syrupy styrene solution compensate for the processing shrinkage in low-shrink (LS) systems (colorable) and low-profile (LP) systems (colorable with difficulty). The fact that the surface color in LP-systems is impaired is of little consequence in, for example, vehicle building, where the molded part must be lacquered in the vehicle color. The thermoplastic additives can be combined with elastomeric modifiers to prevent surface impact crazing.

In the SMZ injection-molding machine (FR), BMC compound with 0.6 to 2.5 mm fibers is metered into a cylinder with a screw that plasticizes the compound without shearing and conveys it into an outer cylinder. After closing the outlet, the screw cylinder works as an injection piston. Large thin-walled profiled SMZ moldings (e.g., car rear-window frames) contain a high portion of unbroken, highly reinforcing long fibers.

In the thick molding compound (TMC) process (US) roving cuttings ordered in parallel strands are fed together with resin paste metered by a pump into the nip of two contra-rotating rolls. The compound is then passed through fast-running take-off rolls onto a nylon carrier film that, together with a covering film, is fed onto a conveyor belt. The resulting rough sheet, up to 5 cm thick and containing 50 to 65% of GF, is cut by hand and laid in the mold.

4.6. Thermosetting Plastics

3. Polyester resin mats and glass-cloth prepregs

Glass fiber mats, non-woven or (for special applications) woven fabric (Table 3.1) impregnated with 20 to 40% resin paste with styrene, in 0.5 to 1 cm thick and up to 160 cm (about 5 ft) wide webs, are wound with separating film onto rolls and delivered to the trade. Supplied in styrene-proof packaging, they can be stored in the cold ($< 20°C$) for two to six months.

The manufacture of moldings in a single working process from impregnation of the resin carrier up to pressing is made difficult by the fact that the maturing of the paste to a compression consistency, using alkaline earth oxides or hydroxides, requires several days. In the interpenetrating thickening process (ITP) (ICI, GB), this time is reduced to several hours by initiating cross-linking with isocyanates. Using magnesium oxide at 60 to 80°C the resins can be thickened to press consistency in a few minutes thus rendering the mass production of large moldings feasible.

Among the varied range of UP-SMC types are C- or A-fiber (mostly hybrid) reinforced. For large area SMC automated presses see Sec. 3.3.4.2. In the IMC (Inmould Coating) technique, 0.1 mm thick PU lacquer coating is sprayed onto the part in a second stroke.

The flow of resin mats makes possible the manufacture of large-surface double-bent parts with cams and ribs such as driver cabs, sliding roofs, motor coverings, rear flaps, bumper bars, gear lever cases, seat shells in commercial vehicles and cars, cable distribution cabinets, machine frames, panel lighting fixtures and large interior fittings in ships and aircraft. For car body parts in the "crumple zone", elastomer-modified, flexible LP-SMC with a low E-modulus (Blendur, DE) is sometimes used. GF products in a corresponding, e.g. vinylester (Sec. 4.6.1.5), matrix have been used for large chemical equipment, acid resistant up to 70°C, e.g. 7 m wide wound pipes for the large smoke scrubbers of power stations. Light structural foam moldings with good impact strength ($d \geq 1 \text{ g/cm}^3$) are obtained by the addition of 1% micro-encapsulated fluoroalkane to the prepreg.

Prepregs reinforced by continuous filament mat (Unipreg, DE, FR) instead of or together with chopped strand mats can be used to advantage for thin-walled, complicated articles.

GF prepregs with light-curing catalyzed VLC resins (p. 350), supplied in rolls or sheets between PVC films (Durodet LH, Palapreg SHZ, DE), have an unlimited shelf life in the dark. After the thermoplastic films are preheated, they can be thermoformed and then cured by short exposure to light. With single-sided, stick-laminated PVC sheets, GF-reinforced apparatus is formed in a single operation plus after-curing with light. Pipes up to 10 mm wall thickness have been dry wound and light cured in one process. VLC prepregs can be filled to 180% of the resin content with UV-transmitting fillers such as $Al(OH)_3$.

Glass-reinforced, prepreg surface mats are pressed as covering layers of highly stressed wood. Adhesive prepregs for the electrical industry are UP resin laminates on flexible carrier layers that first melt and then cure on after-heating. The most highly mechanically stressed glass-cloth prepregs are useful only for flat or uni-directional curved moldings. Similar high mechanical values (750 N/mm^2 bending strength, 200 kJ/m^2 notched impact strength) are desirable for directional forming of cross-wound, filler-free XMC Silenka Mats (NL) with a 65 to 72% glass fiber content.

4.6.2.5. Epoxy resin compounds

ASTM D 3013 does not specify forms of supply for EP compounds. The classification system comprises grades with, in each case, 5-cell limits for

1. Water absorption: 24h, 0.16–3.0%
2. Flexural strength: 42–207 MPa (6–30×10^3 psi)
3. Impact strength: 22–534 J/m (0.41–10 ft-lb/in)
4. Tensile strength: 35–138 MPa (5–20×10^3 psi)
5. Deflection temperature: 70–250°C (158–482°F)

It is supplemented by classes D and F with cells for dielectric and flame retardant behavior.
For practical guide values see Table 4.74.

1. *Dry compounds*

Trade names: Eccomold, Plaskon, Polyset (US); Menzolit, Supraplast (DE); Araldit (CH); Eponac (IT)
Commercial forms: Granules, flakes or rods, also colored.
Epoxide molding compounds have a limited shelf life, depending on the resin curing system used. Because of their excellent flow the compounds can be particularly easily processed by transfer and screw injection molding at low pressures (see Table 3.10).
The moldings are characterized by their high dimensional stability even at high temperatures and, because the processing shrinkage of EP molding compounds is very small, their after-shrinkage is practically nil. As a result of limited water absorption the dimensions and properties are unaffected by variations in climate. Good flow properties enable metallic parts of all sizes to be incorporated in moldings, encapsulated or coated without distortion or cracking. In this connection the strong bonding of the resin to the metal is important.
Applications: Generally used for high-quality precision parts down to the smallest dimensions, particularly with metal inserts as, for example, in the electrical industry as contact carriers and in electronics. They are also used for space

4.6. Thermosetting Plastics

filling armatures and collectors and the encapsulation of wound condensers by injection molding.

2. *Prepregs for composites*

Trade names: Hysol, Poly-Preg, Uni-Flow (US); Elitrex (DE); Fiberite, Fibredux (CH); Syncopreg (NL); Rigidite (DE, US).

Glass-cloth prepregs with EP resins, which are supplied with a 37 to 60% glass content in about 1 m wide rolls, bond together at 120 to 160°C under low pressure to form laminates with a flexural strength of 560 N/mm^2 (80 × 10^3 lb/in^2). Modified EP resins, combining good processability, fracture toughness (also at low temperatures) and stability up to high temperatures, are used for advanced composites reinforced by high-strength aromatic or carbon fibers. Exactly parallel oriented roving prepregs, made of 100 to 200 single fibers unwound from a reeling unit are available as unidirectional tape and cloth prepregs with a good shelf life at room temperature can be cured at 120°C or 180°C. Cured laminates containing 60% v/v C fibers have tensile strengths of > 2000 N/mm^2, moduli of > 150 kN/mm^2 and similar compressive values up to 150°C or 200°C, unaffected by moisture.

Processing methods are compression molding (Sec. 3.3.4.2), filament or tape winding (autoclaving Sec. 3.2.4) or pultrusion (Sec. 3.4.3).

Applications are primary and secondary structural elements for the aircraft and aerospace industries (Sec. 1.8.2). There are helicopters and light aircraft made entirely of such advanced composites, including sandwich structures. EP-GF-prepregs have been used for vehicle leaf springs.

4.6.2.6. Diallyl phthalate (DAP) molding compounds

Trade names: Durez, Parr, Plaskon DAP, Polychem, Poly Dap (US); Neonite (CH).

The resin base for DAP molding compounds consists of prepolymers (p. 356) with catalysts for cross-linking addition polymerization at elevated temperatures.

ASTM D 1636 distinguishes between Type I, high-strength glass fiber reinforced, Type II, general-purpose mineral filled, and Type III, general-purpose synthetic fiber filled. There is a further subdivision into classes A/B, diallyl orthophthalate resin, and C/D, diallyl metaphthalate (i.e., isophthalate) resin, of which B and D are flame retarded. Another differentiation based on grades according to filler shape is rather specialised, as most compounds are tailor-made for particular applications. Those of Type III are less stiff than those of Type I, and classes C/D have a better heat resistance than A/B. For general properties of DAP moldings see Table 4.74.

4.6.3.2. Extruded profiles from PF compounds

There are no difficulties about the *processing* of DAP compounds at 140 to 180°C (300 to 350°F) by compression (10 to 70 N/mm^2), transfer or injection molding (10 N/mm^2 to 50 N/mm^2).

Their applications are primarily for electronic parts in military and aerospace equipment. They have high dimensional stability up to and beyond 200°C (400°F) and their advantageous electrical properties are independent of temperature and weathering under extremes of environmental conditions and climatic variations.

4.6.2.7. Silicone resin molding compounds are made from highly branched methyl or methyl phenyl siloxanes and inorganic fillers. They are molded at 175°C with a pressure of 20 to 30 N/mm^2 for 1 to 3 min. The moldings must be after-heated to achieve optimum properties. Applications include highly temperature-stressed electrical parts.

4.6.2.8. Further high-temperature-resistant thermoset molding compounds based on bis-maleimide or triazine resins and polyimides (Kinel, FR) are dealt with in Sec. 4.8. High-temperature-resistant plastics are in Sec. 4.8.6 and properties in Table 4.82.

4.6.3. Semifinisheds from Thermosetting Plastics

4.6.3.1. Cast resin products

Commercial forms: Panels, rods, pipes, preformed castings crystal-clear and (decoratively) colored for machining into consumer articles.

1. Special purpose phenolic resols with low water content (2-10%), if necessary modified, are cured at moderately raised temperature (60–80°C) within 2 to 5 days without application of pressure. These 'superior' resin semifinisheds are machined. Areas of application are parts for textile machinery (deflection rollers and thread transporting guides), copy milling patterns, billiard balls and skittle balls.

2. Cast semi-finished goods from reactive resins include building and flooring panels with integral decoration and materials for fabricating buttons (Sec. 3.2.4 and 4.6.1.4).

For technical castings see: MMA (Sec. 4.1.1.1.3), UP, EP (Table 4.64), PU (Table 4.70).

4.6.3.2. Extruded profiles from PF compounds

Manufacturing process, Sec. 3.4.3. Properties resemble those of PF-general-purpose moldings (Table 4.72).

4.6. Thermosetting Plastics

Products include insulating pipes, tubes for hanging lamps, support posts with steel tube reinforcement, grips, ram extruded rollers and rolls, e.g. for office equipment.

4.6.3.3. Drawn fiber-reinforced profiles

Hollow and solid profiles drawn in the pultrusion process (Sec. 3.4.3) from rovings impregnated with reactive resin, with 65 to 75% fiber content, with external dimensions up to 5×100 cm are structural elements with extraordinarily high strength/weight ratio (Table 4.75).

Table 4.75. Typical properties for fiber-reinforced composite semifinisheds

Product group		Pultruded tubes and profiles[1]		
Resin/fiber groups		UP/GF	EP/GF	EP/AF
Reinforcement			Rovings	
Resin content	%		ca. 30	
Density	g/cm^3	1.9	2.1	1.4
Water absorption	%	ca. 1	0.2–0.3	0.5
Tensile strength	N/mm^2	700	700	1300
Elongation at break	%	2	2	1.8
Young's modulus in tension	kN/mm^2	35	35	75
Flexural strength in .... room temp.	N/mm^2	800	900	600
at $-$ 55°C	N/mm^2			
$+$ 80°C	N/mm^2			
$+$ 160°C	N/mm^2			
$+$ 250°C	N/mm^2			
Young's modulus in flexion, room temp	kN/mm^2	–	–	–
at $-$ 55°C	kN/mm^2			
$+$ 80°C	kN/mm^2			
$+$ 160°C	kN/mm^2			
$+$ 250°C	kN/mm^2			
Compressive strength	N/mm^2	450	450	270
Interlaminar shear strength	N/mm^2	–	–	–
Service temperature	°C	< 150	< 180	–
Insulation class acc. to aeronautical standards	F °C	H –	– –	– –
Linear thermal expansion	10^{-6} K^{-1}	10	10	0
Thermal conductivity	$\frac{W}{Km}$	0.20	0.24	–

[1] Wacosit [2] Pregnit, cf. Table 4.76, p 385

4.6.3.3. Drawn fiber-reinforced profiles

The main consumer of UP-GF and EP-GF profiles and tubes with continuous service temperatures up to 150°C and 180°C is the electrical industry. In the chemical industry they are suitable for corrosion resistant structures. In mechanical and automotive engineering EP-CF profiles are also used as structural elements for high mechanical and thermal cyclic loads. These, together with Kevlar-reinforced profiles, are extremely lightweight structural materials for aeronautical engineering. For *Pulforming* see Sec. 3.4.3.

High tensile-strength, flexible round profiles with 80% continuous long glass fibers (Polystal) are used as pre-stressed reinforcements in concrete bridges, for mast guys, strain relief elements in optical cables and drawing cables for sewer cleaning equipment.

	Fiber-reinforced plastic sheets[2]						
EP/CF	EP/GF bisphenol type	EP/GF TGDA type	PI/GF	EP/GF	EP/AF	PI/CF	
	Weave			Unidirectional weave	Weave		Unidirectional weave
	35	35	35	32	–	50	40
1.6	1.9–2.0	1.9–2.0	2.1	–	–	1.5	1.6
0.2	0.2	0.2	0.2	0.2	–	0.8	0.8
1400	300	350	350	900	350–500	550/450	1400
0.6	2	2	2	<2	–	–	–
130	23	22	21	35	29	–	–
1500	ca. 300	500/400	500/400	900	350–390	760	1650
					390	760	1440
					240	–	–
					–	720	1520
					–	580	1350
–	20	20	20	35	17–18	57	120
					17–18	59	130
					15	–	–
					–	55	110
					–	43	85
1100	500–600	450–500	450	–	110–170	–	–
–	–	–	–	–	25	57	96
–	≤ 130	155	200	130	generally –55 to +80°C		
–	B	F	C	B			
–	80	150	200	80/100			
0.2	–	–			0	–	
–							

4.6. Thermosetting Plastics

4.6.3.4. Wound fiber reinforced profiles

Wound fiber reinforced profiles can be made with a variety of components, winding angles and cross sections on computer-controlled systems (Sec. 3.2.4) with optimum ratio between low specific weight and high strength values in any load direction. As a typical example may be cited the CF-GF hybrid composite (Pregnit) floor joists in the Airbus A310 which yielded a weight saving of 30% compared with aluminum alloys. Other applications under trial or in mass production include connecting rods, propeller shafts and safety steering columns in automotive engineering, transformer support cylinders and spark-quenching chambers in electrical engineering, radar scaffolds, telescope antennae, grenade launching tubes, and other military uses.

4.6.3.5. Laminates

1. *Industrial laminates*

Laminates for mechanical and electrical purposes (to DIN 7735) are based on hard paper, fabric or mats and are manufactured either as semifinisheds (sheets, tubes, solid rods, flat webs) or as molded articles. Hard paper consists of paper and resin, hard fabric of resin and fabric and hard mats of resin and glass mats. Table 4.77 gives details of composition, type designation and properties.

General trade names[1]: Ferropreg, Gillfab, Insurok, Lamilex, Micarta, Phenolite, Resiten, Spauldite, Statoclad, Textolite (US); Aclait, Birax, Diverrit, Durapol, Ferrozell, Geax, Trolitax (DE); Etronax, Etronit (DM); Resocel, Resofil (CH); Voltis (AT); Warolite, Wartex (BE); Hylam (IN).

Products, properties and applications

Industrial laminates, mainly for electrical requirements, are largely internationally standardized products, coded by *NEMA designation grades* of the National Electrical Manufacturers Association, US. These are generally consistent with ISO 1642, DIN 7735 or the electrotechnical association publications IEC 249 and VDE 318, with the associated test standards (Table 4.76). The laminates are resistant to solvents, lubricants, oils and fuels. Their behavior toward aggressive chemicals depends on the binder. For general properties see Table 4.77.

Phenolic resin laminated paper has long been supplied in various grades as a brown or black insulating working material for heavy current, telephone and radio technology. Light-colored MF paper-covered instrument panels and

---

[1] For the significance of the names for special grades see the same names in chapter 8 "Trade names" list.

Table 4.76. NEMA Designations for Industrial Laminates

NEMA grades	Resin	Rein-forcement	Properties and application	Approximately corresponding types DIN 7735	ISO R 1642
X, XX, XXX	PF	Paper	X gen; XX, electric; XXX, high frequency	HP 2061 to	PF/CP 1 to
XP, XXP, XXXP	PF	Paper	Punchable X types. UL 94-HB XXX-C-FR-2, UL 94-VO	HP 2064	PF/CP 6
C, CE	PF	Cotton fabric	Mechanical, e.g., gears, pulleys, bearings, CE, controlled electrical properties	HGW 2081	
L, LE	PF	Fine weave cotton fabric	Close tolerance machining, punching, threading	HGW 2082/3	PF/CC 1 to PF/CC 4
FR-3	EP	Paper	Punchable, printed circuits, low-loss applications, flame retarded	HP 2361.1	EP/CP 1
FR-6	UP	Glass fiber mat	Printed circuits, flame retarded	HM 2471/2	UP/GM 1
N 1	PF	Nylon staple fiber	Electrical like XXXP, toughness like C	–	–
G 3	PF		High strength, impact; low diss. factor, arc resistant	HGW 2072	PF/GC 1
G 7	SI		humid cond., heat-resistant,	HGW 2572	SI/GC 2
G 9	MF	Continuous woven glass fabric	good electr. prop., strength wet cond., flame retarded.	HGW 2272	MF/GC 1
G 10, FR-4	EP		Particularly good electric properties over a wide range of humidities and temperatures, flame retarded	HGW 2370 to HGW 2374	EP/GC 1 to EP/GC 4
G 11, FR-5	EP				
CEM 1	EP	Glass cloth surfaced laminates: Paper core			
CEM 3	EP	Glass paper	Flame resistant, similar to FR Types		
CEM 5	UP	Glass mat			
GPO 1	UP	Polyesterglass mat low-pressure laminates: Random-laid mat	GP mechanical and electrical	cf. Tables 4.65 (p.347) 4.74 (p.382/383)	
GPO 2	UP	Random-laid mat	Low flammability		
GPO 3	UP	Random-laid mat	Low flammability and tracking resistance		

4.6. Thermosetting Plastics

Table 4.77. Properties of Industrial Laminates According to NEMA Specifications

NEMA types Resin base			X, XX, XXX Phenolic		C Phenolic		G 10, F 4 Epoxy	
reinforcement	Unit SI	U.S.	paper		cotton fabric		glass fabric	
Specific gravity	g/cm³	lb/ft³	1.36-1.32	85-82.5	1.36	85	1.82	114
Bond strength	kg	lb	318-431	700-950	816	1800	907	2000
Tensile strength								
length.	N/mm²	10³ lb/in²	140-105	20-15	70	10	281	40
cross.	N/mm²	10³ lb/in²	112-84	16-12	56	8	246	35
Compression strength								
flat.	N/mm²	10³ lb/in²	253-225	36-32	260	37	422	60
edge.	N/mm²	10³ lb/in²	134-179	19-25.5	165	23.5	246	35
Flexural strength								
length.	N/mm²	10³ lb/in²	176-95	25-13.5	120	17	387	55
cross.	N/mm²	10³ lb/in²	155-58	22-11.8	84	16	316	95
Flexural modulus								
length.	kN/mm²	10⁶ lb/in²	13-9.1	1.8-1.3	7.0	1.0	19	2.7
cross.	kN/mm²	10⁶ lb/in²	9.1-7.0	1.3-1.0	6.3	0.9	16	2.2
Shear strength	N/mm²	10³ lb/in²	84-70	12-10	84	12	134	19
Izod impact								
flat.	J/m	ft-lb/in	214-53	4.0-1.0	171	3.20	374	7.0
edge.	J/m	ft-lb/in	27-19	0.5-0.35	101	1.90	294	5.5
Rockwell hardness	M scale		110-105		103		110	
CORRESPONDING EUROPEAN types:								
Charpy impact:								
unnotched	kJ/m²	–	20		18		100	
notched, flat./edge.	kJ/m²	–	15/4		15/10		–/50	
Max. const. oper. temp.	°C	°F	140	285	129	265	140	285
NEMA temp. indices								
electrical	°C	–	130-140	–	85-115	–	130-170	–
mechanical	°C	–	130-140	–	85-115	–	130-180	–
Coeff. therm. expans.	10⁻⁵ K⁻¹	10⁻⁵ °F⁻¹	2	1	2	1	0.9	0.5
Insulation resistance (96 h/90% RH, 95 °F)	M ohm		XX: 60, XXX: 1000		–		200 000	
Dielectric strength, perpendicular to lam., short-term test								
1/16 in	V/mil		700-650		200		500	
1/8 in	V/mil		500-470		150		400	
Dielectric const., cond. A	at 1 MHz		6.0-5.3		–		5.2	
Dissipation fact., cond. A	at 1 MHz		0.06-0.038		0.10		0.025	
Flame resistant Class UL 94	Scale		HB		HB		VO	
Water absorp.: 1/16 in	%/24 h		6.00-1.40		4.40		0.25	
1/8 in			3.30-0.95		2.50		0.15	
1/2 in			1.10-0.45		1.20		0.10	
Thickness: min.	mm	in	0.13-0.8	0.005-0.031	0.4	0.015	0.10	0.004
max.	mm	in	50	2	200	8	50	2

paper- or fabric-based laminates with MF binder are tracking resistant (KC = 600) construction materials.

Phenolic resin laminated fabrics (or cloth) are highly mechanically stressable laminates used in mechanical engineering. They are machined into low-noise bearing bushes and bushings, slide ways, gear wheels, load-bearing rolls and rollers, bobbin discs, ball-bearing cages and draw and boring tools.

The further *general development of laminates* with UP-, EP- and SI-bonded types (Tables 4.76, 4.77) corresponds with growing electrotechnical demands, in particular the high specifications demanded in the computer and aerospace industries and the demand for heat resistance, fire resistance, insulation properties, tracking resistance and resistance to electrolytic corrosion and precision working. High-quality laminates are also used as substrates for printed circuits. These are supplied with copper foils laminated onto one or both sides and, after printing the circuit with insulating lacquer onto the copper, the metal not covered with lacquer is etched away. Alternatively in the "addition process" on the unlined basic material, a negative of the circuit is printed. The free conductor leads are electroplated with copper. For miniaturized multi-position circuits, bases of up to 0.125 mm thick with intermediate GF-EP resin prepregs, printed on both sides, are laminated. The circuits are connected by plated through holes. Triazine resins (Sec. 4.8.5) with improved fire-resistant properties are binders for such material. For flexible printed circuits PTFE glass-cloth laminates are used. For high temperature resistant laminates (Kerimid) see Table 4.82.

The whole range of applications of "industrial laminates" as engineering plastics can be only partially discussed here. Some of the products mentioned above, with good resistance to high and low temperatures, find increasing application in aircraft structures and as insulation spacers in fusion reactors or superconducting magnet installations, where they operate at very low temperatures. Flame retarded glass fiber laminates with metal layers are used for these purposes.

For machining see Sec. 3.9.2. In cutting and stamping in precision engineering laminates must be accurately heated so that they do not split.

2. *Decorative laminates*

Aminoplastic laminates with decorative surfaces are offered in different qualities:

HPL High pressure laminates, with outer layers based on melamine (mostly single sided but some double sided) and sometimes several core sheets based on phenolic resin, pressed under at least $7\,N/mm^2$;

CPL(CL) Continuous pressure laminates, manufactured in a continuous process under a pressure of (usually) $5\,N/mm^2$;

KF plastic coated chipboard;
KH plastic coated hardboard;
LPL low-pressure laminates

HPL Sheets

Trade names:
For HPL and CPL sheets: Dekodur, Formica, Fundopal, Getalit, Homapal, Hornit, Hornitex, Kellco, MAXI, Perstorp, Resopal, Resopalit, Thermopal-Brilliant, Trespa-Duro.
For external use: Resoplan, Trospa.

Manufacture and availability:
HPL sheets are manufactured in multi-platen presses (12 to 44 daylights) at a temperature of 130–160°C and a pressure of at least 7 (usually 10) N/mm^2. For sheet thicknesses of 0.5 to 1.3 mm, 10 to 22 laminates are pressed per daylight; during this, double sided stainless steel press platens impress the surface structure; with only single sided decor, the stainless steel platens lie between the top sides of 2 sheets while silicone paper is interleaved between the backs of both sheets. The pressing time, including heating and cooling, lasts 50 to 100 minutes, depending on the pressing temperature and coating of the individual daylights. The structure of HPL sheets is as follows:

1. Overlay paper of bleached, high-class sulfate pulp, without pigments, saturated with melamine resin; resin content 65–80%, weight per unit area 17–50 g/m^2.

2. Decorative paper of bleached, high-class sulfate pulp, bulk colored or printed in color, with filler to cover the dark core, saturated with melamine resin; resin content 40–45%, weight per unit area 75–130 g/m^2. In the case where no overlay is used the resin content is 50–55% and the weight per unit area 100–160 g/m^2.

3. Underlay paper (barrier paper), white, used only with particularly bright colors of the decor for additional covering of the dark core, impregnated with melamine resin; resin content 45–50%, weight per unit area 80–100 g/m^2.

4. Core papers of unbleached sulfite pulp, saturated with phenolic resin; resin content ca. 30%, weight per unit area 150–250 g/m^2.

The decorative papers are either dyed in one color or patterned abstractly with multi-colored printing or in imitations of natural materials, the surfaces embossed smooth, matt or structured. Similar laminates are also available with textile, light alloy or true wood veneer outer layers.

The overlay papers, which are frequently omitted, are transparent on pressing; they intensify the color of the decor and improve wear characteristics.

4.6.3.5. Laminates

HPL sheets to DIN 16926 are divided into the following types:

N normal quality
P post formable quality (post formable at a temperature specified by the manufacturer)
F increased fire resistance
C compact laminate, thickness 2.0 to 20 mm (for self supporting applications or glueing to a substrate)
CF compact laminate with increased fire resistance

The type designations are extended by numbers (1 = lowest to 4 = highest quality) for:

1 abrasion loading (number of rotations of an abrasive wheel);
2 impact stress (impact testing equipment simulates hitting a truncated object);
3 scratch stress (increasing loading of a rotating, diamond-tipped scratch device).

From the combination of the numbers for these three properties the designations are obtained for the most important classes:

Application reference numbers for HPL

Application profile	Typical areas of application	Application class*
Specially high abrasion resistance High impact strength Specially high scratch resistance	shop counters floors	434
High abrasion resistance Average to high impact strength Average to high scratch resistance	kitchen work tops cafe tables	333
Average abrasion resistance Low to average impact strength Low to average scratch resistance	kitchen unit fronts shelves vehicles	223
Low abrasion resistance Low impact strength Low scratch resistance	furniture carcasses	111

* With types C and CF every combination has an especially high impact strength.

4.6. Thermosetting Plastics

Further demands and standarization tests (to DIN 53 799) under conditions similar to normal practice concern crack susceptibility (24/27 h at 80/70°C, post-formability for type P, fire performance for the F types (HPL is generally accepted as B 2 to DIN 4102, types with test mark are fire resistant class B 1), mass alteration on climatic change, resistance to glowing cigarettes, hot pots, steam, boiling water, light fastness (minimum grade 6) and stain sensitivity to household chemicals. HPLs have limited resistance to dilute acids and acid salts, chlorine containing bleach solutions, hydrogen peroxide, silver nitrate, iodine and strongly colored tinctures; they are resistant to strong acids.

For rounded edges and other round furniture parts, 1.3 mm thick, sheets, post formable on heating with a bending radius of 10d are standard. The same – preferably fire resistant (B 1 to DIN 4102) – serve for non-toxic, easy to clean, profiled wall coverings in hospitals and homes as well as in ships and aircraft. External facade sheets (Resoplan, Trespa) are weather and fire (B 1) resistant.

CPL sheets

Recently, decorative laminates have been manufactured in a continuous process (CPL or CL sheets or Contiboard). An important element of such a continuous laminate press is a pressure band which winds around a 1300 mm diameter pressure drum which can be heated and has a circumferential velocity of 15 m/min with stepless adjustment. With the pressure band, temperatures of up to 180°C can be used. The pressure is usually about $5 N/mm^2$ and the width 1.30 to 1.40 m.

With a heating zone length of 2 m and rate of advance of 10 to 3 m/min, the pressing times are only 12 to 30 seconds for the manufacture of 0.3 to 0.7 mm thick CPL sheets respectively. For this reason rapid-curing, resin-coated paper films must be used. The core papers are not impregnated with pure phenolic resins but either with melamine resins or phenol/melamine mixtures with the resin content increased by 35–40% compared with films for multi-platen presses.

Continuous laminate manufacture has the following advantages: it is not limited to certain sheet lengths and trim loss only occurs on the longitudinal side of the material path. In conventional multi-platen presses, the pressure plates must be heated and cooled in every pressing cycle, in CPL presses the cool plates lie behind the hot plates so that the hot plates always remain hot and the cold plates cold which saves energy. CPL presses can also be converted from laminate manufacture to direct coating of chipboard and hardboard with decorative film and to laminating on to platens.

Direct coating of wood-based material sheets

Chipboard and hardboard can be directly coated with decorative papers (with decorative surfaces) saturated with melamine resin and used in this form for the manufacture of furniture, in internal building and in vehicle manufacture. This direct coating is now almost exclusively carried out in modern short-cycle presses. Bearing in mind the density of the substrate sheets (chipboard 680–750 kg/m^3, hardboard 850–100 kg/m^3), the pressures used are distinctly lower than for the manufacture of HPL and CPL sheets: for chipboard 2.0–2.5 N/mm^2 for smooth surfaces and 2.5–3.0 N/mm^2 for structured surfaces and for hardboard 3 to 4 N/mm^2. Because of this, plastic coated chip- and hard-board is also called low pressure laminate (LPL). The hot platen temperature is 200 to 222°C, corresponding to 130 to 160°C on the decorative paper and the press cycle time less than 60 seconds.

The test methods given in DIN 53 799 for aminoplastic resin based sheets with decorative surfaces are also valid for chipboard (KF, DIN 68 765) and hardboard (KH, DIN 68 751).

Furniture laminates are often produced by coating chipboard with priming and furniture foils and finishing films. Priming and furniture foils are paper impregnated with urea based resin (resin content 33 to 45%). They are stuck on to chipboard (if necessary after coating with liquid area resin glue). Furniture foils are then also treated with clear lacquer; priming films in contrast are first filled (puttied), polished and then lacquered. Ready-to-use furniture foil is lacquered immediately after impregnation and is therefore easier to process. In order to give the necessary flexibility, impregnating resins based on aminoplastics are elastified with acrylic acid ester copolymers.

Glued laminate wooden materials

Synthetic resin compressed wood
Trade names: Delignit, Lignostone, obo-Festholz, Panzerholz (DE).
Synthetic resin compressed wood can be pressed (to DIN 7707) from red beech veneers and curable resin, preferably phenolic resin, with at least 5 veneers per cm of product at a density of 0.8–1.4 g/cm^3. Depending on whether the veneer is only covered with glue or vacuum impregnated, the resin content can vary between 7 and 35%.

Sheets and molded items are available (see p. 372) with a thickness of 4 to 180 mm, veneers parallel, crossed or star oriented and also round with extensive tangential fiber orientation. Guide values for properties of synthetic resin compressed wood to DIN 7707 in the density range 1.25–1.35 g/m^3 are:

Flexural strength:	150–250 N/mm^2
Flexural E-modulus:	15–25 kN/mm^2
Impact strength:	20–60 kJ/m^2
Dielectric strength	45–70 kV/25
tan δ	0.02
Tracking resistance	500–100

4.6. Thermosetting Plastics

Synthetic resin compressed wood saturated with boron compounds is available for protection against radiation.

Non-densified laminated wood

Non-densified laminated woods are sheets and moldings of at least three laminates which are glued together using curable artificial resins. The most usual of these is plywood (fiber directions of the laminates set crosswise); less common is laminated wood (laminates with fibers parallel).

Depending on the structure, the following types of plywood are distinguished:

a) Veneer plywood (sheets and moldings) in which all layers consist of veneers; veneer sheets of over 12 mm thickness, which consist of 5 or more veneer layers are called multiplex laminates.
b) Wood core plywood consisting of at least 2 cover veneers and a middle layer of wooden bars about 24 mm wide or of wooden bars of a maximum 8 mm thickness.

Plywood is manufactured in hot presses at 100 to 150°C and 0.5 to 2.5 N/mm². The resin content of plywood is usually between 5 and 10%.

Depending on the type of glue used and the consequent resistance to humidity, water and weathering, plywood is divided into the following types for general use (DIN 68 705 part 2):

IF Interior plywood, resistant to general humidity but not weather resistant; urea resin glue.
AW External plywood, limited resistance to weathering; phenolic resin or modified melamine resin glue.

Building plywood (to DIN 68 705 parts 3,4,5) is assigned to a wood material class defined in DIN 68 800 part 2 in an analogous way, i.e. depending on the type of glue:

Wood material class	Resistance of the glue	Maximum permitted wood moisture	Glue
20	not weather fast	15%	Urea resin
100	weather fast	18%	Phenolic resin or melamine resin
100G	weather fast with protection against wood destroying fungi	21%	Phenolic resin or modified melamine resin

Plank (or board) laminates are manufactured by layerwise glueing of boards using cold-curing, two-component synthetic resin glues. They have great importance for load-bearing wood structures to DIN 1052. For resistance against moisture and weathering, resorcinol resin glues or modified melamine resins are used whereas for internal applications urea resins are adequate.

4.6.3.6. GRP sheets and sandwiches

1. All of the following are manufactured for *construction purposes*: sheets, profiles (Sec. 4.6.3.2), sandwich panels with laminated profiled, rigid foamed or honeycomb core, profiled or three-dimensionally formed facade surfacing made by hand or laminated by compression. Binders include UP-, EP- and PF-resins. For properties see Tables 4.74, p. 377; 4.65, p. 347, and 4.77, p. 386.

2. *Light panels and webs* (manufacture, Fig. 3.41)

Composition: UP with 20 to 60% glass fiber mat content and a corresponding density of 1.2 to 2.0 g/cm^3. Natural colors allowing diffused light, with glass content between about 85 and 60% diminishing as the glass content decreases; also transparent and opaque pigment coloring.
Forms of delivery: Transversely corrugated lighting strips in rolls of up to 50 m (164 ft) long and widths of up to 6 m (20 ft). Corresponding length-wise-corrugated and flat translucent sheets in transportable lengths. Standard corrugations fitting corrugated asbestos-cement and steel profiles and many special profiles. Heat-insulating edge-sealed, double skinned, double-corrugated sheets and insulating light panels.
Skylights 1.0 to 1.3 mm (0.04–0.05 in) thick with a glass fiber content of 20 to 25% and about 75% light transmission can be proofed against erosion with MMA resinified covering mats or with PVF film, and can be guaranteed for 10 years with less than 10% loss in light transmission.
Applications: Roofing (for shed roofs also with shadow strips), wall glazing, e.g., in gymnasiums, greenhouses, door panels, separating walls, balcony claddings.

3. *Laminated sheets* for light-transmitting wall panels with a total thickness of 20 to 70 mm; glass fiber-polyester covering layers, e.g., Kalwall, aluminum lattice core (US), Polydet screen plates and core of corrugated Polydet strips (DE).

4.6.3.7. Reinforced thermosetting resin pipes (RTRP)

RTPR Standard Classification ASTM D 2310 provides a complete survey of all possible combinations of pipe structure markings by one or other of the following numbers and letters:

4.6. Thermosetting Plastics

1. Types: (1) filament wound (Sec. 3.2.4), (2) centrifugally cast (Sec. 3.2.3.2), (3) pressure laminated (of prepregs around a mandrel)
2. Grades: (1) EP-GF, (2) UP-GF, (3) PF-GF, (4) UP-Asbestos, (5) EP-Asbestos, (6) PF-Asbestos, (7) FF-GF
3. Classes: (A) no liner, (B) UP, (C) EP, (D) PF resin liner, all non-reinforced; (E) UP, (F) EP, (G) PF liner reinforced, (H) specified thermoplastic liner, (I) Furan (FF) resin liner reinforced
4. Hoop stress hydrostatic design basis (HDB, see Sec. 4.2.1): (A) to (H), 2500 to 12,500 psi (17.2–86.1 MPa) cyclic internal pressure, 25 cycles/min; of (Q) to (Z), 5000–31,500 psi (34.4–217 MPa) static internal pressure.

The standards ASTM D 2996 for 2 to 16 in filament wound pipes and D 2997 for $1\frac{1}{2}$ to 14 in centrifugally cast pipes complete the general standard with some further physical property requirement cells. D 2517 is the approved standard specification for reinforced epoxy-resin-filament wound gas pressure pipes (Type RTPR 11 AV). They have a 2 to 12 in (60–320 mm) nominal diameter and a wall thickness of 0.06 to 0.175 in (1.5 to 4.2 mm); sustained pressure requirements of 1860 to 1005 psi (130 to 70 N/mm^2) for 20 h, 1290 to 638 psi (90–45 N/mm^2) for 1000 h, and further requirements, among others, for the hydrostatic collapse factor for external pressure resistance, according to test method D 2924. Further standards are D 4163 for water conveying systems, D 4162 for sewer and industrial pressure pipes, D 4184 for sewage in gravity flow systems.

Centrifugally cast GF-EP pipes, to meet DIN 16871 (similar to RTRP 21 CX), with DN = 100–500 mm for PIN = 10–14 bar and DN = 40–300 for 25 bar, are designed with a sixfold safety factor against "weeping", i.e., the permeation of liquid below the bursting pressure. DIN 16869 standardizes cast GFR piping for public sewage pipes. DIN 16964 to 16966 deal with different wound UP-GF pipe series and pipe system parts. Such products have a glass fiber content of 30 to 60% with or without internal lining of bisphenol A or vinyl ester resins (Sec. 4.6.1.5) and are also wound on thermplastic or PTFE pipes as internal liners. They are designed in DN = 25–1000 mm (1–40 in) for a PN = 4–16 bar with a six-to-10-fold safety factor.

Forms of supply and applications: EP-GF and UP-GF standardized chemical pipes are delivered in 6 m (20 ft) lengths for all kinds of joints, as required after pre-welding the inner thermoplastic liners or the flange connections. They are suitable for pressure piping in continuous use with corrosive liquids at temperatures of up to 100°C (212°F). With a higher glass content and PTFE inner lining they are usable up to 160°C (320°F). EP-GF wound pipes are also fitted for heating with embedded resistance wires. Extremely thick-walled pipes for working pressures of up to 40 bar are used for deep wells or conveying oil with

DN = 50–100 mm (2–4 in) with threaded-cemented fittings internal and external pressures of up to 200 bar. EP-GF pipes with DN = 50–250 mm (2–10 in) with an insulating layer and an external protective pipe are used in various systems as buried district heating piping with pressurized water at 100 to 130°C (230–266°F).

Large UP-GF pipes up to 4 m (13 ft) in diameter are widely used as free-flowing ducting in water supply and waste-water systems. Earth-covered piping subject to buckling pressures is built up in multi-layers so that a resin-bonded sand layer is introduced between the wound GF core and the outer pipe. GRP waste-gas piping for temperatures from −40 to 150°C (−40 to 302°F) is supplied up to 6 m in diameter, including the associated molded parts. A further application of such wound pipes is in slender engineering structures as, for example, chimneys for aggressive effluent gases or beacons and lighthouses.

4.7. Foams (Summary)

For the manufacture and structure of homogeneous closed and open cell foams with cross-section-invariant bulk density and of structural or integral foam products with bulk densities which decrease towards the centre of the material, see Section 3.1.6). For foam products made from individual polymers see the following pages:

PE	190, 304	PTP	272	UF	338
E/VA	301				
PP	201, 300	PPO	282	MF	336
PS	209, 305	PES	283	UP	352
PVC	231, 332	PEI	284	EP	356, 467
		PMI	319	PU	367
PC	270	PF	336, 442	PI	404

4.7.1. Concepts

According to DIN 7726, foams are cellular materials, always lighter than their framework material. Cells are small hollow spaces which occur during manufacture of the foam.

If the cells are fully surrounded by cell walls the foam is called 'closed celled' (e.g. XPS). Here gas or fluid exchange between the cells is only possible by diffusion. In 'mixed cell' foams (e.g. UF) the cell walls are partially perforated. In 'open-celled' foams the cells have gas-phase connections to each other. In the extreme case only the cross-pieces of the cell walls remain (e.g. MF).

4.7. Foams (Summary)

Structural or integral foams have non-homogeneous density distribution. The lighter foam centre changes continuously into the denser closed outer skin.

As the modulus of elasticity of a piece decreases approximately in line with the bulk density but the stiffness rises with the third power of the wall thickness, lighter but equally stiff parts can be obtained with less material consumed.

Syntactic foams are gas filled polymers which however are not foamed. Their fully closed cells are produced by hollow spheres distributed in the polymer matrix. These foams are very compression resistant with bulk densities to $800 \, kg/m^3$.

Depending on their resistance to deformation under pressure (Pressure for 10% compression, DIN 53 421) a distinction is made between rigid ($>80\,kPa$), semi-rigid ($80–15\,kPa$) and flexible ($<15\,kPa$) foams. With the rigid foams the pressure value rises approximately logarithmically with the bulk density.

The cell structure of hard and brittle foams (e.g. PF) can break down under loading; tough and hard foams (e.g. PVC) are partially, and flexible and elastic foams very fully, reversibly elastically deformed.

4.7.2. Foam Principles

With thermoplastic foams the starting material is the polymer. Continuous manufacture of foam strips, foam films or profiles (PS, PE) always involves an extrusion step.

Thermoplastic foam particles cake on steaming to foam blocks or can be molded. In the Styropor process they are obtained from beads which contain blowing agent and are foamed (PS). In the Neopolen process they arise directly from a melt containing blowing agent by heat removal after extrusion (PE-LD). PE-LLD or PP copolymer granules in aqueous solution are impregnated with blowing agent under pressure and swell to foamed particles after return to normal pressure.

Thermoset macromolecules (PF, MF) arise by cross-linking during the foaming process. This curing can take place in strip foaming equipment (PF, PU, MF), by block molding (PU, PF) or directly in molds suitable for foaming.

The principal types of foam processing are given in Tables 4.71 and 4.72. Further information is contained at the end of Section 4.7).

4.7.3. Blowing Agents

The cells in the polymer matrix are produced using blowing agents. The blowing agent pressure produced under the prevailing processing conditions must be appropriate to the viscosity of the melt or cross-linking mixture. Blowing agents (physical and chemical) are dealt with in Sections 3.1.6.1. and 3.1.6.2. Nucleating agents and pore size regulators are considered in Sec. 3.1.6.3.

4.7.4. Properties

The size of the cells, their type and distribution but above all the polymer itself and its bulk density determine the properties of a foam.
All these affect the mechanical parameters. Thermal and fire performance and resistance to chemicals are determined by the foam substrate. The cell structure is responsible for acoustic and insulating properties. Substrate and cell structure determine the behaviour to water and water vapour.
Properties are summarized in Tables 4.78 and 4.79.
Lightweight foams within the density range of 5 to 100 kg/m^3 (0.3–6.25 lb/ft^3), made of different plastics materials, are of particular importance as rigid thermal insulants (Table 4.78) and as flexible damping and cushioning materials (Table 4.79). These cellular plastics need different *testing methods* to those for bulk plastics. The compressive strength of rigid cellular plastics is defined by the stress at yield point at which the breakdown of the cell structure of brittle rigid foams begins, or by a deformation of 10% for tough rigid foams. For flexible cellular plastics the compression stress at 40% deformation and the compression set after constant deformation at elevated temperature are guiding values for their behavior under load. The most important special standard test methods for rigid and flexible cellular plastics are given in Table 4.80.

4.8. Advanced High-Temperature Plastics

4.8.1. General Limits to Heat Resistance

Thermogravimetric analysis (TGA) gives a measure of the absolute heat resistance of a material. This establishes the weight loss a material undergoes within a short period at a defined high temperature. Figure 4.22 shows examples of such thermograms. With PVC, extensive thermal decomposition begins close to processing temperature. Thermoplastics such as PMMA, PS and PE are fully decomposed at 350 to 400°C. In contrast, the curve for PF indicates that highly cross-linked resins, after a fast initial decomposition at about 300°C, leave a carbonized residue even at temperatures of up to 1000°C. When mineral-filled, they are used in space travel as heat shields, experiencing a certain degree of ablation.
The thermal decomposition temperature and the resulting residue of carbon increase as the proportion of chemically bound hydrogen in the polymer is

4.8. Advanced High-Temperature Plastics

Table 4.78. Rigid Foams

Foam group		Polystyrene			Tough-rigid			Brittle-rigid	
Foam process		Particle foam	Extruder foam without foamed skin	Extruder foam with foamed skin	Polyvinyl-chloride High-pressure foam	Polyether sulfone Block foam	Poly-urethane Block foam without or with covering layer	Phenolic resin	Urea resin Injection foam
Density range	kg/m^3	15-30	30-35	25-60	50-130	45-55	20-100	40-100	5-15
Compression strength	N/mm^2	0.06-0.25	>0.15	>0.2	0.3-1.1	0.6	0.1-0.9	0.2-0.9	0.01-0.05
Tensile strength	N/mm^2	0.15-0.5	0.5	>0.2	0.7-1.6	0.7	0.2-1.1	0.1-0.4	
Shear strength	N/mm^2	0.09-0.22	0.9	1.2	0.5-1.2	–	0.1->1	0.1-0.5	
Bending strength	N/mm^2	0.16-0.5	0.4	0.6	0.6-1.4	0.2	0.2-1.5	0.2-1.0	0.03-0.09
Bending E-modulus	N/mm^2			>15	16-35	3	2-20	6-27	
Thermal conductivity[1]	W/Km	0.032-0.037	0.025-0.035		0.036-0.04	0.05	0.024	0.03-0.02	0.03
Max. service temp.: short-term	°C	100	100		80	210	>150	>250	>100
long-term	°C	70-80	<75		60	180	80	130	90
Water vapor diffusion resistance factor	μ	30-70	100-130	80-300	200->300	9	30-130	30-300	4-10
Water absorption after 7 days immersion in water	Vol.%	2-3	2	<0.5	<1	15	1-4	7-10	>20

[1] Building supervisory authorities prescribe approximately 20% higher thermal conductivity values for the heat insulation of buildings.

4.8.1. General Limits to Heat Resistance

Table 4.79. Semi-rigid to Flexible Elastic Foams

Foam structure		Mainly closed cell					Open cell	
Raw material group		Polyethylene			Polyvinyl chloride		Melamine resin	Polyurethane	
Foam process		Particle foam	Extrusion cross-linked		High-pressure foamed		Web foamed	Block foamed polyester	polyether types
Density range	kg/m³	25-40	30-70	100-200	50-70	100	10.5-11.5	20-45	20-45
Tensile strength	N/mm²	0.1-0.2	0.3-0.6	0.8-2.0	0.3	0.5	0.01-0.15	~0.2	~0.1
Extension at break	%	30-50	90-110	130-200	80	170	10-20	200-300	200-270
Compression resistance	N/mm²	0.03-0.06	0.07-0.16	0.25-0.8	0.02-0.04	0.05	0.007-0.013	0.003-0.006	0.002-0.004
Compression deformation (70 °C, 50%)	%	–	10-4	3	33-35	32	approx. 10	4-20	~4
Shock elasticity	%	40-50	45	–	~	~	–	20-30	40-50
Service temp. range	°C to °C	to 100	70-85	−70-110	−60-50		up to 50	−40-100	
Thermal conductivity	W/mK	0.036	0.04-0.05	0.05	0.036	0.041	0.033	0.04-0.05	
Water vapor diffusion resistance factor	μ	400-4000	3500-5000	15.000-22.000	50-100	–	–	–	–
Water absorption after 7 days immersion in water	Vol.-%	1-2	0.5	0.4	1-4	3	aprox. 1	–	–
Dielectric constant (50 Hz)	ε	1.05	1.1	1.1	1.31	1.45	–	1.45	1.38
Dielectric loss factor (50 Hz)	tan δ	0.0004	0.01	0.01	0.06	0.05	–	0.008	0.003

4.8. Advanced High-Temperature Plastics

Table 4.80. Standard Test Methods for Cellular Plastics

		ASTM	ISO	DIN
Rigid cellular plastics:	Compression properties	D1621	844	53421
	Tensile properties	D1623	1926	53430
	Shear strength	C 272	1922	53427
	Bending test	D 790	1209	53423
Rigid and flexible cellular plastics	Apparent density	D1622	845	53420
	Dimensional stability	D2136	2796	53431
	Apparent thermal conductivity	C 518	2581	52616
	Water vapor transmission	C 355	1663	53429
	Water absorption	D2842	2896	53433
	Burning characteristics	D1692	3582	75200
Flexible cellular plastics	Tensile properties		1798	53571
	Deformation under load		3386	53577
	Compression set	D1546		53572
	Dynamic cushioning performance, impact resilience	D1596	4651	(53573)

reduced. Nitrogen-containing heterocyclic-aromatic cross-linked systems (a in Fig. 4.22; Table 4.87, D) or chain-linked aromatic rings (b in Fig. 4.22, a in Fig. 4.23) are to be considered as "non" flammable – in air – even when without mineral fillers.

The temperature-time limits for long-term service of a plastic material lie below the TGA decomposition zone. They vary with the nature of the exposure and depend on the ambience and the thickness of the material. From experience, the measured continuing temperature vs. log time decomposition follows a straight-line relationship up to 10^5 h (about 10 years). The corresponding measurements of property changes in relation to temperature, to ASTM D 794, ISO 2578, DIN 53446, with storage of test specimens in forced-air-draft ovens for up to three months or one year, lead to figures suitable for comparison with long-term temperature stressing. The internationally recognized *Underwriter Laboratories Temperature Index* gives time-limits at various temperatures for $\frac{1}{32}$, $\frac{1}{16}$, and $\frac{1}{8}$ in (0.8, 1.6, 3.2 mm) thick molded material from which a maximum deterioration of 50% of the original value can be predicted either for electrical values alone, for these and all significant mechanical values or relating to mechanical values excluding notched impact strength (Fig. 4.12, also Sec. 7.3.4).

These indices and other corresponding values of limiting temperatures for continuous service lie at about 60 to 100°C for polyolefins, styrene copolymers,

4.8.1. General Limits to Heat Resistance 401

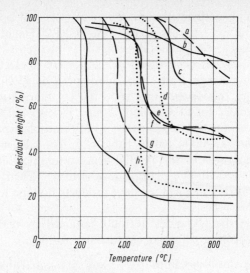

Fig. 4.22. Thermograms of various plastics.
(a) Polybenzimidazole.
(b) Polyphenylene,
(c) Polyimide,
(d) Polyphenylene ether,
(e) Polyoxydiazole,
(f) Polyphenylene sulfide,
(g) Cast phenolic resin,
(h) Polycarbonate,
(i) Polyvinylchloride.

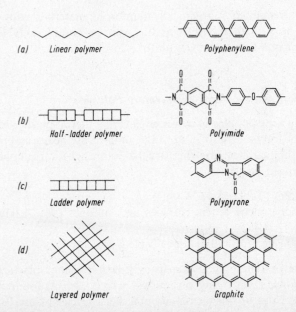

Fig. 4.23. Structures of aromatic-heterocyclic polymer systems.

vinyl polymers, polymethylmethacrylate, polyamides and polyoxymethylene. For inorganic-reinforced grades of PP, PA, PC, PBTP and PPO and most thermoset moldings and laminates the range is 100 to 150°C.

Larger *proportions of inorganic atoms* in the polymer molecule can raise the limiting application temperature. Post-chlorination of PVC to PVC-C (Sec. 4.1.4.3.1) increases this temperature to >100°C; fluorination increases the UL temperature index for PFA to 150°C, for ETFE to 160°C and for PTFE to 180°C. However, the service temperatures of fluoroplastics as measured by other limiting values are up to 260°C (Table 4.28). The same applies to fluorine and silicon rubbers (Sec. 5.1.14/15). High-temperature polymers with unusual atoms, such as those with phosphorous (Sec. 4.1.9.1), arsenic or boron in the molecule, have mostly remained research products because of their high cost and difficulties in synthesis.

Among thermosets the UL index at 150°C attained by many PF and MF molding materials can be raised to 170 to 190°C for, e.g.,

- Phenol aralkyl resin moldings (p. 335 and Xylok in Table 4.82);
- Highly cross-linked, in particular cycloaliphatic epoxide, cast resins, Tables 4.64, 4.66), EP- and EP-novolak-molded material and glass fiber based EP laminates (Tables 4.75, 4.76);
- Cycloaliphatic isocyanate or IPN polyurethane casting resins, Table 4.70);
- Mineral-filled silicones (Sec. 4.6.2.7 and 5.2).

The loss in strength of inorganically reinforced materials with a UL index >100°C, at least with intermittent use, can be considerably less than for light alloys or non-ferrous metals up to >250°C.

4.8.2. New Routes to High-Temperature Polymers

The following are examples of newer engineering thermoplastics with temperature indices of 150 to >200°C that can be molded by the normal processes and are made by linking aromatic ring structures through short aliphatic, oxygen, sulfur or nitrogen bridges:

- Polyarylates (Tables 4.44, 4.45);
- Linear polyarylene ethers, sulfides and sulfones (Sec. 4.1.10), Fig. 4.12 Tables 4.47, 4.48;
- Thermoplastic semi-polyimides (Table 4.49).

The principle of synthesising increasingly heat resistant plastics consists of linking monomers with ring structures, which are predominantly formed from benzene rings or nitrogen-containing heterocycles or are formed during the linking reaction. In this way bi- and multi-functional polymer building

units are formed which on further cross-linking form ladder or network structures (Fig. 4.20).
The difficulties and costs of multi-stage syntheses and of the processing of these materials grow as these become more complex.

4.8.3. Linear Polyarylenes

Linear or branched *polyphenylenes* (Fig. 4.23a, Elmac, H-resins, US) are insoluble and infusible in their final state. Their long-term resistance to corrosion (apart from attack by alkali metals) is maintained up to 200 to 300°C. With the exclusion of oxygen this rises to 400°C.
Poly-p-xylenes (Parylenes, US) thin film or coatings are dielectrics for continuous service up to >200°C. These are formed at cold surfaces on polymerization of the vaporized dimer radical of xylene

[CH₃—⟨O⟩—CH₃] or chlorinated xylene at 600°C under vacuum. Because

they contain $-CH_2-$ groups they resist oxidation only up to about 80°C.

4.8.4. Polyarylene Amides, Aryl Esters and Ethers

The aromatic polyamide (Aramid, p. 408) *poly-m-phenylene-isophthalic acid amide* is supplied in semifinished form for machining (KS Aramid-Vespel, US). It is stable in continuous service in air at temperatures up to 200°C.
Poly-p-hydroxybenzoate (Ekonol, US), from the monomer

HO—⟨O⟩—COOH is a polymer that, apart from ester groups, contains only

aromatic rings. The polymer, melting point 550°C, can be worked by flame-spraying and can be made malleable by additives and shaped into bearing materials. Copolymers with about 40% terephthalate (Ekkcel X7G, US, p. 278) are "self-reinforcing" injection-molding materials. The long, relatively stiff polymer molecules arrange themselves in the melt in thermotropic (p. 408) liquid-crystal phases. Injection moldings from the anisotropic melt are much stronger and stiffer in the direction of flow than at right angles to it (Table 4.49).
Poly-2,6-diphenyl-phenylene ether (Tenax, NL), a member of the PPE group (Table 4.46) but containing only aromatic rings, has a T_g of 235°C and T_m of 480°C. It is spun from organic solvents into fibers for filter cloths and for high-voltage cable-insulating paper.

4.8.5. Bismaleimide-triazine Resins

Bismaleimides (Table 4.81), manufactured by the condensation reaction of e.g. methylenedianiline with maleic anhydride, owe their reactivity to the double bonds at each end of the molecule which can react stepwise with themselves or with co-reactants (Table 4.81, C). Such resins modified, e.g. by allyl-phenyl monomers, are less brittle than homopolymerized bismaleimide but tough and strong at room temperature and elevated temperatures, are under development as matrices for advanced composites.

Triazine resins are an interesting new class in this group.
Dicyanate end groups ($-O-C\equiv N$, cf. isocyanate $-N=C=O$) form triazine nucleus polymers that are cross-linkable by stages and co-curable with bismaleimide according to the equation:

$$3 \text{ NCO}-R-\text{ONC} \longrightarrow -R-C\underset{N}{\overset{N}{\diagdown}}C-R-$$

(R = bisphenol A residue)

As liquid or low viscosity polymers melting at 30 to 130°C, and processible in a variety of ways, they have glass transition temperatures (T_g) in the range 200 to 300°C when cured, with outstanding mechanical, insulating and dielectrical properties. They are used for copper-laminated castings as substrates for printed circuits (Sec. 4.6.3.5.1) and other microelectronics components, as structural materials in the aeronautical industry, insulants for large motors and bearing materials in the automotive industry.

4.8.6. Polyimides

Half-ladder and ladder polymers (Fig. 4.23b), which contain the imide group $O=C-N-C=O$ in heterocyclic-aromatic ring systems, have wide fields of application. (Table 4.81) illustrates pre-products (A) and typical linkages.

The synthesis of pure heterocyclic-aromatic PI by polycondensation (B1 in Table 4.81) requires special processes to obtain infusible and insoluble molding materials because of the required removal of volatile by-products. Solid semi-finished goods, precision-sintered moldings (Vespel) and 7.5 to 125 μm thick electrofilms (Kapton, US) made from such PI are supplied directly by the manufacturer. They can be used continuously from −240°C to +260°C in air and to 315°C in a vacuum. Properties at various temperatures are shown in

4.8.6. Polyimides

Table 4.81. Synthesis of Polyimides and Related Aromatic-Heterocyclic Polymers

(A)
(1) Pyromellitic acid anhydride
(2) Trimellitic acid anhydride
(3) Bismaleimide
(4) Benzidine

(B) Polycondensation with elimination of water from

(B1) 1 + 4 → Polyimide

(B2) 2 + Amino-alcohol → Polyester imide

(C) Stepwise polymerization

of 3 + 4

Polybismaleic imide (thermoplastic intermediate)

Cross-linking reaction (schematic)

(D) Some other important heterocyclics

Hydantoin | Urazol | Paraban acid | Chinoline | Chinoxaline | Benzimidazol

Table 4.82. PI films coated with FEB can be heat sealed. Similar films of the urea derivate Paraban acid (Table 4.81,D) are heat stable up to 155°C.

Semipolyimides (B2 in Table 4.81) are injection-moldable thermoplastics with aliphatic esters or amide groups placed between the long-chain polycondensed imide group (Sec. 4.1.11). Soluble impregnating resins or wire lacquer resins for the electroindustry (Pyre-LM, US; Dobeckan FN, Icdal, Terebec, DE; Rhodeftale, FR) are given a ladder structure by stoving in organic solvents. They can then be continuously exposed at temperatures of up to 200°C.

Precondensates with a preformed polyimide configuration, as in polybismaleic imide (Table 4.81), are curable (by addition polymerization), fusible resins for impregnation of prepregs and wound articles. At 125°C they provide reinforced molding and laminating materials. There are also special foam-forming resins.

Prepolymer trade names: Genon, Kamax, Larc 160, Skybond, Thermimid 600, Upjohn 2080, US; Kerimid, Kermel, Kinel, Nolimid, FR; U 218, CH.

General properties at RT and high temperatures are given in Table 4.82. PIs are resistant to most dilute or weak acids and to almost all organic agents apart from primary and secondary amines and hydrazine. They are attacked by strong alkalis and by ammonia. Hot water or steam cause water absorption, and after drying out there can be a reversible reduction in strength. With continuing exposure to moisture above 100°C cracks can form. PI shows a high resistance to high energy radiation. Its good electrical properties across the entire temperature range have produced significant weight and cost economies for high-temperature electronic and electrical components. Its low coefficient of friction, even in dry surroundings, is even further improved by the addition of graphite, MoS_2 or PTFE.

Solimide polyimide foam is a very light (0.01 g/cm^3) soft and flexible material (compressive strength 0.006 N/mm^2 (40%), compression set 25% (23°/22 h)). From the starting material a powder intermediate is first formed by polycondensation. Water produced as a side product in the reaction and ethanol act as blowing agents during the subsequent heat treatment. Because of its excellent fire resistance the material is used for insulation in aircraft. It was also previously recommended for aircraft seats.

4.8.7. Advanced Heterocyclic Aromatic Polymers

The synthesis of PI-like long-chain ladder polymers with heterocyclic cores, as shown in Table 4.81, D illustrating configuration and nomenclature, is a current field of development. Such polymers are used as impregnating resins for the bases of printed circuits, in microelectronics and as radiation cross-linking "photoresist" lacquers for printed circuits of semiconducting chips in computer technology. They must resist temperatures of about 400°C and other stresses

4.8.7. Advanced Heterocyclic Aromatic Polymers

Table 4.82. High-Temperature Plastics

Trade name polymer	Unit SI	Unit U.S.	Xylok (GB), polyarylalkyl-phenolic resin		Vespel (US), polyimide		Kapton (US), polyimide		Kinel 35 (FR), polyimide		Kerimid 601 (FR), polyimide		QX 13 (GB), polyimide	
product			66 % GF laminate		unfilled molded parts		insulating film		graphite-filled moldings		77 % GF laminate		52 % CF laminate	
Density, RT	g/cm^3	lb/ft^3	1.77	111	1.43	89.3	1.42	88.8	1.55	96.9	1.95	122	1.67	104
Tensile strength:														
−195 °C (−319 °F)	N/mm^2	10^3 lb/in^2			113	16	240	34						
RT	N/mm^2	10^3 lb/in^2	440	63	91	13	180	26			350	50		
200/250 °C (392/482 °F)	N/mm^2	10^3 lb/in^2			46	5.7	120	17						
Flexural strength.														
200/250 °C (392/482 °F)	N/mm^2	10^3 lb/in^2	450	64	120	17			75	11	550/400	78/57	840/630	120/90
316 °C (600 °F)	N/mm^2	10^3 lb/in^2	360	51					50	7.1	400/250	78/36	840/730	120/104
Compression strength.	RT	N/mm^2	10^3 lb/in^2			63	9.0			140	20				
					>280	>40								
Vol. resistivity: 200 °C (392 °F)	ohm cm				10^{10}		10^{10}		—		6·10^{14}-2·10^{15}			
Dielectric constant: 200 °C (392 °F)	ohm cm		4.9		3.4		3.5		—		6.4/5.4			
		1 kHz	4.9		3.1		3.0		—					
Dissipation factor: RT			0.03		0.002		0.003		—		0.014/0.011			
200 °C (392 °F)			0.03		0.002		0.002				0.03/0.01			
Exposure at elevated temperatures			Aged 1000 h at 250 °C. HDT > 330 °C		Fully aged. 50 % loss: at 300 °C in 600 h at 400 °C in 45 h		Zero strength at 815 °C		High temperature-bearing material		Aged 2500 h at 200 °C		Aged 1770 h at 300 °C; 1550 h at 200 °C	
Processing			Compression molding. 175 °C, 35-70 bar; annealing 250 °C		Precision sintering, machining articles		PFEP-coated heat sealable		Compression molding, 200-240 °C, 100-130 bar; injection molding		Laminating, 100-200 °C, 15 bar; annealing, 200 °C			

encountered in the fabrication. Such new syntheses are based on the high reactivity of multi-functional isocyanates (Tables 4.67, p. 357/358, 4.68, p. 360/361). One aim in these syntheses is to produce *open-chain* prepolymers, which are liquid crystals in concentrated solution or as melts. Hence when applied on the substrate, they are pre-oriented in such a way that insoluble and infusible layers are formed by the subsequent cyclizing reaction, producing macromolecular ladders and nets.

4.8.8. Highly Temperature-Resistant and Carbonized Fibers

High-strength, infusible, practically non-flammable fibers are spun from "lyotropic" (cf. Sec. 4.8.4) liquid crystalline oriented solutions of aramids (p. 403, Kevlar, Nomex, US; Twaron, JP, NL) or semi-polyimides (p. 406, Kapton, Kermel F, US). Polybenzimidazole (PBI) fibers (Table 4.81, D with ε_B 28.5% and a moisture regain of 15% at 20°C have been used as asbestos substitute for flameproof, corrosion and radiation resistant, yet comfortable, protective garments for firemen, workmen and astronauts.

According to U.S. specifications aramid fibers must have at least 85% of the aramid groups bound directly to the aromatic ring. Meta-aramid fibers (Nomex US; Technora, JP) are used for fire-proof textiles, high-temperature insulating papers and honeycomb cores. The high-strength para-aramid and similar synthetic fibers are many times more ductile than very stiff carbonized fibers (Table 3.2). "Hybrid" reinforcement with aramid elements alongside S-glass and/or carbon fiber results in highly stressable, specifically light RP composites with optimally well-balanced properties. Aramid long fiber reinforced POM, PA, PPS injection compounds (cf. Sec. 4.1.8.1) less abrasive than GF reinforced types are under development.

The origin of "graphitized" carbon fibers (Fig. 4.23d) from polyacrylonitrile (PAN) precursor is indicated in Table 4.83. The older ladder polymer fibers presented there, Pluton and Orlon, have almost unlimited temperature resistance under short-term mechanical loading: liquid iron can be filtered through a Pluton cloth. But in contrast to carbon fibers they are stable to air oxidation in continuous use only up to 100°C. This is because the ladder reaction in their preparation is so incomplete that many reactive positions remain in the ladder molecule. They provide an example of the difficulties in the syntheses of high-temperature-resistant and thermo-oxidatively stable polymers.

Most ordinary pitches at $\geq 350°C$ are converted into mesophase pitch (MPP), nematic liquid crystalline phases, by dehydrogenation condensation. By spinning such melts, fibers with a high degree of orientation parallel to the shear direction result. By further dehydrogenating carbonization, and subsequent heating to 3000°C, truly graphitized fibers are made from the cheap MPP precursor (Thornel P 100, US; see Table 3.2).

4.8.8. Highly Temperature-Resistant and Carbonized Fibers

Table 4.83. Highly Temperature-Resistant Cyclized and Carbonized Fibers

5. RELATED TOPICS

5.1. Natural and Synthetic Rubber

Natural rubber and those types of synthetic rubbers (1, 2, 3, 7, 8*) that are polymers with conjugated diene monomers contain many unsaturated double bonds along the chains. Only a proportion of them are saturated by traditional vulcanization with sulfur and accelerators to produce soft rubber. The more double bonds remaining in the vulcanizate, the lower is its stability to oxidation and weathering, although its resistance can be much improved by the inclusion of anti-oxidants. Products with a low diene content, which are synthesized by ring-opening polymerization, include poly-trans-1,5-pentene from cyclopentene (this is copolymerized with ethylene in Norsorex powdered rubber) and polyoctene from the partially hydrogenated dimeric butadiene (Vestenamer). The poly-trans-octenamer, which is partially crystalline above room temperature, is used as a property enhancer for many rubber mixes. Polymers with no double bonds require a different sort of vulcanization process; for 10*, 12* and 13* alkaline cross-linking agents are necessary. Oxidative cross-linking with peroxides is generally used for saturated thermoplastics. Radiation cross-linking is also possible. Most types of vulcanization for synthetic rubber can be combined, but it is not possible to combine sulfur and peroxide vulcanization. Rubber-like thermoplastic elastomers which do not need vulcanization for certain service temperatures are dealt with on page 416 *et seq.* Some of them are thermoplastic/rubber composites vulcanized "dynamically" by addition of a curing agent to the melt (Sec. 4.1.2.5).

Mixtures with "reinforcing" active fillers are very common in rubber technology. Carbon black is used for black mixtures and silica for light-colored products. Sometimes fillers are incorporated into the latex. Normal high molecular synthetic rubber types are available as oil-extended rubbers, which are good for processing mixtures. Many rubber compounds also contain specific plasticizers.

The mastication of bales, the customary form of raw material delivery for rubber processing, requires much work. As a result an effort is now under way to make synthetic rubber available as a more easily processable powder or as granules or to produce low molecular weight "liquid rubber" with reactive end groups so that simpler processing techniques using fluid plastic preproducts (Sec. 3.2.2) are available to the rubber industry.

---

* See following pages.

5.1. Natural and Synthetic Rubber

The nomenclature of rubber and latices uses a different standardized system of symbols from that for plastics; for details see DIN ISO 1629. Additionally, many conventional symbols are in use.

1. *Cis-1,4-polyisoprene* is natural rubber (abbreviation NR); the corresponding synthetic products (IR) are Isolene D liquid rubber, Natsyn, (US); Europrene, (IT); and Cariflex, (NL). The synthetic products are not equal to natural rubber in structural strength, elasticity and good processability, but they are purer and more uniform.

2. *Styrene-butadiene rubber* (GR-S, SBR), mostly made with a ratio of about 75 butadiene to 25 styrene by cold polymerization, is most like natural rubber in its behavior. It is used as a component of tire compounds and for cable covering, industrial rubber goods, tubes and profiles and microcellular and foamed rubber. Trade names include: Ameripol, Austrapol, Buna EM, Cariflex S, Carom, Copo, Europrene, Flosbrene (as a liquid rubber), FRS, Gentro, Emaprene, Humex, Intol, ISR, Jetron, Krylene, Krynol, Kyrmix, Nipol, Petroflex, Philprene, Poly bd R 45, Polysar S and Ricon 100 as liquid rubber, Sircis, SKS, Synapren, Synaprene, Ugipol. Solution-polymerized SBR comes as Duradene and Solprene (US). There are also SBR lattices in special grades for external paints and carpet backing. For high styrene containing resins that can be admixed to increase the hardness see Sec. 4.1.1.3.

3. *Cis-1,4-polybutadiene* (BR) gives wear- and cold-resistant compounds, e.g., for tire treads: Ameripol CB, Budene, Buna CB, Cariflex BR, Cis-4, Cisdene, Diene, Duragen, Escorez, Europrene cis, Intene, Plioflex, Polysar, Taktene, Tufsyn.
BR with vinyl-containing groups (Intolene 50) has similar properties to those of SBR. Bunatex HTS, Hycar CTP, Hystl CWG and Ricon 150 are supplied as liquid butadiene rubber, carbon black filled-tire-tread batches also as powder.

4. *Butyl rubber* (GR-I, IIR) is a copolymer. Isobutylene (cf. PIB, Sec. 4.1.1.2.3) as the major component provides its good chemical and aging resistance and low gas permeability while the minor component, isoprene, is needed for vulcanizability into flexible, stable rubber products such as car tubes, heating bellows and roofing membranes: Enjay-butyl, Hycar-butyl, Petrotex-butyl, Polysar-butyl, Soca-butyl. Chlorine-containing (Esso-butyl HT) and bromine-containing (Bromobutyl) butyl rubbers excel through their easy processability, good aging resistance and low air permeability.

5.1. Natural and Synthetic Rubber

Table 5.1. Comparative Data for Natural and Synthetic Rubbers

No. in text	Type of rubber	Density, un-vulc. g/cm³	Tensile strength Vulc., un-filled N/mm²	Tensile strength Vulc., rein-forced N/mm²	Extension at break %	Service range (°C) from	Service range (°C) to	Resistance* to Oxidation	Resistance* to Mineral oil	Resistance* to Organic solvents	Resistance* to Water, acids, alkalis
–	Natural rubber	0.93	22	28	600	–60	60	4	5	6	3
1	Cis-1,4-polyisoprene	0.93	1	24	500	–60	60	4	5	6	3
2	Styrene butadiene	0.94	5	25	500	–30	70	3	4	6	3
3	Cis-1,4-polybutadiene	0.94	2	18	450	–60	90	3	5	6	3
4	Butyl rubber	0.93	5	21	600	–30	120	2	5	6	2
5	Ethylene-propylene terpolymers	0.86	4	25	500	–50	120	1	5	6	2
6	Ethylene-VAC co-polymers	0.98	5	18	500	–30	120	1	4	5	5
7	Polychloroprene	1.25	11	25	400	–30	90	2	2	4	2
8	Nitrile rubber	1.00	6	25	450	–20	110	2	1	3	4
9	Urethane rubber	1.25	20	30	450	–30	100	2	1	2	3
10	Polysulfide rubber	1.35	2	8	300	–50	120	2	1	1	3
11	Acrylic ester rubber	1.10	4	12	250	–30	140	1	1	5	5
12	Epichlorohydrin rubber	1.27 to 1.36	5	15	250	–40	150	2	1	5	3
13	Chlorosulfonated polyethylene	1.25	18	20	300	–30	120	1	2	4	2
14	Fluororubbers	1.85	2	15	450	–40	190	2	1	3	1
15	Silicone rubbers	1.25	1	10	250	–80	>200	1	1	3	5

*Comparative figures for stability: 1 excellent, 2 very good, 3 good, 4 moderate, 5 generally small or variable depending on attacking medium, 6 unstable.

5. *Ethylene-propylene rubbers (EPM and EPDM)* are atactic copolymers.
   Because EPM – Dutral, Intolan, Vistalon, Buna AP (also in crumb) – is a saturated, ethylene-propylene rubber, it is cross-linkable only with peroxides. There are also thermoplastic elastomers (Sec. 4.1.2.5).
   EPDM – Buna AP, Dutral Ter, Epcar, Epsyn, Esprene, Intolan, Keltan, Mitsui EPT, Nordel, Royalene, Vistalon – is an unsaturated ethylene propylene rubber, which as a terpolymer contains dienes, for example dicyclopentadiene (DCP), ethylidene norbornene (EN) or hexadiene (HX). EPDM can be vulcanized with sulfur. Because the unsaturated positions in the terpolymers lie outside the main chain the good aging characteristics, the resistance to ozone and low temperatures and the good electrical properties

of a saturated olefin remain. EPDM roof webbing and building profiles, etc., are weather resistant in light colors. For SEP=SI/EPDM IPN see Sec. 5.2.

6. *Ethylene-vinylacetate copolymers* (E/VA) with medium acetate content (Sec. 4.1.1.5.1 and 4.1.2.4.1) manifest, as oxidative cross-linked synthetic rubbers, above-average heat resistance in Levapren 450, Vinathene VAE, Elvax and Evathane.

7. *Chloroprene polymers* (CR), are difficult to ignite, have leather-like toughness and are weather and corrosion resistant. They are used for building seal profiles, roof coverings, mine conveyor belts, cable covering and linings. Trade names are Baypren, Butaclor, Denkachloropren, Nairit, Neoprene and Switprene. From solutions, they are stiffened to some extent even without vulcanization by crystallization and are therefore suitable for shear-proof contact adhesives.

8. *Nitrile rubber* (NBR) is a copolymer of butadiene that, with 20 to 50% acrylonitrile content, becomes increasingly oil and fuel resistant, properties on which its special applicability is based: Breon, Butacril, Butakon, Butaprene FR-N, Chemigum, Elaprim, Europrene N, Hycar, Krynac, Marbon, Nilac, Nipol N, Nitrex, Paracril, Perbunan N, SKN, Tylac. Liquid rubbers: Hycar CTBN, MTBN, 1312; Poly bd CN-15.
Combinations with PVC (e.g., Breon Polyblend, Paracril OZO, Perbunan N/VC) are ozone-proof; NBR modified with phenolic resins gives vulcanized, tough, elastic products which are resistant to hot water.
Fully *hydrogenated nitrile rubber* (H-NBR: Therban, Tomac) vulcanized by peroxide is extremely resistant to oil and mechanical stress up to 160°C. It has found appropriate applications in oil drilling equipment.

9. *Urethane rubbers* (AU polyester, EU polyether based) are intermediate products of polyurethane chemistry that are processed like rubber. Trade names are Adiprene C, Elastothane 455, Genthane S, Urepan 640 and Vibrathane which are cross-linked with peroxides. Urepan 600 is cross-linked with isocyanates. Elastomeric polyurethanes are considered further on in Sec. 4.6.18.

10. *Polysulfide rubber* (SR) exhibits outstanding resistance to solvents and aging. It is used for container linings (Thiogutt, Thiokol) and for joint sealing in building and construction. Odor and low strength are disadvantages for other applications.

11. *Acrylic ester rubbers* (ABR, ACM, ANM): Hycar 4021, Paracril OHT, Vamac, modified co- or ter-polymers of acrylic esters, have been used for special automotive seals on account of their resistance to lubricants, aging, heat, ozone and UV.

5.1. Natural and Synthetic Rubber

12. *Epichlorohydrin rubbers*, homopolymer (CO) or copolymer (ECO) with ethylene oxide saturated polyethers, which can be cross-linked by diamines, or terpolymers containing vinyl groups (ETER), can also be cured by sulfur or peroxide (Herclor, Hydrin, Gechron). They are fuel and oil resistant materials used for automotive tubing and other movable parts that remain flexible in the cold and stable at high temperatures.

13. *Chlorosulfonated polyethylene* (CSM) is light colored and resistant to weathering. Self-vulcanizing formulations are used as mineral-filled injection materials or weldable webs for roof skins and linings: Hypalon.

13a. *Chlorinated Polyethylene* (CM) used for industrial articles is somewhat more resistant to heat and hot oil, less brittle at low temperature and is also cheaper to compound than CSM.

14. *Fluorinated rubbers* (CFM, FKM, FPM) with trade names like Fluroel, Kalrez, Tecnaflon, Viton (see Sec. 4.1.5.9) are products whose fields of application are determined by elastomeric behavior over a wide range of temperature and their extremely high chemical resistance. Therefore they are used for seals of all kinds in the chemical, aircraft, aerospace and automotive industries. Production capacities are growing fast, and the newer highly fluorinated types fulfill the specifications for use with aggressive fuels and oils as well as for motor temperatures better than NBR, ACM, CO or ECO in similar automotive applications.
    Atlas, a copolymer of TFE and propylene, has exceptionally high chemical resistance.
    Fluoralkoxy phosphazene polymers (Eypel), with a main chain containing only $-P=N-$ and fluorinated alkoxy side chains, are fully resistant to oxygen and ozone and remain elastic at cryogenic temperatures.
    All halogen-containing rubbers (7, 12, 13, 14) are difficult to ignite.

15. *Silicone rubbers* (M ... Q*, Silopren, Wacker-Siliconkautschuk, GE silicon rubber, ICI silicon rubber, KE rubber, Rhodorsil, Silastene, Silastic, Silastomer, SKT) are electrically high grade, non-toxic and, like many silicones, difficult to wet. Their properties are scarcely affected by dry heat over a wide temperature range, from $-60°C$ ($-100°C$ for special grades) to $180°C$ for continuous exposure or to $300°C$ for short periods. In steam silicone rubber is attacked at $130°C$.
    Hot-curing Si-rubber is vulcanized by peroxides at $200°C$. Applications include oil-resistant, non-toxic, moving and stationary seals, tubes,

---
* for details see DIN ISO 1629

electrical insulation, conveyor belts which are non-stick over a wide temperature range, keyboards for electronic equipment with contact points made of electrically conducting SI rubber.

Solvent-free, paste-forming two-component pre-products with special cross-linking agents are cold curing. They may also be single-component materials cross-linking under the influence of air. They serve as non-after-curing building joint seals that accommodate movement in the structure. Others are used as casting materials for flexible elastic molds and in the electrical sector. Boron-containing silicon rubber gives vulcanizates that are self-welding at room-temperature (self-adhering insulating tapes, Silicor).

5.2. Silanes and Silicones (SI)

Manufacturers: Bayer, Goldschmidt, Hüls, Wacker-Chemie, Rhône Poulenc, Dow Corning, General Electric, Shin Etsu, Toshiba, Toray, etc.

Silanes are silicon-hydrogen compounds and their substitution products.

Special low-molecular-weight silanes, e.g. $Y\text{-}Si\text{-}(OR)_3$ in which Y represents an organic group related to certain polymers, are adhesion promoters used when reinforcing thermoplastics or thermosets with glass fiber and other mineral fillers.

Postcuring of thermoplastics such as polyolefins with vinyl silanes leads to increased temperature resistance and improved electrical insulation.

Silicones are oligomeric or polymeric organo-siloxanes which usually have di- or tri-functional structural elements. They are produced from the corresponding organochlorosilanes by hydrolysis followed by polycondensation.

$$-O-\underset{R}{\overset{R}{\underset{|}{\overset{|}{Si}}}}-O- \qquad -O-\underset{R}{\overset{O}{\underset{|}{\overset{|}{Si}}}}-O-$$

difunctional structural element   trifunctional structural element

Hydrolysis of R_2SiCl_2 produces linear polysiloxanes with the following type of chain elements

$$\left(-O-\underset{R}{\overset{R}{\underset{|}{\overset{|}{Si}}}}-O-\right)_n$$

5.2. Silanes and Silicones (SI)

They are important starting products for the manufacture of silicone oils and silicone rubbers (see Sec. 5.1.15). In a similar way, hydrolysis of $RSiCl_3$ produces three-dimensionally cross-linked silicone resins which cure by condensation to hard, thermoset products when heated. Reaction of CH_3SiCl_3 with alcohols gives low-molecular weight alkylalkoxysilanes and siloxanes. These condense to resin-like products at room temperature.

Characteristic of all types of silicones is their chemical and physiological inertness with physical and good electrical properties which scarcely change from very low to high temperatures. They are water-repellent and anti-adhesive; this is important for their use as separating and lubricating agents. Silicones are also characterized by pronounced surface activity which forms the basis for their application as anti-foaming agents and foam stabilizers.

Polysiloxanes are often suitable for copolymerization with thermoplastics to improve such properties as temperature resistance, processing conditions and flexibility. Thermoplastics can be mixed with about 10% siloxane pre-polymers and a high-temperature-active platinum cross-linking catalyst so that, after processing and tempering, a silicone network permeates the thermoplastics. The Petrarch (Rimplast) process is similar but for rubber or thermoset products. Another variation is the modification of thermosets (epoxy resins) with linear siloxanes with amino functions, producing plastics with higher flexibility.

Cross-linking silicone resins have continuous service temperatures up to 180°C (short-term up to 200°C) when cured and serve as binding agents for molding compounds (Sec. 4.6.2.7), and for glass-fiber laminated molding materials. They are also used as impregnating resins for electrical engineering and electronics, as well as base materials for heat-resistant corrosion-inhibiting paints. When reactively modified they improve the environmental and heat resistance of alkyd, epoxy, phenolic and acrylic resin paints.

Monomer water soluble alkali siliconates or alcohol-naphtha soluble alkylalkoxy-silanes and siloxanes are products for waterproof treatment of construction materials and construction elements. All the above eventually cross-link to yield polyorganosiloxanes, that are bound to mineral construction materials via polar SiO groups.

The oligomeric *silicone oils* exhibit little change in viscosity over a wide temperature range (−60 to +300°C), and have low vapor pressures. They boil without decomposition. They are used as boiling liquids for high-vacuum diffusion pumps, temperature independent lubricants, hydraulic oils, damping fluids, dielectrics, permanent mold release agents, in small quantities as paint additives to prevent the separating out of pigments, and with added silicic acid as anti-foam oils. Water soluble modified silicone oils are important as foam stabilizers for plastics foams. In emulsified form, silicone oils are also used for water repellent, breathing impregnations. Silicone oils are also used in the form of aqueous emulsions.

Silicone pastes have similar uses. Like oils, they serve as a skin-compatible base for medicinal and cosmetic preparations.

Siloxane-modified low molecular weight *silanes* which contain organo-functional groups with affinity for a particular polymer, are bonding agents for reinforcing thermoplastics or thermosets with glass-fibers and with other mineral reinforcing fillers.

6. STANDARDIZATION

6.1. International and National Organizations for Standardization

6.1.1. The International Standardization Organization ISO

Central Headquarters Secretariat: 1–3, rue de Varembé, CH 1211 Geneva 20 is the world clearing house for the development of international standards. Its membership includes 89 countries, 72 represented by national standard institutes as members, of which 29 co-operate in the ISO Technical Committee 61 Plastics (TC/61 secretariat ANSI, for addresses see Table 6.2). A survey on the structure of ISO/TC 61 is given in Table 6.1 and a list of its member bodies in Table 6.1. ISO/TC 61 collaborates with the related
   ISO/TC 138 Plastic pipes, fittings and valves for the transport of fluids;
   ISO/TC 45 Rubber and rubber products.
The International Electrotechnical Commission IEC (same address as ISO, Central Secretariat) works with ISO/TC 61 on electrotechnical applications of plastics materials.

Regional international organizations coordinating standards are

ASAC	Asian Standards Advisory Committee,
ASMO	Arab Organization for Standardization and Metrology,
CEN	Commission Européenne de Normalisation (there are e.g. DIN-EN standards),
COPANT	Pan American Standards Commission,
PASC	Pacific Area Standards Committee,

and for their special domains

CENELEC	Comitée Européenne pour la normalisation electrotechnique,
UEAtc	Union Européenne pour l'Agrément Technique dans la construction.

6.1.2. National Standards Institutes

American National Standards are generally based on standard specifications, test methods, classifications, definitions and practices produced by the
American Society for Testing and Materials (ASTM), 1916 Race St., Philadelphia, PA 19103
They appear in the Annual Book of ASTM Standards,
      Vol. 08.01 to 08.03 Plastics I to III
      Vol. 08.04 Plastic Pipes and Building Products
and are there marked as adopted by ANSI, where that is the case.
NEMA Standard Publications, edited by the

Table 6.1. Structure of ISO/TC 61 Plastics Committee

Subcommittee	Working groups (in numerical order)
1. Terminology	General and special vocabularies, symbols
2. Mechanical properties	Creep behavior, Mechanical properties of thin films, Tensile properties, Indentation hardness, Friction and abraison. Small specimens, Damping solid polymers, Multi-axial impact, Pendulum impact, Temperature dependance, Multi-purpose specimens, Machined specimens
4. Burning behavior	Ignitability, Smoke, Flame propagation, Ad-hoc nomenclature
5. Physical and chemical properties	Optical properties, Permeability and absorption, Viscometry, Thermal analysis, Melt flow rate, Gas chromatography, Ash, Reference materials, Electrical properties, Density, Extractable matter, Melting points, Flow of thermosetting material
6. Aging, chemical and environmental resistance	Exposure to light, Various exposures, Cracking and crazing, Thermal stability, Basic standards
9. Thermoplastic materials	Polyolefins, Styrene polymers, Polyamides, Polymer dispersions, Polycarbonates, Cellulose esters, Polyalkylenes, Test specimens, Polymethacrylates, Polyvinylchloride, Polyoxymethylene
10. Cellular materials	Time temperature deformation, Classification – Specification, Polyolefin foams and dynamic cushioning, New Tests, ISO documents revision
11. Products	Industrial and decorative laminates, Films and sheetings, Adhesives, PMMA sheets
12. Thermosetting materials	Resins and molding materials: Phenolic, Amino, Polyesters, Epoxies, Polyurethanes and others
13. Composites and reinforcement fibers	General standards, Mechanical properties, Glass yarns and fabrics, Rovings, Mats, Chopped Strands, Milled fibers, Prepregs, Carbon fibers

National Electrical Manufacturers Association, 2102 L Street, N. W., Washington, DC 20037, are of importance for the use of thermoplastics and (industrial laminated) thermosetting products for conduits and cable assemblies, ducts and fittings as well as for insulating materials. Some are approved as American National Standards.

6.2. Fields of Standardization

The Underwriters Laboratories Inc., 1285 Walt Whitman Road, Melville, L.I., NY 11747, is an independent non-profit-making corporation engaged in testing for public safety. The UL Safety Standard 94 for fire performance (Sec. 7.3.5) and the UL temperature index (Sec. 4.8.1) for electronic and electrotechnical insulating materials have been internationally recognized.

DIN Deutsches Institut für Normung
For the address see Table 6.2. Beuth Verlag GmbH, the publishers at the same address publish from time to time collections of the standards from the FNK (Fachnormenausschuß Kunststoffe) as sectional "DIN-Taschenbuch" editions in one or several volumes, concerning.
 1 Testing of mechanical, thermal and electrical properties,
 2 Testing of chemical, optical and processing properties,
 3 Resins and molding materials,
 4 Semifinished and finished products,
 5 Pipes, pipeline parts and fittings of thermoplastics.
Many DIN standards have been translated into English. A list in given in the catalogue English Translations of German Standards.
DIN and VDE (Verband Deutscher Elektrotechniker) are together responsible for all topics concerning electrical and mechanical safety performance of plastics insulating materials, the standards being DIN as well as VDE.

The British Standards Institution (BSI, for the address see Table 6.2) produces the British Standards Yearbook, with a classified list of all British standards or sectional lists for individual topics such as
 SL 12 Leather, plastics, rubber.

The ASTM Yearbooks and DIN-Taschenbücher also contain lists of corresponding *ISO documents*. ISO documents can be obtained in full text from ISO headquarters and from the cooperating national standards organizations.

6.2. Fields of Standardization

6.2.1. In the field of Terminology, there are lists of equivalent terms in English, French, Russian (ISO 194-DIN 7730) and definitions of terms relating to materials processing and properties (ASTM D 883-ISO 472-DIN 7732). ASTM D 3918 gives such terms for reinforced plastics pultruded products. A new edition of ISO 472 which will include ISO 194 and several addenda is in preparation.

Symbols and Codes for the abbreviation of terms are given in
– ASTM D 1600 for basic plastics, plasticizers and some monomers

6.2.1. Terminology

Table 6.2. Active member bodies of ISO/TC 61

Symbol	Country	Address
ABNT	Brazil	Rio de Janeiro-RJ, Av. 13 de Maio, No. 13-28 andar
AFNOR	France	92080 Paris la Défense, Tour Europe
ANSI	USA	New York, N.Y.10018, 1430 Broadway
BDS	Bulgaria	1000 Sofia, 21, rue de 6 Septembre
BSI	United Kingdom	London W1A 2BS, 2 Park Street
COSQC	Iraq	Baghdad, Aljadiria, P.O.Box 13032
CSBS	China	Beijing, P.O.Box 820
Cz	Czech Republic	113 47 Praha 1, Václavské namesti 19
DIN	Germany, F.R.	10787 Berlin, Burggrafenstraße 6
GOST	Russia	Moskva 117049, Leninsky Propekt 9
IBN	Belgium	1040 Bruxelles, A. de la Brabanconne 29
IRANOR	Spain	28010 Madrid, Calle Fernandez de la Hoz, 52
IRS	Romania	Bucarest 1, Rue Ilie Pintilie 5
ISI	India	New Delhi 110002, 9 Bahadur Sha Zafar Marg
ISIRI	Iran	Tehran, P.O.Box 15875-4618
JISC	Japan	Tokyo 100, 1-3-1, Kasumigaseki, Chiyoda-ku
KBS	Korea	Kyonggi-do 171-11, 2, Chungang-dong, Kwachon-myon
MSZH	Hungary	1450 Budapest 9, Pf. 24
NNI	Netherlands	26 BG Delft, Klafjeslaan 2
NSF	Norway	0306 Oslo 3, Postboks 7020 Homansbyen
PKNMiJ	Poland	00-139 Warszawa, Ul. Elektoralna 2
SAA	Australia	North Sydney-N.S.W. 2060, 80-86 Arthur Street
SABS	South Africa	Pretoria 0001, Private Bag X191
SCC	Canada	Mississauga, Ontario L5N 1V8, 2000 Argentina Rd.
SFS	Finland	00121 Helsinki, P.O.Box 205
SIS	Sweden	10366 Stockholm, Tegnérgatan 11
SNV	Switzerland	8023 Zürich, Kirchenweg 4
TBS	Tanzania	Dar es Salaam, P.O.Box 9524
UNI	Italy	I-20123 Milano, Piazza Armando Diaz 2

- ISO 1043, part 1 – DIN E 7728 for polymers and their special characteristics (see fold out table preceding p. 1),
- ISO DIS 1043, part 2 for filler and reinforcement materials
- ISO DIS 1043, part 3 – DIN E 7223 for plasticizers (see Table 4.2.4).

6.2. Fields of Standardization

For EDP compatibility no character is allowed in front of the basic symbol in the abbreviations for plastics e.g. the standardized abbreviation for linear low density polyethylene has to be written PE-LLD, not LLDPE. The symbols for copolymers should contain those of the monomers separated by an oblique stroke, e.g. E/VA, not EVA (Sec. 4.1.2.4.1). In those for mixtures (blends, alloys), the symbols for the basic polymers are separated by a plus sign. However, these rules are not always followed, and some symbols are often written in traditional form.

Symbols for rubber DIN ISO 1629, ASTM D 1566, (see 5.1) and those for textiles are based on a different system which is not compatible with that for plastics.

6.2.2. *Standardized test methods* within the scope of the ISO/TC 61 subcommittees 2 to 6 are dealt with in Chapter 7. For special standards for the testing of cellular materials (ISO/TC 61/10) see Table 4.80.

6.2.3. *The standardization of plastics raw and molding materials* was started in the 1920s with "type-tables" specifying properties of the, then, new thermosetting molding materials. These had to be certificated and hall-marked for their application as electrical insulating materials (Sec. 4.6.2.1).

Since then, national standard organizations have developed *standard specifications* with "type numbers" for special grades of molding materials (e.g. Table 4.71) or – for a most comprehensive characterization – by a sequence of codes for generic type, group, class, grade, reinforcements, basic requirements, followed by properties "cell" coded by special tables and – if needed – suffixes to ASTM D 4000 "Standard Guide for Identification for Plastics Materials". The meaning of code letters in that system for e.g. suffixes differs from that in the ISO system. Some ASTM specifications in that book provide examples for the application of that standardization system.

With the development of more and more modified polymers and plastics compounds, primarily in thermoplastics, the attempt to demarcate clearly a limited number of fully specified "standardized types" in a common system independent of source widely has been found to be unfeasible in marketing plastics. This was particularly true of the growing international market. As a result ISO has confined itself to setting up a generally applicable system for *Standard Designations of (Thermo-)Plastic Materials* (ISO document N 435 E). This system does not imply that materials having the same designation in any respect necessarily behave similarly but it has accomplished an internationally harmonized, EDP-compatible pattern for all materials. If necessary it may be used as the basis for further specifications. All these ISO standards (like the long established DIN standards) consist of two parts

6.2.3. The standardization of plastics raw and molding materials

1 Designation,
2 Determination of properties.
The test methods and their parameters (specimen, conditions etc.) in Part Two include those needed for the designatory properties as well as those required for specifications, terms of delivery etc.
Part One gives, besides an optional general "Description block" (e.g. Thermoplastics), the Designation by the "Identity block" consisting of the "ISO-Standard Number Block" and the

Individual Item Block				
data block 1	data block 2	data block 3	data block 4	data block 5

These data blocks comprise the following information:

Data block 1: Symbol of the basic plastic that, for copolymers, can be followed by the nominal content of the prevailing monomer, and, optionally, after a hyphen by one of the letters in Table 6.3a. These letters have different meaning from those in ISO 1043, part 1, which are used only for special characteristics of symbols (see list of abbreviations p. xi), not for designations.

Data block 2: 1st position: information about intended application or method of processing; 2nd to 4th position: optional information up to 3 items, sometimes in a simplified form, all coded according to Table 6.3b.

Data block 3: Designatory properties different for every material (see Table 6.4, p. 427, for thermoplastics). Their values are given by a system of figures (and letters) laid down in the respective material standard. They are indicated by a "cell" system representing the value by the nominal (median) value of a cell. In the ISO/DIN standards tables in Chapter 4 (summarized in Table 6.5) these specified values have been recorded for thermoplastics only. For thermosets the transition from traditional specifications is still taking place.

Data block 4 gives the type and content of filler or reinforcing material coded according to Table 6.3c.

Data block 5 is reserved for codes specified by agreement between interested parties or committees. The content of the block is not covered by the general material standard pattern.

Examples for designations according to the ISO pattern are given on page 426.

6.2. Fields of Standardization

Table 6.3. Meaning of Codes in ISO/DIN-Designations

a) Meanings of letters in data block 1

Letter	Meaning	Letter	Meaning
A		N	novolak
B	block copolymer	P	plasticized
C	chlorinated	Q	mixture
D		R	random copolymer, resol
E	emulsion polymer	S	suspension polymer
F		T	
G	casting polymer	U	unplasticized
H	homopolymer	V	
J	prepolymer	W	
K		X	
L	graft polymer	Y	
M	bulk polymer	Z	

b) Definitions of codes in data block 2

Code	Position 1	Code	Position 2 to 4
A	adhesives	A	processing stabilized
B	blow molding	B	antiblocking
C	calendering	C	colored
D	disc manufacture	D	dry blend or powder
E	extrusion of pipe, profiles and sheet	E	expandable
F	extrusion of film and thin sheeting	F	special burning characteristic
G	general use	G	pellets, granules
H	coating	H	heat aging stabilized
J		J	
K	cable and wire coating	K	metal deactivated
L	monofilament extrusion	L	light and/or weather stabilizer
M	injection molding	M	
N		N	natural (not colored)
P	paste resin	O	no indication
Q	compression molding	P	impact modified
R	rotational molding	R	molding release agent
S	powder coating or sintering	S	lubricated
T	tape manufacture	T	improved transparency
U		U	
V	thermoforming	V	
W		W	stabilized against hydrolysis
X	*no indication*	X	cross-linkable
Y	textile yarns	Y	increased electrical conductivity
Z		Z	antistatic

6.2.3. The standardization of plastics raw and molding materials

Table 6.3. Meaning of Codes in ISO/DIN-Designation (cont.)

c) Definitions of codes in data block 4

Code	Material	Form	Code	Mass content W %
A	asbestos			
B	boron	beads, spheres, balls		
C	carbon*	chips, cutting	5	W < 7.5
D		powder	10	7.5 < W < 12.5
E				
F		fiber	15	12.5 < W < 17.5
G	glass	ground	20	17.5 < W < 22.5
H	hybrid	whisker	25	22.5 < W < 27.5
J			30	27.5 < W < 32.5
K	chalk	knitted fabric	35	32.5 < W < 37.5
L	cellulose*	layer	40	37.5 < W < 42.5
M	mineral, metal*	mat (thick)	45	42.5 < W < 47.5
N		non woven (fabric, thin)	50	47.5 < W < 55
P	mica*	paper	60	55 < W < 65
Q	silicone		70	65 < W < 75
R	aramid	roving	80	75 < W < 85
S	synthetic (organic)*	scale, flake	90	85 < W
T	talcum	cord		
U				
V		veneer		
W	wood*	woven fabric		
X	not specified	*not specified*		
Y		yarn		
Z	others*	others*		

* More detailed information can be found in the particular standard restricted by the wording in this standard.

6.2. Fields of Standardization

Examples for designations according ISO pattern:
A polyethylene material (PE), for extrusion of film (F), with antiblocking agent (B), not colored (N), having a conventional density 0.91 g/cm^3 (18), a melt flow rate condition D 3.5 g/10 min (045), would be designated:

	International Standard number block	Individual item block		
		1	2	3
	ISO 1872/1–	PE,	FBN,	118-D045

International Standard
block 1: symbol
block 2: intended application
        antiblocking agent
        not colored
block 3: density cell
        MFR condition (190/21.6)
        MFR cell

Designation: ISO 1872/1-PE, FBN, 18-D 045
For block 3 see Table 4.2.

A polyamide material (PA66), for injection molding (M), with special burning characteristics (F), and a heat ageing stabilizer (H), having a viscosity number of 140 ml/g (14), a tensile modulus of elasticity of 10 200 N/mm^2 (100), reinforced with glass fibers (GF), 37% (35), would be designated:

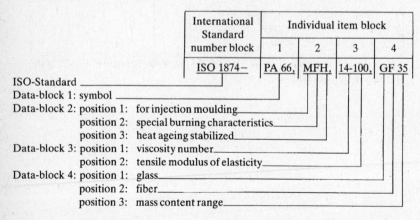

Designation: ISO 1874-PA 66, MFH, 14-100, GF 35
which can be shortened PA 66, GF 35
For block 3 see Table 4.36.

6.2.3. The standardization of plastics raw and molding materials

Table 6.4 Designatory Properties for Thermoplastic Materials

Designatory property	PE	PP	EVA	PVC	PVC-U	PVC-P	PS	SB	SAN	ABS ASA	PA	PC	PMMA	PET PBT
Melt index (MFR)	×	×	×				×	×	×	×		×		
Vicat-temp. (VST)					×		×	×	×	×			×	
Viscosity No (J)				×							×	×	×	×
(Notched) Impact strength		×			×			×		×				
E-Modulus					×			(×)			×	×		
K-Value				(×)										
Density	×					×								
Powder density				×										
Shore hardness						×								
Isotactic index		×												
VAC content			×											
AN content									×	×				
Stress value σ_{100}						(×)								
Torsion stiffness temperature						×								

Table 6.5. Specifications and Designations for Plastic Materials[1])

		ASTM D	table	page	ISO	DIN	table	page
Thermoplastics								
Polyethylenes		1248 / 2952	4.2/4.4 / 4.4	177/179	1872	16776	4.2	177
Ethylene-Vinylacetate Cop.		–	–	–	4613	16778	4.8	193
Crosslinked PE-V		2647	–	190	–	–	–	–
Polypropylenes		2146	4.4	} 179	1873	16774	4.10	197
Polybutene		2581	4.4		–	–	–	–
Reinforced polyolefins		2853	4.4		–	–	–	–
Styrene based polymers	PS	703	4.12	} 208	1622	7741	4.11	} 204
	SAN	4203	4.12		4894	16775	4.11	
	SB	1892	4.12		2897	16771	4.11	
	ABS	1788	4.12		2580	16772	4.11	
*for pipes		3965*	4.12					
	ASA	–	–		6402	16777	4.11	
Reinforced styrene polymers		3011	4.12	208	–	–	–	
PVC Homopolymer		1755	4.19	} 218	} 1060	7746	4.18	217
VC Copolymers		2472	4.19					
Rigid PVC-U compounds		1784	4.22	} 223	1163	7748	4.21	} 221/222
*for pipes		3915*	4.22					
Plasticized PVC-P compounds		2287	4.22		2898	7749	4.21	
Fluorinated polymers	PTFE	1475	4.27	} 236				
	FEP	2116	4.27		–	–	–	–
	ETFE	3159	4.27					
	PVDF	3222	4.27					
	PCTFE	1430	4.27					
	ECTFE	3275	4.27					
Polymethyl methacrylate		788	4.29	241	8257	7745	4.29	241
Acetal resins POM		2133	–	248	–	–	–	248
*calling out spec.		4181*	–	248	–	–	–	
Acetal resins POM, reinforced		2948	–	248	–	–	–	–
Polyamides (Nylons)		789	} 4.35	255	1874	16773	4.36	255
* calling out spec.		4066*						
Polycarbonate		3935	4.40	264	7391	7744	4.40	267
Polyterephthalates		3221	} 4.43	273	7792	16779	–	272
Polyterephthalates, reinforced		3220						
Polyphenylene sulfide		4067	–	282	–	–	–	–
Cellulose Acetate		706	–	} 287	–	} 7742	–	287
Cellulose Acetate Butyrate		707	–		–		–	
Cellulose Propionate		1562	–		–		–	
Ethylcellulose		787	–	288	–	–	–	–
Reaction Resins								
In general (MMA, UP, EP, PUR):								
pre-products and testing						16945	–	342
cast resins		–	–	–	–	16946	4.64	345
GF reinforced laminates						16948	4.65	347
Phenolic resins		–	–	–	–	16916	–	–
Polyester UP-resins		–	–	–	3672	–	4.64	345
Epoxy EP-resins		1763	–	355	3673	–	–	–
Epoxy EP reactants		–	–	–	4597	–	–	355
Thermosetting molding materials								
Phenolics		700 / 3881	4.71 / 4.71	} 373	800	7708/2	4.71	} 373
Melamine/phenolics		–	–	–	4896	} 7708/3	} 4.71	
	MF	704	4.71	} 373	} 2112			
Aminmoplastics								
	UF	705	4.17					
UP based		1201	4.73	376	–	16911	4.73	} 376
EP based		3013	–	379	–	16913[2])	473	
DAP based		1636	–	381	–	–	–	–
Industrial laminates		709	4.76	385	1642	7735	4.76	385
Decorative laminates		(NEMA)	–	–	4586	16926	–	389

[1] The ASTM standards correspond to the ASTM Year Book 1985. Some ISO and DIN documents were (1987) still in the DIS or DP (Draft Standards or Draft Proposals) and DIN E (Entwurf) state. That has been noted in the special tables quoted here.
[2] DIN 16913. part 1 is a general standard for reinforced reactive molding compounds. part 2 deals with the testing of prepregs. part 3 with UP-GF mats.

6.2.4. National and international standardization of semi-finished products

For *pipes and fittings* see
– thermoplastic types Sec. 4.2.1.
– reinforced thermoset types Sec. 4.6.3.5.
For *industrial and decorative laminates* Sec. 4.6.3.5.
There are many general and specialized rules for quality assurance and good practise concerning dimensions, performance, handling and application of semi-finished plastics products e.g. military, naval, aviation, automotive and building construction standards. They still vary greatly between countries and could not be dealt with in this book in detail. The reader will find some references in the sections on the special thermoplastic and thermoset semi-finished products.

7. TESTS AND THEIR SIGNIFICANCE

The testing of plastics includes the following overlapping areas:

(1) *Fundamental and processing data* of polymer preproducts, polymers and molding materials made from them and additives are ascertained by chemical, physical or practical tests on the unshaped material. Corresponding tests during and after processing give information on, for example, curing or decomposition phenomena in the molded material.

(2) *Physical properties of materials* such as density, optical properties, dielectric and insulation behavior, specific heat and thermal conductivity are independent of form and time. Their values are reversibly dependent on ambient influences such as temperature and humidity, insofar as these do not effect changes in the structure of the material. Standard data are determined on appropriate test pieces in a standard atmosphere to ISO 291/DIN 50014/ASTM E 171. This is currently 23/50, i.e, 23°C and 50% relative humidity.

(3) *Data on temperature dependent behavior* are of particular significance. Apart from general research and conventional data relating to changes in state brought about by temperature variations, there are also data relating to the long-term effects of elevated temperature and the burning characteristics of plastics.

(4) *Performance characteristics* reflect predominantly the variable material-dependent effects in the interaction between plastics and the environment. Conventional test methods corresponding to the many types of applications and conditions, which are accelerated and which simulate circumstances of use, only give a qualitative indication of *in-situ* performance.

(5) *The permeability of plastics* to fluids and gases, including water vapor is of special importance for applications as anti-corrosive media, in the packaging and in the building field. It is related to, but different from, the *migration* of substances into or out of plastics.

(6) *The mechanical behavior of plastics products* is more stress, time and temperature dependent than that of other materials as a result of their viscoelastic behavior which follows known laws. Plastics products are also affected by the design of the molding and the conditions of manufacture. Short-term test values for mechanical properties from standard test pieces made under standard conditions are essential for comparison of basic characteristics of different plastics or grades of a given type of plastic.

However, these values cannot be employed in engineering calculations for plastics moldings without at least applying scaling factors (see Sec. 7.7). Industrial parts need refined design methods (see Sec. 7.8).

Plastics test methods are largely defined in national and international standards. For the data groups 1 to 3 mentioned above, which concern the general behavior of materials, values obtained from the standard methods usually agree so well that they are generally valid. For performance characteristics and for permeabilities this applies to qualitative comparisons. Test moldings and preparation and details of testing techniques differ in the various standards.

Average values are contained in the data tables for generic plastics in this book. Deeper evaluation requires a precise statement of the standard governing the test. Mathematical evaluation of measurements over extended periods requires families of curves.

On the basis of many years of experience, test procedures, recording and evaluation of results have been extensively automated and computerized. Testing is required for routine quality control in development and production, for compilation of standardized test data for the designation of product groups and individual products (see Chapter 6) and establishing comprehensive data tables.

The numerous test methods developed historically rather than systematically using many differently prepared specimens and methods which do not correlate completely. For international comparisons of plastics products, it is essential to be able to describe them in standard terms using values determined by the same standardized test methods all over the world. The catalogue of basic values compiled by ISO provides a foundation for such harmonization (ISO 10350). This single-point data catalogue contains a selection of characteristic values which describe the rheological, mechanical, thermal, electrical, and other properties of plastics based on defined test specimen (Fig. 7.1, see next page). It can essentially be used for rapid comparisons and for making an initial selection of plastics and can thus also be taken as a basis for globally harmonized specifications.

Computer Aided Data Bases

In the industrialized countries, some 10,000 grades of thermoplastic and thermoset generic types have been commercialized by more than 200 raw materials producers and compounders. The more or less comprehensive company literature for products can contain up to 200 characteristics, in figures or graphics, that may be of interest to the designer, manufacturer, and user for the development and dimensioning of technical parts. Even if only a limited range of properties is essential in a particular case, searching out the required property

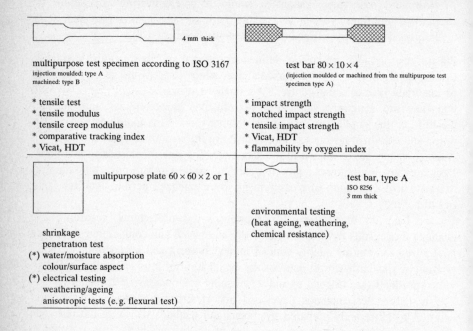

Fig. 7.1. Test specimen to ISO 10350 (single-point data) and ISO 11403 (multi-point data)

profile from the flood of information has become a task which, for the inexperienced user, is pratically insoluble without the aid of computerized data bases.

These contain 50 to 150 basic and dimensioning characteristic values and unquantifiable information for each recorded product, structured to different extents depending on the storage capacity and overall system. Many also provide graphic or tabular representations of functions relevant to processing and applications stored or calculated by the computer according to input formulae. In a computer search operation, which may be directed, for example, towards comparing individual characteristic values or to a product or product group that meets a special characteristics profile, the user first receives a preliminary selection; further searches within such sub-groups narrow the choice and, if the data base is adequate, provide complete information about suitable products including trade names and manufacturer's address.

Beginning in the 80ies, major plastics producers have built up highly structured data bases for their products using basic characteristic values for the most important and significant properties (Campus: Computer Aided Material

7.1.1. Analytical Chemical Methods

*P*reselection by *U*niform *S*tandards), which also permit direct access for CAD (Sec. 1.8.1). In addition to in-house research, particular areas of these data bases can be accessed by customers via on-line remote connection to the manufacturer's central computer or via floppy disks.

In the meantime, major progress has been made on international standardization, and Campus is now to be brought into line with the latest ISO and IEC Standards. This has been made all the more necessary by the fact that the American raw materials producers are currently in the course of converting their measurements to the International Standards. A key step has thus been made towards achieving a uniform solution worldwide. 28 European and American raw materials producers have joined Campus until the mid of 1994. Campus Version 3 not only makes allowance for the new developments in international standardization but also incorporates a new type of user guidance, bringing it into line with the latest developments in hard- and software.

Other data bases for use on a general contractual basis are being built up, and are claimed to provide comprehensive information on all plastics groups; a claim which because of the wealth of data and the limited comparability of different types of data is difficult to substantiate. These include:

- POLSELEC Data Base (Hanser, Kolbergerstr. 22, 81679 Munich, DE),
- PLASCAMS (Plastics Materials Selectors of RAPRA, Shawbury, Shropshire, SV4 4NR, GB),
- PLASPEC (Plastics Technology, 355 Park Avenue South, New York, USA).

The Plastics Selector from D.A.T.A. Inc. (P.O. Box 27875, San Diego, CA 92129, USA), contains up to 50 characteristic data for 9000 thermoplastics and thermosets available on the U.S. market and can be leased on magnetic tape or accessed on-line from the company's central computer. The PROCOP-M data base (Cisigraph, Le Griffon-536, Route de la Seds, 13127 Vitvoller, FR) is part of CAD/CAM-system developed for the automotive and aerospace industries (see Sec. 1.8.2). The Institut für Kunststoffverarbeitung (IKV, Pontstr. 49, 52062 Aachen, DE) offers a data base forming part of the CADFORM System (Sec. 7.8) for computer aided design of injection molded parts.

7.1. Fundamental and Processing Data

7.1.1. Analytical Chemical Methods

Directions for the analytical methods available for the identification of types of plastics can be found in the literature; for example (see also Sec. 2.2):

7.1. Fundamental and Processing Data

Braun/Saechtling, Simple Methods for the Identification of Plastics (Hanser Verlag)

Kraus/Lange/Ezrin, Chemical Analysis for Plastics and Polymers (Hanser Verlag)

Comprehensive treatments for IR Spectral Analysis for the identification of polymers and additives are provided by Hummel/Scholl, Atlas of Polymer and Plastics Analysis (3 vols. – Hanser and Verlag Chemie, Weinheim-New York)

The following standards are given as typical examples of standard test methods for the designation and monitoring of plastics behavior.

ASTM D 1652 – ISO 3001 – DIN 16945.
Determination of epoxide equivalent content of epoxy resins. (DIN 16945 and various individual ISO standards deal with analytical tests for all reaction resins and reactants, see Sec. 4.6.1.4.

ASTM D 834 – ISO 120/172 – DIN 53 707/53 708
Detection and determination of free ammonia in phenol-formaldehyde molded resins.

ASTM D 1303 – ISO 1185 – DIN 53474
Test methods for total chlorine in vinyl chloride polymers and copolymers (and other plastics).

ASTM D 3749/D 3680 – ISO DP 6401 – DIN 53743
Residual vinyl chloride monomer content of polyvinylchloride and copolymers by gas chromatographic headspace technique.

ISO 2561/ISO R 173 (4901)-DIN 53741/53719 (53394)
Determination of residual styrene monomer in polystyrene by gas chromatography/with Wijs solution (in UP resin materials in the molded state)

7.1.2. Data on Flow Behavior of Thermoplastics That Are Dependent on Degree of Polymerization

For the significance of the viscosity numbers obtained from dilute solution of polymers see Secs. 1.4 and 3.1.2, "Characteristic Values". Instructions for the determination are given in ASTM D 2857-ISO DP 1628 and 1191 (PE, PP), 174 (PVC), 1233 (PMMA), 307 (PA), 1228 (PTP), 1157 (CA). DIN 53726 (PVC) and 53727, 53728 deal with the same polymers. The corresponding K values of PVC are given in Table 4.16.

Increasing viscosity numbers or K values lead to a reduction in flow of polymers at processing temperatures. Viscosity numbers are proportional to the chain length of polymers. They can, with certain assumptions, be used to calculate average molecular weights. The *flow rate*, or *melt flow index*, determined according to ASTM D 1238-ISO1133-DIN 53735, is a technological comparative value. The measured value – the amount extruded from the standard plastometer in g/10 min for FR or MFI, in $10\,cm^3/10$ min for MVI – depends on the given material and on the chosen combination of total load and temperature operating on the melt, which should be quoted and is often given as "condition" (Table 7.1). The lower the melt index the less easy flowing the product. This can indicate a high molecular weight polymer. The flow rate is also influenced by fillers present in the compound; therefore only values obtained from similar material groups under like conditions are directly comparable.

Both methods are useful in indicating deterioration resulting from degradation of thermoplastics in processing or use

7.1.3. Processing Data

Apparent density of molding material and bulk factor (ASTM D 1895-ISO R 60, 61, 171-DIN 53466-53468) and pourability (ASTM D 1895-ISO 6186-DIN 53492) provide, from the comparable values obtained from the different standards, the data required for calculation of the space needed in the feed equipment and mold cavities. The cup flow (ASTM 731-DIN 53465) or the spiral flow test (ASTM D 3123) permit qualitative differentiation of soft, medium and hard thermoset molding materials. Similar conventional, but not standardized, spiral flow tests have also been applied for thermoplastics.

Sophisticated torque-rheometers simulate the flow characteristics of thermosets during melting and curing (ASTM 3795) as well as of other processes determined by the viscoelastic behavior of plastics. The rheological properties of polymer melts under processing conditions are measured using dynamic mechanical procedures (ASTM 4440) or by capillary rheometers (ASTM 3795-DIN 54811). In automated molding equipment, such devices, integrated e.g. in a bypass-manner, provide instant process control. Noncontact thickness gauges continuously control extrusion or calendering. The computer records of such data are suitable as quality criteria, and sometimes on-line control is combined with automated separation of rejects.

The use of piezometrically determined dependence of the specific volumes of molding materials on temperature and pressure (pvt-diagrams) as a means for controlling injection molding is considered in Sec. 3.3.2.6.

The significance of processing and mold shrinkage with regard to tolerances and permissible dimensional deviations of moldings is discussed in Sec. 3.3.1.3. Shrinkage and Tolerances.

Table 7.1. Conditions for the Flow Rate Test

Temperature °C	Total Load, Including Piston, g	Approximate Pressure kPa	Approximate Pressure psi	ASTM D 1238	Conditions ISO 1133 Number	Conditions ISO 1133 Code	Test method applied ASTM	Test method applied ISO	Test method applied DIN
125	325	44.8	6.5	A	—	—	PE		
125	2 160	298.2	43.25	B	—	—	PE		
150	2 160	298.2	43.25	C	2	B		EVA	
190	325	44.8	6.5	D	3	C	PE		PE, PP
190	2 160	298.2	43.25	E	4	D	PE, POM	PE, EVA	PE
190	21 600	2982.2	432.5	F	7	G	PE	PE	PE, ABS (200°C)
200	5 000	689.5	100.0	G	8	H	PS, ABS	PS	PS
230	1 200	165.4	24.0	H	11	L	PS, PMMA		PS, PMMA, PA (235°)
230	3 800	524.0	76.0	I	13	N	PS, PMMA		PS, PMMA
265	12 500	1723.7	250.0	J	—	—	CTFE		
275	325	44.8	6.5	K	—	—	PA		PA
230	2 160	298.2	43.25	L	12	M	PP	PP	PP, PA
190	1 050	144.7	21.0	M	5	E	POM		PP, PS, ABS(220°C), PMMA(200°C)
190	10 000	1379.0	200.0	N	6	F	PE		PC
300	1 200	165.4	24.0	O	—	—	PC		
190	5 000	689.5	100.0	P	18	T	PA	PE	PE, PP
235	1 000	138.2	20.05	Q	—	—	PA		
235	2 160	298.2	43.25	R	—	—	PA		
235	5 000	689.5	100.0	S	~20	V	PA	PP (230°C)	PP (230°C)
250	2 160	298.2	43.25	T	—	—	PTP		
310	12 500	1723.7	250.0	U	—	—	PE		
								ABS, SAN: 220°C, 10000 g (19/U)	PA: 230°C, 325 g; 235°C, 325 g PC: 300°C, 5000 g

7.2. General Physical Properties

ASTM-ISO-DIN testing methods are summarized in Table 7.2. For data on *density, optical and thermal properties, melting temperatures* (of semi-crystalline plastics) see Table 7.3.

Table 7.2. Tests for Density, Optical, Electrical, and Thermal Properties

Properties	ASTM	ISO	DIN	Remarks
Density[1]	D 792 D 1505	R 1183	53 479	Hydrostatic balance, Pyknometer, density gradient technique
Refractive Index	D 542	R 489	53 491	Refractometric and microscopic methods, also dispersion, see Table 7.3
Haze and luminous transmittance	D 1003		53 491	Simplified procedure using a hazeometer gives haze in % or ‰
Volume resistivity Surface resistivity Insulation resist.	D 257	IEC 93, 167	53 482 VDE 0303	
Dielectric constant Dissipation factor	D 150	IEC 250	53 483 VDE 0303	Dependent on frequency measured at 50 Hz, 1 kHz, 1 MHz
Dielectric strength	D 149	IEC 243	53 981 VDE 0303	Different methods, do not give comparable values and depend on specimen thickness
Tracking resistance	D 3638	IEC 112	53 480 VDE 0303	Method KC is IEC harmonized, see text
Arc resistance	D 495		53 484 VDE 0303	ASTM: high-voltage, low current, DIN: low-voltage, high current; both not for specifications
Specific heat	C 351	not standardized		See text and Fig. 7.3
Thermal conductivity	C 177		52 612	Steady state, by means of guarded hot plate (*Poensgen*-method) see text
Melting temperature	D 2117	3146	53 736	Optical methods, see text

[1] Density and specific gravity are not always clearly distinguished (at 23°C):
d^{23}, g/cm^3 = sp.gr$^{23/23}$ · 0.9975.

7.2. General Physical Properties

Table 7.3. Characteristic Values of Physical Properties

	Density g/cm³	Refractive index n_{20}^D	Abbe No v_D	Specific heat at 20°C kJ/kg K	Thermal conductivity W/mK	Crystalline melting range °C
Cellulose acetate	1.3	1.50	50	1.5-1.9	0.27	–
Cellulose acetobutyrate	1.19	1.48	61	1.3-1.7	0.31	–
Cellulose propionate	1.21	1.47		1.3-1.7	0.20	–
Epoxide resin, unfilled	1.2	–		1.4		–
Urea molding comp., cellulose filled	1.5	–		1.3	0.36	–
Melamine molding comp., cellulose filled	1.5	–		1.2	0.5	–
Pure phenolic resin	1.2	1.63				–
Phenolic molding comp. mineral filled	1.9	–		1.0	0.7	–
Phenolic molding comp. woodflour	1.4	–		1.3	0.35	–
Polyacetals	1.41	1.48		1.4	0.3	164/167
Polyamide 6	1.13	1.53		1.7	0.29	217/221
Polyamide 66	1.14	1.53		1.7	0.23	265
Polyamide 11	1.04	1.53		1.3	0.23	190
Polyamide 12	1.02	1.53		1.5	0.25	179
Polycarbonate	1.2	1.59	30	1.2	0.21	(220)
Polyester resin unfilled	1.2	1.54/1.60	43	1.2-1.9	0.6	–
Polyethylene LD	0.92			2.1	0.32	105/115
Polyethylene HD	0.96	1.51		1.8	0.4	130/140
Polyimide (film)	1.4	1.78		1.3	0.60	no
Polyisobutylene	0.93	1.50				–
Polymethylmethacrylate	1.18	1.49	58	1.5	0.18	–
PMM copol. with acrylonitrile	1.17	1.51		1.4	0.18	–
Polyphenylene oxide mod.	1.06	–		1.4	0.23	–
Polypropylene	0.91	1.49		1.7	0.22	158/168
Polystyrene	1.05	1.58	31	1.3	0.16	–
PS copol. with acrylonitrile	1.08	1.57	36	1.3	0.18	–
PS copol. with butadiene	1.04	–		1.3	0.18	–
ABS copol.	1.05	1.52		1.5	0.18	–
Polyterephthalate, ethylene	1.37	1.57	27	1.5	0.25	255/258
Polyterephthalate, butylene	1.31	1.55	33	1.5	0.25	225
PVC-S	1.39	1.54		0.9	0.16	–
Polyvinylidene fluoride	1.78	1.42		1.0	0.13	170
Polytetrafluoroethylene	2.2	1.35		1.0	0.23	(327)
PTFE copol. with ethylene	1.75					270

Of the *electrical* values (listed in the tables for particular products) dielectric strength and arc resistance are not invarient but are dependent on the test method used.

The determination of *tracking resistance* (or tracking) at operating voltages below 1 kV (DIN 53480) serves to ascertain deleterious effects of the tracking current caused by conducting dust deposition on plastic surfaces which causes small arcs at the affected positions that carbonize and give rise to short circuiting (for example with phenolics). The resulting volatilization or melting away of thermoplastics can lead to failure of the insulating material. In testing for tracking resistance two platinum electrodes are placed 4 mm apart on a test piece onto which drops of an electrolyte fall. These tests are carried out at different voltage levels. Method KA is obsolete. In methods KB and KC the voltage is determined at which a cut-out switch adjusted to 0.5 A just fails to trigger for a drop rate of 50. Method KC differs from KB in that in place of a solution containing wetting agent, a wetting agent free solution is used; this contains 0.1% NH_4Cl in distilled water. This method can give higher values than KB.

Knowledge of the *specific heat* is necessary for the estimation of the heat required to raise the temperature of a definite amount of a plastic. It is strongly temperature dependent; extensive changes indicate transition regions (Fig. 7.2). Thermal conductivity is generally slightly temperature dependent between 20 and 100°C: in the case of PE it falls in the crystalline region with increasing temperature.

The melting temperature range of semi-crystalline polymers is somewhat dependent on the conditions. A method for the determination of melting point consists of heating a sample 0.04 mm (1.6 mil) or less in thickness using a hot stage unit mounted under a microscope and viewing it between crossed polarizers. When the material melts, the characteristic double refraction from the crystalline aggregates disappears. The point at which it completely disappears is taken as the melting point of the polymer.

7.3. Test Methods for Measuring the Influence of Temperature

7.3.1. The Torsional Pendulum Test

ASTM D 4065-ISO 6721-DIN 53445 (DIN 53520 for elastomers) serves to ascertain continuous and discontinuous changes in the physical state of polymers over a wide temperature range using the torsional pendulum test (cf. Sec. 1.5, Fig. 1.4). In the torsional pendulum apparatus (Fig. 7.3) a rod or ribbon test piece is fixed at the upper end in a temperature controlled chamber while at the lower end it is attached to a light wheel. By striking against a trip lever the system is induced into free damped oscillation the

7.3. Test Methods for Measuring the Influence of Temperature

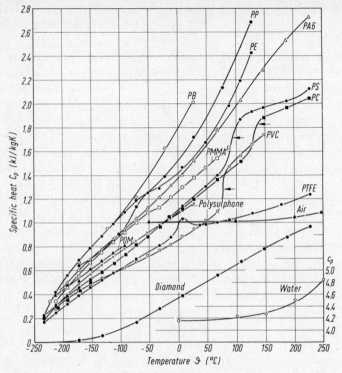

Fig. 7.2. Dependence of specific heat on temperature.

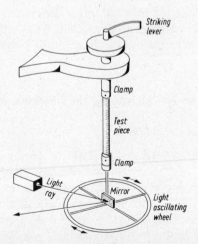

Fig. 7.3. Torsion pendulum test according to ISO 6721.

7.3.1. The Torsional Pendulum Test

decrease in which is recorded. Changes in the shear modulus or maxima of the mechanical loss factor temperature curves, which are calculated from them (Fig. 7.4, and Figures 1.4, 4.1 and 4.9), indicate *state transition temperature ranges*, which are valid for short-term dynamic stressing under the conditions of the test, in particular for the *glass transition temperatures* of thermoplastics and elastomers (Sec. 1.5).

The shear modulus vs temperature curve falls off steeply in the *crystalline melting range* of semi-crystalline thermoplastics (Fig. 7.4). For thermosets,

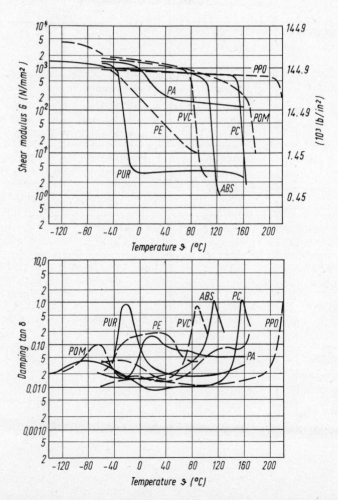

Fig. 7.4. Shear modulus (a) and mechanical loss factor (b) of thermoplastics.

7.3. Test Methods for Measuring the Influence of Temperature

Table 7.4. Determination of Heat Distortion Temperature

	Martens	Vicat	ISO/R 75
Test specification	DIN 53458 VDE 0303/111 43	DIN 53460 VDE 0302/111 43	DIN 53461 ASTM D 648
Test arrangements (measurements in mm)			
Method of loading	Flexing	Penetration of needle, Area: 1 mm^2	Flexing
Load	Flexural stress σ_b 5 N/mm^2	Compression stress σ_d A = 10 N/mm^2 B = 50 N/mm^2*	Flexural stress σ_b A = 1.85 N/mm^2* B = 0.46 N/mm^2
Deformation	Lever-arm depression 6 mm	Penetration depth 1 mm	Bending 0.21 to 0.33 mm** depending on the height of the test specimen
Temperature	Increase 50 K/h	Increase 50 K/h	Increase 2 K/min
Heat exchange medium	Air	Liquid bath, air in particular cases	Liquid bath
Test-piece (l × w × h in mm)	120 × 15 × 10 60 × 15 × 4 50 × 6 × 4	10 × 10 × 3-6.4	l = 110 w = 3.0-4.2 (13) h = 9.8-15
Examples of abbreviated quoted values	$t_{Martens}$ = 125°C	VST/B = 82°C	F_{ISO} = A 95°C

* Preferred method.
** Corresponding to an outer fiber extension of 0.2%.

strength measurements up to high temperatures can give indicators which are more discriminating than this method (Fig. 4.19). The temperature and time dependence of the apparent modulus of rigidity of stiffer plastics in the glassy state is measured by the *stiffness in torsion method*, under static stress, using the Clash and Berg test (ASTM 1043-ISO 458-DIN 53447).

7.3.2. Thermal analytical methods (DIN 51005)

Differential scanning calorimetry or analysis (DSC, DSA, ASTM 3417, 3418) are techniques in which differences in energy inputs or temperatures between a substance and a reference material following a controlled temperature program are recorded. These methods allow exact measurement of heats and temperatures of fusion and crystallisation and other thermal transition processes. For thermogravimetric analysis (TGA) of plastics see Sec. 4.8.1. (Fig. 4.22).

7.3.3. Conventional Single-Value Tests

Conventional single-value tests for cold crack behavior, such as the brittleness temperature test (ASTM D 746/1790-ISO 974) or the drop hammer test for PVC-P films (DIN 53372) are comparable for practical use. Graphs of the temperature dependence of mechanical properties (e.g. Fig. 7.8) or impact resistance (p. 470) vs temperature curves provide a better basis for the designer. *The Vicat softening temperature* (VST) for thermoplastics (ASTM D 1525-ISO 306-DIN 53460) and the heat distortion temperature (HDT test), applicable to all plastics (ASTM D 648-ISO 75-DIN 53461) are used internationally. The *Martens test* (DIN 53458) is an older variation used for testing thermosets under flexural load.

Table 7.4 indicates the principles of these test methods. The single-value temperatures determined by these methods under conventional mechanical loading are not necessarily the upper limits of the application range. The values measured under the various loading conditions A and B (see Table 7.4) provide practical information on the reduction of strength as a function of temperature. For stiff materials, $5.0\,N/mm^2$ stress can be used in the HDT test.

7.3.4. Temperature-Time Limits

Temperature-time limits of the heat aging of plastics without load are determined to ASTM D 794/D 3045-ISO 2578-DIN 53446. Test pieces are stored for between 5000 and 20 000 hours at progressively elevated temperature levels in air ovens and selected properties are examined. The "temperature-time limit" is the temperature at which, after exposure for a predetermined time, the value of

7.3. Test Methods for Measuring the Influence of Temperature

a particular property has fallen to a specified fraction of the initial room-temperature value.

IEC 216, corresponding to VDE 0304 parts 12 and 22, is an international standard for the *thermal resistance of solid insulating materials*. Table 7.5 gives specimen values from this work for several commercial plastics.

For the "*Underwriter Laboratories Temperature Index*" see details of the method in Sec. 4.8.1; index values are given in Sec. 4.8 and also in Fig. 4.12. The temperature-index values given in the above publications allow assessment of thermal resistance for all applications of plastics at elevated temperature, but they do not give property values at their limiting temperatures. Such values fall off more strongly for thermoplastics than thermosets.

The heat storage test, in which moldings are stored under heat for 1–8 hours if thermoplastics (DIN 53497) and 48 hours if thermosets (ASTM D1299-DIN 53498) followed by examination for visible or measurable changes, serves to monitor the manufacture of acceptable moldings. There are some specialized standard methods for the chemical (oxidative) thermal stability of e.g. PE and PP (ASTM D 2445, 3895 – DIN 53383), PVC (ASTM D 2115, 4202 – DIN 53383).

7.3.5. Reaction to Fire Tests

Laboratory scale fire-performance tests cannot give more than a few criteria for selecting plastics because fire hazard depends in a very complex manner on shape, quantity and environment of combustible objects and on other circumstances relating to a particular end use. It is therefore good practice to describe fire performance of plastics by code numbers or letters related to a special test rather than by perhaps misleading descriptions such as "self-extinguishing" or "non-burning". If needed, statements such as "flammable with difficulty" (by a flame) or "ignitable" (by radiation) "*by this test*" might be used.

The *LOI (limiting oxygen index) test* (ASTM D 2863-ISO 4589) gives the lowest percentage of oxygen in an oxygen/nitrogen mixture needed to make a plastics rod in contact with a flame burn like a candle. It can only be used as a screening test because if more than 21% of oxygen is necessary, the test is not realistic. Some plastics with LOI > 21 are ignitable in other conditions.

In the generally used *flammability and flame spread tests* a flame is applied for a short time to the edges and/or on the surface of plastics specimens. Criteria for grading include extinction, after-flame, afterglow time and/or mass burning rate and area of burned material. The articles are tested in a horizontal position (ASTM D 635-ISO 1210) or more severely, vertically (ASTM D 568 for flexible plastics, ASTM D 3713 ignition response index (IRI) and D3801 for solid plastics – DIN 53438 for edge and surface flame action). These standards should be consulted for the different test conditions and coding of results.

Table 7.5. Temperature Indices (for 20 000 or 5000 h) of Some Commercial Plastics with Reference to IEC 216

Property	Dielectric strength E_d (DIN 53481)		Tensile strength σ_B (DIN 53455)		Impact strength a_{zn} (ASTM-D 1822)		Dimensional stability
Limiting value	$\Delta E_d = -50\%$		$\Delta\sigma_B = -50\%$		$\Delta a_{zn} = -50\%$		$S = 0.2\%$
Period of test (h)	$2\times10^4/5\times10^3$		$2\times10^4/5\times10^3$		$2\times10^4/5\times10^3$		$2\times10^4/5\times10^3$
Polymer	Initial value (kV/mm)	°C	Initial value (MPa)	°C	Initial value (kJ/m²)	°C	°C
PA 6 unreinforced ①	31	—	78	90/105	165	(80/85)	(90/115)
PA 6 unreinforced, heat stabilized	52	(125/150)	83	110/130	240	105/120	—
PA 6 30 %w/w GF, heat stabilized	35	125/150	160	145/175	170	110/130	120/145
PA 6 30 %w/w GF, flame resistant	26	135/160	137	130/150	—	—	—
		95/115					
PA 66 unreinforced ①	—	—	90	(90/105)	—	—	—
PA 66 unreinforced, heat stabilized	32	95/115	84	115/135	—	—	—
PA 66 30 %w/w GF, heat stabilized	34	125/165	190	130/170	150	100/140	125/145
PBT unreinforced ②	27	155/180	59	140/160	130	120/135	115/125
PBT 30 %w/w GF	32	155/180	130	140/160	95	135/150	145/165
PBT 30 %w/w GF, flame resistant	30	145/170	120	140/160	85	135/150	145/165
PC unreinforced ③	31	140/(150)	70	135/(150)	560	125/145	115/120
PC unreinforced, flame resistant	31	135/150	70	125/145	470	105/120	120/125
PC 10 %w/w GF, flame resistant	30	140/150	83	135/150	120	130/145	—
PC 30/20 %w/w GF	34	140/150	80/110	135/150	65/120	130/150	140/145

① Durethan, ② Pocan, ③ Makrolon grades.

7.3. Test Methods for Measuring the Influence of Temperature

For surface flame action on sheets which are horizontal or in the 45° position see ASTM D 1433-VDE 0340. Ignition by radiation (ASTM E 162) is applied in the internationally used "Flooring Radiant-Panel Test" (ASTM E 648, French testing see below). ASTM 1929-ISO 881 give standard test methods for the flash- and self-ignition temperatures of plastics.

For *construction and building* the behavior of material in a vertical furnace at $\geq 705°C$ (ASTM E 136-ISO 1182) is used to distinguish between non-combustible (mainly inorganic) and combustible (organic) building materials.

Special arrangements, such as the Steiner tunnel (ASTM E 84), the "Brandschacht" (a vertical combustion chamber for sheets and panels – DIN 4102) and the "Epiradiateur" test chamber (French standard NF P92-501) are used to differentiate between "low" and "normal" (i.e. rather like wood) flammability and flame spreads. Normal flammable materials have to pass tests like those described in the previous paragraphs and by testing of burning particles and sometimes of smoke density. "Easily" flammable materials that fail these tests are not allowed in building. The following govern flammability behavior in building: ASTM E 119 and E 84-DIN 4102; there are, e.g., A, B, C or I, II, III (US), B1, B2, B3 (DE) and M1 to M5 (FR) codes for "low" to "easy" flammability. Testing methods for fire effluents (smoke obscuration – ASTM D 2843-ISO DP 8887, toxicity, corrosivity) are still under development, but these developments are different in particular countries.

The Schramm-Zebrowski *incandescent rod resistance* method (ASTM D 757-ISO/R 181-DIN/VDE 0318 part 2) has been adopted internationally (Method BH in Table 7.6) for *solid electrical insulating materials*. A glowing silicon carbide rod (950°C) is pressed against the specimen for three minutes. BH1 indicates non-ignition and less than 5 mm penetration by the rod. B2 indicates >5 mm penetration in three minutes. If the rod penetrates more than 95 mm in the test time, then B3 gives the burning rate in mm/min.

Methods FH and FV of DIN IEC 707/VDE 0304 part 3 directly correlate with the Bunsen burner flammability test HB and to V-0, V-1, V-2 of the *Underwriters Laboratory Safety Standard 94*, Secs. 2 and 3 (Table 7.6). Test HB is deemed to have been passed if after 30 seconds of flame action a specimen bar, in the horizontal position, has not burned more than 25.4 cm (10 in) in length. Extinction and burning rate are further criteria.

For the V classifications correlated with the thickness of the test specimen the flame is applied twice to the bar in the vertical position.

The criteria are

V-0: <10 s after-flame, <30 s afterglow, no drip of flaming particles, the bar may not have been entirely burned.

7.3.5. Reaction to Fire Tests

Table 7.6. Fire Performance of Solid Insulating Materials to DIN IEC 707/VDE 0304 T.3 and UL 94

Product	Method BH	Method FH	M FV	UL 94
Polyolefines				
PE-LD	BH 3-15 mm/min	FH 3-20 mm/min	–	HB
PE-LD fr	BH 2-20 mm	FH 2-20 mm	FV 2	V 2
PE-HD	BH 3-15 mm/min	FH 3-15 mm/min	–	HB
PE-HD fr	BH 2-20 mm	FH 2-20 mm	FV 2	V 2
PP	BH 3-20 mm/min	FH 3-20 mm/min	–	HB
PP fr	BH 2-20 mm	FH 2-25 mm	FV 2	V 2
Styrene polymers				
PS	BH 3-15 mm/min	FH 3-30 mm/min	–	HB
SAN	BH 3-15 mm/min	FH 3-30 mm/min	–	HB
SB	BH 3-15 mm/min	FH 3-30 mm/min	–	HB
SB fr 2	BH 2-15 mm	FH 2-30 mm	FV 2	V 2
SB fr 3	BH 2-25 mm	FH 2-15 mm	FV 0	V 0
ABS	BH 3-25 mm/min	FH 3-40 mm/min	–	HB
ABS fr	BH 2-25 mm	FH 2	FV 0	V 0
ASA	BH 3-15 mm/min	FH 3-30 mm/min	–	HB
Polyvinyl chloride				
PVC	BH 2-5 mm	–	–	–
Polyacetal				
POM	BH 3-20 mm/min	FH 3-25 mm/min	–	HB
Polyamides				
PA 6	BH 2-15 mm	FH 3-15 mm/min	FV 2	V 2
PA 6 fr	–	–	FV 0	V 0
PA 6-GF	BH 2-60 mm	FH 3-20 mm/min	–	HB
PA 66	BH 2-10 mm	FH 2-20 mm	FV 2	V 2
PA 66 fr	–	–	FV 0	V 0
PA 66-GF fr	BH 2-10 mm	FH 2-15 mm	FV 0	V 0
Polycarbonate				
PC	BH 2	FH 2	FV 2	V 2
PC fr	BH 2	FH 2	FV 0	V 0
Polybutylene terephthalate				
PBT	BH 2-30 mm	FH 3-20 mm/min	–	HB
PBT fr	BH 2-15 mm	–	FV 0	V 0
PBT-GF	BH 2-50 mm	FH 3-15 mm/min	–	HB
PBT-GF fr	BH 2-10 mm	–	FV 0	V 0
Polyphenylene oxide				
PPO, modified	BH 2-25 mm	–	FV 1	V 1
Polyether sulfone				
PESU	–	–	FV 0	V 0
PESU-GF	–	–	FV 0	V 0

fr = With flame retardant treatment.

V-1: < 30 s after-flame, other criteria as for V-1.
V-2: drip of flaming particles, other criteria like those for V-1.
In method 5V for panels, the flame is applied five times at the edges and surface. The criteria are similar to those of V-0. The "recognition" of one of these grades by a *UL yellow card* is combined with supervision of production by the UL "follow-up service". Besides these basic standards there are standards for foamed and cable insulation materials as well as for electrical applications under service conditions.

Further detailed requirements have been included in the *special fire safety standardization* for road vehicles, street-car and railway rolling stock, aircraft, mining and home furnishing.

7.4. Resistance to Environmental Effects

7.4.1. General environmental conditions

The properties of plastics are influenced reversibly by temperature and humidity of the surrounding atmosphere or by immersion in a liquid. In order to predict the service behavior of plastics under various environmental conditions, samples must be *preconditioned* – sometimes for several weeks – and tested in clearly defined atmospheres from $-70°C$ to several hundred degrees, 20% to 93% RH (or immersion) as detailed in ASTM E 41, D 618-ISO 554-IEC 212-DIN 50005, 50013). Normal values (DIN 50014) for simulating humid conditions in the temperate zone and humid and dry conditions of the tropics are expressed in °C/% RH, 23/83, 40/92, 55/20.

The *standard laboratory atmosphere* for *conditioning* and *testing* is generally 23°C/50°% RH but 20°C/65% RH and – in the tropics – 27°C/65% RH are also used (ASTM E 171, ISO 291, 554-DIN 50015). Deviations have to be indicated by duration, temperature, RH of conditioning and of testing conditions, as appropriate.

For several plastics, there are special rules for preparing test pieces, e.g. for testing PA (Sec. 4.1.10) in the dry and in the conditioned state (ASTM D 789-ISO 1110-DIN 53714).

7.4.2. Water Absorption

Water absorption is usually tested by immersion of standardized specimens in cold water. Results of conventional testing to ASTM 570-ISO 62-DIN 53495 are not comparable quantitatively, as specimen dimensions, preconditioning, time of immersion and recording of results are different. This is shown in the following comparison:

7.4.4. Resistance to Fungi, Bacteria and Animals

	Standard	ASTM D 570	DIN 53495
	Time of immersion	24 hours	4 days
	Weight increase:	%	mg
PE-HD		< 0.01	< 0.05
PVC-S-U		0.04	< 4
PF wood flour filled		0.5	150
PA6		1.5	300

Long-term immersion to obtain saturation and estimation of weight loss by extraction of soluble matter, including extraction in boiling water (ASTM D 570-ISO 117-DIN 53457), are needed to obtain precise evaluation of the interaction of plastics and water. Soluble matter is not extracted by exposure of specimens to a humid atmosphere ($20 \pm 2°C$, 92 to 93% RH, DIN 53473).

7.4.3. Light and Weathering Resistance

In testing for resistance to light behind window glass using natural daylight or a xenon lamp (ASTM D 4459-ISO 877/4582-DIN 53388/9) only the lightfastness of the color of the plastics product is generally ascertained. Outdoor exposure for several years (ASTM D 1435-ISO 4607-DIN 53386) is supplemented by accelerated testing for weather resistance with a xenon lamp and rain and temperature cycling (ASTM D 2565-ISO 4892-DIN 53387). As an approximation, a 10-fold acceleration is obtained compared with natural weathering in a temperate zone climate. However the results of such accelerated tests are not strictly comparable with those from natural weathering because of the different environment. Simulated accelerated weathering apparatus should possess an adequate source of UV as this component of solar radiation has a major effect on aging of plastics.

The long outdoor exposure time can be shortened by the use of apparatus in which the test piece is turned in accordance with the solstitial points during the day or by weathering in a tropical climate, which is two to three times more damaging than the mid-European climate. In "Langley units", that is, the global radiation falling on a unit area of the test piece, one year in Florida corresponds to 200 000 Langley and one year in mid-Europe to about 120 000 Langley. Continuous testing of color changes and of significant mechanical or electrical parameters act as indicators of aging in the evaluation of all weathering tests.

7.4.4. Resistance to Fungi, Bacteria and Animals

Test pieces of plastics are innoculated with microorganisms. The test criterion involves measuring their growth and corresponding damage to the plastic (ASTM G 21/22-ISO 846-DIN 53930/32). The inherent resistance of most plastics to microbes can be prejudiced by organic fillers or low molecular weight additives such as fatty acid plasticizers. Cellulose derivatives, PVAC, PVA and UF are only conditionally resistant. Fungicides can protect against fungicidal attack. The harder plastics are, the less they are attacked by *termites* or rodents and thus thermosets show good resistance.

7.4.5. Resistance to Chemicals

Liquids and gases that diffuse into and through polymers (for permeation see Sec. 7.5) can reversibly swell the polymers in the process, or dissolve out components such as plasticizers. Aromatic and chlorinated hydrocarbons are good solvents for most thermoplastics. Alcohols show no more than weak solvent properties. Esters and ketones often have a strongly solvent action (apart from polyolefins, PA and PTFE). Strongly cross-linked thermosets are largely solvent resistant. Polymers without reactive heterogroups in the backbone chain of the macromolecule are the most resistant to irreversible decomposition by acids and bases. Only a few polymers, e.g. PTFE, are completely resistant, without stabilization, to oxidizing agents. The complex effects of physically and/or chemically active media on plastics products are influenced through apparently very small changes in composition as well as in conditions, duration and temperature of the interaction. Further, there are important influences from simultaneous frozen in or application-related mechanical stresses. Immersion tests for seven days at room temperature with standardized test pieces to ASTM D 543-ISO 175-DIN 53476 give only a preliminary indication as to choice of type of plastic. Preferred test liquids include standardized acids, alkalis, oxidizing agents, solvents, detergents, fatty and mineral oils, and fuels. Such tests should be supplemented by extended immersion in individual media under more demanding conditions and also under flexural or tensile stress. Quantitative critical criteria are given in the following table.

Level of significance	Change in weight (%)	Change in dimension (%)	Change in appearance	Change in tensile or flexural strength (%)
Satisfactory	< 0.5–1	< 0.2–1	Slight discoloration	< 10
Special testing needed	1–5	1–3	Discolored	10–30
Unsatisfactory	> 5	> 3	Optically distorted warped, softened, crazed, etc.	> 30

7.4.5. Resistance to Chemicals

Table 7.7. Chemical Resistance of Polymers

1 = Satisfactory, 2 = Special testing needed, 3 = Unsatisfactory,
S = Danger of stress cracking

Polymer	Water	Salt solutions	Resistance to Acids	Alkalis	Oxidizing agents	Solvents
Polyethylene	1	1	1	1	2-3	1-2 S
Polypropylene	1	1	1	1	3	1-2 S
Polybutene-1	1	1	1	1	3	1-3
Polyisobutylene	1	1	1	1	2-3	3
Poly-4-methylpentene	1	1	1	1	2-3	3
Ethylene-propylene copolymers	1	1	1	1	2-3	2-3
EVA	2	1-2	2-3	2	3	2-3
Chlorosulfonated polyethylene	1	1	1	1	1-2	2-3
Polystyrene	1	1	1	1	2-3	3 S
SAN	1	1	2	1	3	3 S
ABS	1	1	2	1	3	3 S
Polystyrene, impact resistant (rubber modified)	1	1	2	1	3	3 S
PVC-U	1	1	1	1	2	1-3
PVC-P	1	1	2	2	2-3	3
Polytetrafluoroethylene and Perfluoroelastomers	1	1	1	1	1	1
Highly fluorinated copolymers, Vinylidenefluoride	1	1	1	1	1	1-2
Polymethylmethacrylate	1	1	2	2	2-3	3 S
Polyoxymethylene	1	1	3	2	3	1-3
Polyamides	2	2	3 S	2	3	1-3 S
Polycarbonate	1	1	2-3	3	3	2-3 S
Polyethylene terephthalates	1	1	2	3	3	1-3 S
Polyphenylene oxide	1	1	2	1	1-2	2-3
Polysulfone	1	1	2-3	2	3	1-3
Polyether imides	1	1	1	2	1-2	1-2
Phenolformaldehyde compounds	1	1	1	3	2-3	1-2
Aminoplastics	1-2	1	1-2	1-3	3	1-2
Furane resins	1	1	1	2	3	1-2
Polyesters, GF reinforced	1-2	1	1-2	3	3	1-2
Epoxy resins	1	1	2-3	2	3	1-3
Polyurethanes	1-2	1-2	3	2	3	1-3
Silicones	1	1	2-3	2-3	3	1-3
Natural rubber	1	1	1-2	1	3	3
Styrene-butadiene rubber	1	1	1-2	1	3	3
Polychloroprene	1	1	1-2	1	2	1-3
Butyl rubber	1	1	1-2	1	2-3	3
Polysulfide-rubber	1	1	1	1	2	1-3

7.4. Resistance to Environmental Effects

The reduction of impact strength and raising or lowering of elongation at break in the range ±50% provide further criteria. In the design of chemical plant, as well as for an assessment of the effect of content, fuels and lubricants on plastics, the extensive experience of raw materials producers should be taken into account. General surveys (Table 7.7) with stability tables containing merely qualitative indicators, such as good/fair/poor, are of limited value.

Creep rupture strength testing of plastics pipes (Sec. 4.2.1) filled with different media is used to ascertain the *resistance factors* for the reduction of the calculated values for plastics piping and equipment when subject to stressing by chemicals.

Premature failure, apart from chemical degradation, can be caused by the *stress cracking effect* of surfactants. After only a short period of exposure some polymers are in danger of stress cracking, for example PE by aqueous wetting agents and detergent solutions, PMMA by fatty acids, and styrene polymers and other thermoplastics by alcohols, which are scarcely effective as solvents (Fig. 7.5, Table 7.8, p 454). In the one-stage Bell telephone test (ASTM D 1693) which has long been used for PE, 10 test-pieces are centrally notched along the length, bent into a U-form and immersed in the test solution at 50°C. The time required for 50% of the test-pieces to show *environmental stress cracking (ESC)* is measured. Another one-stage test for ESC is the Constant Tensile Stress method, in which the tensile strength or creep strain as a function of time (Fig. 7.5) in various media and at different temperatures is determined (ASTM D 2552 – DIN 53449 E, Part 2). In the ball or pin indentation method (ISO 4600 – DIN 53449, Part 1), pre-stresses are applied by the pressing balls or pins of increasing excess diameter into pre-bored holes in test-

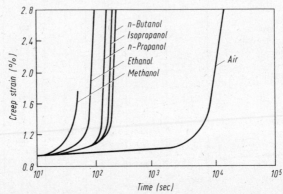

Fig. 7.5. Chemical resistance of high-impact PS to alcohols; creep strain vs. time at 16 N/mm² (2250 psi) stress.

pieces. In the Bend Strip method (ISO DP 4599.2, DIN 53449, Part 3) test strips are fixed onto a bending jig for stretching and are placed for a defined period (a relatively short time) in the testing medium. ECS failure is determined by measuring the residual strength or elongation at break after immersion. There are also stress cracking tests for appropriately filled PE bottles (ASTM D 2561) and helically wrapped strips at 100°C (ASTM D 2951). Table 7.8 summarizes ESC behavior in various media.

7.4.6. Resistance to High-Energy Radiation

This is measured to DIN/VDE 0306 parts 1 and 2, and is so satisfactory (with half values of $> 10^4$ J/kg Table 7.9) that easily decontaminatable plastics can be used in all areas considered safe for people. Many plastics are considered possible reactor construction materials. EP resin laminates and electrical insulation components have proved themselves in practice under long-lasting, high level radiation (> 10 MJ/kg/y). Laboratory equipment, piping and tubes made of PE, and rigid and plasticized PVC last longer than glass in the hot cells of nuclear research establishments. PTFE becomes brittle quickly under the combined effect of radiation and atmospheric oxygen. "Compilations of radiation damage test data" from CERN (CH, Geneva 23) give up to date and comprehensive information on the results of long term tests and practical experience with high-tech plastics.

7.4.7. Resistance to Mechanical Wear, Friction and Sliding

Wear and slide resistance are not physical values of individual materials, but are different for each material pair. The many practical methods for accelerated wear testing often lead to badly reproducible and non-comparable results. In standard test methods loaded plastics surfaces and abrasive discs (e.g., Taber method, ASTM D 1044-ISO 3537-DIN 52347, 53754) or abrasive tapes (e.g., Armstrong apparatus, ASTM 1242 method B-ISO/R 33-DIN 51954, 53516) are rubbed against each other. In ASTM D 1242, method A, with the Olsen Wearometer and similar such as the Böhmer-Scheibe (DIN 52108), a metal disc rotates against the test specimen in the presence of loose granular abrasive.
In testing for resistance to erosion (ASTM D 673-DIN 50352, 53248) a beam of granular grit is allowed, mostly under gravity, to strike the test specimen surface set at an incline. The method is important for assessing the erosion of transparent plastic windows in vehicles by rain and hail as well as piping and plant construction material by moving feedstock. Thickness and volume loss are the parameters measured. The mar resistance of opaque, glossy or trans-

7.4. Resistance to Environmental Effects

Table 7.8. Recommendations for media for testing susceptibility to crazing of various plastics

Plastics	Abbreviation	Crazing-inducing media	Immersion time
Polyethylene	PE	surfactant solution (2%), 50°C surfactant solution (2%), 70°C surfactant solution (5%), 80°C	> 50 h 48 h 4 h
Polypropylene	PP	chromic acid, 50°C	
Polystyrene	PS	n-heptane petroleum, boiling range 50 to 70°C n-heptane:n-propanol (1:1)	
Impact-modified polystyrene	S/B	n-heptane petroleum, boiling range 50 to 70°C n-heptane:n-propanol (1:1) oleic acid	
Styrene-acrylonitrile copolymer	SAN	toluene:n-propanol (1:5) n-heptane carbon tetrachloride	15 min
Acrylonitrile-butadiene-styrene copolymer	ABS	dioctylphthalate toluene:n-propanol (1:5) methanol acetic acid (80%) toluene	 15 min 20 min 1 h
Polymethylmethacrylate	PMMA	toluene:n-heptane (2:3) ethanol n-methylformamide	15 min
Polyvinylchloride	PVC	methanol methylene chloride acetone	 30 min 3 h
Polyacetal (polyoxymethylene)	POM	sulfuric acid (50%), localized wetting	up to 20 min
Polycarbonate	PC	toluene:n-propanol (1:3 to 1:10) carbon tetrachloride caustic soda solution (5%)	3 to 15 min 1 min 1 h
PC/ABS blend	(PC + ABS)	methanol:ethylacetate (1:3) methanol:acetic acid (1:3) toluene:n-propanol (1:3)	
Mod. polyphenylene oxide	PPO/PS	tributyl phosphate	10 min
Polybutylene-terephthalate	PBT	caustic soda solution (1M)	
Polyamide 6	PA 6	zinc chloride solution (35%)	20 min
Polyamide 66	PA 66	zinc chloride solution (50%)	1 h
Polyamide 6–3 (transparent)	PA 6–3–T	methanol acetone	1 min
Polysulfone	PSU	ethyleneglycolmonoethylether ethylacetate 1,1,1-trichloroethane:n-heptane (7:3) methylglycolacetate carbon tetrachloride 1,1,2-trichloroethane acetone	1 min 1 min 1 min
Polyethersulfone	PES	toluene ethylacetate	1 min 1 min
Polyetheretherketone	PEEK	acetone	
Polyarylate	PAR	caustic soda solution (5%) toluene	1 h 1 h
Polyetherimide	PEI	propylene carbonate	

7.4.7. Resistance to Mechanical Wear, Friction and Sliding

Table 7.9. Resistance to high-energy radiation at 20°C (H. Wilski)

Plastics	Irradiation with the exclusion of O_2			Irradiation in air	
	Degradation (A) cross-linking (V)	Gassing mm^3/kg Gy	Half-value dose k Gy	Half-value dose at 500 Gy/h k Gy	Half-value dose at 50 Gy/h k Gy
Thermoplastics[1]					
Polyethylene: PE-LD	V	7	400-1300	180	130
PE-HD	V		60-300	25-95	10-40
highly stabilized PE-V	V		600-1000	780	640
Polypropylene	V	6	30	10-25	6-15
EPDM wire covering cpd.	V		100-1000	200-1000	200-600
Chlorosulfonated PE (CSM)			200-600	650-1000	450-1000
Polystyrene	V	0.05	10 000	590	560
SAN	V		2000		
SB	V	0.2	2000		550
Polyvinylchloride PVC-U	V	10-30	9000		
PVC-P (wire covering cpd.)	V		2000	(135)	(74)
Polytetrafluoroethylene	A			1.4-4.0	1.4-4.0
Polyfluorochloroethylene	A		400		
Polymethylmethacrylate	A	3	200		
Polyoxymethylene	A	9	26	26	
Polyamides	V	2	140-430	85	47
Polycarbonate	A		500		
Polyethylene terephthalate	V	0.4	2100	750	400
Cellulose esters	A	2-3	120-200	160	

Thermosets[2]	Type		at 13 Gy/h
PF, 47% wood flour	31-1518	5 000	2500
PF, 55% asbestos fibers	15-1309	30 000	>1000
MF, 40% asbestos fibers	158	30 000	>1000
UP, 15% GF+56% inorg.	L 1405	>30 000	>1000
EP, cast resin		25 000	
EP, 50% magnesium oxide		30 000	
EP, 50% aluminum oxide		50 000	

Polyimides	not damaged until 40-100 M Gy[3]
Polyamidoazo pyrrolone	not damaged until 100 M Gy[3]
Polyarylether ketone	not damaged until 100 M Gy[3]

[1] Half value dose, elongation at break
[2] Half value dose, flexural strength
[3] Half value dose, from company brochures

parent plastics (ASTM D 673, 1044) is determined by instruments measuring gloss or haze or by an integrating photometer.

Of plastics in the elastic modulus range $10-10^4\,\text{N/mm}^2$, those with a low E-modulus (e.g. elastomers, PE, PA), are more resistant to all types of wear than those with a medium or high E-modulus (e.g., PMMA, PC or thermoset products). The resistance of PMMA and PC to scratching is improved many times by a very hard thin layer of siloxane (Sec. 4.2.6.1 and 4.2.8.1).

Low values of the ratio of frictional force to stress, characterized by the coefficient of friction μ, are advantageous, resulting in good abrasion and slide behavior. PTFE, PA, POM are wear-resistant bearing materials that require no lubricants. Bearings with higher loadings, made of industrial laminates, require lubrication.

Almost equal values of the static or starting coefficient μ_s and the dynamic kinetic or sliding coefficient of friction μ_k prevent the "stickslip" effect. ASTM D 3028 is a standard test method for μ_k of plastics sliding against other substances using a frictiometer with one specimen fixed and the other rotating. ASTM D 1894-DIN 53375 are methods for the measurement of the values of plastic films in relation to one another, one of the important determinants for their processing behavior in the packaging sector.

7.5. Migration and Permeation

7.5.1. General Remarks

Absorption, solubility and diffusion of liquids, volatile substances and gases in polymers and at polymer interfaces are significant processes even when they do not (see Sec. 7.4.5) cause marked changes in the overall behavior of a plastic. Test methods indicate the danger of the *bleeding* of colors from contiguous materials (ISO 183-DIN 53415) and the *migration of plasticizers* (ISO 177-DIN 53405) between plastic films with surfaces in contact under pressure at elevated temperature. The volatility of plasticizers and other additives is measured by the activated carbon absorption test (ASTM D 1203-ISO 176-DIN 53407).

In the migration of a substance across a plastics wall (*permeation*), solubility and diffusion work together in a physically complex way. "Weeping" (or exudation or sweating) of contaminating or dangerous contents of containers or pipes, penetration through plastic liners, evaporation of fuels from tank walls, and the absorption of atmospheric oxygen into the through-put in plastics piping, which can in turn bring about corrosion of metallic elements, are processes that require specifically developed test methods and sometimes simulation by storage for long periods.

The transmission rate of permanent gases and water vapor through films and sheets is the single measurement of most importance for packaging technology. In the measuring apparatus a stream of gas or water vapor at a given partial pressure is directed at one side of a fixed membrane. With gases the permeated quantity is volumetrically or manometrically measured by means of a calibrated capillary system. With water vapor it can also be measured gravimetrically by the weight gain of absorption media such as calcium chloride or silica gel. From the reduction in surface area, thickness and pressure the *permeability coefficient* can be calculated for gases and water vapor at specified humidity conditions. These figures can be used for the comparison of different materials (see Table 4.58). The permeability increases exponentially with temperature. The values for the gas transmission rate (GTR) and the water vapor transmission (WVT) are sensitive to the measuring conditions as well as the nature, condition and structure of the surface. Results for a given polymer from various tests can vary widely. Average values of permeability coefficients in the technical literature (Table 4.58) offer a qualitative comparative scale. For critical applications individual tests can be carried out.

7.5.2. Gas Transmission

Gas transmission is defined (ASTM D 1434-ISO 2556-DIN 53380) as the gas volume reduced to normal conditions which diffuses through $1 m^2$ of the tested material at a given temperature and pressure. The definition in terms of SI units is: 1 GTR = $1 cm^3/24 h\, m^2$ bar, where 1 bar = 0.987 atm. The permeability coefficient P_g in $10^{-12} cm^3$ (NTP) cm s mbar (= $10^{-3} m^2 s^{-1} Pa^{-1}$) is given by the equation

$$P_g = \frac{GTR \times \text{thickness in } \mu m}{8640},$$

or

$$GTR = \frac{8640 P_g}{\text{thickness in } \mu m}$$

Older values derived from the pressure difference in cmHg instead of mbar should be multiplied by 0.0752 to arrive at SI units.
U.S. units are $P_{Barrer} = 10^{-10} cm^3$ (STP)·cm·s·cmHg or permeability in $in^3 \cdot mil/24 h \cdot 100 in^2 \cdot in^2 \cdot atm = 7.4 \cdot 10^{13}$ cm·s·mbar also with cm^3 instead of in^3 ($1 cm^3 = 0.061 in^3$) in the above unit. Basically the P_g values obtained by all the current standard methods with $25 \mu m$ to $50 \mu m$ (1–2 mil) thick films can be compared and can be used for the calculation of the transmission of gases through homogeneous, though not laminated, films and sheets of different thicknesses for given times and surface areas.

7.5.3. Water Vapor Transmission

According to ASTM E 96-ISO 1195-DIN 53122 water vapor transmission is the weight of water vapor diffused through the tested product in unit time at given humidity and temperature. Measuring methods and the atmospheric conditions specified vary between the different standards. Thickness of the material and water vapor transmission are inversely proportional over a limited range. Values measured under different conditions are not generally comparable; for different thicknesses, values can be calculated only to a limited degree.

According to ISO the water vapor transmission q_{wv} is determined gravimetrically with a 40 μm thick sheet in an atmosphere at 23°C/85% RH. The units are g/m^2 24 h.

The permeability coefficient for water vapor is expressed as

$$P_{wv} = \frac{q_{wv} \text{ sheet thickness in } \mu \text{m}}{47.6} \text{ in } 109 \text{ g/cm h mbar}$$

Older forms with values derived from pressure difference in torr instead of mbar should be multiplied by 0.752.

In the US the surface-derived moisture-vapor transmission rate (MVTR) values are often given in g·mil/100 in^2 24 h at 23°C and 37.8°C.

For *vapor barriers* in building construction, the dimensionless number μ indicates how many times higher the vapor transmission resistance of the material is than that of air of the same thickness.

7.6. Mechanical Properties

Standards for testing mechanical properties are given in Table 7.10.

7.6.1. Specimen preparation

Specimens for mechanical testing are prepared according to standardised methods by compression, transfer or injection molding or fabricated from sheets, films and tubes or larger moldings whose properties must be determined. Their surfaces and edges must be so finished that measurements are not invalidated by notch effects. Strength values of injection-molded specimens in the form of a bar can, as a result of orientation stresses, be significantly higher than those of the material pressed in its basic stress-free condition or of an injection-molded test piece that has been thermally freed of stress (ISO 2557, DIN 16770).

7.6.1. Specimen preparation

Table 7.10. Standards for Mechanical Testing

Method	ASTM	ISO	DIN	Remarks
Preparation of specimens				
Injection and compression molding thermoplastics	D 1897, D 1928	293, 294, 2557	16770	ISO 2557, DIN 16770: amorphous thermoplastics specimens with a designed level of shrinkage (for semi-crystallines in development)
Compression molding thermosets	D 796, D 956, D 1896	295	53451, 53457	Transfer injection: ASTM D 1896, D 3881
Machining specimens		2818		
Multi-purpose specimens		3167		
Stress-strain, including creep properties[2]				
Tensile properties	D 638[6]	527	53455	Rigid plastics, for GRP: D 3093, EN 61[3]
	D 2990	899	53444	Tensile creep
	D 2991		53441	Stress relaxation test
	D 882	1184	53445	Thin plastic sheeting and film
Flexural properties	D 790[6]	178	53452	Rigid plastics, for GRP: EN 63
	D 2990	DP 6602		D 2990: tensile, flexural, compression creep
Compression properties	D 695	604	53454	Test specimen prism or cylinder, slenderness ratio 10 to 15
Modulus of elasticity[4]	D 638, D 790, D 695	527, 178, 604	53457	ASTM, ISO: included in the general standards
Fatigue resistance alternating stress tests				
Flexural fatigue test	D 671		53442	Flat specimens, constant amplitude of force (D 671) or strain (53442)
Special tests for GRP			53398	ASTM alternating tensile stress, DIN bending pulsation
Impact resistance				
Izod flexural impact test	D 256	180		Cantilever beam, method A, C, D swing on the same face of the notch, E reversed notch test
Charpy flexural impact test	D 256, method B	179	53453	Horizontal beam, unnotched or notched (D 256 B only notched), impact line midway, directly opposite the notch
Charpy tensile impact test	D 1822[6]		53448	See text, results non-comparable
Multi-axial falling weight (tup) test	D 3029 D 1709 (PE film), D 2444 (pipes)	6603 part 1	53443, part 1	Failure signified by any crack or split
Impact penetration with electronic data recording		6603, part 2	53443, part 2	DIN 53373 for films and thin sheeting
Hardness tests[5]				
Ball indentation hardness	–	2039	53456	Rigid plastics, see text
Rockwell hardness	D 785	2039	–	Rigid plastics, see text
Shore durometer hardness	D 2240	868	53505	Scales A and D, see text
Barcol impression test	D 2583	EN 59[2]	EN 59[2]	Portable impressor used to control the curing degree of GRP-parts

[1] For testing cellular products see Table 4.80, p.400 [2] For environmental stress cracking see p. 454. [3] EN – European Standard, adapted also by DIN. [4] For semi-rigid plastics (PE, Table 4.5, p.182), the shear modulus (DIN 53455, p.439) is sometimes used as a guide value. [5] Hardness tests measuring impression after removal of the load, like methods Brinell (similar to Rockwell), Knoop (ASTM D 1474), Vickers (DIN 50133), both using pyramidal indentors, are seldom used for plastics on account of the effect of plastics elastic recovery. [6] Also M-editions for Metric system.

7.6.2. Stress-Strain Properties under Static Load

7.6.2.1. Short-term tests

The *tensile test* can be applied to all plastics materials to determine their stress-strain behavior. It is therefore one of the most important mechanical tests.

Data from stress deformation tests are determined on various shapes of standardized test specimens with stretching speeds of 1–500 mm/min. These are so chosen that the characteristic values can be found in about one minute of test time. The general shape of the stress-strain diagram from the tensile strength test is shown in Fig. 7.6. It is plotted for the *nominal* tensile stress, that is, the tensile load (W) per unit area of minimum original cross section (A_o) of the test specimen, $\sigma = W/A_o$. The "yield point" stress σ_s at which, through subsequent elongation with necking, the increase of the σ/E curve becomes nil for the first time, is the significant value for partially crystalline materials. With hard, relatively brittle products and also with rubbery products there is no yield point, and the ultimate tensile stress at maximum load is the tensile strength at break (σ_B). There can be little difference between σ_B and the yield rupture strength (σ_R) of semi-rigid or nonrigid plastics. Figs. 4.11 and 7.7 give examples of various curve forms. The slope of the curves increases with the speed of stretching. Figure 7.8 gives examples of the dependence of the strength of thermoplastics on temperature.

The *modulus of elasticity* in tension is conventionally determined as the slope of the tangent to the σ/E curve at the beginning at a standardized rate of strain,

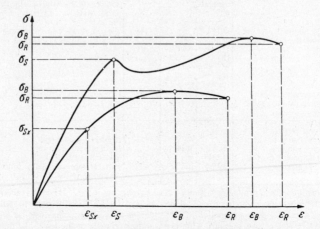

Fig. 7.6. Stress-strain diagram from tensile test. $\sigma_{Sx} = x\%$-extension stress, σ_s = yield stress, σ_B = tensile strength, σ_R = breaking strength, ε_{Sx} = extension at $x\%$ tensile stress, ε_S = extension at yield stress, ε_B = extension at highest force, ε_R = elongation at break.

7.6.2.1. Short-term tests

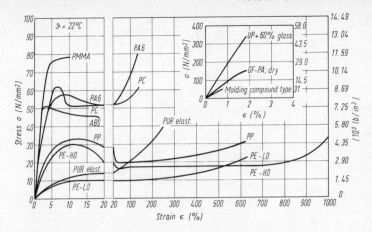

Fig. 7.7. Examples of stress-strain curves from the tensile test.
In considering the run of curves it should be noted that the ordinates of the left hand figure, for reasons of clarity, are plotted at five times the scale of the following figure.

e.g., 1 mm/min. The secant modulus is the ratio of nominal stress to corresponding strain at any specified point on the stress-strain curve. The values of the stress and strain must be quoted along with the secant modulus. Guide values based on tensile test methods of different standards are comparable.

In the *flexural test* for rigid plastics with the usual three-point bending system, the specimen is placed on two supports and the load is applied in the center. Loading at the point of failure when related to the specimen cross section and length between the supports is the *flexural strength at break*. Plastics that are not extremely brittle do not break by deflection at strains of < 5% in the outer fiber. Rather, they show a yield point in the stress-strain diagram. This is recorded as the *flexural yield strength*. The method is not applicable to semi-rigid plastics (e.g., PE-LD), which neither break nor show a yield point in the 5% strain range.

Some methods, e.g., the "Dynstat" (DIN 51230/53435) for small specimens and the flexural modulus method of DIN 53457, use four-point bending systems. The values measured by these systems can be up to 30% lower than those measured by the three-point systems. Therefore flexural moduli measured by ASTM/ISO and the specified DIN methods are not comparable, but flexural strengths measured by three-point bending systems can be compared quite well.

The *short-term compression test* is of little importance for plastics.

7.6. Mechanical Properties

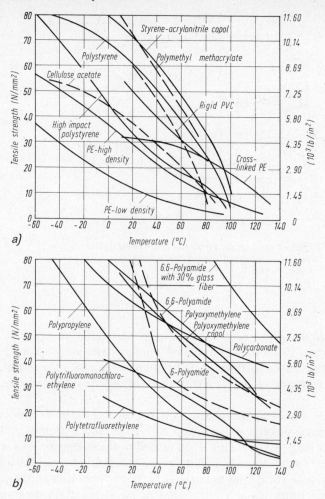

Fig. 7.8. Tensile strength of thermoplastics in short term tests in the range −60°C to 140°C.

7.6.2.2. Creep rupture tests

In the creep rupture test, the test pieces are stressed along a single axis, mainly by weight, in a constant test atmosphere by forces that remain steady during the period of the test. The time-extension curves, $\varepsilon(t, T)$ at several fixed stresses and at a given temperature, as a function of log (time) are designated "*creep curves*". These can be presented as a diagram showing the strain for specified $\sigma_{\varepsilon/t}$ conditions and as isochronous stress-strain curves with the time of stres-

7.6.2.2. Creep rupture tests

sing as the parameter (Fig. 7.9). Figures 7.10a and b are examples of low-strain isochronous stress-strain diagrams with multiple ordinates for different temperatures. The lower limits of the shaded-in areas indicate the $\varepsilon_{F\infty}$ offset yield strains for all temperatures (see paragraph 7.7). From the time after which the test specimen breaks at an applied tension the creep strength $\sigma_{B/t}$, reducing with time, is determined (Fig. 7.11). Complete σ/t diagrams with stress-strain curves from below offset yield strain to break (Fig. 7.12) provide all the data needed for designing for strength at static load and a defined temperature. An important derived quantity for design calculations is the creep modulus ($E_c(t, T) = \sigma/\varepsilon(t, T)$) (Table 7.11). With small creep stresses the stress dependence, in contrast to time dependence, can be neglected. Figures 4.2, 4.13 and 7.13 illustrate creep-rupture tests at normal and at elevated temperatures.

Creep tests for important engineering plastics have been carried out at various temperatures for up to 10^5 h (>1 decade). The data obtained from continuous runs of creep curves on log-time base may, for moderate stresses, be extrapolated by two orders of magnitude of time. Tests of more than 10^3–10^4 h carried out at room temperature, i.e., up to about one year, therefore, make it possible to predict for decades ahead. On the basis of the well-known temperature dependence of creep it is possible to draw similar far-reaching conclusions from shortened creep tests at higher temperatures. *Stress-relaxation* tests are seldom carried out because of their cost. The relaxation modulus ($E_r(t) - \sigma(t)\varepsilon$) differs from the creep modulus only at high stresses. Measurements on creep under multi-directional stress have existed for some

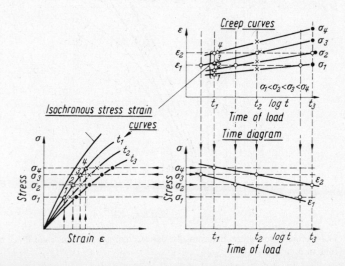

Fig.7.9. Relationship between creep curves and isochronous stress-strain curves.

464 7.6. Mechanical Properties

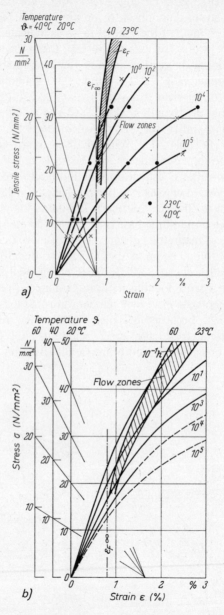

Fig.7.10. Isochronous stress-strain diagrams for various temperatures. (a) rigid PVC, b) PMMA. Parameter time in hours.

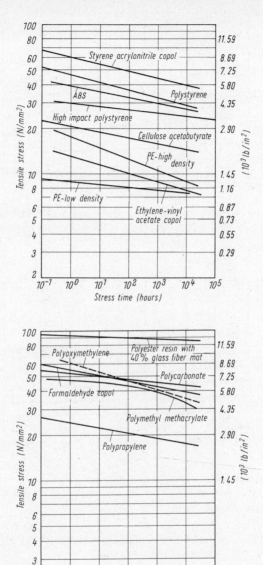

Fig.7.11. Creep rupture strength $\sigma_{B/t}$ of thermoplastics and UP-GF.

7.6. Mechanical Properties

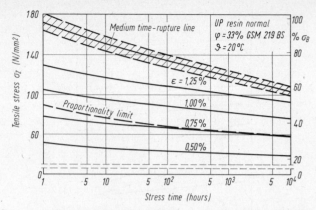

Fig.7.12a. Creep extension stresses and creep rupture strength of GF-UP mat laminates (glass content 33% v/v = 50% w/w) at 20°C. The origin of the ordinates is suppressed.

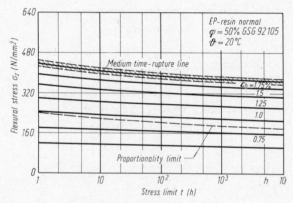

Fig.7.12b. Creep extension stresses and creep rupture strength of EP-GF-textile laminates (glass content 50% v/v = 65% w/w) at 20°C. The scale is different from Fig. 7.12a.

decades from investigations on plastic pipes under internal pressure and these provide the basis for pipe standards (Fig. 4.14).

7.6.3. Long-Term Behavior under Dynamic Stress (Fatigue Tests)

The failure of materials under permanent dynamic stress by mechanical vibration is called *material fatigue*. It can be of striking importance when the material is used in rotating or reciprocating motion and also in stationary but vibrating plastics parts, as for example in vehicles, aircraft and other

7.6.3. Long-Term Behavior under Dynamic Stress (Fatigue Tests)

Table 7.11. Creep Moduli for Thermoplastics

		Test method*	Temperature °C	Test-stress N/mm²	Creep modulus in N/mm² for				
					1 h	100 h	1000 h	10 000 h
PE-HD	High density PE	T	20	5	490	250	180	140
			40	2	330	220	180	140
PP	Polypropylene MFI 230/5 = 1.5	T	22	5	910	590	440	350
			65	2	500	395	300	250
PS	Polystyrene	F	20	10	3100	2400		
SB	−Impact strength a_{n0} = 12-22 kJ/m²	F	20	10	2700	2500		
ABS	−Medium impact strength	T	20	15	3300	2500	1900	800
	−High impact strength	T	20	15	1500	840	600	(370)
PVC	Polyvinylchloride, rigid PVC-S, K ~ 60	T	20	10	3300	3100	2800	1900
			45	10	2200	1200	900	700
PMMA	Polymethylmethacrylate	T	20	10	3200	2900	2500	2000
PA 6	Polyamide 6, moist atmosphere	T	20	6	600	400	300	220
	reinforced with 25% glass fiber	T	23	20	4100	3700	3300	3100
PA 66	Polyamide 66, moist atmosphere	T	20	9	1400	1050	760	500
POM	Acetal copol.	T	20	20	2300	1500	1200	
PPO	Polyphenylene oxide, modified	T	23	21	2300	1800	1700	990
			60	21	2100	1300	1200	
PC	Polycarbonate	T	20	20	2000	1800	1650	1500
			60	21	1800	1300	1200	1000
PET	Polyethylene terephthalate, crystalline	F	20	10	3500	3250	2900	2400
			60	10	3000	1700	1300	1200

* T = tensile modulus, F = flexural modulus.

7.6. Mechanical Properties

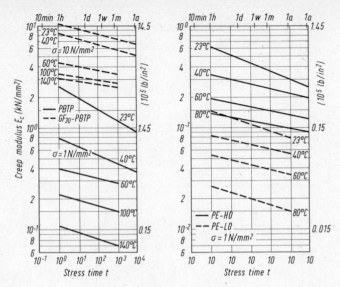

Fig. 7.13. Creep moduli diagrams (PBT, GF$_{30}$-PBT, PE) at various temperatures.

equipment. The range of stress (Fig. 7.14) corresponding to the doubled stress amplitude, characterized as the dynamic load, can be superimposed on the main stress static creep load. Although of great practical importance it is difficult to estimate the combined effect of creep and fatigue (Fig 116b, c, d) by testing and calculation. The simplest case suitable for comparative testing is the "alternating stress load" produced by oscillations with amplitude $\pm\sigma_a(\pm S_a)$ about a main stress of zero (Fig. 7.14a).

For fatigue testing, machines producing sinusoidal flexural oscillations, 1 to 30 Hz are used. In test runs the results are dependent on the alternating stress amplitude and the number of cycles and are brought together in S-N (stress number) or Wöhler curves (Fig. 7.15).

Insofar as the fatigue strength does not tend asymptotically toward a limiting value with a large number of cycles, as it does for example with reinforced thermosets, those that sustain 10^7 cycles of alternating stress without break (or an agreed failure limit) have that stress reported as the fatigue limit.

As a result of the viscoelastic behavior of the polymer matrix of plastics, the response to dynamic stressing is determined not only by crack propagation up to the break but also by the transformation of part of the generated mechanical energy into heat (hysteresis heat). Material- and temperature-dependent mechanical loss factors of plastics can have striking effects on behavior through increased softening resulting from rising temperature with oscillating

7.6.3. Long-Term Behavior under Dynamic Stress (Fatigue Tests) 469

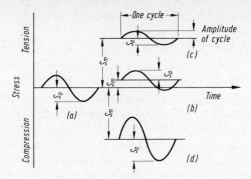

Fig.7.14. Stress-time curves for different values of mean stress (to ASTM D 671). Definitions and terms:

S_{max} and S_{min}: The stresses having the highest and the lowest algebraic values in the stress cycle, tensile stress considered positive, compressive stress negative

Mean stress or steady component of stress:
$$S_m = (S_{max} + S_{min})/2$$
Range of stress: $\quad S_r = S_{max} - S_{min}$
Stress amplitude or variable component of stress:
$$S_a = (S_{max} - S_{min})/2$$
a) Alternating stresses load (Completely reversed stress)
$$S_m = 0 : S_{max} = +S_a, S_{min} = -S_a$$
b) Pulsating stresses load (tension)
$$S_m > 0 : S_{max} = S_m + S_a, S_{min} > S_m$$
c) Pulsating stresses load (tension)
$$S \gg 0 : S_{max} = S_m + S_a, S_{min} = S_m - S_a$$
d) Pulsating stresses load (compression)
$$S_m \ll 0 : S_{max} = S_a - S_m, S_{min} = -(S_a + S_m)$$

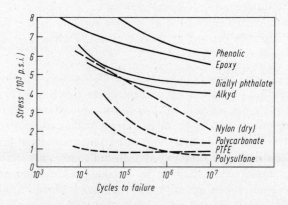

Fig. 7.15. S-N curves of mineral- and glass-filled thermosets and unfilled thermoplastics. Stress is cantilever bending under constant load. Frequency is 18000 cycles/min. Mean stress equals zero.

frequency. Mechanical loss factors of thermoplastics (d in Fig. 1.4) are 0.01 to > 0.1, cf. those of metals about 0.0001).

The nature of the Wöhler (S-N) curves is further influenced by the test conditions in such a way that with peak load held constant the strain amplitude increases; with constant strain amplitude the initial stress reduces. Fatigue tests are only quantitatively useful in strength calculations in as far as their test parameters corresponding to actual requirements are considered and established. For some thermoplastics known ratios between main stress zero fatigue limits and yield tension (Sec. 7.6.2.1) of $\frac{1}{2}$ to $\frac{1}{4}$ can be used as reducing factors for dynamic stresses (Sec. 7.7).

The standard test method for flexural fatigue of plastics ASTM D 671 prescribes alternating stress load conditions (main stress zero), constant amplitude of force and a frequency of 30 Hz (1800 cycles/min). The flexural fatigue test to DIN 53442, with a constant deformation amplitude with time, can be carried out with a variety of oscillating frequencies under alternating stressing.

7.6.4. Impact Resistance Tests

Standards are listed in Table 7.10.

The absorption of impact energy initiates some very complex stress-strain relationships with stress-strain rates $> 10^4$ times higher than those under static load. Elastic deformation or yield, crazing, cracks or splits and further failure propagation to separating or flying broken fragments can follow one another directly after impact.

The type and progress of impact reactions depend among other things on:

> The brittleness of the plastic, which rises with the rate of impact and diminishes with increasing temperature.
> The ability to partly escape the impact by bending of the wall of a thin-walled product or by dissipating the energy of the impact throughout a large volume.
> The nature and location of impact initiation.

Design features that produce stress concentrations (see Fig. 3.13) such as sharp corners, sudden changes of wall thickness, notches and even textured surfaces can cause dangerously impact-sensitive spots. For the service life of a product, initial impact failure points are more important than their propagation to rupture. The traditional *flexural impact tests* indicating only the loss of kinetic energy of a pendulum-type hammer by breaking a specimen under standardized conditions cannot be quantitatively evaluated. However, when properly interpreted, they can be useful, e.g., as screening tests between different grades of the same generic plastic types. They can also be used to characterize impact resistance vs temperature variations. A comparison of *notched and unnotched*

7.6.4. Impact Resistance Tests 471

resistance values of a given plastic gives a hint of its notch sensitivity. But the shape of notching has different effects in different plastics (Fig. 7.16), and the notch impact resistance increases considerably with the depth of the specimen remaining under the notch.

In the conventional *Izod method A* (C for very brittle materials) of ASTM D 256 a rectangular bar mounted as a cantilever is struck by the pendulum-type hammer near the upper end (Fig. 7.17a). The bar is notched with a radius of

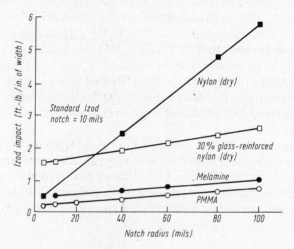

Fig. 7.16. Izod impact vs. notch tip radius.

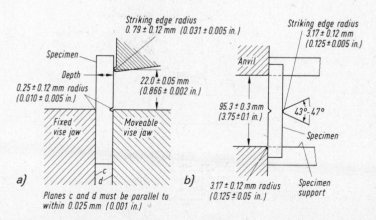

Fig. 7.17. Relationship of vice, specimen, and striking edge to each other; (a) for Izod test methods A and C, (b) for Charpy test method.

7.6. Mechanical Properties

curvature of 0.25 mm ±0.05 mm (0.010 ± 0.002 in). In method D, the notch radius is varied and in E the specimen has a reversed notch to test notch sensitivities. Method B uses standard specimens for the Charpy test.

In the *Charpy method* the specimen is loosely laid horizontally on supports holding it at both ends and then broken by the pendulum swinging against the middle (Fig. 7.17b). From the difference of the impact resistances of a notched (a_k) and unnotched bar (a_n), $a_{\rm rel} = a_k/a_n \times 100\%$ is evaluated as a measure of relative notch sensitivity. Specimens for ASTM D 256 have a general depth of 12.7 mm (0.5 in), and of 10.2 mm at the notch but they have variable widths for particular plastics. Izod-values are given as the average pendulum energy used per unit length of notch in ft lb/in or J/m. In the preferred A 1 Izod method (specimen depth 10 mm, at notch 8 mm) of ISO 180, results are calculated as impact energy per cross-section of the specimen at notch, expressed in kJ/m². The same calculation is applied for the DIN/ISO Charpy impact resistance, however the notch is only about 3 mm in depth. Values determined by the different methods (partly expressed in different units!) cannot be compared quantitatively.

In *tensile impact methods* using a pendulum (Fig. 7.18) the impact is applied to a microtensile specimen. Stress-strain rate, elongation at break and permanent extension can be measured in this way. This method is used for plastics products which do not break, e.g. packaging films, and therefore cannot be tested by flexural impact methods if using notched specimens. The high-speed stress-

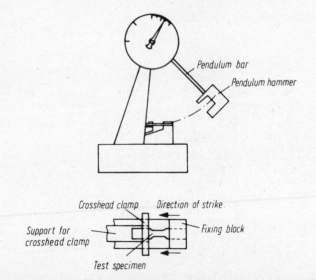

Fig. 7.18. Pendulum hammer equipment set up for the Charpy tensile impact strength test.

strain test ASTM D 2289 makes measurement of ductile-brittle transitions with increasing deformation speed feasible.

The initial failure weight or corresponding failure energy can be determined by *falling weight tests* in which a weighted "tup" with a hemispherical or conical nose is dropped onto flat or curved specimens in defined weight/height ratios. The "main failure energy" required to produce 50% failures is evaluated statistically from about 30 to 40 tests under systematically varied test conditions. Visible initial failure marks are used as criteria. These include crazing, cracking, splitting or (marked) buckling as appropriate.

In the *impact penetration test* a hemispherical penetration head is forced perpendicularly through a clamped specimen at a constant high speed. The force-deformation diagram indicates the damaging and total penetration energy. The falling dart and penetration tests give quite realistic quantitative information for R and D but they are too complicated for routine quality control.

7.6.5. Hardness Tests

Standards are listed in Table 7.10.

The resistance of a solid body to local surface indentation is related to the modulus of elasticity, but as a result of the viscoelastic behavior of plastics conventional hardness tests only give approximately comparable data, depending on the method applied. In the *ball indentation hardness test according to ISO and DIN* the indentation depth of a 5 mm diameter steel ball into a 4 mm thick specimen, measured under load to avoid edge effects, should be between 0.15 and 0.35 mm. One of four different testing forces may be applied preferably for 30 s but alternatively for 10 or 60 s. The ball indentation hardness testing force/surface area under load H in N/mm^2 is indexed by force and time applied, e.g., $H_{132/10}$ means H measured with a force of 132 N during 10 seconds.

Rockwell hardness A numbers generally record the difference between the depth of the impression made by a steel ball under a minor load of 10 kg and a major load of 60 or 100 kg applied for 15 s, with the minor load still operating thereafter, is registered by scales. Each division represents 0.002 mm (0.0008 in) of vertical movement. To differentiate between ranges of hardness there is a progressive series of scales R, L, M, E and K, K being the hardest.

The Shore A and D hardness tests are empirical methods for semi-rigid and elastomeric plastics. They are based on the specialized "Durometer" apparatus with a calibrated spring for applying force to a conically shaped indentor. For the A hardness series the indentor ends with a flat cross-sectional cut, while for D it is more sharply pointed. For the approximate relation between Shore A and D see Table 4.26. For a comparison with ball indentation hardness see Table 7.12.

7.7. Some Formulas for the Calculation of Strength

Table 7.12. Plastics Hardness Test Methods

| Method | | | Examples of material ranges |
Shore A	Shore D	Ball indentation hardness N/mm²	
40–80			Soft to medium soft rubber, very soft flexible PVC
80–90	30–38 40–64 74–90	5.0–8 10–40 60–140	Leather-like material, PVC-P up to Shore A 98, PE-LD, PTFE, CA Indentation hardness range: 50–80: PE-HD, PP, PA, CAB, 100–140: rigid thermoplastics
		140–>200	PS, PMMA, GRP, thermosets

In the Brinell hardness test for metals the area of the indentation of a ball after load is measured. Comparable values: Al, Al alloys 150–1000 N/mm², cast iron, steel 1500–2000 N/mm²

7.7. Some Formulas for the Calculation of Strength

Plastics parts can be designed using strength and elasticity theory if, as for the stressing of metals at high temperatures, the time and temperature dependence of the strength characteristics are taken into account. The equation for the important strength value obtained from creep studies

$$\sigma = E_c(t, T) \cdot \varepsilon(t, T)$$

is generally sufficiently accurate so that data obtained from creep diagrams can be used to design for limits of stress, deformation and life span. Temporary stresses have to be included in the calculations. With adequate recovery intervals between repeated stresses, conditions can be more favorable. Frequently repeated dynamic alternating stressing, requires special factors (Sec. 7.6.3).

With temperature change stressing of plastics parts, the heat-induced movement which is limited by linking or pairing with other materials, must also be considered in the appropriate temperature-tensile-compression stress calculations. For calculations relating to containers and equipment made of rigid PE, PP and PVC, taking into account the graduated safety factor values according to the nature of the stress and also the influence of shape, the "resistance" factors for particular chemicals (Fig. 4.15, and Sec. 7.4.5) and long-term welding factors 0.8 to 0.4 should also be considered.

If only short-term strength values ($\sigma_{B/O}$) for room temperature are available, then it is possible to estimate the failure through break on the basis of general values.

7.7. Some Formulas for the Calculation of Strength

In the equation for the design stress σ_{DS} that can be applied for time t

$$\sigma_{DS} = \sigma_{B/O} \frac{\alpha_t \cdot \alpha_T \cdot \alpha_k \cdot \alpha_u}{S} = \sigma_{B/O} \cdot \frac{A}{S}$$

where the valid reducing factors are

α_t (influence of time), the creep modulus equation

$\dfrac{E_{cxh/20°C}}{Eo}$	x	
	10^4 h	10^5 h
for thermoplastics	0.4	0.35
for GRP mat laminates	0.65	0.59

α_T (influence of temperature)
for thermoplastics (below the transition temperature ranges to be derived from the shear modulus-temperature curves) 0.7, for GRP 0.8 for every 20°C rise in the temperature: thus, for example, 60°C $0.7^2 \sim 0.5$, or $0.8^2 \sim 0.6$.

α_k (notch and influence of mode of preparation, e.g., through orientation) and α_u (environmental effects) must be considered in each case. In such calculations for 10^5 h, general experience shows that $A \approx 0.33$–0.25. The safety factor taken for steady stress is $S \geq 1.5$–2.0 and for alternating stress is ≥ 2. Calculations including terms for reduction and safety factors lead generally to the requirement of the mathematical outcome of an overall safety factor of between 6 and 10. In estimating the load-bearing capacity of a plastic part by this method, the anticipated extensions are usually below critical deformation limits.

Calculation of critical deformation limits is more exact. The flow-limiting extension ε_{Ft}, below which no direct or microscopic changes of the structure of the material, such as stress crazing or stress whitening will appear, tends to a limiting value that is essentially independent of temperature stresses and environmental influences. Values of $\varepsilon_{F\infty}$ are 0.8% for rigid amorphous thermoplastics, 1.5 to 4.0% for partially crystalline thermoplastics and 0.4 to 0.6% for GRP-mat laminates.

With

$$\sigma_{DS} = \frac{\varepsilon_{F\infty} \cdot E_c(t,T)}{S} \cdot A$$

the calculation can be performed if E_c is derived from creep modulus data or if $\varepsilon_{F\infty} \cdot E_c$ is read off directly (for corresponding temperature) from isochronous σ-ε diagrams (see Fig. 7.9). Only the manufacturing and shape influences need to be considered as reducing factors. The known material-independent shape factor can be directly utilized for the notch factor A. The calculation is also applicable for multi-directional stressing, provided that the biggest strain manifested in the plastic part does not exceed the permissible extension. The Poisson ratio of plastics, μ, can be usefully approximated as 0.35. For calculations on instability failure, the Euler crack and buckling formulas can be applied if $E = E_c(\sigma,t,T)$ is included in them. As a safety factor $S \geq 2$ is generally required.

The theoretically derived formulas for strength computation of isotropic polymer materials based on the viscoelastic behavior of homogeneous materials are valid also for inhomogeneous products if the inhomogeneities are small compared with the external dimensions; that is possible for non-directionally filled or reinforced plastics such as mat laminates for stresses parallel with the surfaces, and also for rigid foamed plastics.

For the designed exploitation of the integral sandwich principle for composite products and for the anisotropic strength distribution in directionally reinforced and uni-axial oriented products, special equations are required.

7.8. Computer Aided Design and Manufacturing (CAD/CAM)

The conventional formulas discussed in the previous section for calculating the strength using reducing factors and taking into account the estimated viscoelastic behavior of the polymers, are at best adequate as a basis for the design of simple moldings that are not loaded to their yield limits. The technically and economically optimum exploitation of material characteristics for more complicated moldings requires more powerful computing techniques, such as the application of empirical rules for frequently recurring functional elements (socalled "macros"), the computing of box-shaped parts by superimposition of the computations for base and walls. For problems that have no analytically closed solution, the finite elements method (FEM), an approximation for the analysis of complex three-dimensional geometries in a multi-element network from the node points of connected individual elements, is used. Such computations are impossible without the use of a computer.

The core of *computer modular systems for moldings* is the materials data base which can be directly accessed by the CAD/CAE module of the system to obtain the necessary material characteristic values and functions. With program packages for rheological, thermal and mechanical design, the computation can be linked to that for the injection mold, with the results output as dimensioned design drawings and/or three-dimensional graphical representations of the molding (e.g. Cadmould, IKV; Moldflow, Boronia, Vic./Australia; TMC,

7.8. Computer Aided Design and Manufacturing (CAD/CAM)

Plastic 8, Computer Inst., Milano/Italy; STRIM 100, Cisigraphe - Le-Griffon-536, Route de la Seds, 13127 Vitroless, FR). The data thus obtained, coupled with a CAD standard tools data system are then used for computer controlled mold design.

The same set of data can be used in communication with NC computers for mold manufacture, as well as for generating tool setting data for control of the manufacturing cell (Sec. 3.3.2.6). Extensive networking of computer systems via numerous interfaces leads toward *Computer Integrated Manufacturing* (CIM), which will lead to extensive changes in factory structures, above all in the field of technical plastics applications. The development aims include the integration into the computer network of data exchange between commissioning company and supplier, also on the fields of marketing and quality control.

8. TRADE NAMES FOR PLASTICS AS RAW MATERIALS AND SEMIFINISHED GOODS

8.1. Characteristics of the Register

The following worldwide register combines the commercial names with the name and address of individual suppliers and technical information about the product. It provides comprehensive coverage of trade names for plastics raw materials, molding compounds and important semi-finished products that are available on the market. Obsolete trade and manufacturers' names and trademarks have been omitted.

The list does not include trade names of additives (except for some important plasticizers). Trade names of additives may be found in: Gächter, Müller, "Plastics Additives" (Hanser 1993). Also excluded are trade names of finished products or those merely involving particular features.

The register is based on the "Handelsnamenverzeichnis", which forms part of the author's well-known *Kunststoff-Taschenbuch*. It has been brought up to date by means of a worldwide questionnaire, replied to by all the plastics manufacturers named in that book, as well as by information from firms and general trade literature, including advertisements from different countries, up to the middle of 1995. When commercial names have been registered as trademarks nationally or internationally, the names have been supplemented by ®. However, neither the author nor the publisher has been able to ensure the completeness of these data and cannot be held responsible for the omission of any trademark.

8.2. Arrangement of the Register

There follows after every name the approriate firm and two numbered groups: the first gives the chemical composition of the basic polymer, p. 481 to p. 483; the second indicates the form of supply of the plastic material, p. 484 to p. 486. The arrangement of the following lists is exclusively based on a practical standpoint. The use of the large groups of numbers may appear at first glance to be somewhat complex, but it does make it possible to meet the needs of precise or general description.

Example:

4224		21	
4	Natural product derivative	2	Molding compounds
42	Cellulose derivative	1	Without filler
422	Cellulose ester		
4224	Cellulose acetobutyrate		

Signifies: Cellulose acetobutyrate molding compounds

242		63	
2	Product with cross-linked mixed polymer structure	6	Reinforced rigid plastic
24	Polyester	63	Glass fiber reinforced lighting panels (flat and corrugated)
242	Unsaturated polyester with styrene, or similar, to further polymerize		

Signifies: GRP sheets or webs bound by cured polyester resin crosslinked by polymerization.

       21   113,2

Signifies: The name is generally used for technical phenolic resins (in the broader sense) and for phenolic resin molding compounds.

Insofar as the designation contains only general statements regarding resin and form of delivery it is not to be assumed that the given designation embraces all those products falling withthin the general boundaries; it is much more the case that such facts must be chosen from the dearth of exact knowledge regarding the products.

Arrangements of the Tradename Register
8.2.1 1st Category: Resins

1 **Products with carbon chains**
mainly long-chain thermoplastic polymers and also thermoplastic elastomers.

11	*Polyolefins*
111	Polyethylenes, general
1111	PE-LD, density ≦ 0.925
1111 L	PE-LLD
1112	PE-MD, density 0.926–0.940
1113	PE-HD, density > 0.940
1114	Chlorinated polyethylene
1115	Copols of ethylene (apart from 341)
1116	Cross-linked or cross-linkable (not continuing thermoplastic)
1117	"Thermoplastic elastomers"
112	Polypropylene
1121	Copols of propylene
1122	Chlorinated polypropylene
113	Polybutene
114	Polyisobutene
115	Olefin copols, general
116	Polymethyl pentene
117	Polydicyclopentadien
12	*Polystyrene and related products*
121	Polystyrene
122	Polystyrene modified:
1221	with butadiene and acrylonitrile (ABS polymers)
1222	Copols with acrylonitrile (SAN polymers)
1223	Copols (and polyblends) with butadiene
1224	Polyparamethyl styrene
123	Copols with methyl styrene
124	Copols with acrylonitrile and acrylic ester
125	Other copolymers (apart from 322)
1251	Cross-linked or cross-linkable (not continuing thermoplastic)
1252	With Itaconate
1253	"Thermoplastic elastomers"
126	Polyvinyl carbazole
127	Polyvinyl pyrrolidone
13	*Halogenated polyolefine polymers*
131	Polyvinyl chloride, general, also modified and copols
1311	PVC
1312	Afterchlorinated PVC
1313	Copols of vinyl chloride
1314	Modif. PVC
1315	Thermoplastic elastomers
132	Plasticized PVC and copols
133	Polyvinylidene chloride and copols
134	Polyvinyl dichloride
135	Fluorine-containing polymers
1351	Polytetrafluoroethylene and copols
1352	Polytrifluorochlorethylene and copols
1353	Polyvinyl fluoride and copols
1354	Polyvinylidene fluoride
14	*Polyvinyl ethers*
141	Polyvinyl methyl ether
142	Polyvinyl ethyl ether
143	Polyvinyl isobutyl ether
15	*Polyvinyl esters and derivatives*
151	Polyvinyl acetate
152	Polyvinyl alcohol
153	Polyvinyl formal
154	Polyvinylacetat, also general
155	Polyvinylbutyral
16	*Polyacrylics and polymethacrylics*
161	Polyacrylonitrile (polyvinyl cyanide and copols apart from 164 and 323)
1611	Copols of acrylodinitrile (vinylidene cyanide)
162	Polyacrylic ester
163	Polymethacrylic ester (PMMA)
164	Copols of 161 and 163 (AMMA)
1641	Copols MBS
165	Polyacrylic acid salts
166	Polymethylacrylimide
167	Cyanoacrylate
19	*Polyarylene, etc.*

2 Products with a mixed chain structure

(mostly cross-linking thermosets and reaction resins)
Long-chain thermoplastics and thermopalstic elastomers are designated (*).

21	*Phenoplasts* (by polycondensation with aldehydes)	2441	Phthalyl resins
		2442	Maleic resins
211	Pure phenolic resins		
212	Phenol/cresol and cresylic resins	25	*Resins of mixed structure,* various laquer resins, in particular:
213	Resorcinol resins		
214	Modified phenol resins	251	Lacquer resins synthesis with natural resins
215	Phenol-lignin resins		
216	Furane resins	252	Styrenated alkyd resins
		253	Styrenated drying oils
22	*Aminoplasts* (by polycondensation with aldehydes)	254	Methacrylated alkyd resins
		258	Xylene-formaldehydes resins, mostly modified
221	Urea and thiourea resins		
222	Modified urea resins	259	Aldehyde and ketone resins, mostly modified
223	Melamine resins		
224	Modified melamine resins		
225	Dicyadiamide resins	26	*Polyamides, Polyurethanes,* etc. (by polycondensation or polyaddition)
23	*Polyoxides* etc. (various preparative processes)		
231	Acetal resins*	261	Polyamides*
2311	Polymethylene oxide*	2611	Polyaminoamides, etc.
2312	Copols of polymethylene oxide*	2612	Polyarylamides
		2613	Polyether blockamides
232	Polyethylene oxide*	262	Linear polyurethanes*
233	Epoxy resins	263	Cross-linked polyurethanes or their preproducts
235	Polyphenylene ether, mod.*		
2351	Polyaryl ether*	2631	Polyisocyanates
236	Polysulfone*	2632	Polyesterpolyols
2361	Polyether sulfone*	2633	Polyetherpolyols
237	Polyphenylene sulfide*	2634	Polyaminopolyols
		264	Thermoplastic polyurethane elastomers
24	*Polyesters* (by polycondensation)		
241	Linear polyesters*	265	Polyisocyanurate
2411	Polycarbonates*	266	Polycarbodiimide
2412	Polyterephthalates*		
2413	Thermoplastic elastomers*	27	*Polyimides,* etc. (partly*)
2414	Polyarylates*	271	Polyester imide
2415	Polyester carbonates*	272	Polyacryl imide
242	Unsaturated polyesters to further polymerize (or polymerized) with styrene and the like	273	Polyamide imide
		274	Polyether imide
		276	Polyhydantoin
2421	Vinyl ester resins		
243	Further polymerizable allyl esters	28	*Silicones*
244	Alkyd lacquer resins, also modified with fatty acids	281	Polysiloxane with org. copols

8.2.1 1st Category: Resins

3 **"Vulcanizable rubbers"**
chemically cross-linking on processing to flexible elastomers

31	Products with natural and reclaimed rubber	34	Various polymers
		341	Polyolefins
		3411	Olefin-diente terpolymers
32	Butadiene polymers	342	Chlorsulfonated PE
321	Polybutadiene	343	Ethylene-vinyl acetate copol
3211	Cis-polybutadiene	345	Polyacrylates
3212	Trans-polybutadiene	348	Fluorelastomers
322	Copols with styrene		
3221	Stereospecific copols	35	Chloroprene polymers
323	Copols with acrylonitrile	36	Organic polysulfides
		37	Rubbery elastic cross-linked polyurethane
33	Isoprene polymers		
331	Cis-polyisoprene	38	Silicone rubbers
332	Copols with preponderance of isobutylene (butyl rubber)	39	Chlorhydrin-(co-)polymers

4 **Natural product derivates:**
all products (apart from 421 and 431) are thermoplastic synthetic resins

41	Rubbers derivatives	4224	Cellulose acetobutyrate
411	Chlorinated rubber	4225	Cellulose propionate
412	Rubber hydrochloride	423	Cellulose ether
413	Cyclized rubber	4231	Methyl cellulose
		4232	Ethyl cellulose
42	Cellulose derivatives	4233	Hydroxylated methyl or ethyl cellulose
4211	Vulcanized fiber		
4212	Regenerated cellulose (Rayon, Cellophane)	4234	Carboxylated methyl or ethyl cellulose
422	Cellulose ester	4235	Benzyl cellulose
4221	Cellulose nitrate		
4222	Sec. cellulose acetate	43	Protein derivatives
4223	Cellulose triacetate	431	Casein plastics

5 **Various organic substances**

51	Natural resins	55	Chlorinated paraffins
52	Little-changed natural resins	56	Chlorinated naphthalene and similar products
53	Hydrocarbon resins		
		57	Bitumen- or asphalt-containing products
54	Paraffins		

6 **Inorganic base materials**

61	Polyphosphazene

8.2.2. 2nd Category: Forms of Delivery

1 Raw materials
- 11 *General raw material*
- 110 Thermoplastic elastomers
- 111 Preproducts
- 112 Casting and laminating resins
- 113 Technical resins
- 114 Raw materials for lacquers
- 115 Raw materials for adhesives and glue
- 116 Raw materials for paper and textile refinement
- 117 Foam preproducts
- 118 RIM-system products
- 119 Liquid cristal polymers

- 12 *Raw materials in separate classification*
- 121 Aqueous dispersion
- 122 Pastes (Plastisols and organisols)
- 123 Solutions
- 13 *Products processed in acqueous solution*
- 14 *Products for particular applications*
- 141 Flame spray and centrifugal sintering powder
- 142 Melt-dipping materials
- 143 Lining materials
- 145 Construction sealing materials
- 146 Coating materials
- 147 Dental materials
- 148 Ion-exchange resins
- 149 Cleaning materials
- Intermediate bonding resins
- 150 Coupling agents
- 151 Modicier resins

2 Molding and extrusion compounds
- 21 *Without fillers/reinforcements*
- 22 *With inorganic fillers/reinforcements*
- 221 Not fibrous
- 2211 Quartz
- 2212 Chalk
- 2213 Talcum
- 2214 Mica
- 2217 Glass balls
- 222 Asbestos fibers
- 223 Asbestos string or felt
- 224 Glass fibers
- 225 Glass mat or textiles
- 2251 GR prepregs
- 226 Carbon fibers
- 2261 CF prepregs
- 227 Mineral fibers
- 23 *With finely divided organic fillers*
- 231 Cellulose
- 232 Wood flour
- 24–26 *With structured organic fillers/reinforcements*
- 241 Cellulose fibers
- 242 Cellulose cuttings
- 43 Cellulose (paper) web
- 2431 Cover paper
- 251 Textile fibers
- 2511 Fiber strands
- 2512 Fiber fleece
- 252 Textile cuttings
- 253 Textile webs
- 261 Wood fibers
- 262 Veneer cuttings
- 263 Veneer-webs

3 Profiles and pipes
- 31 Pipes
- 311 Armored
- 312 Wound
- 32 *Tubes*
- 321 Armored
- 33 Rods
- 34 *Profiles with various cross sections*
- 341 Construction profiles
- 3411 Window profiles
- 342 Technical profiles
- 343 Fashion profiles
- 35 *Wires and monofils*
- 36 *Bristles*

8.2.2 2nd Category: Forms of Delivery

4	**Flexible films and sheeting, webs and cuttings**		
41	*Packaging and electrical insulating films*	4321	With inserts
		433	Technical lining and sealing materials
411	Packaging films and tubes		
412	Composite films	434	Construction sealing webs
413	Electrical insulating films	4341	With carrier web
414	Foamed sheets	435	Roof coverings
415	Synthetic paper or cardboard	4351	With carrier web
		436	Technical materials with fabric or fleece
42	*"Plastic" films and artificial leather*		
		437	Laminating films
421	Decorative, garment bags, upholstery, films and sheeting	438	Adhesive carrier films
		44	*Flooring*
422	Self-adhesive films	441	Webs
423	Book-binding material	442	Tiles
424	Artificial leather with fabric backing	443	Tiles filled with absestos
		444	Webs or tiles with carriers
4241	Foamed artificial leather	445	Coating on felt or cork
43	*Webs, technical films and sheetings*		
		45	*Other coverings*
431	Construction site protective webs	451	Tapestrylike wall coverings
4311	With inserts	452	Furniture surfacing material, including multilayered
432	Agricultural and greenhouse films		

5	**Sheets and other unreinforced semifinished goods**		
51	*Sheets, blocks and blanks for machining*	53	*Webs and sheets for automatic thermoforming*
52	*For thermoforming and welding for technical purposes*	54	*Glass-clear products*
		55	*Lighting panels (flat, corrugated, multiwalled*

6	**Rigid reinforced plastics**		
61	*Technical (laminated) products:*	617	With wire netting
611	With paper inserts (hard paper)	618	With saw dust
6111	High-temperature resistant		
612	With textile	62	*Plastic construction sheets*
6121	High-temperature resistant	621	Decorative laminates
613	With veneers (synthetic resin compressed wood)	622	Decorative laminated compressed wood
614	With glass-fiber mat or textile		
6141	Glass fiber reinforced pipes		
615	With asbestos laminates	63	*GRP lighting panels (flat and corrugated)*
616	With mica filter		

7 Various products made by or with synthetic resins

71	*Foamed plastics*	74	*Wood with synthetic resins*
711	Cellular plastics of higher density	741	Impregnated pressed wood
712	Poromerics (air permeable shoe material)	742	Fiber and chipboard
713	Structural foam		
		76	*Composite materials*
		761	Plywood, fiber and chipboard with synthetic resin covering layer
72	*Fibers*		
721	Fleece	762	Plastic-linde metal
		763	Safety glass with plastic interlayer
		766	Sandwich sheets
73	*(Melt-)adhesives, glues and cements*	7661	With plastic foam core
		7662	With plastic resin bound honey-comb core
731	Glue films		
732	Putties		
733	Covers	77	*Melts for casting*

9 Ancillary products for plastics processing

91	Plasticizers
94	Waxes
95	Masterbatches

Abbreviations

a	=	degradable
s	=	for particular applications
cop	=	Copolymer
corr	=	for corrosion protection
el	=	for electrotechnical applications
elc	=	electr. conductive
med	=	for medical purposes
mod	=	modified
se	=	nonflammable
r	=	regenerated material
tra	=	transparent
TE	=	thermoplastic elastomers

8.3 Trade Names

Name	Company (see 8.1, page 479)	Resin (see 8.2.1, page 481)	Forms of delivery (s. 8.2.2, p. 484)
A			
A-C® Polyethylene	Allied Signal Corp. Morristown, NJ 07 962, US	11	11, 2
A-fax	Montell Polyolefins Wilmington, DE 19 894, US	112	11
Absafil®	Akzo Engineering Plastics Inc. Evansville, IN 47 732, US	1221	224, 226
Abscom	–,,–	1	2
Abselex®	Courtaulds PLC, Bradford BD1 1EX W. Yorkshire, GB	1221	53
Absrom	Daicel Chemical Ind. Ltd. Tokyo, JP	1221	141
Ac/Ar®	Allied Signal Corp. Morristown, NJ 07 962, US		
Acell®	BP Chemicals International Ltd. London SW1W OSU, GB	111, 211	713
Acella®	J. H. Benecke GmbH 30007 Hannover, DE	11	422, 437, 721
Acetron®	Polypenco GmbH 51437 Bergisch Gladbach, DE	231	3, 5
Aclacell®	Acla-Werke GmbH 51065 Köln, DE	263	711
Aclaflex	–,,–	263	37, 713
Aclait®	–,,–	212	611, 612
Aclamid®	–,,–	261	2, 31, 33, 51
Aclan®	–,,–	263, 264	11
Aclar	Allied Signal Corp. Morristown, NJ 07962, US	135	41
Aclathan®	Acla-Werke GmbH 51065 Köln, DE	263, 264	2, 71
Aclon®	Allied Signal Corp. Morristown, NJ 07 962, US	1352	2
Aclyn®	Allied Color Ind. Inc. Broadview Heights, OH 44 147, US	1115 s	121, 94
Acme	Acme Chemicals Div. New Haven, CT 06 505, US	233	2 el

8.3 Trade Names

Name	Company (see 8.1, page 479)	Resin (see 8.2.1, page 481)	Forms of delivery (s. 8.2.2, p. 484)
Acorn	Hepworth Building Products Ltd. Stockbridge, Sheffield S30 5 HG, GB	113	31
Acpol	Freeman Chemical Corp. Port Washington, WI 53 074, US	242	111, 2
Acralen®	Bayer AG, 51368 Leverkusen, DE	1115	121
Acrifix®	Röhm GmbH, 64293 Darmstadt, DE	163	115
Acriglas	Acrilex Inc. Jersey City, NJ 07 305, US	163	54, 55
Acriplex®	Röhm GmbH, 64293 Darmstadt, DE	28/223	123
Acrivue	Pilkington Aerospace Inc. Garden Grove, CA 92 641, US	163	51, 54
Acronal®	BASF Aktiengesellschaft 67063 Ludwigshafen, DE	162	114, 121
Acrosol®	–,,–	162	121
Acryalloy V	Mitsubishi Rayon Co. Ltd. Tokyo, JP	163/131	124
Acrycal	Continental Polymers Inc. Compton, CA 90 220, US	163	54
Acrycon	Mitsubishi Rayon Co. Ltd. Tokyo, JP	16	21
Acryester	–,,–	16	111
Acrylafil®	Akzo Engineering Plastics Inc. Evansville, IN 47 732, US	1222	117, 224
Acrylamate	Ashland Chemical Corp. Columbus, OH 43 216, US	242/263	11
Acrylite®	Cyro Industries Woodcliff, NJ 07 675, US	163	21
Acrylivin	Gen. Corp. Polymer Products Newcomerstown, OH 43 832, US	16/135	5
Acryloy	Resolite, Div. Robertson Ceco Co. Zelienople, PA 16 063, US	242/61	6
Acrylux	Tupaj-Technik-Vertrieb GmbH 82538 Geretsried, DE	163	5
Acrylux®	Schock & Co. GmbH 73605 Schorndorf, DE	163	34
Acrypanel®	Mitsubishi Rayon Co. Ltd. Tokyo, JP	163	63

8.3 Trade Names

Name	Company (see 8.1, page 479)	Resin (see 8.2.1, page 481)	Forms of delivery (s. 8.2.2, p. 484)
Acrypet	Mitsubishi Rayon Co. Ltd. Tokyo, JP	163 s	2
Acrysol	Rohm & Haas Co., Research Triangle Park, NC 27 709, US	16	121
Acrysteel®	Aristech Chemical Co. Florence, KY 41 042, US	163 s	54
Acrytex	Röhm GmbH, 64293 Darmstadt, DE	162	116
ACS®	Showa Denko K.K., Tokyo 105, JP	1114/16	2
Acsium	DuPont Co. Inc. Wilmington, DE 19 898, US	3	11
Actilane	SNPE Chemie GmbH 60323 Frankfurt, DE	24	226
Adder®	Tufnol Ltd. Birmingham B42 2TB, GB	21	615
Adell	Adell Plastics Inc. Baltimore, MD 21 227, US	112, 2411, 261	2, 224
Adflex®	Montell Polyolefins Wilmington, DE 19894, US	112, 1121	2
Adheflon®	Elf Atochem 92 091 Paris, La Défense 10, FR	–	150
Adimoll®	Bayer AG, 51368 Leverkusen, DE	–	91
Adiprene®	DuPont Co. Inc. Wilmington, DE 19 898, US	37	111
Admer	Mitsui Petrochemical Ind. Ltd. Tokyo, JP	115, 1117	113
Admex	Ashland Chemical Corp. Columbus, OH 43 216, US	91	
Adsyl®	Montell Polyolefins Wilmington, DE 19894, US	112, 1121	2
Adstif®	–,,–	112, 1121	2
Aecithene	Aeci (Pty) Ltd. Johannesburg 2000, ZA	1111 L	2
Aerodux®	Ciba Geigy AG, 4002 Basel, CH	213	115
Aeroflex®	Anchor Plastics Inc. Great Neck, NY 11 021, US	1111	3
Aerolam®	Ciba-Geigy, Composites Div. Anaheim, CA 9207–2018, US	233	766 s

8.3 Trade Names

Name	Company (see 8.1, page 479)	Resin (see 8.2.1, page 481)	Forms of delivery (s. 8.2.2, p. 484)
Aerotuf	Anchor Plastics Inc. Great Neck, NY 11 021, US	112	3
Aeroweb®	Ciba-Geigy, Composites Div. Anaheim, CA 9207–2018, US	233	f. 7662
Aerowrap	BP Chemicals International Ltd. London SW1W OSU, GB	1113	414
Aflas	3 M Co., St. Paul, MN 55 144, US	1352	110 TE
Aftex	Asahi Glass Co., Tokyo 100, JP	348	11
Agepan	Glunz AG, 66265 Heusweiler, DE	21	761
Agomet®	Degussa AG 60287 Frankfurt, DE	163, 242	73
Agovit®	–,,–	163	73
Ahlstrom RTC	Ahlstrom Glasfibre Ltd. 48 601 Karhula, FI	112	2251
Aicarfen®	Aicar S.A., 08010 Barcelona, ES	21	24
Aipor®	Associazione Italiana Polistirolo Espanso, IT	121	71
Airdec	Isovolta AG 2355 Wiener Neudorf, AT	212	611
Airex®	Airex AG, 5643 Sins, CH	1311, 132, 274	7661
Airline Xtra	Glynwed Int. Inc. Sheldon, Birmingham, GB	1221 mod	31
Airofoam	Airofom AG, 4852 Rothrist, CH	111	71 s, 414 se
Airvol®	Air Products & Chemicals Inc. Allentown, PA 18 105, US	152	11
Aislanpor	Aiscondel S.A., Barcelona 13, ES	121	72
Aisloplastic	–,,–	132	3, 4
Akrylon	PCHZ np, Zilina, CSFR	163	55
Akulon®	DSM, Polymers & Hydrocarbons 6130 AA Sittard, NL	261	2, 22
Akylux	Kaysersberg Packaging 68 240 Kaysersberg, FR	1121	51 corr, el, med
Akyplen	–,,–	112	51 corr, el, med
Akyver	–,,–	2411	55 tra
Alathon®	Oxychem Vinyls Div. Berwyn, PA 19 312, US	1117	2

Name	Company (see 8.1, page 479)	Resin (see 8.2.1, page 481)	Forms of delivery (s. 8.2.2, p. 484)
Albertol®	Hoechst AG 65929 Frankfurt, DE	241, 251	114, 115
Alcantara®	Iganto SpA, Nera Montoro/Terni, IT	263	421 s
Alcryn®	DuPont Co. Inc. Wilmington, DE 19 898, US	343+133	110 TE
Alfane	Atlas Minerals & Chemicals Inc. Mertztown, PA 19 539, US	233	112, 113
Alflow	Ahlstrom Glasfibre Ltd. 48 601 Karhula, FI	112	225
Alftalat®	Hoechst AG 65929 Frankfurt, DE	244, 252	113, 114
Algo-Stat®	Algostat AG, 29201 Celle, DE	121	71
Algoflon®	Ausimont Inc. Morristown, NJ 07 962, US	1351	11, 12, 14, 22
Alkathermic®	Alkudia Empresa para la Industria Madrid 20, ES	1111	432
Alkorflex®	Solvay S.A., 1050 Bruxelles, BE	1311	435
Alkorfol®	–,,–	132, 1354	438, 452
Alkorpack®	–,,–	132	411
Alkorplan®	–,,–	1311	435
ALKOzell	Afelder Kunststoffwerke 31044 Alfeld, DE	1111	71
Alkydal®	Bayer AG, 51368 Leverkusen, DE	244, 252	114
Allacast	Allaco Div., Bacon Industries Inc. Watertown, MA 02 172, US	233	112, 2
Allbond®	–,,–	233	73
Alloprene®	ICI PLC Welwyn Garden City, Herts. AL7 1 HH, GB	411	11
Allotherm®	BASF Lacke+Farben AG 48165 Münster, DE	272	114
Alnovol®	Hoechst AG 65929 Frankfurt, DE	21Novolak	113, 114
Alpex®	–,,–	413	114
Alpha	Alpha Plastics Corp. Pineville, NC 28 134, US	131, 132	2
Alphamid®	Putsch GmbH, 90427 Nürnberg, DE	261	2, 224

8.3 Trade Names

Name	Company (see 8.1, page 479)	Resin (see 8.2.1, page 481)	Forms of delivery (s. 8.2.2, p. 484)
Alpolit®	Hoechst AG 65929 Frankfurt, DE	242	112, 143, 146
Alresat®	Hoechst AG 65929 Frankfurt, DE	2442, 251	113, 114, 115, 116
Alresen®	–,,–	214	114, 115
Alstamp®	Ahlstrom Glasfibre Ltd. 48 601 Karhula, FI	112	2251, 2261
Altair®	Aristech Chemical Co. Florence, KY 41 042, US	163	5
Alteco®	Ceca, La Défense 5 92400 Courbevoie, FR	167	73
Althon®	Occidental Chemical Corp. Dallas, Texas 75380, US	237/1113	2
Altubat®	Elf Atochem 92 091 Paris, La Défense 10, FR	163	55
Altuglas®	Atohaas, 92062 Paris, La Défense 10, FR	163	11, 2
Altulex®	Elf Atochem 92 091 Paris, La Défense 10, FR	163	5, 54
Altulor®	–,,–	163	51
Alveolen®	Alveo AG, 6003 Luzern, CH	112, 1116	71
Alveolit®	–,,–	1116, 112	414, 71
Alveolux®	–,,–	1115, 1116	414, 71
Amberlite®	Rohm & Haas Co., Research Triangle Park, NC 27 709, US	125, 164	148
Ambla®	Wardle Storeys PLC Manningtree, Essex CO11 1NJ, GB	132	424
Amblon®	–,,–	132	4241
Ameripol®	B. F. Goodrich Chemical Group Cleveland, OH 44 131, US	322	11
Amilan®	Toray Industries Inc., Tokyo, JP	261	11, 2, 22
Amilon®	–,,–	261	2
Amodel®	Amoco Performance Products Richfield, CT 06 877, US	2614	2
Amoflo®	Amoco Performance Products Atlanta, GA 30 350, US	1224	9

8.3 Trade Names

Name	Company (see 8.1, page 479)	Resin (see 8.2.1, page 481)	Forms of delivery (s. 8.2.2, p. 484)
Ampacet®	Ampacet Intern. Corp. Mount Vernon, NY 10 550, US	111, 115, 112, 121	111, 2, 9
Ampal®	Raschig AG, 67061 Ludwigshafen, DE	242	2, 22, 23
Ampcoflex	Atlas Plastics Corp. Cape Girardeau, MO 63 701, US	1311	31, 51 corr
Amres®	Georgia-Pacific Atlanta, GA 30 348, US	21, 22, 223	11, 112, 113
Ancorene	Anchor Plastics Inc. Great Neck, NY 11 021, US	1223	3
Ancorex	−„−	1221	3
Andrez	Anderson Development Co. Adrian, MI 49 221, US	1223	11, 14
Andur	−„−	2632, 2633	11
Anjablend	Janßen & Angenendt GmbH 47800 Krefeld, DE	1221/2411, 2411/2412	2
Anjadur	−„−	2412	2
Anjalin	−„−	1221	2
Anjalon	−„−	2411	2
Anjamid	−„−	261	2
Anti-Static	Marley Floors Ltd. Lenham, Maidstone, Kent, ME17 2DE, GB	132	444
Antiflex	Tupaj-Technik-Vertrieb GmbH 82538 Geretsried, DE	163	5
Antron®	DuPont Co. Inc. Wilmington, DE 19 898, US	261	72
Apec HT	Bayer AG, 51368 Leverkusen, DE	2416	2, 21 tra
Apel	Mitsui Petrochemical Ind. Ltd. Tokyo, JP	115 s	124
Apex®	Atlas Fibre Co. Skokie, IL 60 076, US	163 s	54
Aphro Trays	deltaplastic GmbH & Co. KG 27721 Ritterhude, DE	121	414
Aphrolan	−„−	121	414
Apivin	Associated Plastics of Ireland, IR	131	2
Appretan®	Hoechst AG 65929 Frankfurt, DE	151, 162	116, 121, 612, 614

8.3 Trade Names

Name	Company (see 8.1, page 479)	Resin (see 8.2.1, page 481)	Forms of delivery (s. 8.2.2, p. 484)
Appryl®	Appryl SNC, 92 807 Puteaux Cedex, FR	117	11, 2
Apscom®	Akzo Engineering Plastics Inc. Evansville, IN 47 732, US	1, 2, 24	2
Aquaflex	Pantasote Inc. Passaic, NJ 07 055, US	131, 132	433
Aquakeep	Elf Atochem 92 091 Paris, La Défense 10, FR	165	2
Aqualoy®	A. Schulman Inc. Akron, OH 44 309, US	112, 1221, 1222 2361, 237, 2411 2412, 261	2
Aracast®	Ciba Geigy AG, 4002 Basel, CH	233 s	112
Araldit®	–,,–	233, 27	11, 118, 123, 141, 143, 221, 224
Arale	Akzo Engineering Plastics Inc. Evansville, IN 47 732, US	2612	11
Aramid	general name	261	72 se
Arapol	Reichhold Chemicals Inc. Jacksonville, FL 32 245, US	242	112 s, 2
Aratronic®	Ciba Geigy AG, 4002 Basel, CH	233	112, 2
Arcel®	Arco Chemical Co. Newton Square, PA 19 073, US	111/121, 1115	117, 71
Arcoblend	Resinmec S.p.A. 24 040 Pontirolo Nuovo (BG), IT	2415/1221	2
Arcolac	–,,–	1221	2
Arcomid	–,,–	261	2
Arcoplen	–,,–	112	2
Arcosulf	–,,–	2632	2
Arcoter	–,,–	2412	2
Arcoxan	–,,–	2411	2
Ardel®	Amoco Performance Products Richfield, CT 06 877, US	2414	113, 146
Ardylan	Ind. Petroquimicas Argentinas Koppers S.A., Buenos Aires, Argentinien	111	2
Ardylux	–,,–	1222	2

8.3 Trade Names 495

Name	Company (see 8.1, page 479)	Resin (see 8.2.1, page 481)	Forms of delivery (s. 8.2.2, p. 484)
Arenka®	Akzo Fibers Division 6800 AB Arnhem, NL	261 s	72 se
Aristoflex®	Hoechst AG 65929 Frankfurt, DE	15	113, 123, 14
Arjomix	Exxon Chemicals America Houston, TX 77 001, US	112	76 s
Arlen	Mitsui Petrochemical Ind. Ltd. Tokyo, JP	261 s	224
Arlon	Tweed Eng. Plastics, Hartfield, GB	2351	2261
Arloy®	Arco Chemical Co. Newton Square, PA 19 073, US	1225/2411	2, 120
Armaflex	Armstrong World Ind. Inc. Lancaster, PA 17 604, US	323	414, 71
Armaveron	Armaver AG, 4617 Gunzgen, CH	242	6141
Armite	Spaulding Composites Co. Tonawanda, NY 14 150, US	4211	5
Arnar	Ross & Roberts Inc. Stratford, CT 06 497, US	111, 1115, 1311, 132, 263	4, 5
Arnite®	Akzo Engineering Plastics Inc. Evansville, IN 47 732, US	2412	2, 22
Arnitel®	DSM Polymers & Hydrocarbons, 6130 AA Sittard, NL	2413	110 TE
Arofene	Ashland Chemical Corp. Columbus, OH 43 216, US	21	111
Aron	Toa Gosei Chemical Ind. Co. Ltd. Tokyo, JP	131, 1314	2
Aropol	Ashland Chemical Corp. Columbus, OH 43 216, US	242	112, 2, 224
Arotech	–,,–	2613	11
Arotone®	DuPont Co. Inc. Wilmington, DE 19 898, US	2351	11
Arpak	Arco Chemical Co. Newton Square, PA 19 073, US	111	117
Arpro	–,,–	121	117
Arpylene®	TBA Industrial Products Ltd. Rochdale, Lancs., GB	112, 121, 2411, 261	2212, 2213, 222, 224, 226

8.3 Trade Names

Name	Company (see 8.1, page 479)	Resin (see 8.2.1, page 481)	Forms of delivery (s. 8.2.2, p. 484)
Arset	Arco Chemical Co. Newton Square, PA 19 073, US	263/2411	118
Artgranit®	Schock & Co. GmbH 73605 Schorndorf, DE	24	112
Artimer®	–,,–	24	112
Artonyx®	–,,–	24	112
Arylon®	DuPont Co. Inc. Wilmington, DE 19 898, US	2414	2
Asaprene	Asahi Chemical Ind. Co. Ltd. Tokyo, JP	12/32/12	110 TE
Ashlene®	Ashley Polymers Brooklyn, NY 11 219, US	1221, 231, 235, 261, 2411, 2412	2, 21, 2213, 2214, 224
Aslan®	Aslan, 51491 Overath, DE	1113, 132	422 s
Asp®	Tufnol Ltd. Birmingham B42 2TB, GB	21	615
Asplit®	Hoechst AG 65929 Frankfurt, DE	214, 216, 233, 242	732, 733, corr
Assil®	Henkel KGaA, 40589 Düsseldorf, DE	263	71
Asterite®	ICI PLC Welwyn Garden City, Herts. AL7 1 HH, GB	163	112
Astra	Drake (fibres) Ltd. Huddersfield, W. Yorks., GB	112	71
Astra Star	–,,–	112	72
Astralon®	HT Troplast AG 53840 Troisdorf, DE	131, 16	51, 52, 53
Astro Glaze	Commercial Decal Inc. Mt. Vernon, NY 10 550, US	223	452
Astro Turf®	Monsanto Chemical Co., Plastics Div. St. Louis, MO 63 167, US	261	44 s
Astryn	Montall Polyolefins Wilmington, DE 19 894, US	112, 1121	2
Atlac	DSM Resins BV, 8000 AP Zwolle, NL	2421	11, 112, 113
Atlantic	Norplex Oak Inc. La Crosse, WI 54 601, US	1351, 233, 27	614 el
Autan®	Acla-Werke GmbH 51065 Köln, DE	263	5, 711

8.3 Trade Names

Name	Company (see 8.1, page 479)	Resin (see 8.2.1, page 481)	Forms of delivery (s. 8.2.2, p. 484)
Autothane®	Kemira Polymers Ltd. Stockport, Cheshire SK 12 5BG, GB	263	711
Avalon®	ICI Polyurethanes 3078 Kortenberg, BE	2632, 2633	117/711
Avimid®	DuPont Co. Inc. Wilmington, DE 19 898, US	27	1
Avotone®	−„−	2351	1
Avron®	ICI PLC Welwyn Garden City, Herts. AL7 1 HH, GB	116	51
Axpet®	Axxis N.V., 8700 Tielt, BE	2412	54
Axxis®-PC	−„−	2411	54
Axxis®-Sunlife	−„−	2411	55
Azdel®	Azdel Inc., Shelby, NC 28 150, US	112 s	224, 614, 76
Azfab®	−„−	2412	2251
Azloy®	−„−	2411/235, 2412	124
Azmet®	−„−	2412	111
B			
Bakelite®	Bakelite GmbH 58642 Iserlohn, DE	21, 22, 242	2
Bakelite®	−„−	211–214, 223, 216	111–117, 12, 123, 13, 141, 146, 73
Bamberko	Claude Bamberger Molding Compounds Carlstadt, NJ 07 072, US	16	149
Bapolan	Bamberger Polymers Inc. New Hyde Park, NY 11 042, US	121, 1221, 1222	2
Bapolene	−„−	111, 1111 L, 1112, 112	2
Bapolon	−„−	261	41
Barex®	BP Chemicals International Ltd. London SW1W OSU, GB	1611	11
Baricol	BCL, Bridgewater, Somerset, TA6 4PA, GB	1115/152	412

8.3 Trade Names

Name	Company (see 8.1, page 479)	Resin (see 8.2.1, page 481)	Forms of delivery (s. 8.2.2, p. 484)
Baroflex	American Barmag Corp. Charlotte, NC 28 217, US	–	413
Basenol®	BASF Aktiengesellschaft 67063 Ludwigshafen, DE	2633	123
Basonat®	–„–	2631	114
Basotect®	–„–	223	71 se
Bauder PUR	Paul Bauder GmbH & Co. 70499 Stuttgart, DE	263, 265	71
Bayblend®	Bayer AG, 51368 Leverkusen, DE	1221/124/2411	124
Baybond®	–„–	263	121
Baydur®	–„–	2632, 2633	118/71
Bayfill®	–„–	2632, 2633	117/71
Bayfit®	–„–	263	117/71
Bayflex®	–„–	2632, 2633	118, 117/71
Bayfol®	–„–	2411	4
Baymer®	–„–	2632, 2633	117/71
Baymod®	–„–	115, 1221, 125, 264, 343	9
Baymoflex	–„–	124/345	437
Baynat®	–„–	263	117/53
Baypreg®	–„–	263	225
Baypren®	–„–	35	11, 115
Baysilone®	–„–	28	11
Baysport®	–„–	263	146
Baystal®	–„–	322	121
Baytec®	–„–	263	117/71
Baytherm®	–„–	2632, 2633	117/71
Bear®	Tufnol Ltd. Birmingham B42 2TB, GB	21	612
Beckocoat®	Hoechst AG 65929 Frankfurt, DE	263	114
Beckopox®	–„–	233	112, 114, 115, 143, 146
Beckurol®	–„–	222	114

8.3 Trade Names

Name	Company (see 8.1, page 479)	Resin (see 8.2.1, page 481)	Forms of delivery (s. 8.2.2, p. 484)
Bectran®Pk	BASF Lacke+Farben AG 48165 Münster, DE	263	112
Bectron®Pl	BASF Lacke+Farben AG 48165 Münster, DE	27	114
Bedacryl	Cray Valley Products Int. Farnborough, Kent, BR6 7EA, GB	16	114
Bedesol	–„–	25	114
Beetle®	BIP Chemicals Ltd. Warley, West Midl., B69 4PG, GB	21, 22, 2412, 2413, 261, 27	2, 22
Begra	Begra GmbH & Co. KG 13407 Berlin, DE	131	2
Bekaplast	Steuler Industriewerke GmbH 56203 Höhr-Grenzhausen, DE	111, 112, 131, 1354	52 s, corr
Benecor®	J. H. Benecke GmbH 30007 Hannover, DE	132	421, 424
Benefol®	–„–	132	434, 435
Benelit®	–„–	1311, 132	452, 53
Beneron®	–„–	1221	53
Benova®	–„–	132	421
Benvic®	Solvay S.A., 1050 Bruxelles, BE	1311, 132	2, 11
Bergacell®	Theodor Bergmann Kunststoffwerk GmbH 76560 Gaggenau, DE	4222	2
Bergadur®	–„–	2412	2
Bergaflex®	–„–	1253	2
Bergamid®	–„–	261	2
Bergaprop®	–„–	112	2
Berlene®	Gurit-Worbla AG 3063 Ittigen-Bern, CH	11	43, 52, 53
Bestfight®	Toho Rayon, Div. Beslon, Tokyo, JP	5	226
Betamid	Putsch GmbH, 90427 Nürnberg, DE	261	2, 224
Bexel	Wardle Storeys PLC Manningtree, Essex, CO11 1NJ, GB	–	52
Bexfilm®	Bexford Ltd., Manningtree, Essex CO11 1NL, GB	2412, 4222, 4223	4

8.3 Trade Names

Name	Company (see 8.1, page 479)	Resin (see 8.2.1, page 481)	Forms of delivery (s. 8.2.2, p. 484)
Bexloy	DuPont Co. Inc. Wilmington, DE 19 898, US	115 s, 2412 s, 261 s	110 TE
Bexphane®	Moplephan UK, Brantham, GB	112	4
Bextrene	Wardle Storeys PLC Manningtree, Essex, CO11 1NJ, GB	–	52
Biafol	TVK Tisza Chemical Combine 3581 Leninvaros, HU	112	41
Biapen	Chemolimpex, 1805 Budapest, HU	235	2
Bicor®	Mobil Chemical Co., Films Dept. Macedon, NY 14 502, US	112	41, 412
Bifan	Showa Denko K.K., Tokyo 105, JP	112	411, 412
Bimoco®	DSM Italia s.r.l., 22 100 Como, IT	242	224
Bio-Net	Norddeutsche Seekabelwerke AG 26954 Nordenham, DE	1113	35, 221
Bioceta®	Tubize Plastics S.A., 1360 Tubize, BE	4222 s	2 a
Biodrak	Antonios Drakopoulos S.A. Athen, Griechenland	121, 1311, 163	5, 55, 6
Biopol®	ICI PLC Welwyn Garden City, Herts. AL7 1 HH, GB	241 s	11, a
Biopolymer	Biopolymers Ltd., Dassenberg, ZA	1223	2 tra
Bipeau	Elf Atochem 92 091 Paris, La Défense 10, FR	131	31/7661
Bisoflex®	BP Chemicals International Ltd. London SW1W OSU, GB	–	91
Bisol®	–,,–	151	11
Blak-Stretchy®	The Perma-Flex Mold Co. Inc. Columbus, OH 43 209, US	237	111
Blak-Tufy®	–,,–	237	111
Blane	Blane Polymers Div. Vista Chemical Mansfield MA 02 048, US	132	22
Blavin	–,,–	131	2
Blaze Master	B. F. Goodrich Chemical Group Cleveland, OH 44 131, US	134	2
Blendex®	GE Speciality Chemicals Parkersburg, WV 26 102, US	1221	111

8.3 Trade Names

Name	Company (see 8.1, page 479)	Resin (see 8.2.1, page 481)	Forms of delivery (s. 8.2.2, p. 484)
Blendur®	Bayer AG, 51368 Leverkusen, DE	233/242/263	11, 112, 113
Blu-Sil®	The Perma-Flex Mold Co. Inc. Columbus, OH 43 209, US	237	111, 112, 2
Boltamask	Gen. Corp. Polymer Products Newcomerstown, OH 43 832, US	131	41
Boltaron	–,,–	124, 131, 134	4, 5
Bondfast	Sumitomo Chemical Co. Ltd. Tokyo 103, JP	1121	73
Bondstrand®	Ameron PCD Brea, CA 92 621, US	233, 242	6141
Bondwave	Flexible Reinforcements Ltd. Clitheroe, Lancaster, GB	131	247/436
Bonosol	Ernst Jäger & Co. OHG 40599 Düsseldorf, DE	163	123
Bri-Nylon®	DuPont Co. Inc. Wilmington, DE 19 898, US	261	72
Bricling	BCL, Bridgewater, Somerset, TA6 4PA, GB	1111, 1111 L	411
Brilen	Brilen SA, Barbastro/Huesca, ES	2412	72
Brithene	–,,–	1111, 1111 L	411
Bromobutyl®	Polymer Corp. Ltd. Reading, PA 19 603, US	332 s	11
Budene®	Goodyear Tire & Rubber Co. Akron, OH 44 316, US	3211	11
Buflon®	Solvay S.A., 1050 Bruxelles, BE	132	451
Bultex	Recticel Foam Corp., Morristown Div. Morristown, TN 37 816–1197, US	2633	71
Buna NB 186	Buna AG, 06258 Schkopau, DE	323+131	11
Buna® *AP*	Bayer AG, 51368 Leverkusen, DE	1121	11
Buna® *SL,VSL*	–,,–	1223	11
Buna® *VI*	–,,–	–	11
Buna®*CB*	–,,–	32	11
Bunatex®	Hüls AG, 45764 Marl, DE	322	121
Butacite®	DuPont Co. Inc. Wilmington, DE 19 898, US	155	43
Butaclor®	Distugil, 92408 Courbevoie, FR	32, 35	11, 2

8.3 Trade Names

Name	Company (see 8.1, page 479)	Resin (see 8.2.1, page 481)	Forms of delivery (s. 8.2.2, p. 484)
Butaprene	Firestone Synthetics Rubber and Latex Co., Akron, OH 44 301, US	322, 323	11
Butofan®	BASF Aktiengesellschaft 67063 Ludwigshafen, DE	1223	121
Butvar®	Monsanto Chemical Co., Plastics Div. St. Louis, MO 63 167, US	155	11
Butylex®	Nordmann Rassmann GmbH & Co. 20459 Hamburg, DE	1111+332, 1113+332	1
Bytac®	Norton Performance Plastics Wayne, NJ 07 470, US	1351, 1352	437
C			
C-Flex®	Shell International Co. Ltd. London, SE1 7NA, GB	12/341/12	110 TE
Cabelec®	Cabot Corp. Billerica, MA 01 821, US	1111, 1113, 112, 1115, 121, 131	2 el
Cablon-Flex	–„–	263	2 med
Cabocell	Cabon Plastics Corp. Newark, NJ 07 102, US	4224	32
Cadon®	Monsanto Chemical Co., Plastics Div. St. Louis, MO 63 167, US	1225	21
Calibre®	The Dow Chemical Company Midland, MI 48641-2166, US	2411	117, 2, 224
Calmica	Isovolta AG 2355 Wiener Neudorf, AT	233	616
Calmicaflex	–„–	233	616
Calmicaglas	–„–	233	616
Calthane	Cal Polymers Inc. Long Beach, CA 90 813, US	264	112, 115
Cambrelle®	ICI PLC, Fibres Div. Harrogate, HG2 8QN, GB	2412, 261	721
Canevasit	Von Roll Isola 4226 Breitenbach, CH	212	612
Cantrece®	DuPont Co. Inc. Wilmington, DE 19 898, US	261	35, 72
Canusaloc	Shrink Tubes & Plastics Ltd. Redhill, Surrey, RH1 2 LH, GB	1115	31 s

8.3 Trade Names 503

Name	Company (see 8.1, page 479)	Resin (see 8.2.1, page 481)	Forms of delivery (s. 8.2.2, p. 484)
Capilene	Carmel Olefins Ltd. Haifa 31 014, IL	112	2
Capron®	Allied Signal Corp. Morristown, NJ 07 962, US	261	2
Caradate®	Shell International Co. Ltd. London, SE1 7NA, GB	2631	117
Caradol®	–,,–	2633	117
Caraplas	Caraplas, Dublin, IR	2412	53, 54
Carb-o-life	Carbolux S.p.A. 05 027 Nera Montoro (Terni), IT	2411	55
Carbaicar®	Aicar S.A., 08010 Barcelona, ES	22	24
Carboflex	Ashland Chemical Corp. Columbus, OH 43 216, US	5	226
Carbofol®	HT Troplast AG 53840 Troisdorf, DE	1111, 1113	433, 434, 435, 436
Carboglass	Carbolux S.p.A. 05 027 Nera Montoro (Terni), IT	2411	55
Carbolux®	–,,–	2411	55
Carbowax®	Union Carbide Corp. Danbury, CT 06 817, US	232	94
Cardon	Advanced Elastomer Systems L.P. St. Louis, MO 63 167, US	1225	224
Cardura®	Shell International Co. Ltd. London, SE1 7NA, GB	2632	11, 117
Cariflex TR®	–,,–	12/32/12	110 TE
Cariflex®	–,,–	3211, 322, 331	2
Caril®	–,,–	121/235	71
Carilon®	–,,–	–	2
Carina®	–,,–	1311	11, 2
Caripak	Shell International Co. Ltd. London, SE1 7NA, GB	2412	11, 2
Carlex®	Carlon, Cleveland, OH 44 122, US	111, 131	31
Carlona®	Montell Polyolefins Wilmington, DE 19894, US	112, 1121	2
Carodel	ICI America Inc. Wilmington, DE 19 897, US	2414 s	11
Carom®	Chemisches Kombinat, Borzesti, RO	322	1

8.3 Trade Names

Name	Company (see 8.1, page 479)	Resin (see 8.2.1, page 481)	Forms of delivery (s. 8.2.2, p. 484)
Carp®	Tufnol Ltd. Birmingham B42 2TB, GB	21	612
Carpran	Allied Signal Corp. Morristown, NJ 07 962, US	261	4
Carrilen	Rio Rodano S.A., Madrid 20, ES	322	2
Carta®	Isola Werke AG, 52353 Düren, DE	21, 223, 233	312, 611, 612
Cartonplast	Antonios Drakopoulos S.A. Athen, Griechenland	112	5
Casco-Resin®	Borden Chemical Co. Geismar, LA 70 734, US	221	115
Cascomelt®	–„–	1115	115, 73
Cascophen®	–„–	213, 214	112, 115
Cascorez®	–„–	151	115
Cashmilon	Asahi Chemical Ind. Co. Ltd. Tokyo, JP	161	72
Casocryl®	Elf Atochem 92 091 Paris, La Défense 10, FR	163	5
Casoglas®	–„–	163	763 s
cast-film	Karl Dickel & Co. KG 47057 Duisburg, DE	112	411
Castomer	Baxenden Chemical Co. Ltd. Accrington, Lancs. BB5 2SL, GB	263	112
Catalloy®	Montell Polyolefins Wilmington, DE 19 894, US	112	124, 120
Cebian	Daicel Chemical Ind. Ltd. Tokyo, JP	1222	11, 2
Celanex®	Hoechst Celanese Corp. Engineering Plastics Div. Chatham N.J. 07928, US	2412	2, 2217, 224
Celcon®	Hoechst Celanese Corp. Engineering Plastics Div. Chatham N.J. 07928, US	2312	2, 224
Cellasto®	Elastogran GmbH 49440 Lemförde, DE	263	711
Cellidor®	Albis Plastic GmbH 20531 Hamburg, DE	4224, 4225	11, 141, 224
Cello M	BCL, Bridgewater, Somerset, TA6 4PA, GB	4222 s	41

8.3 Trade Names

Name	Company (see 8.1, page 479)	Resin (see 8.2.1, page 481)	Forms of delivery (s. 8.2.2, p. 484)
Cellobond®	BP Chemicals International Ltd. London SW1W OSU, GB	21, 222, 223, 233, 242	112, 113, 115, 117, 2
Cellolam®	UCB N.V. Filmsector 9000 Gent, BE	42	412
Cellolux	La-Es s.p.a., IT	4222	2
Celmar®	Courtaulds Chemicals & Plastics Ltd. Spondon, Derby DE2 7BP, GB	112	614 corr
Celoron®	Budd Co., Polychem. Div. Phoenixville, PA 19 460, US	261	2, 321
Celsir	Sirlite Srl., 20 161 Milano, IT	222	11
Celstran®	Hoechst AG 65929 Frankfurt, DE	112, 231, 237, 2412, 261, 263	22 s
Celuform	Caradon Celuform Ltd. Aylesford, Maidstone, Kent ME20 75X, GB	131	34/713
Celulon®	Unitex Ltd., Knaresborough, North Yorks., HG5 OPP, GB	263	711
Celuvent	Caradon Celuform Ltd. Aylesford, Maidstone, Kent ME20 75X, GB	131	34/713
Celvin®	Courtaulds Chemicals & Plastics Ltd. Spondon, Derby DE2 7BP, GB	132	4, 5
Ceno®	Carl Nolte, 48268 Greven, DE	132, 1111, 1111 L	436
Centrex Q836	Monsanto Chemical Co., Plastics Div. St. Louis, MO 63 167, US	1222/124	2
Centrex®	–,,–	124	21
Cereclor	ICI PLC, Welwyn Garden City, Herts. AL7 1 HH, GB	–	91
Cestidur	Cestidur Industries S.A. 01360 Balan, FR	1113 s	31, 33, 342, 52, 61
Cestilene	–,,–	1113 s	31, 33, 342, 52, 61
Cestilite	–,,–	1113 s	31, 33, 342, 51, 61
Cetamoll®	BASF Aktiengesellschaft 67063 Ludwigshafen, DE	91	

8.3 Trade Names

Name	Company (see 8.1, page 479)	Resin (see 8.2.1, page 481)	Forms of delivery (s. 8.2.2, p. 484)
Cetex	Ten Cate Composites bv, NL	2361, 261, 274	2251, 2261
Cevian	Daicel Chemical Ind. Ltd. Tokyo, JP	1222	11
Chemfluor®	Norton Performance Plastics Wayne, NJ 07 470, US	135, 1354	31, 33, 5, 61
Chemigum®	Goodyear Tire & Rubber Co. Akron, OH 44 316, US	322, 323	11
Chemigum® SL	–,,–	37	112
Chemlink	Sartomer Inc., Exton, PA 19 341, US	–	9
Chemlon	Chemlon AS, Humenne, CSFR	261	72
Chempex	Golan Plastics Products Jordan Valley 15145, IL	1116	31 s
Chempol	Freeman Chemical Corp. Port Washington, WI 53 074, US	2631	114
Chissonyl	Chisso Corp., Tokyo, JP	151	11
Cibamin®	Ciba Geigy AG, 4002 Basel, CH	221, 223	114
Cibatool®	–,,–	233	51
Cisamer	Indian Petrochemicals Corp. 391 346 Gujarat State, IN	321	2
Cisdene	American Synthetic Rubber Corp. Louisville, KY, US	3211	11
Cisrub	Indian Petrochemicals Corp. 391 346 Gujarat State, IN	3211	11
Citax®	Henkel KGaA, 40589 Düsseldorf, DE	1	73
Citroflex®	Pfizer International Inc. New York, NY 10017, US	91	
Civic®	Neste Oy Chemicals, 02 151 Espoo, FI	242	11
Clarifoil®	Courtaulds Chemicals & Plastics Ltd. Spondon, Derby DE2 7BP, GB	4222	41
Clarino	Kuraray, JP	263	424/721
Claryl	Rhône-Poulenc Films, 92 080 Pa La Défense, Cedex 6, FR	2412	41 s
Clarylene	–,,–	2412/111	412 s
Clarypac	–,,–	131	412
Clearen	Denki Kagaku Kogyo, Tokyo 100, JP	1223	2
Clearflex®	ECP EniChem Polimeri srl. 20 124 Mailand, IT	1111, 1111 L, 115	11, 113

8.3 Trade Names

Name	Company (see 8.1, page 479)	Resin (see 8.2.1, page 481)	Forms of delivery (s. 8.2.2, p. 484)
Clearlac	Mitsubishi Rayon Co. Ltd. Tokyo, JP	125	2
Clearseal	Columbus Coated Fabrics Columbus, OH 43 216, US	131	42
Cleartuf	Goodyear Tire & Rubber Co. Akron, OH 44 316, US	2412	2
Clocel®	Baxenden Chemical Co. Ltd. Accrington, Lancs. BB5 2SL, GB	263	117
Clysar	DuPont Co. Inc. Wilmington, DE 19 898, US	11	411
Coathylene®	Plast Labor SA, 1630 Bulle, CH	1111, 1111 L, 1113, 1112, 1115, 112, 2413, 264	121, 14
Cobex	Wardle Storeys PLC Manningtree, Essex CO11 1NJ, GB	1311	51, 52, 53, 55
Cobocell	Cobon Plastics Corp. Newark, NJ 07 102, US	4224	32
Cobothane	–,,–	1115	32
Cobovin	–,,–	132	321
Cole	Montell Polyolefins Wilmington, DE 19 894, US	112/1221/ 1222, 12	2
Collacral®	BASF Aktiengesellschaft 67063 Ludwigshafen, DE	127, 165	115, 116, 123
Collimate	Mitsubishi Monsanto Chemical Co. Tokyo, JP	1253	224
Colo-Fast®	Recticel Foam Corp., Morristown Div. Morristown, TN 37 816–1197, US	263	118, 4
Comalloy	Comalloy Intern. Corp. Nashville, TN 37 027, US	112/2411, 261	22
Combidur®	Gebr. Kömmerling Kunststoffwerke GmbH, 66954 Pirmasens, DE	1314	3411
Combithen®	Wolff Walsrode AG 29655 Walsrode, DE	111, 26	411, 412
Combitherm®	–,,–	111, 26	411, 412
Comco	Commercial Plastics and Supply Corp. Cornwells Heights, PA 19 020, US	11, 13, 16, 261	43, 53

8.3 Trade Names

Name	Company (see 8.1, page 479)	Resin (see 8.2.1, page 481)	Forms of delivery (s. 8.2.2, p. 484)
Commax	Tecknit, Cranford, NJ 07 016, US	28+Ni	5 el
Comp	Putsch GmbH, 90427 Nürnberg, DE	112	2213
Compimide	Boots Comp. PLC Nottingham NG2 3AA, GB	276	11, 111
Compodic F	Dainippon Ink. & Chemicals Inc. Tokyo 103, JP	261 s	22
Comshield®	A. Schulman Inc. Akron, OH 44 309, US	112, 1221, 1222, 2361, 237, 2411, 2412, 261	2
Comtuf®	–,,–	112, 1221, 1222, 2361, 237, 2411, 2412, 261	2
Conacure®	Conap Div. of WFI Olean, NY 14 760, US	–	9
Conapoxy	–,,–	233	113, el, 115
Conaspray®	–,,–	263	111
Conathane	–,,–	263, 264	11, 112, 14
Conductherm	Isovolta AG 2355 Wiener Neudorf, AT	233	616
Conductofol	–,,–	233	616
Contactfoam	Sentinel GmbH 73441 Bopfingen, DE	111	711
Conolite®	Pioneer Valley Plastics Inc. Bondsvillers, MA 01 009, US	242	614
Contafel	Isovolta AG 2355 Wiener Neudorf, AT	233	614
Contapreg	–,,–	233	614
Contaval	Isovolta AG 2355 Wiener Neudorf, AT	233	611, 612, 614
Copolene	Asahi Chemical Ind. Co. Ltd. Tokyo, JP	1115 s	11
Cordoglas	Ferro Corp. Evansville, IN 47 711, US	13	436
Cordopreg®	–,,–	242	2251
Corducell	Nemitz, Kunststoff-Additive GmbH 48338 Altenberge, DE	–	95

Name	Company (see 8.1, page 479)	Resin (see 8.2.1, page 481)	Forms of delivery (s. 8.2.2, p. 484)
Cordulen	Nemitz, Kunststoff-Additive GmbH 48338 Altenberge, DE	–	95
Corduma®	–,,–	–	95
Cordura	DuPont Co. Inc. Wilmington, DE 19 898, US	261	72
Cordustat	Nemitz, Kunststoff-Additive GmbH 48338 Altenberge, DE	–	95
Coremat	Firet B.V., 3900 AA Veenendaal, NL	2412	721
CoRezyn	IMI-Tech Corp. Elk Grove Village, IL 60 007, US	242, 2421	112
Corial®*Grund*	BASF Aktiengesellschaft 67063 Ludwigshafen, DE	161, 162	121
Corian®	DuPont Co. Inc. Wilmington, DE 19 898, US	163	62
Corkelast	Edilon B. V., Haarlem, NL	262, 263	112/23
Corlar	DuPont Co. Inc. Wilmington, DE 19 898, US	233	2251, 614 el
Cornex® *CMR*	Teijin Ltd., Osaka 541, JP	2612	11
Coroplast	Coroplast Inc., Irving, TX 75 038, US	112	5
Coroplast®	Coroplast Fritz Müller 42279 Wuppertal, DE	11, 12, 13, 24, 26, 38	3, 4
Correx	Cordek Ltd. Billinghurst, W. Sussex, GB	112	7661
Corrolite®	Reichhold Chemicals Inc. Jacksonville, FL 32 245, US	2421	11
Corvic®	ICI PLC, Welwyn Garden City, Herts. AL7 1 HH, GB	131	1
Cosmax	Asahi Chemical Ind. Co. Ltd. Tokyo, JP	16	4
Courtelle®	Courtaulds PLC, Bradford BD1 1EX W. Yorkshire, GB	161	72
Courthene®	Courtaulds Chemicals & Plastics Ltd. Spondon, Derby DE2 7BP, GB	1111	4, 5
Courtoid®	–,,–	4222	5
Cova	Forbo-CP, Cramlington, North. NE23 8AQ, GB	132	4
CR-Compound	Borealis Compounds AB 12 613 Stockholm, SE	35	11 se

8.3 Trade Names

Name	Company (see 8.1, page 479)	Resin (see 8.2.1, page 481)	Forms of delivery (s. 8.2.2, p. 484)
Crastin®	DuPont Co. Inc. Wilmington, DE 19 898, US	2412	2
Crelan®	Bayer AG, 51368 Leverkusen, DE	16, 2632	114
Cremonil®	La Nuova Cremonese 26 025 Pandino (CR)	261	2
Crestomer®	Scott Bader Co. Ltd. Wollaston, Wellingborough, Norths. NN9 7RL, GB	242	112, 113, 114
Crilux	Critesa S.A., Barcelona, ES	163	51, 53, 55
Cristalite®	Schock & Co. GmbH 73605 Schorndorf, DE	16	112
Cristamid	Elf Atochem 92 091 Paris, La Défense 10, FR	261	11
Cropolamid	SCM Chemicals Corp. Baltimore, MD 21 202, US	26	114
Crow®	Tufnol Ltd. Birmingham B42 2TB, GB	21	612
Crylor®	Rhône-Poulenc 92 097 Paris, La Défense 2, FR	161	72
Cryovac®	Cryovac Div. W.R. Grace & Co. Duncan, SC 29 334, US	111, 1311, 133, 261	411, 412
Crystalene	Crystal-X Corp. Darby, PA 19 023, US	111	4
Crystic Fireguard®	Scott Bader Co. Ltd. Wollaston, Wellingborough, Norths. NN9 7RL, GB	242	113
Crystic Impel®	–,,–	242	224, 2251
Crystic Impreg®	Scott Bader Co. Ltd. Wollaston, Wellingborough, Norths. NN9 7RL, GB	242	2251
Crystic®	–,,–	242	112, 113
CSM-Compound	Borealis Compounds AB 12 613 Stockholm, SE	342	11 se
Cumar	Neville Chemical Co. Pittsburgh, PA 15 225, US	19	114, 115
Curon	Reeves Brothers Canada Ltd., Toronto, Ontario M8W 2T2, CA	2631	71

8.3 Trade Names

Name	Company (see 8.1, page 479)	Resin (see 8.2.1, page 481)	Forms of delivery (s. 8.2.2, p. 484)
Cuticulan®	Odenwald-Chemie GmbH 69250 Schönau, DE	1111	43
Cutilan®	−„−	1113	43
Cutipylen	−„−	112	41
Cuvolt	Isovolta AG 2355 Wiener Neudorf, AT	212	611, 612
CX-Serie	Unitika Ltd., Osaka, JP	261	2 tra
Cyanacryl®	American Cyanamid Corp. Wayne, NJ 07 470, US	345	2
Cyanaprene®	−„−	264	11
Cyandrothane	−„−	2631 s	121
Cyanolit Crystal	Panacol-Elosol GmbH 6000 Frankfurt/M., DE	167	73
Cycolac®	GE Plastics Pittsfield, MA 01 201, US	1221	117, 2
Cycoloy®	−„−	1221+2411	124
Cycom®	Cyanamid Aerospace Products Ltd. Wrexham, Clwyd LL13 9UF, GB	233	2251, 2261
Cycovin®	GE Plastics Pittsfield, MA 01 201, US	1221/131	2
Cymel®	American Cyanamid Corp. Wayne, NJ 07 470, US	223	2
Cyrolite	Cyro Industries Woodcliff, NJ 07 675, US	163	2
Cyrolite®	Röhm GmbH, 64293 Darmstadt, DE	1641	21
Cytop	Asahi Glass Co., Tokyo 100, JP	135 s	123 tra
Cytor	American Cyanamid Corp. Wayne, NJ 07 470, US	2633	110 TE

D

d-c-fix®	Konrad Hornschuch AG 74679 Weissbach, DE	132	422
Dacron®	DuPont Co. Inc. Wilmington, DE 19 898, US	2412	72
Daiamid®	Daicel Chemical Ind. Ltd. Tokyo, JP	261	2
Daicel	−„−	422	2

8.3 Trade Names

Name	Company (see 8.1, page 479)	Resin (see 8.2.1, page 481)	Forms of delivery (s. 8.2.2, p. 484)
Daiel	Daikin Kogyo Co. Ltd., Osaka, JP	348	11
Daiflon	−„−	1352	11, 2
Daiso	Osaka Soda Co. Ltd., Osaka, JP	243	111, 2
Daiso DAP	Daikin Kogyo Co. Ltd., Osaka, JP	243	111
Daisolac®	Osaka Soda Co. Ltd., Osaka, JP	1114	11
Daltocel®	ICI Polyurethanes 3078 Kortenberg, BE	2633	117
Daltolac®	−„−	2633	117
Daltorez®	−„−	2632, 2633	117, 120
Daltotherm	−„−	263	117
Danar 1000®	Dixon Industries Corp. Bristol, RI 02 809, US	274	4
Danat®	−„−	274	413 s
Danulon	Viscosefaserfabrik Nyergesujfalo, HU	261	35
Daotan®	Hoechst AG 65929 Frankfurt, DE	244, 26	114, 121, 123
Daplen®	PCD Polymere GmbH, 4021 Linz, AT	112	2
Dapon®	FMC Corp. Philadelphia, PA 19 103, US	243	111, 22
Daran	W. R. Grace & Co., Organic Chemicals Lexington, MA 02 173, US	133	733
Daratak	−„−	151	11, 121
Darex	−„−	322	11
Daron	DSM Resins BV, 8000 AP Zwolle, NL	242	11, 112, 113
Dartek®	DuPont Co. Inc. Wilmington, DE 19 898, US	261	411
Darvic®	Wardle Storeys PLC Manningtree, Essex CO11 1NJ, GB	1311	51, 52, 53, 55
Dayplas	Dayton Plastics Inc. Dayton OH 45 419, US	1351	3
Decarglas	Degussa AG 60287 Frankfurt, DE	2411	55
Decelith®	ECW-Eilenburger Chemie Werk AG i.GV. 04838 Eilenburg, DE	1311, 1313, 1314, 132	2 el, med, se, s, tra, 51, 52, 53, 54, 33, 34

8.3 Trade Names

Name	Company (see 8.1, page 479)	Resin (see 8.2.1, page 481)	Forms of delivery (s. 8.2.2, p. 484)
Declar	DuPont Co. Inc. Wilmington, DE 19 898, US	2351	1
Deconyl	Plascoat Int. Ltd. Sheerwater, Woking, Surrey, GB	261	2
decospan®	André & Gernandt 69434 Hirschhorn, DE	223	761
Degadur®	Degussa AG 60287 Frankfurt, DE	16	123, 143, 145, 146
Degalan®	–,,–	162, 163	114, 116
Degaroute	–,,–	163	112
Deglas	Isovolta AG 2355 Wiener Neudorf, AT	24	614
Deglas®	Degussa AG 60287 Frankfurt, DE	163	51, 53, 54, 55
Dehoplast	A. & E. Schmeing 57399 Kirchhundem, DE	1113, 112	33, 43, 433, 51
Dekadur	Deka Rohrsysteme 35228 Dautphetal, DE	1311, 1314, 1312	31 tra
Dekadur-C	–,,–	1312	31
Dekalen H	–,,–	1113	31
Dekaprop®	–,,–	112, 1121, 237	31 se
Dekasab	–,,–	1221	31
Dekazol	–,,–	4224	31
dekodur®	André & Gernandt 69434 Hirschhorn, DE	223	621
dekorial®	–,,–	223	621
Dekorit F®	Raschig AG, 67061 Ludwigshafen, DE	211	51 corr
Dekorit M®	–,,–	211	51
Delifol®	DLW AG, Deutsche Linoleum-Werke 74301 Bietigheim-Bissingen, DE	132	435, 4351
Delignit®	Blomberger Holzindustrie, B. Hausmann GmbH & Co. KG 32817 Blomberg, DE	21	613
Dellatol	Bayer AG, 51368 Leverkusen, DE	–	91
Dellit	Von Roll Isola 4226 Breitenbach, CH	212	611

8.3 Trade Names

Name	Company (see 8.1, page 479)	Resin (see 8.2.1, page 481)	Forms of delivery (s. 8.2.2, p. 484)
Delmer	Asahi Chemical Ind. Co. Ltd. Tokyo, JP	163	111
Delpet	–,,–	163	11, 2
Delrin®	DuPont Co. Inc. Wilmington, DE 19 898, US	2311	2
Delta-Folie	Ewald Dörken AG 58313 Herdecke, DE	111	4311
Deltra	Porvair P.L.C., King's Lynn, Norfolk PE30 2 HS, GB	262	712, 437
Denka ER	Denki Kagaku Kogyo, Tokyo 100, JP	343	111
Denka LCS	–,,–	131/323	119
Denka Malecca	–,,–	27 Cop	21, 22
Denka® Arena	–,,–	2412	2
Denkastyrol	–,,–	12	2
Denkavinyl	–,,–	131, 1314	2
Densite®	General Foam Corp. Paramus, NJ 07 652, US	263	711 se
Depron	Hoechst Diafoil GmbH 65174 Wiesbaden, DE	121	414
Desmalkyd®	Bayer AG, 51368 Leverkusen, DE	244	114
Desmobond®	Miles Chemical Corp. Pittsburgh, PA 15 205, US	233/263	115
Desmocap®	Bayer AG, 51368 Leverkusen, DE	262, 263	111
Desmocast®	–,,–	263	112
Desmocoll®	–,,–	2632	111f., 731
Desmodur®	–,,–	2631	111
Desmoflex®	Bayer AG, 51368 Leverkusen, DE	263	114
Desmolac®	–,,–	262	116
Desmopan®	–,,–	264	2
Desmophen®	–,,–	2632, 2633, 162	111
Desmotherm®	–,,–	263	114
Destex	DSM Resins BV, 8000 AP Zwolle, NL	2412	224, 2251
Dexcarb®	Dexter Corp. Windsor Locks, CT 06 096, US	2411/263	2

8.3 Trade Names

Name	Company (see 8.1, page 479)	Resin (see 8.2.1, page 481)	Forms of delivery (s. 8.2.2, p. 484)
Dexel®	Courtaulds Chemicals & Plastics Ltd. Spondon, Derby DE2 7BP, GB	4222	2
Dexlon®	Dexter Corp. Windsor Locks, CT 06 096, US	112/261	2, 224
Dexpro®	–„–	112/261	224
Diaclear®	Mitsubishi Chemical Corp. Tokyo, JP	261 s	11
Diafoil	Hoechst Diafoil GmbH 65174 Wiesbaden, DE	2412	41
Diakon®	ICI PLC, Welwyn Garden City, Herts. AL7 1 HH, GB	163	2
Diamid®	Daicel Chemical Ind. Ltd. Tokyo, JP	261	2
Diamiron	Mitsubishi Plastics Inc. Tokyo, JP	261	411, 417
Diapet	Mitsubishi Rayon Co. Ltd. Tokyo, JP	1221	2
Diaprene	Advanced Elastomer Systems L.P. St. Louis, MO 63 167, US	112+3411	110 TE
Diarex®	Mitsubishi Monsanto Chemical Co. Tokyo, JP	121, 1223	2
Diaron	Reichhold Ltd. Mississauga, ON L4Z 1S1, CA	223	111
Diawrap	Mitsubishi Plastics Inc. Tokyo, JP	132	411
Dieglas	Glastic Corp. Cleveland, OH 44 121, US	242	224
Dielektrite	Industrial Dielectrics Inc. Noblesville, IN 46 060, US	242	112, 224
Diene 1000	Firestone Synthetics Rubber and Latex Co., Akron, OH 44 301, US	3221	11
Diofan®	BASF Aktiengesellschaft 67063 Ludwigshafen, DE	133	121
Diorez®	Kemira Polymers Ltd. Stockport, Cheshire SK12 5BR, GB	2632	111
Diprane®	–„–	263	111, 112, 14
Disflamoll®	Bayer AG, 51368 Leverkusen, DE	–	91
Dispercoll	–„–	163	121

8.3 Trade Names

Name	Company (see 8.1, page 479)	Resin (see 8.2.1, page 481)	Forms of delivery (s. 8.2.2, p. 484)
Divinycell	Divinycell International AB 31222 Laholm, SE	131/261 s	71
DLW-EPDM	DLW AG, Deutsche Linoleum-Werke 74301 Bietigheim-Bissingen, DE	3411	43
DLW-Hypalon	–„–	1114/342	43
Dobeckan FN®	BASF Farben+Fasern AG 22309 Hamburg, DE	271	114 el
Dobeckan®	BASF Lacke+Farben AG 48165 Münster, DE	242	112
Dobeckot®	–„–	233	112
Doeflex	Doeflex Ind. Ltd. Redhill, Surrey RH1 2NR, GB	112	4, 2212/2213/ 53
Dolan®	Hoechst AG 65929 Frankfurt, DE	161	72
Dolanit 10	–„–	16	72
Dolphon®	John C. Dolph Co., Monmouth Junction, NJ 08 852, US	233, 242	2 el
Dolph's®	–„–	–	–
Doplan	Südwestdeutsche Sperrholzwerke E. Dold, 77694 Kehl, DE	–	761
Doplex	–„–	–	761
Dorfix	Egyesült Negyimüvek 1172 Budapest, HU	216	113
Dorix®	Bayer AG, 51368 Leverkusen, DE	261	72
Dorlastan®	–„–	37	35, 72
Dorlyl	Dorlyl 92100 Boulogne Billancourt, FR	131	2
Dorolac	Egyesült Negyimüvek 1172 Budapest, HU	21	113, 114
Doroplast	–„–	211, 212	232
Dowlex®	The Dow Chemical Company Midland, MI 48641-2166, US	1111L	2
Draimoco®	DSM Italia s.r.l., 22 100 Como, IT	242	224
Drakafoam	British Vita Co. Ltd. Manchester M24 2D3, GB	2633	71
Dralon®	BASF Aktiengesellschaft 67063 Ludwigshafen, DE	161	72

8.3 Trade Names

Name	Company (see 8.1, page 479)	Resin (see 8.2.1, page 481)	Forms of delivery (s. 8.2.2, p. 484)
Driscopipe	Phillips Petroleum Chemicals NV 3090 Overijse, BE	111	31
Dry-Stat	Web Technologie Oakville, CT 06 779, US	2412	146
Dryflex®	Perstorp AB, 28 480 Perstorp, SE	1253	11
Dryton XL	Monsanto Chemical Co. Akron, OH 44 314–9914, US	322, 3411	110 TE
Dualoy®	Ciba-Geigy, Composites Div. Anaheim, CA 9207–2018, US	233	312
Dularit	Henkel KGaA, 40589 Düsseldorf, DE	233	115
Dunova®	Bayer AG, 51368 Leverkusen, DE	16	72 s
Duoflex	Röhrig & Co., 30453 Hannover, DE	111/132 se	32, 34
Duplothan	H. Hützen GmbH 41747 Viersen, DE	263	71
Duplotherm	–„–	263	71
Durabit	Durabit Bauplast GmbH & Co. KG 4050 Traun, AT	1113, 115+ Bitumen	432, 433, 435
Duracap	B. F. Goodrich Chemical Group Cleveland, OH 44 131, US	131	1
Duracarb®	PPG Industries Inc. Pittsburgh, PA 15 272, US	2632 s	11
Duracon®	Polyplastics, Tokyo, JP	2312	2
Duracryn®	DuPont Co. Inc. Wilmington, DE 19 898, US	–	110 TE
Duragen	Goodyear Tire & Rubber Co. Akron, OH 44 316, US	3211	11
Duragrid	London Artid Plastics, London, GB	112	442
Dural	Alpha Plastics Corp. Pineville, NC 28 134, US	1314	2
Duralex®	–„–	1314, 264	2
Duraloy®	Hoechst AG 65929 Frankfurt, DE	231/2412	2, 224

Name	Company (see 8.1, page 479)	Resin (see 8.2.1, page 481)	Forms of delivery (s. 8.2.2, p. 484)
Duramix®	Isola Werke AG, 52353 Düren, DE	242	224
Duranex®	Polyplastics, Tokyo, JP	2412	2
Durapipe	Glynwed Int. Inc. Sheldon, Birmingham, GB	1221, 131	31
Durapol®	Isola Werke AG, 52353 Düren, DE	242	224, 225, 614
Durapox®	–„–	233	22
Durapreg®	–„–	242	2251, 614
Duraver® E-Cu	–„–	223, 233, 28, 233	312, 614, 6141, 614-Cu
Durax®	–„–	21, 22, 242	22, 23, 24, 25
Durayl	American Filtrona Corp. Richmond, VA 23 233, US	163	2
Durel®	Hoechst AG 65929 Frankfurt, DE	2114	2, 224
Durelast®	Kemira Polymers Ltd. Stockport, Cheshire SK12 5BR, GB	264	73
Durestos®	TBA Industrial Products Ltd. Rochdale, Lancs., GB	21, 233	6111, 6121
Durethan®	Bayer AG, 51368 Leverkusen, DE	261, 261+32	2, 11, 224
Durette	Durette-Kunststoff GmbH & Co. KG 52304 Düren, DE	1311	341
Durex	Oxychem, Durez Div. Tonawanda, NY 14 120, US	242	11
Durez®	Occidental Chemical Corp. Dallas, Texas 75380, US	21, 242, 243	113, 2, 22
Duripor	Binné & Söhne 25421 Pinneberg, DE	121	71
Durocron	Mitsubishi Rayon Co. Ltd. Tokyo, JP	16	11
Durodet®	Mitras Kunststoffe GmbH 92637 Weiden, DE	242, 112	61, 64, 2251
Duroform Composite	Duroform GmbH & Co. KG 56357 Miehlen, DE	242, 2421	224, 225, 2251 r, el, elc, se
Duroftal®	Hoechst AG 65929 Frankfurt, DE	244	114, 123
Durolite®	Mitras Kunststoffe GmbH 92637 Weiden, DE	242	61, 2251

Name	Company (see 8.1, page 479)	Resin (see 8.2.1, page 481)	Forms of delivery (s. 8.2.2, p. 484)
Durolon	Policarbonatos do Brasil S.A. 06412–140 Barueri-Sao Paulo, BR	2411	2
Duropal	Duropal-Werk E. Wrede GmbH & Co KG, 59717 Arnsberg, DE	223	621
Durophen®	Hoechst AG 65929 Frankfurt, DE	214	113, 114
Durostone®	Röchling Haren KG 49733 Haren, DE	214, 233, 242, 2421	312, 342, 51, 614, 6141
Duroxyn®	Hoechst AG 65929 Frankfurt, DE	233 Ester	114
Dutral®	Montell Polyolefins Wilmington, DE 19894, US	1117, 1121	2
Dutralene®	–„–	1117	2
Dyflor®	Hüls AG, 45764 Marl, DE	1354	1
Dylark®	Arco Chemical Co. Newton Square, PA 19 073, US	122, 1225	2, 224
Dylene®	–„–	121	2
Dylite	–„–	264	110 TE
Dymetrol®	DuPont Co. Inc. Wilmington, DE 19 898, US	241	110 TE
Dynacoll®	Hüls AG, 45764 Marl, DE	241	115
Dynamar®	3 M Co., St. Paul, MN 55 144, US	348	11, 9
Dynapol®	Hüls AG, 45764 Marl, DE	241	114
Dynapor®	–„–	211	117
Dyneema®	DSM, 6160 AP Geelen, NL/ Toyobo Co. Ltd., Osaka, JP	111 s	72
Dynodren	Dyno Industrier A.S., Oslo 1, NO	131	31
Dynofen	–„–	21	114
Dynoform	–„–	212	2
Dynomin	–„–	222, 224	114
Dynopon	Dyno Industrier A.S., Oslo 1, NO	233	114
Dynorit	–„–	221	115, 116
Dynos®	HT Troplast AG 53840 Troisdorf, DE	4211	–
Dynosol	Dyno Industrier A.S., Oslo 1, NO	21	113, 115

Name	Company (see 8.1, page 479)	Resin (see 8.2.1, page 481)	Forms of delivery (s. 8.2.2, p. 484)
Dynotal	Dyno Industrier A.S., Oslo 1, NO	241, 244	123
Dynoten	–„–	111	41
Dynova®	–„–	21	117
Dytherm®	Arco Chemical Co. Newton Square, PA 19 073, US	1225	117, 2
Dytron	Advanced Elastomer Systems L.P. St. Louis, MO 63 167, US	264	110 TE

E

Name	Company	Resin	Forms of delivery
EACM-Compound	Borealis Compounds AB 12 613 Stockholm, SE	345	2
Eastar	Eastman Chemical Int. Kingsport, TN 37662, US	241	2
Easypoxy	Conap Div. of WFI Olean, NY 14 760, US		
Eccofoam	Emerson & Cuming Inc. Canton, MA 02 021, US	233	117
Eccogel	–„–	233	112 el
Eccomold	–„–	233	2
Eccoseal	–„–	233	112
Eccosorb	W. R. Grace & Co., Organic Chemicals Lexington, MA 02 173, US	263	71 s
Eccothane	–„–	263	112, 146
Ecdel®	Eastman Chemical Intern. Kingsport, TN 37 662, US	2412 s	110 TE
Ecocryl	Elf Atochem 92 091 Paris, La Défense 10, FR	1225	11, 12
Ecofelt	Chemie Linz AG, 4040 Linz, AT	112	721
Ecolo F	Mitsubishi Petrochemical Co. Ltd. Tokyo, JP	1121	22
Econol	Sumitomo Chemical Co. Ltd. Tokyo 103, JP	2412	119
Ecostarplus	DSM Compounds UK Ltd. Eastwood, Lancs., GB	1	a
Ecostarplus	Frost & Sullivan 6000 Frankfurt/M., DE	1	a
Edenol®	Henkel KGaA, 40589 Düsseldorf, DE	–	91

8.3 Trade Names

Name	Company (see 8.1, page 479)	Resin (see 8.2.1, page 481)	Forms of delivery (s. 8.2.2, p. 484)
Edistir®	ECP EniChem Polimeri srl. 20 124 Mailand, IT	121	2
Editer	–„–	1221	2, 224
Efroit	Ernst Frölich GmbH 37520 Osterode, DE	1311, 132	2, 32
Efweko	Degussa AG 60287 Frankfurt, DE	263	12
Egelen	Egeplast, Werner Strumann GmbH & Co. 48282 Emsdetten, DE	111, 1113	31
Egerit®	Gehr-Kunststoffwerk GmbH 68219 Mannheim, DE	1311	3
Ekonol	Carborundum Corp. Niagara Falls, NY 14 302, US	2414 s	11, 2
Ektar®	Eastman Chemical Intern. Kingsport, TN 37 662, US	237	224
Elamed	Chemitex-Elana, Torun, PL	2412	2
Elan	Putsch GmbH, 90427 Nürnberg, DE	112	2213, 2211
Elana	Chemitex-Elana, Torun, PL	2412	72
Elapor	EMW-Betrieb, 65582 Diez, DE	1116, 263	71 s
Elaslen®	Showa Denko K.K., Tokyo 105, JP	1114	11
Elast-o-Fluor®	Norton Performance Plastics Wayne, NJ 07 470, US	1351	32
Elastalloy	GLS Plastics Woodstock, IL 60 098, US	1	11, 120
Elastan®	Elastogran GmbH 49440 Lemförde, DE	263	117 s
Elaster	Nippon Zeon Co. Ltd., Tokyo, JP	1311+323	110 TE
Elastocoat®	Elastogran GmbH 49440 Lemförde, DE	263	146
Elastodrain	Zin Co Dachsysteme 72622 Nürtingen, DE	31	43 r
Elastoflex®	Elastogran GmbH 49440 Lemförde, DE	263	117
Elastofoam®	–„–	263	713 s
Elastolen®	–„–	263, 264	123, 146, 120
Elastolit®	–„–	263	117

8.3 Trade Names

Name	Company (see 8.1, page 479)	Resin (see 8.2.1, page 481)	Forms of delivery (s. 8.2.2, p. 484)
Elastollan®	Elastogran GmbH 49440 Lemförde, DE	264	110 TE
Elaston	Chemitex, Cellviskoza, Torun, PL	263	72
Elastonat®	Elastogran GmbH 49440 Lemförde, DE	263	117
Elastopal®	–,,–	263	7
Elastopan®	–,,–	263	117
Elastophen®	–,,–	263	117
Elastopor®	–,,–	263	117
Elastopreg®	–,,–	1	225 s
Elastopren®	Record-Kunststoffwerke GmbH 79232 March, DE	263	71
Elastosil®	Wacker-Chemie GmbH 81737 München, DE	38	112
Elastotec®	Elastogran GmbH 49440 Lemförde, DE	264	120
Elastotherm®	–,,–	263	117
Elastuf	Goodyear Tire & Rubber Co. Akron, OH 44 316, US	2413	2
Elasturan®	Elastogran GmbH 49440 Lemförde, DE	263	117
Elate	Akzo Engineering Plastics Inc. Evansville, IN 47 732, US	2631	111
Elaxar®	Shell International Co. Ltd. London, SE1 7NA, GB	12/341/12	110 TE
Electrafil®	Akzo Engineering Plastics Inc. Evansville, IN 47 732, US	11, 12, 13, 23, 24, 274	226 el
Electroglas	Glasflex Corp. Stirling, NJ 07 980, US	163	31, 33, 55
Elektroguard®	The Perma-Flex Mold Co. Inc. Columbus, OH 43 209, US	28	–
Elektroplast	Egyesült Negyimüvek 1172 Budapest, HU	21	224, 611
Elisol	Werner Hahm GmbH & Co. KG 42109 Wuppertal, DE	132	32 el
Elit	Chemitex-Elana, Torun, PL	2412	224

8.3 Trade Names

Name	Company (see 8.1, page 479)	Resin (see 8.2.1, page 481)	Forms of delivery (s. 8.2.2, p. 484)
Elite HH®	Monsanto Chemical Co., Plastics Div. St. Louis, MO 63 167, US	1224	11
Elitel	Chemitex-Elana, Torun, PL	2412	110 TE
Elitrex®	AEG Isolier- und Kunststoff GmbH 34123 Kassel, DE	242	2251, 2261
Elix®	Monsanto Chemical Co., Plastics Div. St. Louis, MO 63 167, US	122	151
Elkalite	Elkaplast, Bruxelles, BE	1221/1354	76
Elkoflex®	Elkoflex Isolierschlauchfabrik 10553 Berlin, DE	244	32 el
Elkosil®	–„–	28, 38	32 el
Elkotherm®	–„–	24, 263	32 el
Elmit	Mitsui Petrochemical Ind. Ltd. Tokyo, JP	112/261, 115 s/261	124
Elmo®	BASF Lacke+Farben AG 48165 Münster, DE	244	114 el
Elmotherm®	–„–	27	114 el
Eltex®	Solvay S.A., 1050 Bruxelles, BE	1112, 1113	11, 2
Eltex® P	–„–	112, 1121	11, 2
Elvacite®	DuPont Co. Inc. Wilmington, DE 19 898, US	163	114, 115, 2
Elvaloy®	–„–	111/131, 131 s	124
Elvamide®	–„–	261	114, 115, 2
Elvanol®	–„–	152	111
Elvax®	–„–	1115	94
Elvon®	–„–	15	94
Emblem	Unitika Ltd., Osaka, JP	133+151	411 s
Emflon	Pallflex Products Corp. Putnam, CT 06 260, US	135	413
Emi-X®	LNP Plastics Nederland BV 4940 Raamsdonksveer, NL	11	110 TE
Empee®	Monmouth Plastics Inc. Freehold, NJ 07 728, US	1113, 112, 1127	124, 2 sc
Emu®-Pulver	BASF Aktiengesellschaft 67063 Ludwigshafen, DE	1223, 125	111, 113
Enalon	Enalon Plastics Ltd. Tonbridge, Kent TN9 2BE, GB	2	61

8.3 Trade Names

Name	Company (see 8.1, page 479)	Resin (see 8.2.1, page 481)	Forms of delivery (s. 8.2.2, p. 484)
Enathene	Quantum Chemical Corp. Cincinnati, OH 42 249, US	111/162	11
Endur®	Rogers Corp., Rogers, CT 06 263, US	71	
Enplex®	Kanegafuchi Chemical Ind. Co. Ltd. Osaka, JP	1221/131	124
Ensolite	Uniroyal, Mishawaka, IN 46 544, US	1311	71 s
Envirez	PPG Industries Inc. Pittsburgh, PA 15 272, US	242	113
Enviroplastic	Planet Packaging Technologies San Diego, CA, US	232	11 a
Epcar®	B. F. Goodrich Chemical Group Cleveland, OH 44 131, US	3411	11
EPDM Semicon	Borealis Compounds AB 12 613 Stockholm, SE	3411	11 el
EPDM-Compound	–,,–	3411	11
Eperan	Kanegafuchi Chemical Ind. Co. Ltd. Osaka, JP	111, 112	71
Epibond®	Ciba Geigy AG, 4002 Basel, CH	233	115
Epichlomer®	Osaka Soda Co. Ltd., Osaka, JP	3411	11
Epidian	Ciech Chemikalien GmbH 00013 Warschau 1, PL	233	112, 113
Epikote®	Shell International Co. Ltd. London, SE1 7NA, GB	233	112, 113, 114, 115
Epo-tek	Epoxy Technology Inc. Billerica, MA 01 821, US	233	11
Epocast®	Ciba Geigy AG, 4002 Basel, CH	233	112, 2
Epocryl®	Ashland Chemical Corp. Columbus, OH 43 216, US	233 s	111
Epodil L®	Anchor Plastics Inc. Great Neck, NY 11 021, US	53	11
Epodite	Showa High Polymer Co. Ltd. Tokyo, JP	233	11
Epoflex®	Von Roll Isola 4226 Breitenbach, CH	233	442 el
Epolene®	Eastman Chemical Intern. Kingsport, TN 37 662, US	111 s, 112 s	94

8.3 Trade Names 525

Name	Company (see 8.1, page 479)	Resin (see 8.2.1, page 481)	Forms of delivery (s. 8.2.2, p. 484)
Epolite	Hexcel Corp. Chemical Div. Chatsworth, CA 93 111, US	233	112, 113
Epon®	Shell International Co. Ltd. London, SE1 7NA, GB	233	11
Eponol®	–,,–	2633	110 TE
Epophen®	Borden Chemical Co. Geismar, LA 70 734, US	233	112, 115
Eposet	Hardman Inc. Belleville, NJ 07 109, US	233	11
Eposir®	Sirlite Srl., 20 161 Milano, IT	233	111
Eposyn	Copolymer Rubber & Chemical Corp. Baton Rouge, LA 70 821, US	341	11
Epotal®	BASF Aktiengesellschaft 67063 Ludwigshafen, DE	111	121
Epotuf®	Reichhold Chemicals Inc. Jacksonville, FL 32 245, US	233	114
Epovoss	FAW Jacobi AB, SE	233	112, 113
Epoxical®	United States Gypsum Co. Chicago, IL 60 606, US	233	112, 114
Epsyn 70 A®	Monsanto Chemical Co. Akron, OH 44 314–9914, US	3411	11
Era®	Gustav Ernstmeier GmbH & Co. KG 32049 Herford, DE	132, 263	42, 43
Eraclear®	ECP EniChem Polimeri srl. 20 124 Mailand, IT	1111 L	2
Eraclene®	–,,–	1115	2
Ercom®	Ercom Composite Recycling GmbH 76437 Rastatt, DE	242	2262
Ercusol®	Bayer AG, 51368 Leverkusen, DE	16	121
Eref®	Solvay S.A., 1050 Bruxelles, BE	261/112	224
Ergeplast®	Roga KG Dr. Loose GmbH & Co. 50389 Wesseling, DE	111, 131, 132	32, 34
Ertacetal®	Erta-Plastics N.V., 8880 Tielt, BE	231	31, 33, 342, 51
Ertalon®	–,,–	261	31, 33, 342, 51
Ertalyte®	–,,–	2412	31, 33, 342, 51
Esall	Sumitomo Chemical Co. Ltd. Tokyo 103, JP	112/1253	2

Name	Company (see 8.1, page 479)	Resin (see 8.2.1, page 481)	Forms of delivery (s. 8.2.2, p. 484)
Esbrite®	Sumitomo Chemical Co. Ltd. Tokyo 103, JP	121	2
Escalloy®	A. Schulman Inc. Akron, OH 44 309, US	112, 1221, 1222, 2361, 237, 2411, 2412, 261	2
Escor®	Exxon Chemicals America Houston, TX 77 001, US	1115	94
Escorene alpha	–,,–	1111 L	2
Escorene Micro	–,,–	111	141
Escorene ultra	–,,–	1115	2
Escorene®	–,,–	1111, 1113, 1115,	111, 146, 21
Escorez®	–,,–	53	11
Eska®	Mitsubishi Rayon Co. Ltd. Tokyo, JP	163	72 s
Eslon FFU	Sekisui Chemical Co. Ltd. Osaka 530, JP	263	614/71
Espesol	Sumitomo Seika Chemicals Co. Ltd. Osaka, JP	23	12
Espet	Toyobo Co. Ltd., Osaka 530, JP	2412	2, 411
Esprene®	Sumitomo Chemical Co. Ltd. Tokyo 103, JP	3411	11
Esprit	Caradon Celuform Ltd. Aylesford, Maidstone, Kent ME20 75X, GB	131	34/713
Estaloc®	B. F. Goodrich Chemical Group Cleveland, OH 44 131, US	264	110 TE/224
Estane®	–,,–	264	110 TE
Estar	Mitsui Toatsu Chemicals Inc. Tokyo, JP	242	11
Este®	Max Steier GmbH & Co. 25337 Elmshorn, DE	11, 132	431, 432
Esteform	Chromos – Ro Polimeri 4100 Zagreb, Croatia	242	22
Estemix	–,,–	242	22
Esteral	Makhteshim Chemical Works Ltd. Beer-Sheva, IL	242	11, 114

Name	Company (see 8.1, page 479)	Resin (see 8.2.1, page 481)	Forms of delivery (s. 8.2.2, p. 484)
Estyrene®	Nippon Steel Chemical Co. Ltd. Tokyo, JP	1225	2
ET-Polymer	Borealis Compounds AB 12 613 Stockholm, SE	111	110 TE
ET-Semicon	–,,–	111/332	117
Ethofil®	Akzo Engineering Plastics Inc. Evansville, IN 47 732, US	1113	224
Ethylux®	Westlake Plastics Co. Lenni, PA 19 052, US	111	3, 4, 5
Etinox	Aiscondel S.A., Barcelona 13, ES	131	2
Etronax	Elektro-Isola A/S, 7100 Vejle, DK	212, 233	612, 6121
Etronax G	–,,–	212, 223, 233, 235, 242, 28	614, 6141
Etronit	Elektro-Isola A/S, 7100 Vejle, DK	212, 223, 233	611, 6121
Eucarigid®	Manufactures de Cables électriques et de Caoutchouc S.A., Eupen, BE	1311, 1312	31
Eurelon®	Witco GmbH, 59180 Bergkamen, DE	261, 2611	112, 113, 114, 115
Euremelt®	–,,–	26	115
Eurepox®	–,,–	233	112, 113, 114, 115
Euresyst®	–,,–	233	11
Eurocell	Europlastic Pahl & Pahl GmbH & Co, 40472 Düsseldorf, DE	263	71
Eurodrain	Hegler Plastik GmbH 97714 Oerlenbach, DE	1311	31, 32
Euroflex M	Scheuch GmbH & Co. KG 64367 Mühltal, DE	11+2412+Al	412
Europan	Europlastic Pahl & Pahl GmbH & Co, 40472 Düsseldorf, DE	263	71
Europhan®	4P Folie Forchheim GmbH 91301 Forchheim, DE	1311	411
Europlastic®	Europlastic Pahl & Pahl GmbH & Co, 40472 Düsseldorf, DE	2632, 2633	71
Europlex®	Röhm GmbH, 64293 Darmstadt, DE	23	4
Europrene®	ECP EniChem Polimeri srl. 20 124 Mailand, IT	12/32/12, 12/33/12	110 TE, 120

Name	Company (see 8.1, page 479)	Resin (see 8.2.1, page 481)	Forms of delivery (s. 8.2.2, p. 484)
Eutan®	Acla-Werke GmbH 51065 Köln, DE	263	51
Evaco®	Neste Oy Chemicals, 02 151 Espoo, FI	1115 Cop	2
Evaflex	DuPont-Mitsui Polychemical Co. Ltd. Tokyo, JP	1115	11, 2
Eval®	Quantum Chemical Corp. Cincinnati, OH 42 249, US	1115 s	11, 411, 412
Evalastic	Alwitra KG, Klaus Göbel 54296 Trier, DE	112+3411	4351
Evalon	–„–	1115/131	435
Evatane®	Elf Atochem 92 091 Paris, La Défense 10, FR	1115	111, 2
Evatate®	Sumitomo Chemical Co. Ltd. Tokyo 103, JP	1115	2
Evazote®	BP Chemicals International Ltd. London SW1W OSU, GB	1115	71
Everlite®	Everlite A/S, Skaevinge, DK	1, 1311	61, 614
Evoprene®	Evode Plastics Ltd. Syston, Leicester LE7 8PD, GB	21, 2212	2
Exact	Exxon Chemicals America Houston, TX 77 001, US	1111 L	115, 731, 733
Excelon	Armstrong World Ind. Inc. Lancaster, PA 17 604, US	132	442
Excelprint®	Scott Bader Co. Ltd. Wollaston, Wellingborough, Norths. NN9 7RL, GB	125	121
Exnor®	Norton Performance Plastics Wayne, NJ 07 470, US	131	34/71
Exolite®	Cyro Industries Woodcliff, NJ 07 675, US	163, 2411	55
Expancel	Expancel, Casco Nobel AB 85 013 Sundsvall, SE	161	117
Extir®	ECP EniChem Polimeri srl. 20 124 Mailand, IT	121	117/2
Extrel®	Exxon Chemicals America Houston, TX 77 001, US	112	411
Exulite	Cyro Industries Woodcliff, NJ 07 675, US	163	55

Name	Company (see 8.1, page 479)	Resin (see 8.2.1, page 481)	Forms of delivery (s. 8.2.2, p. 484)
Exxtraflex®	Exxon Chemicals America Houston, TX 77 001, US	11	41
Eymyd	Ethyl Corp. Baton Rouge, LA 70 801, US	1, 2	2251
Eypel F®	–,,–	348 s	11
F			
F ...	Makhteshim Chemical Works Ltd. Beer-Sheva, IL	233	9 se
Fabelnyl	Tubize Plastics S.A., 1360 Tubize, BE	261	2, 22
Fabeltan®	–,,–	264	2
Fablon®	Forbo-CP, Cramlington, North. NE23 8AQ, GB	111, 132	4
Fabtex®	Clopay Corp. Plastics Products Div. Cincinnati, OH 45 202, US	111	42, 45
Faradex	DSM, Polymers & Hydrocarbons 6130 AA Sittard, NL	112, 1221, 1221/2411	2 el
Fardem	Fardem Ltd., Louth, Lincs., GB	111	4
Farfen	Fabbrica Adesivi Resine S.p.A. Cologno Monzese, IT	214	113, 123, 73
Feinmicaglas	Isovolta AG 2355 Wiener Neudorf, AT	233	616
Felor®	DuPont Co. Inc. Wilmington, DE 19 898, US	261	35
Femso®	Franz Müller & Sohn, Femso Werk 61440 Oberursel, DE	11, 1115, 1117, 12, 1253, 131, 132, 1354, 231, 2413, 261, 264, 2412	3
Fenlac	AMC-Sprea S.p.A., 20 101 Milano, IT	21	113
Fenochem	Chemiplastics S.p.A. 20 151 Mailand, IT	2121	2
Fenoform	Chromos – Ro Polimeri 4100 Zagreb, Croatia	212	2
Ferobestos	Tenmat Ltd., Trafford Pk., Manchester M17 1RU, GB	21	615

8.3 Trade Names

Name	Company (see 8.1, page 479)	Resin (see 8.2.1, page 481)	Forms of delivery (s. 8.2.2, p. 484)
Feroform	Tenmat Ltd., Trafford Pk., Manchester M17 1RU, GB	212, 28	61
Feroglas	–,,–	21, 242	614
Ferrene®	Ferro Corp. Evansville, IN 47 711, US	115	22
Ferrex®	–,,–	1121	22
Ferro-Flex®	–,,–	112+3411	110 TE
Ferroflex®	Ferrozell GmbH 86199 Augsburg, DE	212, 214, 223, 233, 242, 28	312, 33, 436, 438, 6111, 6121
FerroLene	Ferro Eurostar, 95470 St. Landre, FR	112, 1127	110 TE, 2
Ferroplast®	Ferrozell GmbH 86199 Augsburg, DE	214	2261
Ferropreg	Ferro Corp., Composites Div. Los Angeles, CA 90 016, US	21, 233, 242, 27	2221, 2251, 2261
Ferrostat	Ferro Eurostar, 95470 St. Landre, FR	1	2 el
Ferrotron	Polypenco GmbH 51437 Bergisch Gladbach, DE	1351 s	3, 5
Ferrozell®	Ferrozell GmbH 86199 Augsburg, DE	212, 223, 233, 242, 28	214, 225, 611, 612, 614, 6141, 615
FF ...	Fränkische Rohrwerke GmbH + Co. 97486 Königsberg, DE	111, 1116, 1122, 113, 121, 1311, 261	31(71)
FF-Kabuflex	–,,–	1113	31 el
FF-pordrän	–,,–	121	71
FF-therm	–,,–	1116, 113	31
Fib	Putsch GmbH, 90427 Nürnberg, DE	112	224
Fibercast®	Fibercast Co. Sand Springs, OK 74 063, US	233, 242	6141
Fiberesin®	Fiberesin Industries Inc. Oconomowoc, WI 53 066, US	223	621
Fiberfil TN	DSM RIM Nylon vof 4202 YA Maastricht, NL	261	2
Fiberform	Fiberesin Industries Inc. Oconomowoc, WI 53 066, US	21	611

8.3 Trade Names

Name	Company (see 8.1, page 479)	Resin (see 8.2.1, page 481)	Forms of delivery (s. 8.2.2, p. 484)
Fiberloc®	B. F. Goodrich Chemical Group Cleveland, OH 44 131, US	131	224, 614
Fiberod	Polymer Composites Inc. Winona, MN 55 987, US	112, 1221, 231, 2411, 2412, 237, 263	22 s
Fiberstran	Akzo Engineering Plastics Inc. Evansville, IN 47 732, US	1	224
Fibredux®	Ciba-Geigy, Composites Div. Anaheim, CA 9207–2018, US	2	2261
Fibrelam®	–„–	233, 242	7662
Fibrolux	Fibrolux, 65719 Hofheim, DE	242	3
Fibron®	Fibron GmbH 75015 Bretten, DE	242, 614, 2251, 34, 51	225, 614
Filmon	Nyltech Italia Srl 20 020 Ceriano (MI), IT	261	411
Fina X	Isofoam S.A., 7170 Manage, BE	121	71
Finaclear®	Fina Chemicals 1040 Bruxelles, BE	32, 322	2
Finalloy®	–„–	–	22
Finaprene®	–„–	32, 322	2
Finapro®	–„–	112, 1121	2
Finathene®	–„–	1111, 1112, 1113	2
Firewall FRB	Coroplast Inc., Irving, TX 75 038, US	112	5 se
Fish-paper	Weston Hyde Products (EVC) Hyde, Cheshire SK14 4EJ, GB	4211	5
Flakeglas	Owens Corning Fiberglas Corp. Toledo, OH 43 659, US	242	225
Flakeline	–„–	242	143
Flamulit	Herberts GmbH 50943 Köln, DE	1111	141
Flex	Röhrig & Co., 30453 Hannover, DE	132	32, 34
Flex-O-Crylic	Flex-O-Glass Inc. Chicago, IL 60 051, US	163	55
Flex-O-Film	–„–	1115 s, 4224, 4225	411, 53

Name	Company (see 8.1, page 479)	Resin (see 8.2.1, page 481)	Forms of delivery (s. 8.2.2, p. 484)
Flexel	B. F. Goodrich Chemical Group Cleveland, OH 44 131, US	–	110 TE
Flexibel	Felten & Guilleaume, 51063 Köln, DE	2, 27	225, 2261
Flexifilm	Tredegar Film Products B.V. Kerkrade, NL	111	43 s
Flexipol	Flexible Products Co. Marietta, GA 30 061, US	263	11
Flexirene	ECP EniChem Polimeri srl. 20 124 Mailand, IT	1111, 1111 L	113, 2
Flexline®	Elf Atochem 92 091 Paris, La Défense 10, FR	2412	35
Flexocel®	Baxenden Chemical Co. Ltd. Accrington, Lancs. BB5 2SL, GB	263	117
Flexom	Sommer B.T.P. Dtschl. GmbH 6000 Frankfurt/M. 60, DE	1115/13, 132	43
Flexomer®	Union Carbide Corp. Danbury, CT 06 817, US	1111 L	11
Flexvin®	Techno-Chemie Kessler & Co. GmbH 64546 Mörfelden-Walldorf, DE	112, 1127, 1315, 132, 264, 343	32, 321
Flo-Blen	Sumitomo Seika Chemicals Co. Ltd. Osaka, JP	11	11
flo-foam	Flo-pak GmbH 89542 Herbrechtingen, DE	263	117
flo-pak	–„–	121	71 s
Flo-Thene	Sumitomo Seika Chemicals Co. Ltd. Osaka, JP	11	11
Flo-Vac	–„–	11	11
Flomat	DSM Compounds UK Ltd. Eastwood, Lancs., GB	242	2251
Flomat	Frost & Sullivan 6000 Frankfurt/M., DE	242	2251
Floratroop	Hegler Plastik GmbH 97714 Oerlenbach, DE	1121	31, 32
Florit	Mayer Enterprises Ltd., Coating Dept., Tel Aviv, IL	415/53	
Fluobond	James Walker & Co. Ltd. Woking, Surrey GU22 8AP, GB	1351	33, 4, 5

8.3 Trade Names

Name	Company (see 8.1, page 479)	Resin (see 8.2.1, page 481)	Forms of delivery (s. 8.2.2, p. 484)
Fluon®	ICI PLC, Welwyn Garden City, Herts. AL7 1 HH, GB	1351	11, 121
Fluorel®	3 M Co., St. Paul, MN 55 144, US	348	1
Fluorocomp®	LNP Plastics Nederland BV 4940 Raamsdonksveer, NL	135	22 s
Fluoroloy	Fluorocarbon Anaheim, CA 92 803, US	1351	11
Fluoromelt	LNP Plastics Nederland BV 4940 Raamsdonksveer, NL	1351, 1352, 1354	224, 226
Fluorosint®	Polypenco GmbH 51437 Bergisch Gladbach, DE	1351	32, 33, 42, 51
Fluran®	Norton Performance Plastics Wayne, NJ 07 470, US	348	32 med
Foamex®	Airex AG., 5643 Sins, CH	1311	51
Foamosol	Watson Standard Co. Pittsburgh, PA 15 238, US	131	117
Foamspan®	A. Schulman Inc. Akron, OH 44 309, US	112, 1221, 1222, 2361, 237, 2411, 2412, 261	2
Folitherm	4P Folie Forchheim GmbH 91301 Forchheim, DE	112	43
Fomox	Bayer AG, 51368 Leverkusen, DE	s	–
Fomrez®	Witco Corp. Organics Div. New York, NY 10 017, US	2632	117
Foraflon®	Elf Atochem 92 091 Paris, La Défense 10, FR	1351, 1354	2
Foramine	Reichhold Chemicals Inc. Jacksonville, FL 32 245, US	221	73
Forasite	Reichhold Chemicals Inc. Jacksonville, FL 32 245, US	214	73
Forbon	NVF Container Div. Hartwell, GA 30 643, US	4211	5
Forco®	4P Folie Forchheim GmbH 91301 Forchheim, DE	112	411
Forex®	Airex AG., 5643 Sins, CH	1311	51
Formaldafil®	Akzo Engineering Plastics Inc. Evansville, IN 47 732, US	231	224

8.3 Trade Names

Name	Company (see 8.1, page 479)	Resin (see 8.2.1, page 481)	Forms of delivery (s. 8.2.2, p. 484)
Formica®	Formica Corporation, Wayne, NJ 07 470	223	621
Formion®	A. Schulman Inc., Akron, OH 44 309, US	1117	2
Formosir®	Sirlite Srl., 20 161 Milano, IT	21	113
Formula Format	Marley Floors Ltd., Lenham, Maidstone, Kent, ME17 2DE, GB	132	444
Formula P	Putsch GmbH, 90427 Nürnberg, DE	112	2
Formvar®	Monsanto Chemical Co., Plastics Div. St. Louis, MO 63 167, US	153	113, 115
Forsan®	Kaucuk n. Vlt., 27 852 Kralupy, CSFR	1221	2
Fortiflex®	Solvay S.A., 1050 Bruxelles, BE	1113	2, 11
Fortilene®	–„–	112, 1121	2, 11
Fortron®	Hoechst AG, 65929 Frankfurt, DE	237	21, 22
Fosta-Tuf-Flex®	Huntsman Chemical Corp. Chesapeake, VA 23 320, US	1223	2
Fostafoam®	–„–	121	117
Fostalite®	–„–	121	2
Fostarene®	–„–	121	11
Foundrez®	Reichhold Chemicals Inc. Jacksonville, FL 32 245, US	21	113
FPM-R	Borealis Compounds AB 12 613 Stockholm, SE	348	11 se
FR-PET®	Teijin Chemicals Ltd., Tokyo, JP	2412	224
Franklin Fibre	Franklin Fibre-Lamitex Corp. Wilmington, DE 19 899, US	4211	31, 33, 5
Freemix	DSM Compounds UK Ltd. Eastwood, Lancs., GB	242	224
Freemix	Frost & Sullivan 6000 Frankfurt/M., DE	242	224
frianyl	Frisetta GmbH 79677 Schönau, DE	261	2
friatherm	Friatec AG 68229 Mannheim, DE	1116, 1312	31
Fric®	Oy Wijk & Höglund AB Vase, FI	111	41

Name	Company (see 8.1, page 479)	Resin (see 8.2.1, page 481)	Forms of delivery (s. 8.2.2, p. 484)
Friedola	Friedola Gebr. Holzapfel GmbH&Co. KG 37276 Meinhard, DE	112, 132	4, 5
Frisetta	Frisetta GmbH 79677 Schönau, DE	261	2
Fudowlite	Fudow Chemical Co. Ltd. Tokyo, JP	21, 22, 223, 242, 243	2
Fürkaform	Regeno-Plast, 42697 Solingen, DE	2312 r	2
Fürkalan	–,,–	1221 r	2
Fürkamid	–,,–	261 r	2
Füron	–,,–	121 r	2
Fulcon	Sakai Kasei Kogyo Co. Ltd. Osaka, JP	131	411 tra
Fulton®	LNP Plastics Nederland BV 4940 Raamsdonksveer, NL	135/231	2
Fundopal	Funder Ind. Ges.mbH. 9300 St. Veit a. d. Glan, AT	223	621
FurCarb	QO Chemicals Inc. Oak Brook, IT 60 251, US	216	112, 113
Furesir®	Sirlite Srl., 20 161 Milano, IT	216	112
Furnidur®	Hoechst Diafoil GmbH 65174 Wiesbaden, DE	1313	41, 452, 53
furnit®	Konrad Hornschuch AG 74679 Weissbach, DE	1311	437
Furset	Raschig AG, 67061 Ludwigshafen, DE	216	113
G			
Gabotherm®	Thyssen Polymer GmbH 81671 München, DE	113	31
Gaflon®	Plastic Omnium S.A. Levallois-Perret, FR	1351	3
Galirene	Carmel Olefins Ltd. Haifa 31 014, IL	121	2
Garbeflex®	Elf Atochem 92 091 Paris, La Défense 10, FR	–	91
Gardglas	American Filtrona Corp. Richmond, VA 23 233, US	163	4, 5

8.3 Trade Names

Name	Company (see 8.1, page 479)	Resin (see 8.2.1, page 481)	Forms of delivery (s. 8.2.2, p. 484)
Gealan	Gealan Werk Fickenscher GmbH 95145 Oberkotzau, DE	132	31, 341
Geax®	AEG Isolier- und Kunststoff GmbH 34123 Kassel, DE	21	611
Geberit®	Geberit International 8640 Rapperswil, CH	1113	31
Gecet	GE Plastics Pittsfield, MA 01 201	235 s	71
Gedexcel®	Elf Atochem 92 091 Paris, La Défense 10, FR	12	117
Gehr	Gehr-Kunststoffwerk GmbH 68219 Mannheim, DE	111, 112, 12, 1221, 131, 1354, 2351, 2361, 263, 274	31, 33, 342, 5
Gekaplan	Göppinger Kaliko GmbH 73054 Eislingen, DE	132	435
Geloy®	GE Plastics Pittsfield, MA 01 201, US	1222(+345)	124
GenaKor®	Hoechst AG 65929 Frankfurt, DE	1, 2, 3	143, 433
Genesis	Novacor Chemicals Ltd. Calgary, Alberta T2P 2 H6, CND	11, 121	110 TE
Geniplex 80	Rhein Chemie Rheinau GmbH 68219 Mannheim, DE	–	–
Genopak®	Hoechst AG 65929 Frankfurt, DE	1311	41, 35
Genotherm®	–,,–	1311	411, 53
Gensil®	GE Silicones Waterford, NY 12 188, US	38	2
Geolast®	Advanced Elastomer Systems L.P. St. Louis, MO 63 167, US	112/323	110 TE
Geon®	B. F. Goodrich Chemical Group Cleveland, OH 44 131, US	131, 1312, 132	11, 117, 141
Gerodur	Gerodur AG 8717 Benken SG, CH	111, 112, 131	31
Gesadur	G. H. Sachsenröder 42285 Wuppertal, DE	212	612

Name	Company (see 8.1, page 479)	Resin (see 8.2.1, page 481)	Forms of delivery (s. 8.2.2, p. 484)
Getadur	Westag & Getalit AG 33378 Rheda-Wiedenbrück, DE	223	761
Getaform	–„–	223	452
Getalan	–„–	223	761
Getalit	–„–	223	452, 621
Getaplex	–„–	223	761
Gilco	Gilman Brothers Co. Gilman, CT 06 336, US	111, 12, 1223	41, 5
Gillcoat	M. C. Gill Corp. El Monte, CA 91731, US	223	621
Gillfab	–„–	16, 21, 233, 242, 27, 28	2251, 61, 614
Gillfoam	–„–	21, 242	71
Gillite	–„–	21, 242	76
Gislaved®	Gummifabriken Gislaved AB 33200 Gislaved, SE	132	411, 413, 421, 423, 432, 433, 435
Glad®	Union Carbide Corp. Danbury, CT 06 817, US	111	411
Glaskyd	American Cyanamid Corp. Wayne, NJ 07 470, US	243	2
Glasotext®	AEG Isolier- und Kunststoff GmbH 34123 Kassel, DE	233	614
Glasrod	Glastic Corp. Cleveland, OH 44 121, US	242	61
Glasspack	ECP EniChem Polimeri srl. 20 124 Mailand, IT	121	411 s
Glastic®	Glastic Corp. Cleveland, OH 44 121, US	233, 242	224, 614
Glendion®	ECP EniChem Polimeri srl. 20 124 Mailand, IT	2639, 2633	17
Glitex	Sybron Chemicals Inc. Birmingham, NJ 08 011, US	132	32
Glutofix®	Hoechst AG 65929 Frankfurt, DE	4231	73
Glutolin®	–„–	423	11
Godiflex	Godiplast Kunststoffgranulate GmbH, 66636 Tholey, DE	1315	2

Name	Company (see 8.1, page 479)	Resin (see 8.2.1, page 481)	Forms of delivery (s. 8.2.2, p. 484)
Godigum	Godiplast Kunststoffgranulate GmbH, 66636 Tholey, DE	1314	2
Godiplast	–„–	1311, 132	2
Gölzalit	Pipelife International Holding 06369 Weissandt-Gölzau, DE	131	31
Gölzathen	–„–	111	31, 411, 43, 5
Gohsenol	Nippon Synthetic Chemical Ind. Co. Ltd., Osaka, JP	152	11
Gohsenyl	–„–	151	11
Golan Profiles	Golan Plastics Products Jordan Valley 15145, IL	1311	341
Gore-tex®	W. L. Gore Ass. Inc. Elkton, MD 21 921, US	–	712
Gotalene®	Plast Labor SA, 1630 Bulle, CH	1111	141
Gra-Tufy®	The Perma-Flex Mold Co. Inc. Columbus, OH 43 209, US	237	111
Grafil®	Hysol Div., Dexter Corp. Seabrook, NH 03 874, US	–	226
Gran	TBA Industrial Products Ltd. Rochdale, Lancs., GB	261	224
Granlar®	Montedison S.p.A., 20 121 Milano, IT	2415	119
Granulit®	Gurit-Worbla AG 3063 Ittigen-Bern, CH	28	11
Green-Sil	The Perma-Flex Mold Co. Inc. Columbus, OH 43 209, US	28	111
Greenflex®	ECP EniChem Polimeri srl. 20 124 Mailand, IT	1115	113
Gremodur	Gremolith AG, 9602 Bazenheid, CH	212	113
Gremolith	–„–	431	33, 51
Gremopal	–„–	242	112
Gremothan	–„–	263	117
Gril-tex®	Ems-Chemie AG, 7013 Domat/Ems, CH	261 s	116, 73
Grilamid 7ELY	–„–	2613	120
Grilamid TR 55	–„–	261	2 tra, 224
Grilamid®	–„–	261, 2613	11, 141, 224
Grilene	–„–	2412	72

8.3 Trade Names

Name	Company (see 8.1, page 479)	Resin (see 8.2.1, page 481)	Forms of delivery (s. 8.2.2, p. 484)
Grilesta	Ems-Chemie AG, 7013 Domat/Ems, CH	241	73
Grillodur®	Fibron GmbH 75015 Bretten, DE	242	225, 34, 63
Grilon®	Ems-Chemie AG, 7013 Domat/Ems, CH	261, 2613	21, 224, 35, 120
Grilon®ELY	–„–	2613	120
Grilonit®	–„–	233	114, 146
Grilpet®	–„–	2412	21, 224
Grisuten®	Märkische Faser AG 14723 Premnitz, DE	2412	72
Grivory®	Ems-Chemie AG, 7013 Domat/Ems, CH	2412+261	110, 110 TE
Gumiplast	Saplast S.A. 67 100 Straßburg-Neuhof, FR	132	2
Gurimur®	Gurit-Worbla AG 3063 Ittigen-Bern, CH	11, 13	43, 45
Gurit®	–„–	11, 12, 13	41, 42, 43, 45, 52, 53, 54
Guron	Koepp AG 65375 Oestrich-Winkel, DE	1111	712 porous
Gymlene®	Drake (fibres) Ltd. Huddersfield, W. Yorks., GB	112	72

H

Name	Company	Resin	Forms of delivery
H-Resin®	Hercules Powder Company Inc. Wilmington, DE 19 894, US	19	14
Hagulen	Hagusta GmbH, 77867 Renchen, DE	1113	31
Hakathen	Haka, 9202 Gossau SG 1, CH	111, 1112, 1113, 112, 113	31
Halar®	Ausimont Inc. Morristown, NJ 07 962, US	135	11, 14
Haloflex®	ICI PLC Welwyn Garden City, Herts. AL7 1 HH, GB	1114, 162	11, 114
Halon®	Ausimont Inc. Morristown, NJ 07 962, US	1351	2, 224

8.3 Trade Names

Name	Company (see 8.1, page 479)	Resin (see 8.2.1, page 481)	Forms of delivery (s. 8.2.2, p. 484)
Halweftal	Hüttenes-Albertus 40505 Düsseldorf, DE	244	114
Halwemer	–,,–	16	114
Halweplast	–,,–	244	114
Halwepol	–,,–	241	114
Halwepox	–,,–	233	114
Halwetix	–,,–	24	114
HAP	Colorant GmbH 65555 Limburg, DE	135	2
Harden	Toyobo Co. Ltd., Osaka 530, JP	261	411
Harex®	Resopal GmbH 64828 Groß-Umstadt, DE	212, 242	224, 612
Haysite	Haysite Corp. Erie, PA 16 509, US	242	224, 225
HD Acoustic	Marley Floors Ltd., Lenham, Maidstone, Kent, ME17 2DE, GB	132	444
HD Hitech	–,,–	132	444
HD Vinyl	–,,–	132	444
HDPEX	Borealis Compounds AB 12 613 Stockholm, SE	1116	11
Heglerflex	Hegler Plastik GmbH 97714 Oerlenbach, DE	111, 112, 1311	31, 32
Heglerflex med	–,,–	111	31, 32 med
Heglerplast	–,,–	1311	31, 32
Hekaplast	–,,–	1113	31, 32
Helidur	A. G. Petzetakis SA, 10210 Athen, GR	111, 1313	312 s
Heliflex	–,,–	13	311, 321
Helioflex®	Papeteries de Belgique, Bruxelles, BE	111	411, 412
Helioplast®	–,,–	112	411
Heliothen®	–,,–	1	412
Heliovir®	–,,–	131	41
Hemit	Garfield Molding Co. Garfield, NJ 07 026, US	9	27

Name	Company (see 8.1, page 479)	Resin (see 8.2.1, page 481)	Forms of delivery (s. 8.2.2, p. 484)
Hep O	Hepworth Building Products Ltd. Stockbridge, Sheffield S30 5 HG, GB	113	31
Hercoflex®	Hercules Powder Company Inc. Wilmington, DE 19 894, US	–	91
Hercolyn®	–„–	52	91
Herex®	Airex AG, 5643 Sins, CH	1311	7661
Herox®	DuPont Co. Inc. Wilmington, DE 19 898, US	261	35, 72
Hetron®	Ashland Chemical Corp. Columbus, OH 43 216, US	2421	11 se
Hexcel	Hexcel Corp. Chemical Div. Chatsworth, CA 93 111, US	21, 233, 242, 27	7662
Hexcelite	–„–	242	225
Hexene	TVK Tisza Chemical Combine 3581 Leninvaros, HU	1112, 1113	2
Heydeflon	Chemiewerk Nünchritz GmbH 01612 Nünchritz, DE	1351	3, 5
Hi-Carbolon	Asahi Nippon, JP	–	226
Hi-Fax	Montell Polyolefins Wilmington, DE 19 894, US	111 s	11, 2
Hi-Glass	–„–	112	224
Hi-Selon	Nippon Synthetic Chemical Ind. Co. Ltd., Osaka, JP	151	411
Hi-Therm®	John C. Dolph Co., Monmouth Junction, NJ 08 852, US	321	146 el
Hi-Zex®	Mitsui Petrochemical Ind. Ltd. Tokyo, JP	1112, 1113	2
Hidens	Nissan Chemical Industries Ltd. Tokyo, JP	1113	141 s
Hiflex®	Montell Polyolefins Wilmington, DE 19 894, US	112, 1121	2
Hiflon®	Hindustan Fluorocarbons Ltd. Andrah Pradesh, Medak, IN	1351	11, 121, 2
Higram®	Montell Polyolefins Wilmington, DE 19 894, US	112, 1121	71
Hilex®	Courtaulds Chemicals & Plastics Ltd. Spondon, Derby DE2 7BP, GB	1113	4, 5

8.3 Trade Names

Name	Company (see 8.1, page 479)	Resin (see 8.2.1, page 481)	Forms of delivery (s. 8.2.2, p. 484)
Hiloy®	A. Schulman Inc. Akron, OH 44 309, US	112, 1221, 237, 1222, 2361, 2411, 2412, 261	2
Hilube®	Akzo Engineering Plastics Inc. Evansville, IN 47 732, US	1222, 231, 2351, 2411, 261, 263	224 s
Himet	Himac Inc., Danbury, CAT 06 811, US	1122	41 s
Himiran	DuPont-Mitsui Polychemical Co. Ltd. Tokyo, JP	1115	2
Hishi plate	Mitsubishi Plastics Inc., Tokyo, JP	1311	51, 52
Hishi Tube	–,,–	132	41
Hishi-metal	–,,–	132	51, 52
Hishirex	–,,–	1311	41
Hitafran	Hitachi Chemical Co. Ltd., Tokyo, JP	216	113
Hitanol	–,,–	121	11
Hivalloy®	Montell Polyolefins Wilmington, DE 19 894, US	1115 s	124
HM 50	Teijin Ltd., Osaka 541, JP	261 s	72
Hobas-Rohre	Armaver AG, 4617 Gunzgen, CH	242	6141
Hoechst-Wachse®	Hoechst AG 65929 Frankfurt, DE	11, 4, 5	94
Hofalon	Hornitex Werke Gebr. Künnemeyer 32805 Horn-Bad Meinberg, DE	–	742
Homanit®	Homanit GmbH & Co. KG 37412 Herzberg, DE	742, 761	
Homapal®	Homapal Plattenwerk GmbH & Co. KG, 37412 Herzberg, DE	223	621, 761
Hornex®	Vulkanfiberfabrik Ernst Krüger + Co. KG., 47608 Geldern, DE	4211	5
Hornit	Hornitex Werke Gebr. Künnemeyer 32805 Horn-Bad Meinberg, DE	223	621
Hornitex MB®	–,,–	223	761
Hostacom	Hoechst AG 65929 Frankfurt, DE	112	22
Hostaflam®	–,,–	5	9

Name	Company (see 8.1, page 479)	Resin (see 8.2.1, page 481)	Forms of delivery (s. 8.2.2, p. 484)
Hostaflex®	Hoechst AG 65929 Frankfurt, DE	1313/143	11, 114
Hostaflon®	–,,–	135	11, 121, 2
Hostaform®	–,,–	2312	2, 224
Hostalen PP3100 schwarz 12	–,,–	112/341	2 r
Hostalen PP®	–,,–	112, 1121	2, 2213, 224, 2217
Hostalen®	–,,–	111	2, 224
Hostalit®	–,,–	1311, 1312, 1313	11, 2
Hostalub®	–,,–	11, 4, 5	9
Hostanox®	–,,–	5	91
Hostaphan®	Hoechst Diafoil GmbH 65174 Wiesbaden, DE	111+133+ 2412+Al	411, 412, 413
Hostapher	Hoechst AG 65929 Frankfurt, DE	112	11
Hostapren®	–,,–	1114/1117	11
Hostastab®	–,,–	5	9
Hostastat®	–,,–	5	9
Hostatron®	–,,–	5	9
Hostavin®	–,,–	5	9
Hot-Hard	Dexter Corp. Specialty Coating Div. Waukegan, IL 60085, US	27	114 el
howelon®	Konrad Hornschuch AG 74679 Weissbach, DE	132	421, 423
Hutex®	A. Huppertsberg KG 65520 Bad Camberg, DE	132	32, 34
Hy Comp	Hysol Div., Dexter Corp. Seabrook, NH 03 874, US	27	3, 5
Hy-Bar	BCL, Bridgewater, Somerset, TA6 4PA, GB	111, 1111 L, 1115, 112	412
Hycar®	B. F. Goodrich Chemical Group Cleveland, OH 44 131, US	323, 332	11
Hydrex®	Reichhold Chemicals Inc. Jacksonville, FL 32 245, US	242	11

Name	Company (see 8.1, page 479)	Resin (see 8.2.1, page 481)	Forms of delivery (s. 8.2.2, p. 484)
Hydrin®	B. F. Goodrich Chemical Group Cleveland, OH 44 131, US	39	11
Hyflo MC 18	Hysol Div., Dexter Corp. Seabrook, NH 03 874, US	233	11, 2 tra
Hyflon®	Ausimont Inc. Morristown, NJ 07 962, US	135	11, 12, 14
Hygel	W. R. Grace & Co., Organic Chemicals Lexington, MA 02 173, US	263	117
Hylak®	Hylam Ltd., Hyderabad 18, IN	21, 242	112, 113, 2
Hylam	–„–	21	611, 612
Hylar®	Ausimont Inc. Morristown, NJ 07 962, US	1354	11, 14
Hypalon®	DuPont Co. Inc. Wilmington, DE 19 898, US	1114/342	11
Hyperlast®	Kemira Polymers Ltd. Stockport, Cheshire SK12 5BR, GB	263	112, 14
Hypol®	W. R. Grace & Co., Organic Chemicals Lexington, MA 02 173, US	263 s	117
Hyrizon®	Aristech Chemical Co. Florence, KY 41 042, US	–	71
Hysol	Hysol Div., Dexter Corp. Seabrook, NH 03 874, US	233	2251
Hytemp	Nippon Zeon Co. Ltd., Tokyo, JP	323	11
Hytrel®	DuPont Co. Inc. Wilmington, DE 19 898, US	2413	110 TE
Hyvis®	BP Chemicals International Ltd. London SW1W OSU, GB	114	12
Hyzod	Sheffield Plastics Inc. Sheffield, MA 01 257–0428, US	2411	53

I

Name	Company	Resin	Forms of delivery
Icdal®	Hüls AG, 45764 Marl, DE	271	114
Icosolar	Isovolta AG 2355 Wiener Neudorf, AT	24	43
Igoform	Faigle AG Igoplast, 9434 Au, CH	11+231	2
Igopas®	–„–	112, 1113, 135, 231, 261	31, 33, 5

8.3 Trade Names   545

Name	Company (see 8.1, page 479)	Resin (see 8.2.1, page 481)	Forms of delivery (s. 8.2.2, p. 484)
Illandur®	Ems-Polyloy GmbH 64823 Groß-Umstadt, DE	242	11, 2, 22, 224
Illen	–„–	2411, 2412	110 TE
Illenoy	–„–	2413	11, 2
Illmid	Illbruck GmbH Schaumstofftechnik 51381 Leverkusen, DE	27	71
Illtec	–„–	223	71
Imidex®	GE Plastics Pittsfield, MA 01 201, US	271	114 el
Imipex®	–„–	271	11
Impet®	Hoechst AG 65929 Frankfurt, DE	2412	2, 2217, 224
Implex®	Rohm & Haas Co. Philadelphia, PA 19105, US	16	2
Impolex®	ICI PLC, Welwyn Garden City, Herts. AL7 1 HH, GB	242	112, 225
Impranil®	Miles Chemical Corp. Pittsburgh, PA 15 205, US	263	111, 115
Imprenal®	Raschig AG, 67061 Ludwigshafen, DE	21, 212, 214	113, 114, 123
Imprez®	ICI PLC, Welwyn Garden City, Herts. AL7 1 HH, GB	53	114/116
Inbord	Isovolta AG 2355 Wiener Neudorf, AT	212	611, 612
Incoblend	Zipperling Kessler & Co. 22926 Ahrensburg, DE	1313 s	12 s
Indothene	Indian Petrochemicals Corp. 391 346 Gujarat State, IN	1111, 1111 L	2
Indovin	–„–	131	2
Infolite TM	Hoechst AG 65929 Frankfurt, DE	1	35 s
Inklurit®	BASF Aktiengesellschaft 67063 Ludwigshafen, DE	221	117
Innovex®	BP Chemicals International Ltd. London SW1W OSU, GB	1111 L	11
Insert	DSM Resins B.V. 8000 AP Zwolle, NL	242	111
Insul F	Mateson Chemical Corp. Philadelphia, PA 19125, US	263	2

8.3 Trade Names

Name	Company (see 8.1, page 479)	Resin (see 8.2.1, page 481)	Forms of delivery (s. 8.2.2, p. 484)
Insularc	Franklin Fibre-Lamitex Corp. Wilmington, DE 19 899, US	6	
Insultrac	Industrial Dielectrics Inc. Noblesville, IN 46 060, US	242	112 s
Insurok®	Spaulding Composites Co. Tonawanda, NY 14 150, US	212, 223	61, 62
Intec	Intec Ltd. Plymouth, Dev. PL6 8 LA, GB	261	321
Intene®	International Synthetic Rubber Co. Ltd., Southampton SO9 3AT, GB	321	11
Intol®	–,,–	322	11
Intolan®	–,,–	3411	11, 2
Intrasol®	Kibbuz Ginegar, IL	1111	432
Intrex	Sierracin Corp. Sylmar, CA 91 342, US	2412	413
Intrile	International Synthetic Rubber Co. Ltd., Southampton SO9 3AT, GB	323	11, 4, 32
Iotek	Exxon Chemicals America Houston, TX 77 001, US	1115 s	11
Ipethene	Carmel Olefins Ltd. Haifa 31 014, IL	111	2
Irganod	Ciba Geigy AG, 4002 Basel, CH	13/16	9
Irgastab®	–,,–	–	9
Irodur	Morton International Inc. Chicago, IL 60 606, US	2631	111
Irogran	–,,–	264	2
Irophen®	Morton International Inc. Chicago, IL 60 606, US	2632	115
Irostic®	–,,–	263	115
Irracure®	Reichhold Chemicals Inc. Jacksonville, FL 32 245, US	1116	2 el
Iso-Genopak®	Hoechst AG 65929 Frankfurt, DE	131	43 s
Iso-Glasnetz	Isovolta AG 2355 Wiener Neudorf, AT	231	224
Isocord®	–,,–	214	61 el

Name	Company (see 8.1, page 479)	Resin (see 8.2.1, page 481)	Forms of delivery (s. 8.2.2, p. 484)
Isofix-M	Fränkische Rohrwerke GmbH + Co. 97486 Königsberg, DE	131	31 el
Isofoam®	Witco Corp. Organics Div. New York, NY 10 017, US	2631	117
Isoglas	Isovolta AG 2355 Wiener Neudorf, AT	24	614
Isolama	L.M.P. S.p.A., 10156 Turin, IT	121	414
Isolant®	–,,–	121	414
Isolene D	Hardman Inc. Belleville, NJ 07 109, US	33	11
Isomat®	Chemie Linz AG, 4040 Linz, AT	121	71
Isonom	Isovolta AG 2355 Wiener Neudorf, AT	233	614
Isopak	Great Eastern Resins, Taiwan	1221	1
Isophen	Isovolta AG 2355 Wiener Neudorf, AT	212	113
Isopreg	–,,–	233	614
Isosan	Great Eastern Resins, Taiwan	1222	1
Isoschaum®	Schaum-Chemie Wilhelm Bauer GmbH & Co. KG, 45001 Essen, DE	221	71 s
Isoseal	Isovolta AG 2355 Wiener Neudorf, AT	233	612
Isospan	–,,–	233	616
Isothane®	Recticel Foam Corp., Morristown Div. Morristown, TN 37 816–1197, US	263	71 se
Isoval	Isovolta AG 2355 Wiener Neudorf, AT	212	611, 612
Isphen	Repsol Quimica, ES	112, 1121	2
Itamid	Ciech Chemikalien GmbH 00013 Warschau 1, PL	5	
Itenite®	Iten Industries Ashtabula, OH 44 004, US	21, 223, 4211	61, 612, 614
Iupiace	Mitsubishi Gas Chem. Comp. Inc. Tokyo, JP	235	2
Iupilon	–,,–	2411	2, 41, 5
Iupital	–,,–	2312	2
Ixan®	Solvay S.A., 1050 Bruxelles, BE	133	11, 121

8.3 Trade Names

Name	Company (see 8.1, page 479)	Resin (see 8.2.1, page 481)	Forms of delivery (s. 8.2.2, p. 484)
Ixef®	Solvay S.A., 1050 Bruxelles, BE	2612	224
Ixol®	–„–	2633	117 se
J			
Jackodur	Gefinex Jackon GmbH, 33803 Steinhagen, DE	21	41
Jägalux	Ernst Jäger & Co. OHG 40599 Düsseldorf, DE	16	114
Jägalyd	–„–	233, 2441	114
Jägapol	–„–	24	114
Jägotex	–„–	16	121
Jayflex	Exxon Chemicals America Houston, TX 77 001, US	–	91
Jeffamine	Texaco Chemical Co. Bellaire, TX 77 401, US	2634	111
Jekrilan	J. K. Synthetics Ltd. Kota (Rajarthan), IN	16	72
Jonylon®	BIP Chemicals Ltd. Warley, West Midl., B69 4PG, GB	261	21, 22
Julon	Jung-Wehbach GmbH 57548 Kirchen, DE	261	31
K			
K-Flex	Kureha Chemical Industry Co. Ltd. New York, NY 10 170, US	133	411
K-Resin®	Phillips Petroleum Chemicals NV 3090 Overijse, BE	122, 1223	2
KaCepol®	Kali-Chemie Akzo GmbH 30173 Hannover, DE	263	117
Kadel®	Amoco Performance Products Richfield, CT 06 877, US	2351	2
Kaifa®	Beijing Chemical Ind. R. & D. Corp. Hongkong, HK	2412	2
Kaladex	ICI PLC Welwyn Garden City, Herts. AL7 1 HH, GB	1118	41
Kalar	Hardman Inc. Belleville, NJ 07 109, US	332	2

Name	Company (see 8.1, page 479)	Resin (see 8.2.1, page 481)	Forms of delivery (s. 8.2.2, p. 484)
Kalen	Emil Keller AG 9220 Bischofszell/TG, CH	111	31
Kalen®	Kalenborn Kalprotect 53560 Vettelschoss, DE	1113	5(31)
Kalene	Hardman Inc. Belleville, NJ 07 109, US	332	11
Kalex	–,,–	264, 37	11
Kalidur	Emil Keller AG 9220 Bischofszell/TG, CH	1211, 131	31, 52
Kalit	–,,–	1311	33
Kaliten	–,,–	1113	31
Kalrez®	DuPont Co. Inc. Wilmington, DE 19 898, US	348	33, 34, 4
Kamax®	Rohm & Haas Co. Philadelphia, PA 19105, US	1642	2, 21 tra
Kanalite®	Creators Ltd. Woking, Surrey GU21 5RX, GB	132	312 s
KaneAce®	Kanegafuchi Chemical Ind. Co. Ltd. Osaka, JP	1221, 1641	14
Kanebian	–,,–	152	72
Kaneka	–,,–	1312	2
Kaneka CPVC	–,,–	134	11
Kaneka Telalloy	–,,–	125/163	124
Kanelite	–,,–	121	414, 71
Kanevinyl®	–,,–	1315	110 TE
Kapex®	Airex AG., 5643 Sins, CH	264	7661
Kapton®	DuPont Co. Inc. Wilmington, DE 19 898, US	27	412, 413
Karboresin	Hoechst AG 65929 Frankfurt, DE	53	114, 115
Kardel®	Union Carbide Corp. Danbury, CT 06 817, US	121	411
Kartex	Fabbrica Adesivi Resine S.p.A. Cologno Monzese, IT	151	11, 115, 73
Kartothene	Plastona Ltd. Subs. John Waddington, GB	11, 112	53

8.3 Trade Names

Name	Company (see 8.1, page 479)	Resin (see 8.2.1, page 481)	Forms of delivery (s. 8.2.2, p. 484)
Kasobond	Kaso-Chemie GmbH & Co. KG 49733 Haren, DE	263	115
Kasothan	–,,–	263	2
Kauramin®	BASF Aktiengesellschaft 67063 Ludwigshafen, DE	223	73
Kauranat®960 *flüssig*	–,,–	2631	73
Kauresin®	–,,–	212, 213	123, 73
Kaurit®	–,,–	221	73
Kauropal®	–,,–	242	73
Kayfax	Toa Gosei Chemical Ind. Co. Ltd. Tokyo, JP	27	117 se
KaZepol®	Kali-Chemie Akzo GmbH 30173 Hannover, DE	263	117
Keebush®	APV Kestner Ltd. Greenhithe, Kent, GB	21	2, 61 corr
Keeglas	–,,–	21, 242	614 corr
Kel-F®	3 M Co., St. Paul, MN 55 144, US	1352, 1353	11, 114
Kelanex	Hoechst AG 65929 Frankfurt, DE	2412	2, 224
Kelburon®	DSM, Polymers & Hydrocarbons 6130 AA Sittard, NL	112/3411	110 TE
Keldax®	DuPont Co. Inc. Wilmington, DE 19 898, US	1115	221 s
Kellco	Novopan-Keller AG 4314 Kleindöttingen, CH	223	621
Kelon®	Lati S.P.A. 21040 Vedano Olona, IT	261	22
Kelprox	DSM, Polymers & Hydrocarbons 6130 AA Sittard, NL	112/3411	110 TE
Kelrinal®	–,,–	1114/1117	11
Keltaflex®	–,,–	11	9
Keltan®	–,,–	112+3411	110 TE
Kemamide®	Witco Corp. Organics Div. New York, NY 10 017, US	9	
Kematal®	Hoechst AG 65929 Frankfurt, DE	2312	2, 224

Name	Company (see 8.1, page 479)	Resin (see 8.2.1, page 481)	Forms of delivery (s. 8.2.2, p. 484)
Kemester CP®	Witco Corp. Organics Div. New York, NY 10 017, US	9	
Kemid®	Norton Performance Plastics Wayne, NJ 07 470, US	274	4
Kemipur	Kemipur-Polyurethane-Systeme GmbH, Solymár, HU	263	2, 11
Kenflex	Kenrich Petrochemicals Inc. Bayonne, NJ 07 002, US	53	91
Kenplast	–,,–	53	91
Kepital	Korea Eng. Plastics Co, Seoul, Korea	231	2
Keraphen	Keramchemie GmbH 56427 Siershahn, DE	212	2251, 2261
Kerimid®	Rhône-Poulenc Specialités Chim 69 006 Lyon, FR	27	112
Keripol®	Phoenix AG, 21079 Hamburg, DE	242	222, 224, 231, 232, 251
Kermel®	Rhône-Poulenc S.A. 92 408 Courbevoie, FR	272	111
Kerni®	Oy Finlayson AB, Tampere 10, FI	132	424
Kevlar®	DuPont Co. Inc. Wilmington, DE 19 898, US	261 s	72
KF Film	Kureha Chemical Industry Co. Ltd. Tokyo, JP	1353	4 el
KF Piezo Film	–,,–	1353	4 el
KF Polymer	–,,–	1354	11, 41
Kialite	Heuvelmans B.V., Tilburg, NL	233, 242, 2421	6141, 76
Kibisan	Chi Mei Industrial Co. Ltd. Tainan, Shien, Taiwan	1222	2
Kinel®	Rhône-Poulenc Specialités Chim 69 006 Lyon, FR	27	22
Kite®	Tufnol Ltd. Birmingham B42 2TB, GB	21	611
Klea	ICI PLC Welwyn Garden City, Herts. AL7 1 HH, GB	5	9
Kleer Kast®	Kleer Kast Inc. Kearny, NJ 07 032, US	163	55

8.3 Trade Names

Name	Company (see 8.1, page 479)	Resin (see 8.2.1, page 481)	Forms of delivery (s. 8.2.2, p. 484)
Kleiberit®	Klebchemie M. G. Becker GmbH+Co. KG, 76356 Weingarten, DE	151, 162, 343, 35	73
Klemite	Garfield Molding Co. Garfield, NJ 07 026, US	223	2
Klingerflon	Rich. Klinger GmbH 65510 Idstein, DE	1351	3, 4, 5
Klucel®	Hercules Powder Company Inc. Wilmington, DE 19 894, US	4233	2
Kobe-Lite	Shin-Kobe Electric & Machinery Co. Ltd., Tokyo 160, JP	61 el	
Koblend®	ECP EniChem Polimeri srl. 20 124 Mailand, IT	111/1223	11/411
Kodacel®	Eastman Chemical Intern. Kingsport, TN 37 662, US	4223	42, 53
Kodapak® PET	–,,–	2412	2
Kodar®	–,,–	2412	2
Kodel®	–,,–	2412	72
Kö-Profile	Gebr. Kömmerling Kunststoffwerke GmbH, 66954 Pirmasens, DE	132	34
Kömabord Ce®	–,,–	1311	713
Kömacel®	–,,–	1311	713, 5
Kömadur®	–,,–	1314	5
Kömalen®	–,,–	121 se, 1311	5
Kömapan	–,,–	131	34, 342
Kömapor	–,,–	131	71
Kömatex	–,,–	131	71
Kohinor®	Pantasote Inc. Passaic, NJ 07 055, US	13, 1311, 132	11, 2, 4
Koit	Koitwerk Herbert Koch GmbH & Co. 83253 Rimsting, DE	132	435
Kollerdur®	Scott Bader Co. Ltd. Wollaston, Wellingborough, Norths. NN9 7RL, GB	263	113
Kollernox®	–,,–	233	113
Kollidon®	BASF Aktiengesellschaft 67063 Ludwigshafen, DE	127	1

Name	Company (see 8.1, page 479)	Resin (see 8.2.1, page 481)	Forms of delivery (s. 8.2.2, p. 484)
Konlux	G. Roggemann GmbH 49504 Lotte, DE	242	55
Koplen	Kaucuk n. Vlt., 27 852 Kralupy, CSFR	121	117
Korad®	Polymer Extruded Products Newark, NJ 07 105, US	16	437, 452
Koreforte®	BASF Aktiengesellschaft 67063 Ludwigshafen, DE	214	113
Koresin®	–„–	214	113
Koretack®	–„–	214	113
Korez	Atlas Minerals & Chemicals Inc. Mertztown, PA 19 539, US	211	123
Koroseal	B. F. Goodrich Chemical Group Cleveland, OH 44 131, US	132	2
Korton®	Norton Performance Plastics Wayne, NJ 07 470, US	1352	41
Korvex®	–„–	135	32 s, el
Kostil®	ECP EniChem Polimeri srl. 20 124 Mailand, IT	1222	2, 224
Koylene®	Indian Petrochemicals Corp. 391 346 Gujarat State, IN	112, 1121	2
Kralastic®	Uniroyal/Sumitomo Chemical Co. Ltd., Tokyo 103, JP	1221/131	124
Krasten	Kaucuk n. Vlt., 27 852 Kralupy, CSFR	121	2
Kraton D®	Shell International Co. Ltd. London, SE1 7NA, GB	12/32	110 TE
Kraton G®	–„–	12/341	110 TE
Krehalon	Krehalon Industrie BV, Deventer, NL	133	411
Krene®	Union Carbide Corp. Danbury, CT 06 817, US	1313	4 med
Kronospan	Kronospan Ltd. Chirk, Wrexham, GB	223	761
Krylene®	Bayer AG, 51368 Leverkusen, DE	1223	–
Krynac®	–„–	323	11
Krynol®	–„–	1223	–
Krystaltite®	Allied Signal Corp. Morristown, NJ 07 962, US	131	411
Kunstharz AP, SK	Hüls AG, 45764 Marl, DE	259	114

Name	Company (see 8.1, page 479)	Resin (see 8.2.1, page 481)	Forms of delivery (s. 8.2.2, p. 484)
Kunsto®-ABS	Kunstoplast-Chemie GmbH 61440 Oberursel, DE	1221	2
Kunstolen®	–,,–	111, 112	2
Kunstomid®	–,,–	261	2
Kunstonyl®	–,,–	131	122, 2
Kunstyrol®	–,,–	121	2
Kureha	Kureha Chemical Industry Co. Ltd. Tokyo, JP	123, 1311, 1354	111, 2
Kydene	Rohm & Haas Co. Philadelphia, PA 19105, US	131/16	2
Kydex®	–,,–	1313	2, 52, 53
Kynar®	Elf Atochem 92 091 Paris, La Défense 10, FR	1354	11, 121, 41 el
Kynol	Carborundum Corp. Niagara Falls, NY 14 302, US	21	226
Kyowalite	Kyowa Gas Chemical Co. Ltd. Tokyo, JP	61 s	54 s

L

Name	Company	Resin	Forms of delivery
Lacovyl	Elf Atochem 92 091 Paris, La Défense 10, FR	131	122, 2
Lacqrene	–,,–	12	11, 2
Lacqtene®	–,,–	1111, 1113, 1111 L	2
Lacros	SCM Chemicals Corp. Baltimore, MD 21 202, US	251	114
Ladene®	Saudi Basic Industries Corp. (Sabic) Riyadh 11 422, Saudi Arabien	1111 L, 1113, 12, 131, 223	117, 2
laif®	Konrad Hornschuch AG 74679 Weissbach, DE	263	721
Lamigamid®	G. Schwartz GmbH & Co. 46509 Xanten, DE	111, 231, 261	31, 33, 51
Lamilux®	Lamilux-Werk GmbH 95111 Rehau, DE	242	55, 614, 63
Laminex®	G. Schwartz GmbH & Co. 46509 Xanten, DE	212	612
Lamipor®	Lamilux-Werk GmbH 95111 Rehau, DE	–	7661

8.3 Trade Names

Name	Company (see 8.1, page 479)	Resin (see 8.2.1, page 481)	Forms of delivery (s. 8.2.2, p. 484)
Lamitex	Franklin Fibre-Lamitex Corp. Wilmington, DE 19 899, US	21, 233, 242, 28	611, 612, 614 el
Larc	U.S. Polymeric Santa Ana, CA 92 707, US	27	2
Larflex	Lati S.P.A. 21040 Vedano Olona, IT	112/3411	2, 21, 22
Laril	–,,–	235	2, 21, 22
Laripur	Larim S.p.A. 20 019 Settimo Milanese (MI), IT	264	111, 2
Larodur®	BASF Aktiengesellschaft 67063 Ludwigshafen, DE	162	114
Laroflex®	–,,–	1313	114
Laromer®	–,,–	162	114
Laropal®	–,,–	222, 259	114
Laros	SCM Chemicals Corp. Baltimore, MD 21 202, US	251	114
Larton	Lati S.P.A. 21040 Vedano Olona, IT	237	21, 22
Lastane	–,,–	262	2, 21, 22
Lastiflex	–,,–	1221/132	2, 21
Lastil	–,,–	1222	2, 21, 22
Lastilac	–,,–	1221/2411	124, 2, 21, 22
Lastirol	–,,–	122	2, 21, 22
Lasulf	–,,–	236	2, 21, 22
Latamid	–,,–	261	2, 21, 22, 224
Latamid FE	–,,–	261	113, 224
Latan	Lati S.P.A. 21040 Vedano Olona, IT	2312	2, 21, 22
Latekoll®	BASF Aktiengesellschaft 67063 Ludwigshafen, DE	165	123
Latene	Lati S.P.A. 21040 Vedano Olona, IT	1113, 112, 1121	2, 21, 22
Later	–,,–	2412	2, 21, 22
Latilon	–,,–	2411	2, 21, 22
Latilub	–,,–	261, 231	113, 224

Name	Company (see 8.1, page 479)	Resin (see 8.2.1, page 481)	Forms of delivery (s. 8.2.2, p. 484)
Latishield	Lati S.P.A. 21040 Vedano Olona, IT	2411, 2412, 26, 261	113, 224
Latistat	–,,–	2411, 2412, 26, 261	113, 224
Lauramid®	Albert Handtmann GmbH & Co. 88400 Biberach, DE	261	111/51
Lavella®	Sondex AB, Malmö, SE	1311	34, 341
Leben®	Dainippon Ink. & Chemicals Inc. Tokyo 103, JP	131	11
Lekutherm®	Bayer AG, 51368 Leverkusen, DE	233	111, 112, 113
Lemac	Borden Chemical Co. Geismar, LA 70 734, US	151	11
Lemaloy	Mitsubishi Chemical Corp. Tokyo, JP	23/261	2
Lemapet	–,,–	235	113, 6111
Lennite®	Westlake Plastics Co. Lenni, PA 19 052, US	2412	224, 6111
Lenser®	Lenser Kunststoff-Preßwerk GmbH+Co., KG, 89250 Senden, DE	111, 121, 135	5
Lenzing Modal	Lenzing AG, 4860 Lenzing, AT	4212	72
Lenzing P84	–,,–	27	72
Lenzing Profilen	–,,–	1351	72
Lenzing PTFE	–,,–	1351	72
Lenzingtex	–,,–	11	411, 431, 432, 435
Lenzingtex Alu	–,,–	111	412, 43
Leona	Asahi Chemical Ind. Co. Ltd. Tokyo, JP	261	2 s
Leotel	Nichimen + Co., Tokyo, JP	261	2
Lerille	Courtaulds PLC, Bradford BD1 1EX W. Yorkshire, GB	2412	72
Leukorit®	Raschig AG, 67061 Ludwigshafen, DE	21	51
Levapren®	Bayer AG, 51368 Leverkusen, DE	1115, 343	11, 115, 91
Levasint®	–,,–	1115	141
Lewatit®	–,,–	–	148
Lexan®	GE Plastics Pittsfield, MA 01 201, US	2411	2, 4, 5

8.3 Trade Names

Name	Company (see 8.1, page 479)	Resin (see 8.2.1, page 481)	Forms of delivery (s. 8.2.2, p. 484)
Lexan® PPC	GE Plastics Pittsfield, MA 01 201, US	2415	11
Lexgard®	–„–	2411	5
Lighter	Inca International s.p.a. 75 010 Pisticci (MT), IT	2412	1, 2
Lightlon®	Sekisui Chemical Co. Ltd. Osaka 530, JP	1111	413, 71
Lignoform®	Kunststoffwerk Voerde 58256 Ennepetal, DE	1311	232
Lignostone®	Lignostone Ter Apel B.V. 9560 AB Ter Apel, NL	21, 22	613, 622
Lindolen®	A. u. E. Lindenberg GmbH 51436 Bergisch Gladbach, DE	11, 112	232, 52
Linklon	Mitsubishi Chemical Corp. Tokyo, JP	1116	21
Lipaton®	Hüls AG, 45764 Marl, DE	1223	121
Lipolan®	–„–	1223	121
Liquiflex H	Krahn Chemie GmbH, 20457 Hamburg, DE	263	111
Lisa®	Bayer AG, 51368 Leverkusen, DE	1/24	2, 54 s
Litac	Mitsui Toatsu Chemicals Inc. Tokyo, JP	1221, 1222	11, 2
Litex®	Hüls AG, 45764 Marl, DE	1223	121
Lithene®	Revertex Chemicals Ltd. Harlow, Essex CM20 2AH, GB	321	111
Llumar	Martin Processing, Film Div. Martinsville, VA 24112, US	2412	41
Lomod®	GE Plastics Pittsfield, MA 01 201, US	2413	110 TE
Lotader®	Elf Atochem 92 091 Paris, La Défense 10, FR	115 s	11
Lotrene®	Sofrapo/EniChem 92 080 Paris, La Défense 2, FR	1111 L, 1115	2, 41
Lotrex®	–„–	1111 L	2
Lotryl	Elf Atochem 92 091 Paris, La Défense 10, FR	1115	11, 1117
Loxiol®	Henkel KGaA, 40589 Düsseldorf, DE	–	91

Name	Company (see 8.1, page 479)	Resin (see 8.2.1, page 481)	Forms of delivery (s. 8.2.2, p. 484)
Lubonyl	N. Lundbergs Fabriks AB Fristad, SE	1116, 1311	31
Lubricomp	ICI PLC Welwyn Garden City, Herts. AL7 1 HH, GB	2411	2
Lubrilon®	A. Schulman Inc. Akron, OH 44 309, US	112, 1221, 237, 2411, 2412, 2361, 261, 1222	2
Lucalen®	BASF Aktiengesellschaft 67063 Ludwigshafen, DE	1115	115
Lucalor®	Elf Atochem 92 091 Paris, La Défense 10, FR	1312	11
Lucarex	–„–	131	2
Lucel	Standard Polymers Lake Grove, NY 11 755, US	2311	2
Lucite®	DuPont Co. Inc. Wilmington, DE 19 898, US	163	1, 2
Lucky	Standard Polymers Lake Grove, NY 11 755, US	1221	2
Lucobit	BASF Aktiengesellschaft 67063 Ludwigshafen, DE	1115+Bitumen	112, 434, 4351
Lucorex®	Elf Atochem 92 091 Paris, La Défense 10, FR	131	2
Lucryl®	BASF Aktiengesellschaft 67063 Ludwigshafen, DE	163	21
Ludopal®	–„–	242	114
Luflexen®	–„–	111	2
Luhydran	–„–	16	114
LUMAsite	American Acrylic Corp. West Babylon, NY 11 704, US	163, 242	614, 63
Lumax	Standard Polymers Lake Grove, NY 11 755, US	1221/2412	124
Lumiflon®	ICI PLC, Welwyn Garden City, Herts. AL7 1 HH, GB	135	146
Lumitol®	BASF Aktiengesellschaft 67063 Ludwigshafen, DE	162	114

Name	Company (see 8.1, page 479)	Resin (see 8.2.1, page 481)	Forms of delivery (s. 8.2.2, p. 484)
Lupan	Standard Polymers Lake Grove, NY 11 755, US	1222	224
Luperox	Elf Atochem 92 091 Paris, La Défense 10, FR	–	9
Luphen®	BASF Aktiengesellschaft 67063 Ludwigshafen, DE	2632	114
Lupolen®	–,,–	111, 1111, 1111 L, 1112, 1113, 1115	21
Lupos	Standard Polymers Lake Grove, NY 11 755, US	1221	224
Lupox	–,,–	2412	2
Lupox TE	–,,–	2411/2412	110 TE
Lupoy	–,,–	2411/1224	124
Lupragen®	Elastogran GmbH 49440 Lemförde, DE	263	117
Lupranat®	–,,–	2631	117
Lupranol®	–,,–	2633	117
Lupraphen®	–,,–	2632	117
Luran S®	BASF Aktiengesellschaft 67063 Ludwigshafen, DE	124	21
Luran®	–,,–	1222	21
Luranyl®	–,,–	1223/235	21, 22
Lusep	Standard Polymers Lake Grove, NY 11 755, US	237	2
Lustran®	Monsanto Chemical Co. Plastics Div. St. Louis, MO 63 167, US	1221, 1222	11, 2
Lustran® *Elite AMS*	–,,–	123/161	2
Lutofan®	BASF Aktiengesellschaft 67063 Ludwigshafen, DE	1313	123
Lutonal®- *Marken*	–,,–	14	114
Lutraflor	Fränkische Rohrwerke GmbH + Co. 97486 Königsberg, DE	2412	31
Luvatol	Lehmann & Voss & Co. 20354 Hamburg, DE	24	95

Name	Company (see 8.1, page 479)	Resin (see 8.2.1, page 481)	Forms of delivery (s. 8.2.2, p. 484)
Luviskol®	BASF Aktiengesellschaft 67063 Ludwigshafen, DE	127	1
Luvitherm®	–,,–	1311	411, 413
Luvobloc	Lehmann & Voss & Co. 20354 Hamburg, DE	11	16
Luvocom®	–,,–	23, 27	64
Luvofilm	–,,–	11	16
Luvoflex®	–,,–	264	2
Luvogard	–,,–	11, 24, 2412	16
Luvomag	–,,–	5	16
Luvomelt	–,,–	11	16
Luvopor	–,,–	11, 131, 2441, 242	16
Luvosil	–,,–	11	16
Luwax	BASF Aktiengesellschaft 67063 Ludwigshafen, DE	11	94
Luwipal®	–,,–	223	114
Luxacryl	Tupaj-Technik-Vertrieb GmbH 82538 Geretsried, DE	163	5
Lycra®	DuPont Co. Inc. Wilmington, DE 19 898, US	37	35, 72
Lynx®	Tufnol Ltd. Birmingham B42 2TB, GB	21	612
Lytex	Quantum Chemical Corp. Cincinnati, OH 42 249, US	233	2261
Lytron®	Monsanto Chemical Co., Plastics Div. St. Louis, MO 63 167, US	121, 122	113
Lyvertex®	Brochier S. A. 69 152, Decines Charpieù, FR	2612	72
M			
Macromelt	Henkel Corp. Ambler, PA 19 002, US	261	73
Macromer®	Sarnatech-Xiro AG 3185 Schmitten, CH	1253	2
Macroplast®	Henkel KGaA, 40589 Düsseldorf, DE	263, 35	73

Name	Company (see 8.1, page 479)	Resin (see 8.2.1, page 481)	Forms of delivery (s. 8.2.2, p. 484)
Macrynal®	Hoechst AG 65929 Frankfurt, DE	16	114
Madurit®	–„–	223, 224	112, 113, 116
Magnacomp®	LNP Plastics Nederland BV 4940 Raamsdonksveer, NL	261 s	22
Magnamite	Sumika-Hercules Wilmington, DE 19 894, US	5	226
Magnoval	Isovolta AG 2355 Wiener Neudorf, AT	212	611, 612
Magnum®	The Dow Chemical Company Midland, MI 48641-2166, US	1221	2
Maicro®	Savid S.p.A., 22 100 Como, IT	233	114
Makrofol®	Bayer AG, 51368 Leverkusen, DE	2411	413 el
Makrofol® LT	–„–	2411	436 s
Makrolon®	–„–	2411	2, 224
Makrolon®	Röhm GmbH, 64293 Darmstadt, DE	2411	51, 55
Makrolon® longlife	–„–	163/2411	51, 55
Malecca	Denki Kagaku Kogyo, Tokyo 100, JP	27 Cop.	11, 2
Mantopex	Golan Plastics Products Jordan Valley 15145, IL	1116	31
Maprenal®	Hoechst AG 65929 Frankfurt, DE	223, 224	114
Maraglas	Acme Chemicals Div. New Haven, CT 06 505, US	233	112 el
Maranyl®	ICI PLC, Welwyn Garden City, Herts. AL7 1 HH, GB	261	21, 224
Maraset	Acme Chemicals Div. New Haven, CT 06 505, US	233	112
Margard®	GE Plastics Pittsfield, MA 01 201, US	2411	5
Maricon	Riken Vinyl Ind. Co. Ltd., Tokyo, JP	1221/131	2
Marlex®	Phillips Petroleum Chemicals NV 3090 Overijse, BE	1112, 1113	11, 21
Marley Elite	Marley Floors Ltd., Lenham, Maidstone, Kent, ME17 2DE, GB	132	444

Name	Company (see 8.1, page 479)	Resin (see 8.2.1, page 481)	Forms of delivery (s. 8.2.2, p. 484)
Marley Premier Collection	Marley Floors Ltd., Lenham, Maidstone, Kent, ME17 2DE, GB	132	444
Marleyflex	–„–	132	444
Marleyflor Plus	–„–	132	444
Marvec	Brett Martin Ltd. Antrim BT 368 RE, IR	131	5
Marvyflo	–„–	1314	141
Marvylan	Limburgse-Vinyl Maatschapij Tessenderlo, NL	1311	2
Marvylex	Limburgse-Vinyl Maatschapij Tessenderlo, NL	1315	1
Marvyloy	–„–	1314	124 se
Mater-Bi	Novamont Italia Srl. 20 123 Mailand, IT	11	23 a
Max-Platte®	Isovolta AG 2355 Wiener Neudorf, AT	212	621
Maxprene	Sumitomo Seika Chemicals Co. Ltd. Osaka, JP	33	11
Mecanyl-Rohr®	Mecano-Bundy GmbH 69123 Heidelberg, DE	2411	2 med
Mektal®	Rogers Corp., Rogers, CT 06 263, US	21	124
Melacel	Pallflex Products Corp. Putnam, CT 06 260, US	223	452
Meladurol®	SKW Stickstoffwerke Piesteritz GmbH 06886 Lutherstadt Wittenberg, DE	224	114, 116
Melaform	Chromos – Ro Polimeri 4100 Zagreb, Croatia	223	2
Melaicar®	Aicar S.A., 08010 Barcelona, ES	223	24
Melamite	Pioneer Valley Plastics Inc. Bondsvillers, MA 01 009, US	223	621
Melan®	Henkel KGaA, 40589 Düsseldorf, DE	223	113
Melana®	Chemiefaserwerk Savinesti Piatra Neamt, RO	161	72
Melbrite®	Montedison S.p.A., 20 121 Milano, IT	223	231
Meldin®	Dixon Industries Corp. Bristol, RI 02 809, US	27	5, 6

Name	Company (see 8.1, page 479)	Resin (see 8.2.1, page 481)	Forms of delivery (s. 8.2.2, p. 484)
Melfeform	Chromos – Ro Polimeri 4100 Zagreb, Croatia	224	2
Melinar®	ICI PLC, Welwyn Garden City, Herts. AL7 1 HH, GB	2412	11
Melinex®	–„–	2412	41, 43
Melinite®	–„–	2412	2
Meliodent®	Bayer AG, 51368 Leverkusen, DE	163	147
Melit®	Sirlite Srl., 20 161 Milano, IT	223	112, 113
Melmex®	BIP Chemicals Ltd. Warley, West Midl., B69 4PG, GB	223	2
Melmorite	M/s Rattanchand Harjasrai Pvt. Ltd. Faridabad, IN	221, 223	2
Melochem	Chemiplastics S.p.A. 20 151 Mailand, IT	223	22, 23
Melopas®	Raschig AG, 67061 Ludwigshafen, DE	223, 224	22, 23
Melsir	Sirlite Srl., 20 161 Milano, IT	223	2
Melsprea®	AMC-Sprea S.p.A., 20 101 Milano, IT	223	2
Menzolit	Menzolit-Werke, Albert Schmidt GmbH, 76703 Kraichtal, DE	242, 233	22, 23, 24, 25, 26, 73 s
Meraklon	Moplefan S.p.A., 20124 Milano, IT	1121	35, 72
Merlon®	s. Makrolon	2411	2 se
Merporal	Makhteshim Chemical Works Ltd. Beer-Sheva, IL	242	11, 112, 113, 114
Mesamoll®	Bayer AG, 51368 Leverkusen, DE	–	91
Metallogen®- Metadur	Metallogen GmbH 4630 Bochum-Wattenscheid, DE	233, 263	146 s
Metallon®	Henkel KGaA, 40589 Düsseldorf, DE	233, 242	115, 73
Metamarble®	Teijin Chemicals Ltd., Tokyo, JP	16/2411, 1221/ 2411	2
Methafil	Akzo Engineering Plastics Inc. Evansville, IN 47 732, US	116	22
Metton®	Montell Polyolefins Wilmington, DE 19 894, US	53	118
Metylan®	Henkel KGaA, 40589 Düsseldorf, DE	4231	11

8.3 Trade Names

Name	Company (see 8.1, page 479)	Resin (see 8.2.1, page 481)	Forms of delivery (s. 8.2.2, p. 484)
Metzoplast®	Metzeler Schaum GmbH (British Vita PLC) 87700 Memmingen, DE	111, 112, 112+341, 12, 163, 235, 1221/2411	2, 721/76
Micares®	Micafil AG, 8048 Zürich, CH	263	112, 113
Micarta®	Westinghouse Electric Corp. Manor, PA 15 665, US	223	621
Microflex®	Clopay Corp. Plastics Products Div. Cincinnati, OH 45 202, US	111	4
Microlen	Sekisui Chemical Co. Ltd. Osaka 530, JP	111	71
Microlite	Web Technologie Oakville, CT 06 779, US	2412	413 s
Microthene®	Quantum Chemical Corp. Cincinnati, OH 42 249, US	1111, 1111 L, 1115	11, 141
Millathane HT	Notedome Ltd. Coventry CV3 2RQ, GB	263	11
Min	Putsch GmbH, 90427 Nürnberg, DE	112	227
Mindel®	Amoco Performance Products Atlanta, GA 30 350, US	2351, 263/ 1221	2, 224
Minicel®	Sekisui Chemical Co. Ltd. Osaka 530, JP	112	71
Minlon®	DuPont Co. Inc. Wilmington, DE 19 898, US	261	224
Mipofolie®	Aslan, 51491 Overath, DE	1311, 132	422, 437
Mipolam®	HT Troplast AG 53840 Troisdorf, DE	1311, 11	441, 442, 444
Mipoplast®	–,,–	131	432, 433
Mirason®	Mitsui Polychemical Co. Ltd. Tokyo, JP	1111 L, 1111, 1112	11, 2
Mirex	Mitsui Toatsu Chemicals Inc. Tokyo, JP	21	113
Mirvyl	Rio Rodano, S.A., Madrid 20, ES	131	11
Mitsubishi Polyethylene®	Mitsubishi Chemical Corp. Tokyo, JP	111	11, 2, 3, 4, 5, 6
Modar®	ICI PLC, Welwyn Garden City, Herts. AL7 1 HH, GB	163 s	118

Name	Company (see 8.1, page 479)	Resin (see 8.2.1, page 481)	Forms of delivery (s. 8.2.2, p. 484)
Modular	Polyú Italiana s.r.l. 20 010 Arluno, Milano, IT	2411	55
Moform	Chemische Betriebe a. d. Waag, CSFR	222	11
Moldesite®	AMC-Sprea S.p.A., 20 101 Milano, IT	21	23
Mollan®	Chemie Linz AG, 4040 Linz, AT	91	
Moltopren®	Bayer AG, 51368 Leverkusen, DE	2632, 2633	71
Monarfol	Billerud AB Abt. Nya Produkter Säffle, SE	1111	43
Mondur	Miles Chemical Corp. Pittsburgh, PA 15 205, US	263, 2631	117, 71
Monocast®	Polypenco GmbH 51437 Bergisch Gladbach, DE	261	3, 5
Monoplex	C.P. Hall Comp. Chicago, IL 60638, US	91	
Monosol®	Mono-Sol Div. Chris Craft Ind. Inc. Gary, IN 46403, US	152, 4233, 4234, 4231	41, 411, 412, 432, 437
Monothane	Synair Corp. Chattanooga, TN 37 406, US	263	112
Montac®	Monsanto Chemical Co., Plastics Div. St. Louis, MO 63 167, US	2611, 271	110 TE, 73
Moplan®	Montell Polyolefins Wilmington, DE 19 894, US	112 s	2
Moplefan®	–,,–	112	411, 412, 413
Moplen Ro®	–,,–	1113	2
Moplen-EP®	–,,–	1115	2
Moplen®	–,,–	112, 1121	11, 2
Moplex®	–,,–	112	2 s
Morthane	Morton International Inc. Chicago, IL 60 606, US	264	115, 111 f. 73
Mowilith®	Hoechst AG 65929 Frankfurt, DE	151, 16	114, 115, 116, 121
Mowiol®	–,,–	152	11
Mowital®	Hoechst AG 65929 Frankfurt, DE	155	112, 114
Multi-Flow®	Norplex Oak Inc. La Crosse, WI 54 601, US	233, 27	2251

Name	Company (see 8.1, page 479)	Resin (see 8.2.1, page 481)	Forms of delivery (s. 8.2.2, p. 484)
Multican®	Unitex Ltd., Knaresborough, North Yorks., HG5 OPP, GB	263	112, 146
Multifil®	Chemie Linz AG, 4040 Linz, AT	112	72 se, s
Multilon	Teijin Chemicals Ltd., Tokyo, JP	1121/2411	2 se
Multranol	Miles Chemical Corp. Pittsburgh, PA 15 205, US	2633	11
Multrathane	–,,–	2631	111
Multron	–,,–	2632	11
Muratherm®	Murtfeldt GmbH & Co. KG 44309 Dortmund, DE	–	6121, 31, 33, 51
Murdopol®	–,,–	261 s	33, 51
Murlubric®	–,,–	261 s	33, 51, 342
Murtex®	–,,–	21	612
Mylar®	DuPont Co. Inc. Wilmington, DE 19 898, US	133/2412	412
Myoflex	Von Roll Isola 4226 Breitenbach, CH	233/261	4341/611
Myosam	–,,–	233/26	4341/616 el

N

Name	Company	Resin	Forms of delivery
Nabutene	Ets. G. Convert 01100 Oyonnax, FR	1221, 124	5
Nafion®	DuPont Co. Inc. Wilmington, DE 19 898, US	135	4 s
Nakan	Elf Atochem 92 091 Paris, La Défense 10, FR	131	2
Naltene	Ets. G. Convert 01100 Oyonnax, FR	1111	5
Nandel	DuPont Co. Inc. Wilmington, DE 19 898, US	161	72
NAP resin	Kanegafuchi Chemical Ind. Co. Ltd. Osaka, JP	2412	11
Naprene	Ets. G. Convert 01100 Oyonnax, FR	112	5
Narmco®	BASF Structural Materials Inc. Charlotte, NC 28 217, US	23	2261/6121

Name	Company (see 8.1, page 479)	Resin (see 8.2.1, page 481)	Forms of delivery (s. 8.2.2, p. 484)
NAS	Novacor Chemicals Inc. Plastics Div. Leominster, MA 01 453, US	125	2
Natsyn®	Goodyear Tire & Rubber Co. Akron, OH 44 316, US	331	11
Naugahyde®	Uniroyal, Mishawaka, IN 46 544, US	132	436
Naugapol®	–„–	322	11
Naxoglas	Ets. G. Convert 01100 Oyonnax, FR	163	55
Naxoid	–„–	4222	5
Naxolene®	–„–	1223	5
Naxols	–„–	244	11
Naxorese	–„–	2442	11
Necirès®	Nevcin Polymers B.V. 1420 AD Uithoorn, NL	53	114, 115
Necofene®	Ashland Chemical Corp. Columbus, OH 43 216, US	235	2
Nelco	New England Laminates Corp. Inc. Walden, NY 12 586, US	233	2251, 614 el
Neo-zex	Mitsui Petrochemical Ind. Ltd. Tokyo, JP	1111, 1111 L, 1112	141, 2
Neocryl®	Polyvinyl Chemie BV Holland 5140 AC-Waalwijk, NL	16	114, 115
Neoflon®	Daikin Kogyo Co. Ltd., Osaka, JP	1351, 1352	121
Neofract®	Elastogran GmbH 49440 Lemförde, DE	263	117 med
Neolith	Fabbrica Adesivi Resine S.p.A. Cologno Monzese, IT	151	11, 121, 123, 13, 73
Neonit®	Ciba Geigy AG, 4002 Basel, CH	223, 233, 243	224 s
Neopolen®	BASF Aktiengesellschaft 67063 Ludwigshafen, DE	112	71
Neoprene	DuPont Co. Inc. Wilmington, DE 19 898, US	35	11
Neoprex	Fabbrica Adesivi Resine S.p.A. Cologno Monzese, IT	221	13, 73
NeoRez®	Polyvinyl Chemie BV Holland 5140 AC-Waalwijk, NL	263	146

Name	Company (see 8.1, page 479)	Resin (see 8.2.1, page 481)	Forms of delivery (s. 8.2.2, p. 484)
Neox	Matsushita Electric Works Ltd. Mie, JP	242	224
Nepol	Neste Oy Chemicals, 02 151 Espoo, FI	112	224
Net-O-Fol	Billerud AB Abt. Nya Produkter Säffle, SE	111	4311
Neuthane	Notedome Ltd. Coventry CV3 2RQ, GB	2632, 2633	111
Nevchem®	Nevcin Polymers B.V. 1420 AD Uithoorn, NL	53	114, 115, 121
Nevex®	–,,–	53	114, 115
Nevillac®	–,,–	53	114, 115
Nevroz	–,,–	53	14
New-PTI	Mitsui Toatsu Chemicals Inc. Tokyo, JP	27	11, 21, 224, 226
News®	Neste Oy Chemicals, 02 151 Espoo, FI	1111 L	1, 2
Niax®	Union Carbide Corp. Danbury, CT 06 817, US	2632	117
Nibren-Wachs®	Bayer AG, 51368 Leverkusen, DE	56	94
Nika Temp®	Nippon Carbide Ind. Co. Inc. Tokyo, JP	1312	11, 2
Nikalet	–,,–	223	2
Nipeon	Nippon Zeon Co. Ltd., Tokyo, JP	131	11, 2
Nipoflex	Toyo Soda Mfg. Co. Ltd., Tokyo, JP	1115	11, 2, 73
Nipol	Nippon Zeon Co. Ltd., Tokyo, JP	32	11
Nipolit®	Chisso Corp., Tokyo, JP	131, 1314	11, 2
Nipolon	Toyo Soda Mfg. Co. Ltd., Tokyo, JP	1111	11, 2
Nipren®	Nuova Italresina 20 027 Rescaldina (Milano), IT	112	31
Nirlene®	–,,–	1111	31
Nivion	ECP EniChem Polimeri srl. 20 124 Mailand, IT	261	72
Nivionplast	ECP EniChem Polimeri srl. 20 124 Mailand, IT	261	224
Noblen®	Mitsubishi Chemical Corp. Tokyo, JP	112	11, 2, 4, 5
Nokrythane®	Akzo Engineering Plastics Inc. Evansville, IN 47 732, US	261	112

8.3 Trade Names

Name	Company (see 8.1, page 479)	Resin (see 8.2.1, page 481)	Forms of delivery (s. 8.2.2, p. 484)
Nolimold	Rhône-Poulenc Films, 92 080 Paris La Défense, Cedex 6, FR	272	2
Nomelle	DuPont Co. Inc. Wilmington, DE 19 898, US	161 s	72
Nomex®	Euro-Composites S.A. Echternach, LU	163/21	611
Nopinol®	Raschig AG, 67061 Ludwigshafen, DE	211	113
Nor-Pac	Norddeutsche Seekabelwerke AG 26954 Nordenham, DE	1113	35, 221
NorCore	Norfield Corp. Danbury, CT 06 810, US	112, 1113, 121, 1221, 2415	7662
Nord-ht	Nordchem S.p.A. 33035 Martignacco (Ud), IT	1312	2
Nordcell	–,,–	1311	117/21
Nordel®	DuPont Co. Inc. Wilmington, DE 19 898, US	112/3411	110 TE
Nordvil	Nordchem S.p.A. 33035 Martignacco (Ud), IT	1311	2
Norflex®	Norddeutsche Seekabelwerke AG 26954 Nordenham, DE	121	43, 54
Norplex®	Norplex Oak Inc. La Crosse, WI 54 601, US	212, 223, 233, 265, 27	231, 6111, 612, 614
Norpol	Jotun Polymer AS 3235 Sandefjord, NO	242	112, 6
Norsorex®	Elf Atochem 92 091 Paris, La Défense 10, FR	3 s	11
Nortuff®	Quantum Chemical Corp. Cincinnati, OH 42 249, US	1113/112	22
Norvinyl	Norsk Hydro A/S., Oslo 2, NO	131	11
Noryl GTX	GE Plastics Pittsfield, MA 01 201, US	235/261	2
Noryl®	–,,–	235	2, 4
Nosaflex®	Von Roll Isola 4226 Breitenbach, CH	24/28	4341/616 el
Nourythane®	Akzo Engineering Plastics Inc. Evansville, IN 47 732, US	261	112
Novaclad	Sheldal Inc. Northfield, MN 55 057, US	27	6111 el

8.3 Trade Names

Name	Company (see 8.1, page 479)	Resin (see 8.2.1, page 481)	Forms of delivery (s. 8.2.2, p. 484)
Novacor	Novacor Chemicals Ltd. Calgary, Alberta T2P 2 H6, CA	121	2
Novadur®	Mitsubishi Chemical Corp. Tokyo, JP	2412	21, 2213, 224
Novamid®	–,,–	261	21, 224, 31, 33
Novapol LL	Novacor Chemicals Ltd. Calgary, Alberta T2P 2 H6, CA	1111 s, 1111 L	2
Novarex®	Mitsubishi Chemical Corp. Tokyo, JP	2411	2, 224, 54
Novatec®	–,,–	152, 261	422, 61
Novatron®	Polypenco GmbH 51437 Bergisch Gladbach, DE	2412	3, 5
Novex	BP Chemicals International Ltd. London SW1W OSU, GB	1111	2
Novodur®	Bayer AG, 51368 Leverkusen, DE	1221	11, 224
Novoid	Ets. G. Convert 01100 Oyonnax, FR	4221	4
Novolak	general name	21	111
Novolen®	BASF Aktiengesellschaft 67063 Ludwigshafen, DE	1121	21, 2213, 224
Novomikaband®	Elektrotechn. Werke Hennigsdorf, DE	212, 28	61-Cu
Novomikaflex®	–,,–	28	614+616
Novomikanit®	–,,–	28	616
Novon	Novon Products Div. Morris Plains, NJ, US	11	23 a
Nucrel®	DuPont Co. Inc. Wilmington, DE 19 898, US	111, 1115/16	2
Nupol®	Freeman Chemical Corp. Port Washington, WI 53 074, US	163	112
Nupreg®	Scott Bader Co. Ltd. Wollaston, Wellingborough, Norths. NN9 7RL, GB	242	2251
Ny-Kon®	LNP Plastics Nederland BV 4940 Raamsdonksveer, NL	261	22
Nycoa	Nylon Corp. of America Manchester, NH 03 103, US	261 s	2
Nydur®	Bayer AG, 51368 Leverkusen, DE	261	2

Name	Company (see 8.1, page 479)	Resin (see 8.2.1, page 481)	Forms of delivery (s. 8.2.2, p. 484)
Nylafil	Akzo Engineering Plastics Inc. Evansville, IN 47 732, US	261	117, 224
Nylaflow®	Polypenco GmbH 51437 Bergisch Gladbach, DE	26	31, 33
Nylane®	Rhône-Poulenc 92 097 Paris, La Défense 2, FR	111+261	412
Nylatrack®	Polypenco GmbH 51437 Bergisch Gladbach, DE	261	342
Nylatron®	–,,–	261	32, 33, 34, 42, 51
Nylon	general name	261	–
Nypel®	Allied Signal Corp. Morristown, NJ 07 962, US	261	2
Nyrim®	DSM RIM Nylon vof 4202 YA Maastricht, NL	261 s	118

O

Name	Company	Resin	Forms of delivery
obo-Festholz	Otto Bosse, 31655 Stadthagen, DE	61	613
obomodulan	–,,–	51	263
OC-Plan 2000	Odenwald-Chemie GmbH 69250 Schönau, DE	1115+Bitumen	43
Oekolex-G	Gurit-Worbla AG 3063 Ittigen-Bern, CH	11, 12, 24	41, 42, 52, 53, 54
Oekolon G	–,,–	11, 12	41, 42, 52, 53, 54
Ohmoid	Wilmington Fibre Specialty Co. New Castle, DE 19 720, US	21, 223	61
Oilamid®	Nylontechnik Licharz GmbH 53757 St. Augustin, DE	261	31, 33, 51
Oilex	Schüder Oilex KG 56412 Nentershausen, DE	2311, 2312, 261	31, 33, 51
Olapol®	Philippine GmbH & Co. KG 56112 Lahnstein, DE	2633	71
Oldoflex	Büsing & Fasch GmbH & Co. 26123 Oldenburg, DE	263	111
Oldopal®	–,,–	242	112, 114
Oldopren	–,,–	264	14

8.3 Trade Names

Name	Company (see 8.1, page 479)	Resin (see 8.2.1, page 481)	Forms of delivery (s. 8.2.2, p. 484)
Oldopur	Büsing & Fasch GmbH & Co. 26123 Oldenburg, DE	263	111
Oltvil®	Chemisches Kombinat, Pitesti, RO	131	2
Omniplast	Alphacan Omniplast GmbH 35627 Ehringshausen, DE	111, 121, 131	434
Ondex	Adriaplast S.p.A. 34074 Monfalcone, IT	132	55
Ondex®	Solvay S.A., 1050 Bruxelles, BE	1311	55
Onduline	Onduline S.A., Paris 17, FR	1311	55
Ongrodur®	BorsodChem Rt. 3702 Kazincbarcika, HU	131	52
Ongrofol	–„–	131	41
Ongrolit	–„–	131, 132	2
Ongromix	–„–	131	11
Ongronat	–„–	2632	11
Ongropur	–„–	26	11
Ongrovil	–„–	131	11
Ontex	Dexter Corp. Windsor Locks, CT 06 096, US	111/1115	11, 120
Opalen	UPM Walki Pack 37 601 Valkeakoski, FI	261/111	412
Opax®	–„–	111/261	412
Opet	Owens Brockway Plastics Toledo, OH 43 666, US	2412 s	111
Oppalyte®	Mobil Polymers US Inc. Norwalk, CT 06 856, US	112	412
Oppanol®-B	BASF Aktiengesellschaft 67063 Ludwigshafen, DE	114	11
Opssalak	Opssa 08850 Gava (Barcelona), ES	221	2
Opssalit	–„–	121	2
Opssalkyd	–„–	242	2
Opssamin	–„–	22, 224	2
Opssapol	–„–	242	224
Opti...	Fränkische Rohrwerke GmbH + Co. 97486 Königsberg, DE	131	31

8.3 Trade Names

Name	Company (see 8.1, page 479)	Resin (see 8.2.1, page 481)	Forms of delivery (s. 8.2.2, p. 484)
Opto	ICI Fiberite Winona, MN 55 987, US	233	2
Optum®	Ferro Corp. Evansville, IN 47 711, US	11	124
Or-on	Mayer Enterprises Ltd., Coating Dept., Tel Aviv, IL	132	436
Orel®	DuPont Co. Inc. Wilmington, DE 19 898, US	2412	36
Orevac®	Elf Atochem 92 091 Paris, La Défense 10, FR	11	115, 120
Orgalloy	–,,–	112/261	11, 2
Orgamide®	–,,–	261	2, 224
Orgasol	–,,–	261	14
Orgater	Palmarole sarl 68480 Bettlach, FR	241	2
Orit	Mayer Enterprises Ltd., Coating Dept., Tel Aviv, IL	132	423
Orlon®	DuPont Co. Inc. Wilmington, DE 19 898, US	161	72
Ornamenta	Tarkett Pegulan AG 67227 Frankenthal, DE	132	444
Oroglas®	Atohaas, 92062 Paris, La Défense 10, FR	163	54
Oromid	Nyltech Italia Srl 20 020 Ceriano (MI), IT	261	2
Orthane®	Ohio Rubber Co. Denton, TX 76 201, US	2633	1111
OSMOpane®	Ostermann & Scheiwe GmbH & Co. 48155 Münster, DE	131	341, 3411, 342
Osstyrol	A. Hagedorn & Co. AG 49078 Osnabrück, DE	1221, 1222, 1223	5
Otipol	Bayer AG, 51368 Leverkusen, DE	2411	35
OxyBlend	Occidental Chemical Corp. Dallas, Texas 75 380, US	131	2
Oxycal	–,,–	1221, 131	4
OxyClear	–,,–	131	111
Oxyshield	Allied Signal Corp. Morristown, NJ 07 962, US		

8.3 Trade Names

Name	Company (see 8.1, page 479)	Resin (see 8.2.1, page 481)	Forms of delivery (s. 8.2.2, p. 484)
Oxyshield	Allied Signal Corp. Morristown, NJ 07 962, US	–	–
Oxytuf®	Occidental Chemical Corp. Niagara Falls, NY 14 303, US	1315	2
P			
P 84	Lenzing AG, 4860 Lenzing, AT	27	72
P-Flex	Putsch GmbH, 90427 Nürnberg, DE	1221	2, 224
Pacrosir	Sirlite Srl., 20 161 Milano, IT	16	123
Pacton	Wardle Storeys PLC Manningtree, Essex CO11 1NJ, GB	1314	51, 52, 53, 55
Pagholz®	PAG Preßwerk AG, 45356 Essen, DE	21	16, 622
Paja	Paja Kunststoffe Jaeschke 51503 Rösrath, DE	1111	411
Palamoll®	BASF Aktiengesellschaft 67063 Ludwigshafen, DE	–	91
Palapet	Kyowa Gas Chemical Co. Ltd. Tokyo, JP	163	111, 2
Palapreg®	BASF Aktiengesellschaft 67063 Ludwigshafen, DE	242	224/2251
Palatal®	–,,–	242	11, 112
Palatinol®	–,,–	–	91
Pallaflon®	Schieffer GmbH & Co. KG 59557 Lippstadt, DE	1351	3, 5
Pallanorm®	Schieffer GmbH & Co. KG 59557 Lippstadt, DE	237, 2412	6111
Pamflon®	Norton Pampus GmbH 47877 Willich, DE	1351	2r
Panaflex®	3 M Co., St. Paul, MN 55 144, US	132	436
Pandex	Dainippon Ink. & Chemicals Inc. Tokyo 103, JP	264	2
Panlite®	Teijin Chemicals Ltd., Tokyo, JP	2411	2, 224
Pantalast®	Pantasote Inc. Passaic, NJ 07 055, US	1315	110 TE
Pantarin	Koepp AG 65375 Oestrich-Winkel, DE	263	71

8.3 Trade Names

Name	Company (see 8.1, page 479)	Resin (see 8.2.1, page 481)	Forms of delivery (s. 8.2.2, p. 484)
Panzerholz®	Blomberger Holzindustrie, B. Hausmann GmbH & Co. KG 32817 Blomberg, DE	21	613
Papertex	Lenzing AG, 4860 Lenzing, AT	111	411, 412
Papia	Mikuni Lite Co., Osaka, JP	112	23/62
Paracril®	Uniroyal, Mishawaka, IN 46 544, US	323, 131/323	11
Paraglas®	Degussa AG 60287 Frankfurt, DE	163	51, 53, 54, 55
Paralac®	Cray Valley Products Int. Farnborough, Kent, BR6 7EA, GB	244	114
Paraloid®	Rohm & Haas Co. Philadelphia, PA 19105, US	1223, 16, 164, 322	11, 112, 114
Paraplex®	–„–	233, 242, 244	112, 91
Parapol®	Exxon Chemicals America Houston, TX 77 001, US	111	1
Paraprene	Hodogaya Ltd., JP	264	2
Parcloid	Parcloid Chemical Co. Ridgewood, NJ 07 451, US	1313	121, 122
Parketfoam	Sentinel GmbH 73441 Bopfingen, DE	111	711
Parlon®	Hercules Powder Company Inc. Wilmington, DE 19 894, US	411	11
Parylene	Union Carbide Corp. Danbury, CT 06 817, US	19	1
Patix	Chemische Betriebe a. d. Waag, CSFR	242	11
Pattex®	Henkel KGaA, 40589 Düsseldorf, DE	35	73
Paulownia	Mitsui Toatsu Chemicals Inc. Tokyo, JP	112	414
Pavex®	Pavag AG/SA, Nebikon, CH	112	41, 43
PCI...®	PCI Augsburg GmbH 86159 Augsburg, DE	16, 163, 233, 24, 263, 28	146, 145, 732
Pe-vo-lon	Räder-Vogel, Räder+Rollenfabrik GmbH, 21109 Hamburg, DE	261	2, 224
Pebax®	Elf Atochem 92 091 Paris, La Défense 10, FR	2613	110 TE
Pecolit®	Pecolit Kunststoffe GmbH & Co KG 67105 Schifferstadt, DE	242	614, 63, 766

Name	Company (see 8.1, page 479)	Resin (see 8.2.1, page 481)	Forms of delivery (s. 8.2.2, p. 484)
Pedigree	P. D. George Co. St. Louis, MO 63 147, US	242	11, 225
PEEK	ICI PLC, Welwyn Garden City, Herts. AL7 1 HH, GB	2351	11, 2
Peerless Insulation	NVF Container Div. Hartwell, GA 30 643, US	4211	fishpaper
Pegulan	Tarkett Pegulan AG 67227 Frankenthal, DE	132	44
Pegutan	–,,–	132	433, 434, 435
Pekatop®	Bayer AG, 51368 Leverkusen, DE	163	147
Pekatray®	–,,–	163	147
Pekevic	Neste Oy Chemicals, 02 151 Espoo, FI	113	2
Pekutherm	Unitemp, Coloma, MI 49 038, US	–	149
Pelprene®	Toyobo Co. Ltd., Osaka 530, JP	264	1, 2
PEN-resin	NKK Corp., Tokyo, JP	1123	11
Pennlon®	Dixon Industries Corp. Bristol, RI 02 809, US	11	3, 5
Pentacite	Reichhold Chemicals Inc. Jacksonville, FL 32 245, US	251	114
Pentaclear	Klöckner-Pentaplast GmbH 56410 Montabaur, DE	1311	411
Pentadur®	–,,–	1311	421, 53
Pentafood	–,,–	3111	411, 412, 53
Pentaform	–,,–	3111	411, 53
Pentalan	–,,–	1311	43, 432, 452, 53
Pentapharm	–,,–	3111	411, 412, 53
Pentaplus	Klöckner-Pentaplast GmbH 56410 Montabaur, DE	3111	43, 431
Pentaprint	–,,–	3111	411, 412, 53
Pentatec	–,,–	236, 237, 2411, 261	4
Pentatherm®	–,,–	1311	43
Perbunan®	Bayer AG, 51368 Leverkusen, DE	323	11
Perfluorogum	Daikin Kogyo Co. Ltd., Osaka, JP	348	11, 2
Pergut®	Bayer AG, 51368 Leverkusen, DE	411	114

Name	Company (see 8.1, page 479)	Resin (see 8.2.1, page 481)	Forms of delivery (s. 8.2.2, p. 484)
Perl	Van Besouw BV Kunststoffen 5050 AA Goirle, NL	11	433
Perlon®	Perlon Warenzeichenverband	261	72
Permair	Porvair P.L.C. King's Lynn, Norfolk PE30 2 HS, GB	262	437, 712
Permaloc	Ma-Bo AS, 1760 Berg, NO	131	312 s
Permapol	ECP EniChem Polimeri srl. 20 124 Mailand, IT	236	112 el
Peroxidol®	Reichhold Chemicals Inc. Jacksonville, FL 32 245, US	–	91
Perspex®	ICI PLC, Welwyn Garden City, Herts. AL7 1 HH, GB	163	51, 53, 55
Perstorp Compounds	Perstorp AB, 28, 480 Perstorp, SE	21, 22, 223	2
Petlite	Goodyear Tire & Rubber Co. Akron, OH 44 316, US	2412	72
Petlon	Miles Chemical Corp. Pittsburgh, PA 15 205, US	2412	2 se, 224
Petpac®	Hoechst AG 65929 Frankfurt, DE	2412	14
Petra®	Allied Signal Corp. Morristown, NJ 07 962, US	2412	2, 22
Petrothene®	Quantum Chemical Corp. Cincinnati, OH 42 249, US	1111, 1111 L, 1112, 1113	2, 2212, 226
Petsar®	Bayer AG, 51368 Leverkusen, DE	261/2412	2
Pevikon®	Norsk Hydro A/S., Oslo 2, NO	131	11, 2
Pexgol	Golan Plastics Products Jordan Valley 15145, IL	1116	31
Phenmat	DSM Compounds UK Ltd. Eastwood, Lancs., GB	212	2251
Phenmat	Frost & Sullivan 6000 Frankfurt/M., DE	212	2251
Phenodur®	Hoechst AG 65929 Frankfurt, DE	21	112, 113, 114, 2
Phenolite	NVF Container Div. Hartwell, GA 30 643, US	21, 223, 233, 242, 28	61

8.3 Trade Names

Name	Company (see 8.1, page 479)	Resin (see 8.2.1, page 481)	Forms of delivery (s. 8.2.2, p. 484)
Phenorit®	Lautzenkirchener Kalksandsteinwerk GmbH, 66440 Blieskastel, DE	21	71
Phenox	Glunz AG, 66265 Heusweiler, DE	21	761
Philan®	Philippine GmbH & Co. KG 56112 Lahnstein, DE	37	5
Phoenolan®	Phoenix AG, 21079 Hamburg, DE	37	3, 4, 711
Phophazene	Firestone Synthetics Rubber and Latex Co., Akron, OH 44 301, US	348	2
Phtalopal®	BASF Aktiengesellschaft 67063 Ludwigshafen, DE	2632	114
Piadurol®	SKW Stickstoffwerke Piesteritz GmbH 06886 Lutherstadt Wittenberg, DE	221, 224	114, 116
Piamid	–,,–	224	114, 116
Piapox®	–,,–	2611	112, 114, 115
Pierson	Pierson Industries Palmer, MA 01 069, US	1115	411, 412, 432, 438
Pioloform®	Wacker-Chemie GmbH 81737 München, DE	154	11
Planolen®	Rethmann Plano GmbH 48356 Nordwalde, DE	1111, 1113, 112	2 r
Planomid®	–,,–	261	2 r, 2212, 2217, 224
Planox	Glunz AG, 66265 Heusweiler, DE	21	761
Plaper®	Mitsubishi Monsanto Chemical Co. Tokyo, JP	1223	41 s
Plascoat®	Plascoat-Systems Ltd. Farnham, Surrey GU9 9NY, GB	112	2 se
Plaskon®	Plaskon Electronic Inc. Philadelphia, PA 19 105, US	21, 22, 233, 242, 243, 261	2
Plastalloy	Akzo Engineering Plastics Inc. Evansville, IN 47 732, US	236	224
Plastazote®	BXL Plastics Ltd. London, SW1W OSU, GB	111	71 el
Plasticell	Permali Gloucester Ltd. Gloucester, GB	1311	713
Plasticon	KTD Plasticon GmbH 46339 Dinslaken, DE	242	614, 6141

8.3 Trade Names

Name	Company (see 8.1, page 479)	Resin (see 8.2.1, page 481)	Forms of delivery (s. 8.2.2, p. 484)
Plastigen®	BASF Aktiengesellschaft 67063 Ludwigshafen, DE	222	114
Plastilit®	Solvay S.A., 1050 Bruxelles, BE	1311	31
Plastin®	4P Folie Forchheim GmbH 91301 Forchheim, DE	1113	411
Plastiroll	Hermann Wendt GmbH 12103 Berlin, DE	1221, 1311, 132	31, 311
Plastoflex	Elkoflex Isolierschlauchfabrik 10553 Berlin, DE	132	32 el
Plastolein®	Quantum Chemical Corp. Cincinnati, OH 42 249, US	233	91
Plastomoll®	BASF Aktiengesellschaft 67063 Ludwigshafen, DE	–	91
Plastopal®	–„–	222	114
Plastopil	Mayer Enterprises Ltd., Coating Dept., Tel Aviv, IL	111, 132	432
Plastopreg®	Reichhold Chemicals Inc. Jacksonville, FL 32 245, US	242	224, 225
Plastor	Mayer Enterprises Ltd., Coating Dept., Tel Aviv, IL	132	4 a
Plastotex	Elkoflex Isolierschlauchfabrik 10553 Berlin, DE	132, 4212	32 el
Plastothane®	Morton International Inc. Chicago, IL 60 606, US	264	2
Plastotrans®	4P Folie Forchheim GmbH 91301 Forchheim, DE	1111	411
Platamid®	Elf Atochem 92 091 Paris, La Défense 10, FR	261	121, 123
Plathen®	–„–	111	73
Platherm®	–„–	2412	73
Platilon®	–„–	112, 261, 264	41, 411, 43
Platon®	–„–	2412, 261	35
Plenco®	Plastics Engineering Company Sheboygan, WI 53 082, US	21, 223, 224, 242, 243	113, 2, 22, 23
Plex®	Röhm GmbH, 64293 Darmstadt, DE	16	115, 73

Name	Company (see 8.1, page 479)	Resin (see 8.2.1, page 481)	Forms of delivery (s. 8.2.2, p. 484)
Plexacryl	Röhm GmbH, 64293 Darmstadt, DE	–	–
Plexalloy®	–,,–	1641	2
Plexar®	Quantum Chemical Corp. Cincinnati, OH 42 249, US	1113, 1115	115
Plexidon®	Röhm GmbH, 64293 Darmstadt, DE	16	147
Plexifix®	–,,–	163 s	149
Plexiglas®	–,,–	163	2, 31, 33, 54, 55
Plexigum®	–,,–	16	111, 113, 114
Plexileim®	–,,–	165	116, 123
Plexilith	–,,–	163	112
Pleximid®	–,,–	1642	2, 21 tra
Pleximon®	–,,–	16	111, 91
Plexisol®	–,,–	164	116, 123
Plexit®	–,,–	163	112, 146, 147
Plexitex®	–,,–	61	116
Plextol®	–,,–	16	121
Pliocord VP 107	Goodyear Tire & Rubber Co. Akron, OH 44 316, US	127	12
Pliofilm	–,,–	412	4
Plioflex	–,,–	322	11
Pliolite®	–,,–	322	11, 115
Pliovic	–,,–	131	114
Pluracol®	BASF Wyandotte Corp. Parsippany, NJ 07 054, US	2633	117
Pluragard	BASF Wyandotte Corp. Parsippany, NJ 07 054, US	263	146
Pluronic	–,,–	2633	117
Plyamin	Reichhold Chemicals Inc. Jacksonville, FL 32 245, US	221	113, 114
Plyamul	–,,–	151	115
Plyocite	–,,–	21	2431
Plyophen	–,,–	211	113
Plytron®	ICI PLC, Welwyn Garden City, Herts. AL7 1 HH, GB	1, 2	225

Name	Company (see 8.1, page 479)	Resin (see 8.2.1, page 481)	Forms of delivery (s. 8.2.2, p. 484)
PMI-Resin	Mitsubishi Co. Ltd. Tokyo, JP	2416	2, 21 tra
Pocan®	Bayer AG, 51368 Leverkusen, DE	2412	2
Pokalon®	Lonza-Folien GmbH 79576 Weil am Rhein, DE	2411	43 s
Polathane	Polaroid Chemicals Assonet, MA 02 702, US	264	111
Policril	Fabbrica Adesivi Resine S.p.A. Cologno Monzese, IT	16	115
Polidux	Aiscondel S.A., Barcelona 13, ES	121	2
Polinat®	Raschig AG, 67061 Ludwigshafen, DE	263	113
Politarp®	ICI PLC, Welwyn Garden City, Herts. AL7 1 HH, GB	111	433
Politen®	Iten Industries Ashtabula, OH 44 004, US	24	34, 611, 612, 614
Polivar®	Polivar S.p.A., Rom, IT	163	33, 51, 55
Polnac®	AMC-Sprea S.p.A., 20 101 Milano, IT	242	113
Poly bd	Elf Atochem 92 091 Paris, La Défense 10, FR	31, 32	111
Poly Pearl	Chi Mei Industrial Co., Ltd. Tainan, Shien, Taiwan	121	117
Poly-Net®	Norddeutsche Seekabelwerke AG 26954 Nordenham, DE	1111, 1113, 112	35, 54
Poly®129u	Polyú Italiana s.r.l. 20 010 Arluno, Milano, IT	2411	55
Polyathane®XPE	Polaroid Chemicals Assonet, MA 02 702, US	264	2
Polybond®	BP Chemicals International Ltd. London SW1W OSU, GB	1113, 1121, 16	9
Polycar	Polyon-Barkai, Kibbutz Barkai M. P. Menashe 37 860, IL	132	41
Polycarbafil®	Akzo Engineering Plastics Inc. Evansville, IN 47 732, US	2411	117, 224, 226
Polychem®	Budd Co., Polychem. Div. Phoenixville, PA 19 460, US	21, 223, 233, 243	22
Polyclad®	GE Plastics Pittsfield, MA 01 201, US	2411	63

8.3 Trade Names

Name	Company (see 8.1, page 479)	Resin (see 8.2.1, page 481)	Forms of delivery (s. 8.2.2, p. 484)
Polyclear	Hoechst AG, 65929 Frankfurt, DE	2412 s	21
Polyclip	Polyú Italiana s.r.l., 20 010 Arluno, Milano, IT	2411	55
Polycoat®	EMW-Betrieb, 65582 Diez, DE	263	146
Polycom®	Huntsman Chemical Corp., Chesapeake, VA 23 320, US	112	2212, 2213
Polycomp®	LNP Plastics Nederland BV, 4940 Raamsdonksveer, NL	135	22
Polycril®	Irpen S.A., Barcelona 13, ES	163	53, 55
Polycup	Polyú Italiana s.r.l., 20 010 Arluno, Milano, IT	2411	55
Polydene®	Scott Bader Co. Ltd., Wollaston, Wellingborough, Norths. NN9 7RL, GB	133	116, 121
Polydet®	Mitras Kunststoffe GmbH, 92637 Weiden, DE	242	61, 63
Polyfelt®	Chemie Linz AG, 4040 Linz, AT	112	721
Polyfill	Polykemi AB, 27 100 Ystad, SE	112	224
Polyflam®	A. Schulman Inc., Akron, OH 44 309, US	111, 112, 1221, 1223, 237, 2411/1221	2 se
Polyflon	Daikin Kogyo Co. Ltd., Osaka, JP	1351	11, 12, 2
Polyfluron®	Dr. Schnabel GmbH, 65549 Limburg, DE	1351	31, 32, 43
Polyfoam Plus	Lin Pac Insulation Products, Hartlepool, GB	121	71
Polyfort®	A. Schulman Inc., Akron, OH 44 309, US	111, 1113, 112, 1121, 1222	2 s
Polygard MR	Polytech Inc., Owensville, M 65 066, US	2411 s	5
Polyglad	GE Plastics, Pittsfield, MA 01 201, US	2411	63
Polyimidal®	Raychem Corp., Menlo Park, CA 94 025, US	27	112 el
Polylac	Chi Mei Industrial Co. Ltd., Tainan, Shien, Taiwan	1221	2

Name	Company (see 8.1, page 479)	Resin (see 8.2.1, page 481)	Forms of delivery (s. 8.2.2, p. 484)
Polylam	Polyon-Barkai, Kibbutz Barkai M. P. Menashe 37 860, IL	11	437
Polylam AF	–,,–	11	731
Polylite®	Reichhold Chemicals Inc. Jacksonville, FL 32 245, US	242	11
Polyloy®	Ems-Polyloy GmbH 64823 Groß-Umstadt, DE	261+111	11, 2, 224, 2217, 226
Polymag®	A. Schulman Inc. Akron, OH 44 309, US	111, 112, 237, 2412, 261	2
Polyman®	–,,–	1221, 1222, 1225, 124, 163, 1641, 1221/261, 2411/1221	2 se
Polymar	Hammersteiner Kunststoffe GmbH 41836 Hückelhoven, DE	132	436
Polymeg	QO Chemicals Inc. Oak Brook, IT 60 251, US	2633	117
Polymist®	Ausimont Inc. Morristown, NJ 07 962, US	1351	11, 14
Polyorc	Uponor, Stourton, Leeds, LS10 1UJ, GB	111, 1311	31
Polyox®	Union Carbide Corp. Danbury, CT 06 817, US	232	111
Polyphon	Steinbacher 6383 Erpfendorf/Tirol, AT	263	71
Polypur®	A. Schulman Inc. Akron, OH 44 309, US	264	2
Polyrex	Chi Mei Industrial Co. Ltd. Tainan, Shien, Taiwan	121	2
Polyrite	Polyply Inc. Amsterdam, NY 12 010, US	242	112, 225
Polysar® Bromobutyl	Bayer AG, 51368 Leverkusen, DE	332	11
Polysar® Butyl	–,,–	332	11
Polysar® Chlorobutyl	–,,–	332	11
Polysar® EPM/EPDM	–,,–	1121	11

8.3 Trade Names

Name	Company (see 8.1, page 479)	Resin (see 8.2.1, page 481)	Forms of delivery (s. 8.2.2, p. 484)
Polysar®S/SS	Bayer AG, 51368 Leverkusen, DE	1223	11
Polyset	Morton International Inc. Chicago, IL 60 606, US	233	22
Polysizer	Showa High Polymer Co. Ltd. Tokyo, JP	152	116
Polysol®	–„–	151	121
Polystat®	A. Schulman Inc. Akron, OH 44 309, US	111, 112, 122, 1221, 2411, 2411/1221	2
Polystone®	Scobalitwerk Wagner GmbH 56626 Andernach, DE	1113, 112	342, 51, 52, 53
Polystron	Svenska Polystyrenfabriken AB, SE	121, 1223	2
Polystron	B. F. Goodrich Chemical Group Cleveland, OH 44 131, US	131	2 el
Polythan	Steinbacher 6383 Erpfendorf/Tirol, AT	263	71
Polytherm	4P Folie Forchheim GmbH 91301 Forchheim, DE	1311	43
Polytrope®	A. Schulman Inc. Akron, OH 44 309, US	1127	2
Polyvin®	–„–	1311, 1314	2
Polyviol®	Wacker-Chemie GmbH 81737 München, DE	152	11
Ponal®	Henkel KGaA, 40589 Düsseldorf, DE	151	73
Poolliner	Van Besouw BV Kunststoffen 5050 AA Goirle, NL	132	434
Porelle®	Porvair P.L.C. King's Lynn, Norfolk PE30 2 HS, GB	262	437
Poret®	EMW-Betrieb, 65582 Diez, DE	263	71 s
Poroband	Isovolta AG 2355 Wiener Neudorf, AT	233	616
Porofol	–„–	233	614
Poromat	–„–	233	614
Poron®	Rogers Corp., Rogers, CT 06 263, US	132, 263	71
poronor	Fränkische Rohrwerke GmbH + Co. 97486 Königsberg, DE	121	32/71

8.3 Trade Names

Name	Company (see 8.1, page 479)	Resin (see 8.2.1, page 481)	Forms of delivery (s. 8.2.2, p. 484)
Porovlies	Isovolta AG, 2355 Wiener Neudorf, AT	233	616
Porvair®	Porvair P.L.C., King's Lynn, Norfolk PE30 2 HS, GB	262	712
Porvent®	Porvair P.L.C., King's Lynn, Norfolk PE30 2 HS, GB	1113, 112	43, 712
porzyl	Fränkische Rohrwerke GmbH + Co., 97486 Königsberg, DE	121	5/71
Poticon	Biddle-Sawyer Corp., New York, NY 10 121, US	2312, 261, 2412	22
Poval	Denki Kagaku Kogyo, Tokyo 100, JP	152	11
Powersil	Wacker-Chemie GmbH, 81737 München, DE	38	112
Pre-Elec	Premix Oy, 05201 Rajamäki, FI	11, 12, 13, 19, 23, 24, 26	95 el
Pregnit	August Krempel Soehne GmbH & Co., 71655 Vaihingen, DE	233, 26, 27	342, 612, 614
Premi-Dri®	Premix Inc., North Kingsville, OH 44 068, US	242	2 el
Premi-Glas	–,,–	242	224, 2251
Premi-Ject	–,,–	242	224
Pres-Rite	M/s Rattanchand Harjasrai Pvt. Ltd. Faridabad, IN	221, 223	2
Pressal-Leime®	Henkel KGaA, 40589 Düsseldorf, DE	223	73
Prester	Soc. Provencale de Résines Appliquées, Sauveterre, FR	242	11
Presto	Reynolds Metals Co., Richmond, VA 23261–7003, US	1111, 1111 L	411
Prestocol	Soc. Provencale de Résines Appliquées, Sauveterre, FR	24	73
Prevex®	GE Plastics, Pittsfield, MA 01 201, US	235	2, 224
Prezenta	Märkische Faser AG, 14723 Premnitz, DE	42	72
Prima®	European Vinyls Corp. (EVC), 1160 Bruxelles, BE	1314	2
Primal®	Rohm & Haas Co., Philadelphia, PA 19105, US	16	116, 121

Name	Company (see 8.1, page 479)	Resin (see 8.2.1, page 481)	Forms of delivery (s. 8.2.2, p. 484)
Primef®	Solvay S.A., 1050 Bruxelles, BE	237	22
Primocel	HPP-Profile GmbH Primoplast 21629 Neu-Wulmsdorf, DE	131	7661
Prinem	Isovolta AG 2355 Wiener Neudorf, AT	233	72
Pro-fax®	Montell Polyolefins Wilmington, DE 19 894, US	112, 1121, 1127	2
Pro-Seal	Products Research & Chemical Corp. Woodland Hills, CA 91 365–4226, US	233	112
Probimage®	Ciba Geigy AG, 4002 Basel, CH	233	11
Probimer®	–,,–	233	114 el, s
Probimid®	–,,–	27	113 el, s
PROCO®	Hüni + Co. 88046 Friedrichshafen, DE	135	146
Procom®	BASF Aktiengesellschaft 67063 Ludwigshafen, DE	112, 1121	22
Procor	Mobil Chemical Co. Pittsford, NY 14 534, US	16/112	412
Prodoral®	T. I. B. Chemie (Shell) 68219 Mannheim, DE	233, 263	111, 146 el
Prodorit®	–,,–	233	146
Profil®	Akzo Engineering Plastics Inc. Evansville, IN 47 732, US	112+3411	110 TE, 22
Profilon	Hoechst Celanese Corp. New York, NY 10 036, US	2412	41
Propafilm®	ICI PLC, Welwyn Garden City, Herts. AL7 1 HH, GB	112	411, 412
Propafoil®	–,,–	112	4 s
Propaply®	–,,–	112	411
Propiofan®	BASF Aktiengesellschaft 67063 Ludwigshafen, DE	1313	121
Proponite	Borden Chemical Co. Geismar, LA 70 734, US	112	412
Propylex®	Courtaulds Chemicals & Plastics Ltd. Spondon, Derby DE2 7BP, GB	112	52, 53
Propylex®	Nordmann Rassmann GmbH & Co. 20459 Hamburg, DE	112+3411	11

Name	Company (see 8.1, page 479)	Resin (see 8.2.1, page 481)	Forms of delivery (s. 8.2.2, p. 484)
Propylux®	Westlake Plastics Co. Lenni, PA 19 052, US	112	3, 4, 5
Protax	Hercules Powder Company Inc. Wilmington, DE 19 894, US	3411	11
Protefan®	T. I. B. Chemie (Shell) 68219 Mannheim, DE	132	117/122
Pryphane®	Rhône-Poulenc Films 69 398 Lyon, Cedex 3, FR	112	411
Pulse®	The Dow Chemical Company Midland, MI 48641-2166, US	1221, 2411	124
Pure-CMC	The Perma-Flex Mold Co. Inc. Columbus, OH 43 209, US	263	111
Purenit	Puren-Schaumstoff GmbH 88662 Überlingen, DE	263	71
Purez	ICI Polyurethanes 3078 Kortenberg, BE	2632	111
Pyralin®	DuPont Co. Inc. Wilmington, DE 19 898, US	272	1
Pyratex®	Bayer AG, 51368 Leverkusen, DE	15 s	121
Pyro-Chek®	Ferro Corp. Evansville, IN 47 711, US	121	2 se
Pyrofil	Mitsubishi Rayon Co. Ltd. Tokyo, JP	161 s	72 s
Pyroguard	Recticel Foam Corp., Morristown, Div. Morristown, TN 37 816–1197, US	263	71 se
Q			
Q200.5	Polypenco GmbH 51437 Bergisch Gladbach, DE	122	33, 51
QuaCorr®	QO Chemicals Inc. Oak Brook, IT 60 251, US	216	112, 21
Qualiprint®	Scott Bader Co. Ltd., Wollaston, Wellingborough, Norths. NN9 7RL, GB	125	121
Quarite®	Aristech Chemical Co. Florence, KY 41 042, US	16	2211, 5
Quelflam®	Baxenden Chemical Co. Ltd. Accrington, Lancs. BB5 2SL, GB	265	117 se

8.3 Trade Names

Name	Company (see 8.1, page 479)	Resin (see 8.2.1, page 481)	Forms of delivery (s. 8.2.2, p. 484)
R			
Rabalon®	Mitsubishi Chemical Corp. Tokyo, JP	1253	11
Radaflex	Dash Cable Industries (Israel) Ltd., IL	131/323	4
Radel®	Amoco Performance Products Richfield, CT 06 877, US	236	22, 411 el
Radil	Radici Novacips S.p.A. 24 020 Villa d'Ogna (BG), IT	112	41
Radilon	–,,–	261	2, 224
Raditer B	–,,–	2412	2
Radlite®	Azdel Inc., Shelby, NC 28 150, US	1	721
Ralupol®	Raschig AG, 67061 Ludwigshafen, DE	242	2, 22
Rapok®	–,,–	214	113
Rau ...	Rehau AG & Co., 95111 Rehau, DE	11, 12, 13, 23, 2411, 26, 28, 3, 4	3, 5
Ravemul®	ECP EniChem Polimeri srl. 20 124 Mailand, IT	151	12
Raventer®	–,,–	1313	1, 2
Ravepox®	–,,–	233	11
Ravinil®	–,,–	1311, 1313	11
Rayopp®	UCB/Sidac, 1620 Drogenbos, BE	112	411, 412
Reclyn R 134 a	Hoechst AG 65929 Frankfurt, DE	135 s	135 CFC alt.
Recticel	Recticel NV SA., 1150 Brüssel, BE	263	71
Recyclen	Recyclen Kunststoffprodukte GmbH 74706 Osterburken, DE	1, 2	5
Redux®	Ciba-Geigy, Composites Div. Anaheim, CA 9207–2018, US	153, 21	73
Reedex-F	Berndt Rasmussen 3460 Birkerod, DK	112	43
Reevane	Reeves Brothers Canada Ltd., Toronto, Ontario M8W 2T2, CA	263	424
Regulus	Mitsui Toatsu Chemicals Inc. Tokyo, JP	27	4

Name	Company (see 8.1, page 479)	Resin (see 8.2.1, page 481)	Forms of delivery (s. 8.2.2, p. 484)
Relatin®	Henkel KGaA, 40589 Düsseldorf, DE	4234	11
Reliapreg®	Ciba-Geigy, Composites Div. Anaheim, CA 9207–2018, US	121, 233	2251, 7662
Relon®	Chemiefaserwerk Savinesti Piatra Neamt, RO	261	72
Reny	Mitsubishi Gas Chem. Comp. Inc. Tokyo, JP	261(2)	2, 224
Reoplast®	Ciba Geigy AG, 4002 Basel, CH	–	91
Repak	AB Akerlund & Rausing, SE	11r	2
Repete	Goodyear Tire & Rubber Co. Akron, OH 44 316, US	2412	11re
Repolem	Elf Atochem 92 091 Paris, La Défense 10, FR	1225	11, 12
Reproflon®	Mikro-Technik GmbH 63927 Bürgstadt, DE	1351	11, 2, 31, 33, 34, 433, 51, r
Repsol	Repsol Quimica, ES	111, 112	2
Res-I-Glas	Micafil AG, 8048 Zürich, CH	16, 242	614
Resamim®	Hoechst AG 65929 Frankfurt, DE	221	113, 114, 115
Resart® PC	Resart GmbH, 55120 Mainz, DE	2411	51, 52, 53, 54, 55
Resartglas®	–,,–	163	51, 52, 53, 54, 55
Resicast®	DSM, Polymers & Hydrocarbons 6130 AA Sittard, NL	263	112
Resifix	Raschig AG, 67061 Ludwigshafen, DE	216	113
Resifoam®	DSM, Polymers & Hydrocarbons 6130 AA Sittard, NL	263	117
Resimene®	Cargill Inc., Lynwood, CA 90 262, US	223, 224	112, 113, 114
Resinite®	Borden Chemical Co. Geismar, LA 70 734, US	131	41
Resinol®	Raschig AG, 67061 Ludwigshafen, DE	21	22, 23
Resinoplast	Resinoplast, Reims, FR	131	2
Resiplast®	Raschig AG, 67061 Ludwigshafen, DE	211, 212, 214 211	113
Resistit-Perfekt®	Phoenix AG, 21079 Hamburg, DE	1411	434
Resistit®	–,,–	35	4351

8.3 Trade Names

Name	Company (see 8.1, page 479)	Resin (see 8.2.1, page 481)	Forms of delivery (s. 8.2.2, p. 484)
Resiten®	Iten Industries Ashtabula, OH 44 004, US	21, 223, 233, 28	611, 6111, 612, 614
Resocel®	Micafil AG, 8048 Zürich, CH	212	611
Resofil®	–,,–	212	612
Resolam®	–,,–	233	611, 612
Resopal®	Rethmann Plano GmbH 48356 Nordwalde, DE	223	621
Resopalan®	–,,–	223, 242	452 s
Resopalit®	–,,–	–	761
Resoplan®	–,,–	223	621
Resoplast	Ciba Geigy AG, 4002 Basel, CH	–	91
Resoweb®	Micafil AG, 8048 Zürich, CH	233/2412	612
Resproid	Goodyear Tire & Rubber Co. Akron, OH 44 316, US	132	424
Rest Easy	BASF Structural Materials Inc. Charlotte, NC 28 217, US	263	71
Resticel	Sirlite Srl., 20 161 Milano, IT	21	117
Restil	Montedison S.p.A., 20 121 Milano, IT	1222	2
Restiran	–,,–	1221	11, 2
Restiran ML	–,,–	1221	11
Restirolo	–,,–	121, 122	2
Resydrol®	Hoechst AG 65929 Frankfurt, DE	16, 21, 22, 24	114
Retiflex®	Montedison S.p.A., 20 121 Milano, IT	112	4
Retipor®	EMW-Betrieb, 65582 Diez, DE	263	71 s
Rexene®	Rexene Products Co. Dallas, TX 75 244, US	1111, 1111 L, 1115, 112, 1121	2
Rexolite®	Brand-Rex Co. Willimantic, CT 06 226, US	1251	5 el
Rextac®	Rexene Products Co. Dallas, TX 75 244, US	1111 L	2
Reynolon	Reynolds Metals Co. Richmond, VA 23261–7003, US	1311, 261	411
Rezibond	Chromos – Ro Polimeri 4100 Zagreb, Croatia	21	2

8.3 Trade Names

Name	Company (see 8.1, page 479)	Resin (see 8.2.1, page 481)	Forms of delivery (s. 8.2.2, p. 484)
Rezolin	Hexcel Corp. Chemical Div. Chatsworth, CA 93 111, US	233	11
Reztex	Erez Thermoplastic Products M.P. Ashkelon 79150, IL	111, 132	43
Rhenoflex®	Hüls AG, 45764 Marl, DE	1312	11, 113, 114, 115
Rhenofol®	Braas Flachdachsysteme GmbH 61440 Oberursel, DE	132	433, 434
Rhenogran Geniplex 80	Rhein Chemie Rheinau GmbH, 68219 Mannheim, DE	–	–
Rhenoverit®	Rhenowest, Kunststoff- u. Spanplattenwerk GmbH 56759 Kaisersesch, DE	223	761
Rhepanol®	Braas Flachdachsysteme GmbH 61440 Oberursel, DE	114	435, 434
Rhino Hyde	Cargill Inc., Lynwood, CA 90 262, US	263	5
Rhodeftal®	Rhône-Poulenc 92 097 Paris, La Défense 2, FR	272	11
Rhodester® CL	–,,–	–	119
Rhodiastab®	–,,–	–	9
Rhodopas®	–,,–	121, 1223, 124, 151	121
Rhodorsil®	–,,–	281, 38	11
Rhodoviol®	–,,–	152	11, 2
Rhonacryl	Chemotechnik Abstatt GmbH & Co. 74232 Abstatt, DE	16	112
Rhonaston	–,,–	233	112
Riacryl®	Rias A/S, 4000 Roskilde, DK	163	5
Riblene®	ECP EniChem Polimeri srl. 20 124 Mailand, IT	1111, 1115	113, 241
Richform	Richmond Technology Inc. Redlands, CA 92373, US	124	5
Ricolor®	Holztechnik GmbH-Ricolor 95336 Mainleus, DE	223	621, 761
Rigidex®	BP Chemicals International Ltd. London SW1W OSU, GB	1113	2
Rigidite	BASF Structural Materials Inc. Charlotte, NC 28 217, US	23	2261/6121

8.3 Trade Names

Name	Company (see 8.1, page 479)	Resin (see 8.2.1, page 481)	Forms of delivery (s. 8.2.2, p. 484)
Rigidsol	Watson Standard Co. Pittsburgh, PA 15 238, US	131	121, 122
Rigilene®	Stanley Smith & Co. Plastics Ltd. Isleworth, Middx. TW7 7AU, GB	111	5
Rigipore®	BP Chemicals International Ltd. London SW1W OSU, GB	121	117
Rigiwall	Gen. Corp. Polymer Products Newcomerstown, OH 43 832, US	131/16	51
Rigolac	Showa High Polymer Co. Ltd. Tokyo, JP	242	112, 113
Rilsan®	Elf Atochem 92 091 Paris, La Défense 10, FR	261	11, 116, 21, 224, 35
Rimline®	ICI Polyurethanes 3078 Kortenberg, BE	2633, 2634	118
Rimplast	LNP Plastics Nederland BV 4940 Raamsdonksveer, NL	28 s	118
Rimplast®	Hüls AG, 45764 Marl, DE	11, 24, 26, 264+38	118, 21 s, 120
Ripolit	Rilling & Pohl KG 70469 Stuttgart, DE	111, 112, 1314	5
Ripoxy	Takeda Chemical Industries Ltd. Osaka, JP	16/233	113
Riteflex BP	Hoechst AG 65929 Frankfurt, DE	2412	110 TE
Robadur	Leripa, AT	1113	2
Robalon	–„–	1113	2
Roblon	Roblon A/S, 9900 Frederikshavn, DK	224, 41, 413, 414, 616, 72	
Rocel®	Courtaulds Chemicals & Plastics Ltd. Spondon, Derby DE2 7BP, GB	4222	4, 5
Rodrun	Unitika Ltd., Osaka, JP	2415	119
Roga®	Roga KG Dr. Loose GmbH & Co. 50389 Wesseling, DE	111, 131, 132, 422, 28	32, 34
Röco	Röhrig & Co., 30453 Hannover, DE	1311(71) se	31, 33, 34, 342
Röcothene	–„–	111 se	32, 33, 34, 342
Rohacell®	Röhm GmbH, 64293 Darmstadt, DE	166	71
Rohafloc®	–„–	161	121, 123, 2

Name	Company (see 8.1, page 479)	Resin (see 8.2.1, page 481)	Forms of delivery (s. 8.2.2, p. 484)
Rohagit®	Röhm GmbH, 64293 Darmstadt, DE	165	121, 123
Rohatex®	–„–	16	116
Rolan	Chemiefaserwerk Savinesti Piatra Neamt, RO	161	72
Romicafil®	Micafil AG, 8048 Zürich, CH	233	413
Romicaglas®	–„–	233, 26, 28	614
Romicapreg®	–„–	233	413
Ronfalin®	DSM, Polymers & Hydrocarbons 6130 AA Sittard, NL	1221	2
Ronfaloy®	–„–	1221/3411, 1221/263, 1212/2411	2 el
Ropet®	Rohm & Haas Co. Philadelphia, PA 19105, US	2412/16	2
Ropol®	Chemisches Kombinat, Borzesti, RO	1111	1, 2
Ropoten	–„–	111	2
Rosevil®	–„–	1311	2
Rosite	Rostone Corp. Lafayette, IN 47 903, US	242	224, 225
Roskydal®	Bayer AG, 51368 Leverkusen, DE	244	114
Rotothene	Rototron Corp. Babylon, NY 11 704, US	1111, 1111 L, 1112	14
Rotothon	–„–	112	141
Roxan®	Roxan Folien GmbH 70469 Stuttgart, DE	1311, 132	423, 452, 53
Roy®...	J. H. Benecke GmbH 30007 Hannover, DE	132, 262	421, 424, 4241
Royalene®	Polycast Technology Corp. Stamford, CT 06 904, US	3411	2
Royalex	Royalite Thermopl. Div. Uniroyal Technology Corporation Mishawaka, IN 46 546–0568, US	1221, 1311	7661
Royalite®	–„–	1221/131	341, 342, 353, 52, 53
Royalstat	–„–	1221/131	413
Royaltherm	Polycast Technology Corp. Stamford, CT 06 904, US	3411 s	11

8.3 Trade Names

Name	Company (see 8.1, page 479)	Resin (see 8.2.1, page 481)	Forms of delivery (s. 8.2.2, p. 484)
Roylar®	B. F. Goodrich Chemical Group Cleveland, OH 44 131, US	264	2
Rozylit	Romatin AG, CH	163	2
Rubinate	ICI America Inc. Wilmington, DE 19 897, US	–	11
Rucoflex	Ruco Polymer Corp. Hicksville, NY 11 802, US	2632	14
Rucothane	–„–	263, 264	11
Rütamid®	Bakelite GmbH 58642 Iserlohn, DE	261	2, 224
Rütaphen®	–„–	211–214, 216, 21	111–117, 12, 123, 13, 141, 146, 73, 2
Rütapox®	Bakelite GmbH 58642 Iserlohn, DE	233	111, 112, 113, 114, 115, 121, 122, 123, 13, 141, 142, 143, 144, 145, 146
Rütapur®	–„–	263	112 el, 115
Rulon®	Dixon Industries Corp. Bristol, RI 02 809, US	1351	61
RX	Rogers Corp., Rogers, CT 06 263, US	212, 243	2
Rynite®	DuPont Co. Inc. Wilmington, DE 19 898, US	2412	1, 2, 224
Ryton®	Phillips Petroleum Chemicals NV 3090 Overijse, BE	237	11, 2, 22 el
Ryulex®	Dainippon Ink. & Chemicals Inc. Tokyo 103, JP	121/2411	2
S			
S-dine®	Sekisui Chemical Co. Ltd. Osaka 530, JP	167	73
S-lec®	–„–	155	11, 4f. 763
Saduren®	BASF Aktiengesellschaft 67063 Ludwigshafen, DE	224	113
Safecoat	Chemie Linz AG, 4040 Linz, AT	112	435/721

8.3 Trade Names

Name	Company (see 8.1, page 479)	Resin (see 8.2.1, page 481)	Forms of delivery (s. 8.2.2, p. 484)
Safetred	Marley Floors Ltd., Lenham, Maidstone, Kent, ME17 2DE, GB	132	444
Safetred Universal	–,,–	132	444
Saflex®	Monsanto Chemical Co., Plastics Div. St. Louis, MO 63 167, US	155	41/763
Salvex	Mitsubishi Chemical Corp. Ltd. Tokyo, JP	11	22 s
Samicaflex® Si	Von Roll Isola 4226 Breitenbach, CH	28	4351/616
Samicanit®	–,,–	233	616
Samicatherm®	–,,–	233	616
Sanprene	Sanyo Chemical Ind. Ltd. Kyoto, JP	263	111
Sanrex®	Mitsubishi Monsanto Chemical Co. Tokyo, JP	1222	2
Santoclear®	–,,–	121	411
Santolite®	Advanced Elastomer Systems L.P. St. Louis, MO 63 167, US	2351	2
Santoprene®	–,,–	112+3411	110 TE
Sapedur	Saplast S.A. 67 100 Straßburg-Neuhof, FR	13	2
Sapelec	–,,–	131	2 el
Sarafan	Karl Dickel & Co. KG 47057 Duisburg, DE	111, 1111 L, 112, 1311	411
Sarleim	Elf Atochem 92 091 Paris, La Défense 10, FR	221	73
Sarlink	DSM, Polymers & Hydrocarbons 6130 AA Sittard, NL	1117	21
Sarnafil	Sarna Kunststoff AG 6060 Sarnen, CH	11, 1113, 132	4341, 4351
Sarnatex	–,,–	132	4351, 436
Sasolen	Sasol Polymers Johannesburg, ZA	112	2
Saterflex	Saplast S.A. 67 100 Straßburg-Neuhof, FR	1253	2

Name	Company (see 8.1, page 479)	Resin (see 8.2.1, page 481)	Forms of delivery (s. 8.2.2, p. 484)
Satinflex®	Clopay Corp. Plastics Products Div. Cincinnati, OH 45 202, US	111	4
Savilit SPP	DSM, Polymers & Hydrocarbons 6130 AA Sittard, NL	242	2251
Saxerol®	ECW-Eilenburger Chemie Werk AG i.GV., 04838 Eilenburg, DE	122, 1221	51, 52, 53
Saxolen®	–,,–	111	51, 52, 53
Scarab®	BIP Chemicals Ltd. Warley, West Midl., B69 4PG, GB	221	2
Schuladur®	A. Schulman Inc. Akron, OH 44 309, US	2412	2
Schulaform®	–,,–	2311	2
Schulamid®	–,,–	261	2 se, 224
Schwarzafol	Polymer und Filament GmbH 07407 Rudolstadt, DE	2412	2
Schwarzamid	–,,–	261	2
Sclair-Tak	DuPont Co. Inc. Wilmington, DE 19 898, US	1111 L/-	9
Sclair®	–,,–	111, 1111 L, 1112	2
Sclairfilm®	–,,–	111	411
Sclairlink®	DuPont Co. Inc. Wilmington, DE 19 898, US	111	14
Sclairpipe®	–,,–	111	31
Scobalit®	Scobalitwerk Wagner GmbH 56626 Andernach, DE	242	63
Scolefin®	Buna AG, 06258 Schkopau, DE	1113	2
Scona	–,,–	1117	9
Sconaran	–,,–	242	112
Sconarol®	–,,–	1222	2
Sconater®	–,,–	1221, 1253	11, 22
Sconatex®	–,,–	133	121
Scotchcast	3 M Co., St. Paul, MN 55 144, US	233	112
Scotchpak	–,,–	2412	411
Scotchpar	–,,–	2412	411
Scotchply	–,,–	21, 233	224, 225

Name	Company (see 8.1, page 479)	Resin (see 8.2.1, page 481)	Forms of delivery (s. 8.2.2, p. 484)
Scovinyl	Buna AG, 06258 Schkopau, DE	131	11f., 122
Sculpture	Marley Floors Ltd., Lenham, Maidstone, Kent, ME17 2DE, GB	132	444
Scuranate®	Rhône-Poulenc S.A. 92 408 Courbevoie, FR	2631	11
Sea-Lok®	Blomberger Holzindustrie, B. Hausmann GmbH & Co. KG 32817 Blomberg, DE	1311	34, 51
SEBS-Compound	Borealis Compounds AB 12 613 Stockholm, SE	1127	110 TE, el
Seemilite	Saurastra Electrical & Metal Ind. Pty. Ltd., Bombay 2, IN	21	11, 2
Selar®	DuPont Co. Inc. Wilmington, DE 19 898, US	1115/261, 261	1, 21 s
Selectrofoam	PPG Industries Inc. Pittsburgh, PA 15 272, US	263	117
Selectron®	–„–	242	11, 112
Semicon®	Borealis Compounds AB 12 613 Stockholm, SE	111/332	117
Semper	Schütte-Lanz, 50321 Brühl, DE	21	16, 622
Senocryl	Senoplast Klepsch & Co. 5721 Piesendorf, AT	163	54
Senosan®	Senoplast Klepsch & Co. 5721 Piesendorf, AT	121, 1221, 241	53
Sequel	D & S Plastics Int. Auburn Hills, MI 48 326, US	11	2
Serfene	Morton International Danvers, MA 01 923, US	1354	14
Serinil	Rio Rodano, S.A., Madrid 20, ES	131	2
Setilithe®	Tubize Plastics S.A., 1360 Tubize, BE	4222	2
Sevinil	Rio Rodano, S.A., Madrid 20, ES	131	2
SG laminat	Saar-Gummiwerk GmbH 66687 Wadern-Böschfeld, DE	3411	434
SG tan	–„–	3411	43
SG tyl	–„–	332	433
Sheldal	Sheldal Inc. Northfield, MN 55 057, US	2412, 27	413

Name	Company (see 8.1, page 479)	Resin (see 8.2.1, page 481)	Forms of delivery (s. 8.2.2, p. 484)
Shellvis	Shell International Co. Ltd. London, SE1 7NA, GB	12/341/12	110 TE
Shimoco®	DSM Italia s.r.l., 22 100 Como, IT	242	225
Shindex®	ICI PLC, Welwyn Garden City, Herts. AL7 1 HH, GB	263	71
Shinko-Lac®	Mitsubishi Rayon Co. Ltd. Tokyo, JP	1221	2
Shinkolite®	–,,–	163	112, 2, 54
Sho-Allomer®	Showa Denko K.K., Tokyo 105, JP	112	2
Sholex®	–,,–	1111, 1113	2
Sicalit	Mazzucchelli Vinyls S.R.L. 21 043 Castiglione Olona/Varese, IT	4222	2
Sicoamide	Radici Novacips S.p.A. 24 020 Villa d'Ogna (BG), IT	261	2
Sicobox	Mazzucchelli Vinyls S.R.L. 21 043 Castiglione Olona/Varese, IT	1311	415
Sicodex	–,,–	1311	5
Sicofarm	–,,–	1311	41 med
Sicoffset	–,,–	1311	415
Sicolene	–,,–	111, 131	4
Sicoplast	Mazzucchelli Vinyls S.R.L. 21 043 Castiglione Olona/Varese, IT	131	4
Sicoprint	–,,–	1311	415
Sicoreg	–,,–	1311	415
Sicovimp	–,,–	1311/134	412
Sicovinil	–,,–	1311, 1313	36, 4, 5
Sicron®	ECP EniChem Polimeri srl. 20 124 Mailand, IT	131, 132	2
Sigrafil®	Hoechst AG 65929 Frankfurt, DE	261 s	226
Sika Norm	Sika AG. Flexible Waterproofing 8048 Zürich, CH	1114/342	433
Sikaplan	–,,–	132	43
Silacron	Schock & Co. GmbH 73605 Schorndorf, DE	16	112
Silamid	Chemische Betriebe a. d. Waag, CSFR	261	2

Name	Company (see 8.1, page 479)	Resin (see 8.2.1, page 481)	Forms of delivery (s. 8.2.2, p. 484)
Silastic®	Metallogen GmbH 4630 Bochum-Wattenscheid, DE	38	11
Silbione®	Rhône-Poulenc Silicones 92 5207 Neuilly sur Seine, FR	281, 38	112, 113, 114, 145, 146, 147
Silmar	Sohio Chemical Co. Cleveland, OH 44 115, US	242	112, 224
Silopren®	BASF Aktiengesellschaft 67063 Ludwigshafen, DE	38	11
Silres	Wacker-Chemie GmbH 81737 München, DE	38	112, 114
Siltem®	GE Plastics Pittsfield, MA 01 201, US	263+28	2
Siluminite	Tenmat Ltd., Trafford Pk., Manchester M17 1RU, GB	21	614, 615 el
Simona®	Simona AG Kunststoffwerke 55606 Kirn, DE	1113, 112, 131, 1354	31, 33, 51, 52, 53, 54
Simona®	–,,–	1352	51, 52
Simona®	–,,–	2412	54
Sinalloy	Montell Polyolefins Wilmington, DE 19 894, US	112	2
Sinaplast	Aluminium Walzwerke GmbH 78201 Singen, DE	132	762
Sinkral®	ECP EniChem Polimeri srl. 20 124 Mailand, IT	1221	2
Sintimid®	Hochleistungskunststoff GmbH Reutte, AT	273	1, 5, 72
Sintrex®	Airex AG, 5643 Sins, CH	121	51
Sinvet®	ECP EniChem Polimeri srl. 20 124 Mailand, IT	2411	21, 224
Siraldehyd	Sirlite Srl., 20 161 Milano, IT	221	111
Siralkyd	–,,–	25	114
Siramide	–,,–	2611	111
Sirban®	–,,–	323	11
Sircel®	–,,–	121	117
Sircis®	–,,–	321(1)	11
Sirel®	–,,–	322	11

8.3 Trade Names

Name	Company (see 8.1, page 479)	Resin (see 8.2.1, page 481)	Forms of delivery (s. 8.2.2, p. 484)
Sirester®	Sirlite Srl., 20 161 Milano, IT	242	11
Sirfen®	–„–	212	113, 2
Sirit®	–„–	221	113
Sirminol®	–„–	22	114
Siroplan	Hegler Plastik GmbH 97714 Oerlenbach, DE	1311	31, 32
Siroplast	–„–	1113	31, 32
Sirowell	–„–	1311	31, 32
Skai®	Konrad Hornschuch AG 74679 Weissbach, DE	132	4241
Skailan®	–„–	263	4241, 712
Skybond®	Monsanto Chemical Co., Plastics Div. St. Louis, MO 63 167, US	27	112
Sniamid®	Nyltech Italia Srl 20 020 Ceriano (MI), IT	261	2, 224
Sniatal®	–„–	231	2
Snowpearl	Nihon Polystyrene Co. Ltd. Kawasaki, JP	121	117
Soarblen	Nippon Gohsei Chemical Ind. Co. Ltd., Osaka, JP	1115	11, 2
Soarlex	Nippon Gohsei Chemical Ind. Co. Ltd., Osaka, JP	1115	11, 2
Soarnol®	Elf Atochem 92 091 Paris, La Défense 10, FR	1125/152	11 s
Sobral®	Scott Bader Co. Ltd., Wollaston, Wellingborough, Norths. NN9 7RL, GB	244	123, 146
Soflex®	Von Roll Isola 4226 Breitenbach, CH	132	32 el
Softlex®	Nippon Petrochemicals Co. Tokyo, JP	1115	11
Softlite	Gilman Brothers Co. Gilman, CT 06 336, US	1115 s	414, 5/71
Softlon®	Sekisui Chemical Co. Ltd. Osaka 530, JP	1116	414, 71
Solarflex	Pantasote Inc. Passaic, NJ 07 055, US	1114	43

8.3 Trade Names

Name	Company (see 8.1, page 479)	Resin (see 8.2.1, page 481)	Forms of delivery (s. 8.2.2, p. 484)
Solef®	Solvay S.A., 1050 Bruxelles, BE	1354	11, 2
Solidur®	Solidur Deutschland GmbH & Co. KG 48691 Vreden, DE	1113	3, 42, 5
Solidur® 1000	–,,–	111	33, 342, 51
Solidur® DS	–,,–	111	33, 342, 51
Solidur® super	–,,–	111	33, 342, 51
Solimide	IMI-Tech Corp. Elk Grove Village, IL 60 007, US	27	71
Solithane	Morton International Inc. Chicago, IL 60 606, US	2631	112, 114
Soluphene	Elf Atochem 92 091 Paris, La Défense 10, FR	221	73
Solvic®	Solvay S.A., 1050 Bruxelles, BE	1311, 1313	11, 2
Sonit	Chemie-Werk Weinsheim GmbH 67547 Worms, DE	263	71
Sonite	Smooth-On Inc. Gillette, NJ 07 933, US	233, 237, 264	113, 2
Soplasco	American Filtrona Corp. Richmond, VA 23 233, US	131	4, 5
Sorane	ICI Polyurethanes 3078 Kortenberg, BE	263	712
Sorbothane	ICI Polyurethanes 3078 Kortenberg, BE	263	37
Spandal	Baxenden Chemical Co. Ltd. Accrington, Lancs. BB5 2SL, GB	263	7661
Spandofoam	–,,–	263	71
Sparlux®	Solvay S.A., 1050 Bruxelles, BE	2411	55
Spauldite	Spaulding Composites Co. Tonawanda, NY 14 150, US	212, 223, 233, 4211	61, 612, 614
Spaulrad	–,,–	27	6, 614
Spectra 900®	Allied Signal Corp. Morristown, NJ 07 962, US	111 s	72
Spectra-Visions	Planox B.V. 5705 AA Helmond, NL	132	451, 71
Spectrum	Royalite Thermopl. Div. Uniroyal Technology Corporation Mishawaka, IN 46 546–0568, US	1111, 1112, 1113, 112, 1121	341, 342, 353, 52, 53

8.3 Trade Names

Name	Company (see 8.1, page 479)	Resin (see 8.2.1, page 481)	Forms of delivery (s. 8.2.2, p. 484)
Spherilene	Daeilim Industrial Co., Korea	11	11
Spheripol®	Montell Polyolefins. Wilmington, DE 19 894, US	11	11
Spilac®	Showa Denko K.K., Tokyo 105, JP	2421	123
Spiral-bauku	Troisdorfer Bau- u. Kunststoff GmbH 51674 Wiehl, DE	11	312
Spralkyd	Soc. Provencale de Résines Appliquées, Sauveterre, FR	25	113
Spreacol®	AMC-Sprea S.p.A., 20 101 Milano, IT	221	73
Sprelacart®	Sprela-Schichtstoff GmbH 03130 Spremberg, DE	223	621
Sprelaform	–,,–	242	452
Sprigel	Soc. Provencale de Résines Appliquées, Sauveterre, FR	242	112
Springvin	Eurohose, Stroud, GB	132	321 tra
Sprunglow	Ross & Roberts Inc. Stratford, CT 06 497, US	111, 132	4
Sriver	Soc. Provencale de Résines Appliquées, Sauveterre, FR	242	146
Sta-Form	Georgia-Pacific Atlanta, GA 30 348, US	221	11
Stabar®	ICI PLC, Welwyn Garden City, Herts. AL7 1 HH, GB	2351, 2361	413
Stabilox	Henkel KGaA, 40589 Düsseldorf, DE	–	9
Stabiol	–,,–	–	9
Staflene	The Nisseki Plastic Chemical Co. Ltd Tokyo, JP	1113, 1115	2
Staloy	DSM, Polymers & Hydrocarbons 6130 AA Sittard, NL	1221/261	2
Stamylan P®	–,,–	112	11, 2
Stamylan®	–,,–	111	11, 2
Stamylex®	–,,–	1111 L	11, 2
Stamyroid	–,,–	112 s	11
Stamytec®	–,,–	111	11, 2
Stanyl®	–,,–	261 s	11, 2

8.3 Trade Names

Name	Company (see 8.1, page 479)	Resin (see 8.2.1, page 481)	Forms of delivery (s. 8.2.2, p. 484)
Stapron®	DSM, Polymers & Hydrocarbons 6130 AA Sittard, NL	1225, 1221+ 2411, 1221+ 261, 1222+3411	124, 2
Star L	Ferro Eurostar, 95470 St. Landre, FR	1	2 s
Star Xlam	–,,–	261	2213, 2217, 224
Star XX	–,,–	261	2
Star-C	–,,–	237, 2411, 2412, 261, 263	226
Star-Flam	–,,–	111, 112, 121, 261	2 se
Staralloys	–,,–	1	1
Staramid	–,,–	261	2213, 2217, 224
Starex	Cheil Industries, Seoul, Korea	1221	2
Starglas	Ferro Eurostar, 95470 St. Landre, FR	1113, 121, 1221, 1222, 2351, 237, 231, 2411, 2412	22
Starpylen	–,,–	112	2213, 222, 224
Stat-Kon®	LNP Plastics Nederland BV 4940 Raamsdonksveer, NL	112, 1352, 2411	11, 226 el
Stat-Kon® Z	–,,–	235	2
Statoil TPE	Statoil Den norske stats oljeselskap a.s., 3960 Stathelle, NO	1117	110 TE
Staufen®	ICI PLC, Welwyn Garden City, Herts. AL7 1 HH, GB	131	4
Steier®	Max Steier GmbH & Co. 25337 Elmshorn, DE	111, 131	4
Steierpack	–,,–	111	4
Steierplast	–,,–	132	4
Stereon®	Firestone Synthetics Rubber and Latex Co., Akron, OH 44 301, US	12/32/12	110 TE
Steriweb	Wihuri Oy Wipak, 15561 Nastola, FI	111/261, 112/261	412

8.3 Trade Names

Name	Company (see 8.1, page 479)	Resin (see 8.2.1, page 481)	Forms of delivery (s. 8.2.2, p. 484)
Sternite®	Elf Atochem 92 091 Paris, La Défense 10, FR	121, 1221	112, 2
Sterocoll	BASF Aktiengesellschaft 67063 Ludwigshafen, DE	16, 162	121, 123
Sterpon®	Ets. G. Convert 01100 Oyonnax, FR	242	11
Stevens Urethane	JPS Elastomeric Corp. Northampton, MA 01 061–0658, US	264	32, 33, 34, 342, 411, 43, 5
Strapan	Chemie Linz AG, 4040 Linz, AT	111	76
Stratimat®	Stratinor S.A., 87000 Limoges, FR	242	61, 2251
Stratoclad	Spaulding Composites Co. Tonawanda, NY 14 150, US	21	614
Stren®	DuPont Co. Inc. Wilmington, DE 19 898, US	261	36
Strippex	Borealis Compounds AB 12 613 Stockholm, SE	1115	11 el
Stycast	Emerson & Cuming Inc. Canton, MA 02 021, US	233	22
Stylac	Asahi Chemical Ind. Co. Ltd. Tokyo, JP	1221	11
Stypol®	Freeman Chemical Corp. Port Washington, WI 53 074, US	242	112, 113, 225
Styrafil®	Akzo Engineering Plastics Inc. Evansville, IN 47 732, US	121, 1223	224
Styritherm	Kulmbacher Spinnerei AG Kunststoffwerk Mainleus 95336 Mainleus, DE	121	71
Styroblend	BASF Aktiengesellschaft 67063 Ludwigshafen, DE	122	2
Styrocell	Shell International Co. Ltd. London, SE1 7NA, GB	121	117
Styrodur®	BASF Aktiengesellschaft 67063 Ludwigshafen, DE	121	71, 51
Styrofan®	–,,–	121	121
Styrofill®	–,,–	121	71
Styroflex®	Norddeutsche Seekabelwerke AG 26954 Nordenham, DE	121	35, 413, 43

Name	Company (see 8.1, page 479)	Resin (see 8.2.1, page 481)	Forms of delivery (s. 8.2.2, p. 484)
Styrolux	Westlake Plastics Co. Lenni, PA 19 052, US	122	3, 5
Styrolux®	BASF Aktiengesellschaft 67063 Ludwigshafen, DE	125	21
Styromull®	–,,–	121	71 s
Styron®	The Dow Chemical Company Midland, MI 48641-2166, US	12	11, 2
Styronal®	BASF Aktiengesellschaft 67063 Ludwigshafen, DE	1223	121
Styronol	Norton Performance Plastics Wayne, NJ 07 470, US	121	5
Styroplus®	BASF Aktiengesellschaft 67063 Ludwigshafen, DE	1223	124
Styropor®	–,,–	121	71
Sucorad	Huber & Suhner AG, 9100 Herisau, CH	1116	31
Sulfil	Akzo Engineering Plastics Inc. Evansville, IN 47 732, US	1351/236, 237	224
Sumiepoch®	Sumitomo Chemical Co. Ltd. Tokyo 103, JP	1115 s	115
Sumiflex®	–,,–	1315	2
Sumigraft®	–,,–	1115/132	2
Sumika Flex®	–,,–	1115, 1117	11, 2
Sumikadel	–,,–	16/152	11
Sumikagel	–,,–	115	4
Sumikathene®	–,,–	1111, 1111 L, 1112, 1113, 1115	2
Sumilit®	–,,–	131	11
Sumilite FST	–,,–	2361	413
Sumipex®	–,,–	163	2, 411
Sunlet	Mitsui Petrochemical Ind., Ltd. Tokyo, JP	112	224
Sunloid KD	Tsutsunaka Plastic Ind. Co. Ltd. Osaka 541, JP	1314	5
Sunloid PC	–,,–	2415	5
Sunloid®	–,,–	131	5

Name	Company (see 8.1, page 479)	Resin (see 8.2.1, page 481)	Forms of delivery (s. 8.2.2, p. 484)
Sunprene®	A. Schulman Inc. Akron, OH 44 309, US	1314	2
Suntec	Asahi Chemical Ind. Co., Ltd. Tokyo, JP	1113	11, 2
Supaboard	Caradon Celuform Ltd., Aylesford, Maidstone, Kent ME20 75X, GB	131	34/713
Supaliner	–,,–	131	34/713
Supastik	–,,–	51	73
Supec®	GE Plastics Pittsfield, MA 01 201, US	237	224
Super® Dylan	Arco Chemical Co. Newton Square, PA 19 073, US	1113	2
Superex®	Mitsubishi Monsanto Chemical Co. Tokyo, JP	1225	2
Superkleen®	Alpha, Div. Dexter Plastics Newark, NJ 07 105, US	132 s	2
Supernaltene	Ets. G. Convert 01100 Oyonnax, FR	1113	5
Superohm®	A. Schulman Inc. Akron, OH 44 309, US	1116	2 el
Superpolyorc	Uponor, Stourton, Leeds, LS10 1UJ, GB	1311	31
Supra-Carta®	Isola Werke AG, 52353 Düren, DE	21, 233	611
Supra-Carta®-Cu	Isola Werke AG, 52353 Düren, DE	21, 233	611-Cu
Supramid	VIS Kunststoffwerk GmbH 77652 Offenburg, DE	261	3, 51
Suprane®	Rhône-Poulenc Films, 92 080 Paris La Défense, Cedex 6, FR	1113	413
Supraplast®	Süd-West-Chemie GmbH 89231 Neu-Ulm, DE	21, 22, 233, 242	113, 2, 2251, 2431
Suprasec®	ICI Polyurethanes 3078 Kortenberg, BE	2631, 265, 266	117, 71
Suprathen	Hoechst AG 65929 Frankfurt, DE	1111	411, 43
Supratherm	–,,–	132	111
Suprel	Vista Chemical Co. Houston, TX 77 079, US	12/131/161	2 se

Name	Company (see 8.1, page 479)	Resin (see 8.2.1, page 481)	Forms of delivery (s. 8.2.2, p. 484)
Supronyl	Hoechst AG, 65929 Frankfurt, DE	261	412, 43
Sur-Flex	Flex-O-Glass Inc., Chicago, IL 60 051, US	1115 s	41
Suritex	Isovolta AG, 2355 Wiener Neudorf, AT	28	616
Surlyn®	DuPont Co. Inc., Wilmington, DE 19 898, US	1115 s	146, 4
Sustamid®	Röchling Sustaplast KG, 56112 Lahnstein, DE	261	3, 4, 5
Sustarin®	–,,–	231, 2311	3, 5
Sustatec®	–,,–	2351, 2361	3, 5
Susteel	Tosoh Susteel, Nagoya, JP	237	2
Sustodur®	Röchling Sustaplast KG, 56112 Lahnstein, DE	2412	3, 5
Sustonat®	–,,–	2411	3, 5
Suva	DuPont Co. Inc., Wilmington, DE 19 898, US	135 s	CFC alt.
Suwide®	Planox B. V., 5705 AA Helmond, NL	132, 26	421, 424, 451
Swan®	Tufnol Ltd., Birmingham B42 2TB, GB	21	611
Syfan	Syfan B.O.P.P. Films, 85140 D.N. Hanegev, IS	112	411
Sylomer	Getzner Chemie, AT	263	711
Symalen®	Symalit AG, 5600 Lenzburg 1, CH	34	31, 52
Symalit®	–,,–	11, 34	31, 342, 51, 52, 614
Syncomat®	N.V. Syncoglas S.A., Zele, BE	242	225
Syncopreg	–,,–	233, 242	224, 225
Synergy	N. V. Allied Corp. Int. S. A., 3030 Heverlee, BE	235/261	110 TE, s
Synlon	Synlon Limited, Elland, W. Yorkshire HX5 9DZ, GB	11	342
Synlon	–,,–	261	21, 2213, 2217, 224
Synlon	–,,–	262	21, 224, 2217

Name	Company (see 8.1, page 479)	Resin (see 8.2.1, page 481)	Forms of delivery (s. 8.2.2, p. 484)
Synocryl	Cray Valley Products Int. Farnborough, Kent, BR6 7EA, GB	15	112
Synocure	–,,–	15	11
Synolac	–,,–	242, 244	112
Synolite®	DSM Resins BV 8000 AP Zwolle, NL	242	112, 113
Synotex®	–,,–	413	11
Syntac	W. R. Grace & Co., Organic Chemicals Lexington, MA 02 173, US	27	71
Synthacryl®	Hoechst AG 65929 Frankfurt, DE	16	114, 121
Synthite®	John C. Dolph Co., Monmouth Junction, NJ 08 852, US		
Synthopan	Synthopol Chemie, 21614 Buxtehude, DE	242	11, 112
Synthoplex®	Röhm GmbH, 64293 Darmstadt, DE	16	116, 123
Systanat®	Elastogran GmbH 49440 Lemförde, DE	2631	117
Systol S®	–,,–	2632	117
Systol T®	–,,–	2633	117
T			
Ta-adin	Mayer Enterprises Ltd., Coating Dept., Tel Aviv, IL	132	4241
Ta-or	–,,–	132	4241
Taff-a-flex®	Clopay Corp. Plastics Products Div. Cincinnati, OH 45 202, US	11, 112	4
Taffen®	Exxon Chemicals America Houston, TX 77 001, US	112	225
Taflite	Mitsui Toatsu Chemicals Inc. Tokyo, JP	125 s	2
Tafmer	Mitsui Petrochemical Ind. GmbH 40211 Düsseldorf, DE	1117	2
Taktene®	Bayer AG, 51368 Leverkusen, DE	3211	11
Taradal	BAT taraflex, 69170 Tarare, FR	132	442
Taraflex	–,,–	132	441, 444

8.3 Trade Names

Name	Company (see 8.1, page 479)	Resin (see 8.2.1, page 481)	Forms of delivery (s. 8.2.2, p. 484)
Taralay	BAT taraflex, 69170 Tarare, FR	132	441
Tarflen	Stickstoffwerke Tarnow, Tarnow, PL	135	11
Tarnamid T	–,,–	261	2
Tarnoform	–,,–	235	11
Tauride®	Veritex B.V., 7300 AA Appeldoorn, NL	132	451
Tauro-pren	Taurus Gummiwerke, HU	112, 1127	11
Taylorclad	Synthane-Taylor Corp. La Verne, CA 91 750, US	233	614
Taylorite	–,,–	4211	3, 5
Teamex	DSM, Polymers & Hydrocarbons 6130 AA Sittard, NL	1111	2
Technoduct®	Techno-Chemie Kessler & Co. GmbH 64546 Mörfelden-Walldorf, DE	112, 1127, 1315, 132, 264, 343	32, 321
Technoflex	Hegler Plastik GmbH 97714 Oerlenbach, DE	261	31, 32
Technoply	Howe Industries Inc. Van Nuys, CA 91 402, US	233, 27	614
Technora®	Teijin Ltd., Osaka 541, JP	2612	72
Technyl®	Nyltech Italia Srl 20 020 Ceriano (MI), IT	261	2
Techster®	Rhône-Poulenc Specialités Chim 69 006 Lyon, FR	2412	2 tra
Techtron®	Polypenco GmbH 51437 Bergisch Gladbach, DE	237	3, 5
Tecnoflon®	Ausimont Inc. Morristown, NJ 07 962, US	348	11, 12
Tecnoprene	ECP EniChem Polimeri srl. 20 124 Mailand, IT	112	2, 224
Tecoflex	Thermo Electron Corp. Waltham, MA, GB	263	11, 114, 2 med
Tecolite®	Toshiba Chemical Product Co. Ltd. Tokyo, JP	212	2, 23
Tediflex	ECP EniChem Polimeri srl. 20 124 Mailand, IT	263	117
Tedilast	–,,–	263	117

8.3 Trade Names

Name	Company (see 8.1, page 479)	Resin (see 8.2.1, page 481)	Forms of delivery (s. 8.2.2, p. 484)
Tedimon®	ECP EniChem Polimeri srl. 20 124 Mailand, IT	2631	113
Tedirim	–,,–	263	118
Tedistac	Stac, Erstein Gare, FR	263, 2631, 2633	117
Teditherm	ECP EniChem Polimeri srl. 20 124 Mailand, IT	263	117
Tedlar	DuPont Co. Inc. Wilmington, DE 19 898, US	1353	41
Tedur	Albis Plastic GmbH 20531 Hamburg, DE	237	2
Teflon®	DuPont Co. Inc. Wilmington, DE 19 898, US	135	11, 2, 4, 7, 72
Tefzel®	–,,–	1115/1351	2
Tegit	Garfield Molding Co. Garfield, NJ 07 026, US	212	22
Tego-Tex®	Th. Goldschmidt AG, 45127 Essen, DE	211, 223	2431
Tegocoll®	Th. Goldschmidt AG, 45127 Essen, DE	233	73
Tegophan AC®	–,,–	163, 242	4 s
Tegophan UP®	–,,–	242	4 s
Tehadur	Tehalit-Kunststoffwerk GmbH 67716 Heltersberg, DE	1311	31
Teklan	Courtaulds PLC, Bradford BD1 1EX W. Yorkshire, GB	61 mod	72 se
Tekmilon®	Mitsui Polychemical Co. Ltd. Tokyo, JP	111 s	72 s
Tekudur	Tekuma Kunststoff GmbH, DE	2412	2
Tekuform	–,,–	231	2
Tekulac	–,,–	1221	2
Tekulan	–,,–	2411	2
Tekumid	–,,–	261	2
Tekusan	–,,–	1222	2
Telalloy	Kanegafuchi Chemical Ind. Co. Ltd. Osaka, JP	163/125	124

Name	Company (see 8.1, page 479)	Resin (see 8.2.1, page 481)	Forms of delivery (s. 8.2.2, p. 484)
Telcar®	Teknor Apex Co. Pawtucket, RI 02 862, US	112+3411	110 TE
Telcon	Telcon Plastics Ltd. Orpington, Kent BR6 6BH, GB	11, 12, 13	3, 4, 5
Telcoset	–„–	233	14
Telcothene	–„–	111	14
Telcovin	–„–	1314	14
Telene	B. F. Goodrich Chemical Group Cleveland, OH 44 131, US	53	118
Telstrene	Telcon Plastics Ltd. Orpington, Kent BR6 6BH, GB	121	414
Tempalloy®	A. Schulman Inc. Akron, OH 44 309, US	112, 1221, 1222, 2411, 2412, 2361, 237, 261	2
TempRite®	B. F. Goodrich Chemical Group Cleveland, OH 44 131, US	1312, 134	2 se, 11
Tenac	Asahi Chemical Ind. Co. Ltd. Tokyo, JP	2311	2
Tenax®	Akzo Fibers Division 6800 AB Arnhem, NL	161 s	226
Tenex®	Teijin Chemicals Ltd., Tokyo, JP	4224	2
Tenite®	Eastman Chemical Intern. Kingsport, TN 37 662, US	111, 1111 L, 112, 1121, 2412, 4222, 4224/4225	11, 116, 141, 2, 224, 713
Tensar	Netlon, Blackburg, GB	1113	433
Tensiltarpe	BCL, Bridgewater, Somerset, TA6 4PA, GB	111	43
Tepcon®	Tepco, Taiwan	2312	2
Teramide	P. D. George Co. St. Louis, MO 63 147, US	271	14 el
Terate	Hercules Powder Company Inc. Wilmington, DE 19 894, US	2412	11
Terathane®	DuPont Co. Inc. Wilmington, DE 19 898, US	2633	11
Terblend® S	BASF Aktiengesellschaft 67063 Ludwigshafen, DE	124, 2411	124

8.3 Trade Names

Name	Company (see 8.1, page 479)	Resin (see 8.2.1, page 481)	Forms of delivery (s. 8.2.2, p. 484)
Tercarol	ECP EniChem Polimeri srl. 20 124 Mailand, IT	2632	117
Terebec®	BASF Lacke+Farben AG 48165 Münster, DE	271	114
Terene	Chemicals & Fibres of India Ltd. Bombay, IN	2412	72
Tergal®	Rhône-Poulenc Fibres 69 003 Lyon, FR	2412	72
Teriber®	Sosiedad Anonima de Fibras Artificiales, Barcelona, ES	2412	72
Terluran®	BASF Aktiengesellschaft 67063 Ludwigshafen, DE	1221	21
Terlux®	–,,–	1221	21
Termovil	Fabbrica Adesivi Resine S.p.A. Cologno Monzese, IT	151	11, 121, 123, 33, 51
Termovir	Fiap, Milano, IT	1311	41
Terocor®	Teroson GmbH 69123 Heidelberg, DE	263	71
Teroform®	Teroson GmbH 69123 Heidelberg, DE	43, 5	
Terokal®	–,,–	233, 3	73
Terolan®	–,,–	131, 3	145
Terosol®	–,,–	61	122
Terostat®	–,,–	345, 36, 37	145
Terotop®	–,,–	167	73
Terphane®	Rhône-Poulenc Films 69 398 Lyon, Cedex 3, FR	2412	41, 413
Terthene	–,,–	111+2412	412
Terylene®	ICI PLC, Fibres Div. Harrogate, HG2 8QN, GB	2412	72
Tesamoll®	BDF Beiersdorf AG 20253 Hamburg, DE	37	422 s
Tetnet	Billerud AB Abt. Nya Produkter Säffle, SE	412	4
Tetoron®	Teijin Chemicals Ltd., Tokyo, JP	2412	72
Tetradur	Tetra-Dur-Kunststoff-Produktion GmbH, 21220 Seevetal, DE	242	224

8.3 Trade Names

Name	Company (see 8.1, page 479)	Resin (see 8.2.1, page 481)	Forms of delivery (s. 8.2.2, p. 484)
Tetrafil®	Akzo Engineering Plastics Inc. Evansville, IN 47 732, US	2412	2, 224
Tetralene	Tetrafluor Inc. El-Segundo, CA 90 245, US	1111 s	2
Tetralon	–,,–	1351	22
Tetraloy	ICI PLC, Welwyn Garden City, Herts. AL7 1 HH, GB	135	22
Tetraphen	Georgia-Pacific Atlanta, GA 30 348, US	214 s	11
Tetronic	BASF Wyandotte Corp. Parsippany, NJ 07 054, US	2633	117
Texalon®	Texapol Corp. Bethlehem, PA 18 017, US	261	2, 224
Texicote®	Scott Bader Co. Ltd. Wollaston, Wellingborough, Norths. NN9 7RL, GB	24	113, 146
Texicryl®	–,,–	12, 16	121
Texigel®	–,,–	16	121
Texin®	Miles Chemical Corp. Pittsburgh, PA 15 205, US	2411/264	110 TE
Texipol®	Scott Bader Co. Ltd. Wollaston, Wellingborough, Norths. NN9 7RL, GB	121	
Texon	Bonded Laminates Ltd. London E3 5NP, GB	223	621
Texrim®	Texaco Chemical Co. Bellaire, TX 77 401, US	–	118
Textolite	GE Plastics Pittsfield, MA 01 201, US	21, 223, 233, 242, 28	61
Thanate	Texaco Chemical Co. Bellaire, TX 77 401, US	2631	11
Thanol	–,,–	2632	11
Thelan	Thelen & Co. 53757 St. Augustin, DE	37	433, 51, 71, 713
Thelon	Interplastic Werk AG, 4600 Wels, AT	132	441, 442
Thelotron	–,,–	132	441, 442 el
Therban®	Bayer AG, 51368 Leverkusen, DE	323 s	11

8.3 Trade Names

Name	Company (see 8.1, page 479)	Resin (see 8.2.1, page 481)	Forms of delivery (s. 8.2.2, p. 484)
Thermaflow	Evode Plastics Ltd. Syston, Leicester LE7 8PD, GB	21, 2212	2
Thermalux®	Westlake Plastics Co. Lenni, PA 19 052, US	236	33, 5
thermassiv®	Schock & Co. GmbH 73605 Schorndorf, DE	16	34
Thermex-1®	A. Schulman Inc. Akron, OH 44 309, US	112, 1221, 237, 261, 2361, 2411, 2412	2
Thermid	National Starch & Chemical Corp. Bridgewater, NJ 08 807, US	27	2
Thermoclear®	GE Plastics Pittsfield, MA 01 201, US	2411	55
Thermocomp®	LNP Plastics Nederland BV 4940 Raamsdonksveer, NL	1, 23, 2351, 2411, 26, 274	2213, 224, 226, 2217, se
Thermodet®	Mitras Kunststoffe GmbH 92637 Weiden, DE	122, 1221, 1222	53
Thermofilm® IR	Polyon-Barkai, Kibbutz Barkai M. P. Menashe 37 860, IL	1115	41
Thermogreca	Polyú Italiana s.r.l. 20 010 Arluno, Milano, IT	2411	56
Thermolast K	Gummiwerk Kraiburg 84478 Waldkraiburg, DE	1253	2
Thermolast®	–„–	1253	2
Thermonda	Polyú Italiana s.r.l. 20 010 Arluno, Milano, IT	2411	55
Thermoprene	Evode Plastics Ltd. Syston, Leicester LE7 8PD, GB	323, 3411	110 TE
Thoprene	Thoprene Co., Yokkaichi, JP	237	2
Thor	Borden Chemical Co. Geismar, LA 70 734, US	216, 222	113
Thornel	Amoco Performance Products Atlanta, GA 30 350, US	5	226
Timbrelle®	ICI PLC, Fibres Div. Harrogate, HG2 8QN, GB	261	72
Timbron	Plexite India Pot Ltd., Bombay (Glynwed Int. Birmingham), IN	12	76

Name	Company (see 8.1, page 479)	Resin (see 8.2.1, page 481)	Forms of delivery (s. 8.2.2, p. 484)
Tipolen	TVK Tisza Chemical Combine 3581 Leninvaros, HU	1111	2
Tipox	–,,–	233	113
Toghpet®	Mitsubishi Rayon Co. Ltd. Tokyo, JP	2412	22
Tolonate®	Rhône-Poulenc S.A. 92 408 Courbevoie, FR	2631, 265	111
Tone Polymers	Union Carbide Corp. Danbury, CT 06 817, US	241	2 a
Tonen	Tonen Sekiyukagaku K. K. Tokyo, JP	112	226
Topamid	Ton Yang Nylon Co. Seoul, Korea	261	2
Topan	Glunz AG, 66265 Heusweiler, DE	–	742
Toporex	Mitsui Toatsu Chemicals Inc. Tokyo, JP	121	2
Torayca®	Toray Industries Ltd., Osaka 530, JP	261	2261
Torayfan	Toray Plastics America North Kingstown, IR 02 852, US	112	415
Toraylina	–,,–	237	11
Torlen	Elana-Werke, Torun, PL	2412	72
Torlon®	Amoco Performance Products Atlanta, GA 30 350, US	272	2, 22
Toughlon	Idemitsu Petrochemical Co. Ltd. Tokyo, JP	2411	2
Toyolac®	Toray Industries Inc., Tokyo, JP	1221, 1222	2 s
TPX®	Mitsui Petrochemical Ind. Ltd. Tokyo, JP	116	21 tra, 22
Tradlon	Fluor Plastics Inc. Philadelphia, PA 19 134, US	27	413
Transparene	Neste Oy Chemicals, 02 151 Espoo, FI	2411	2
Transparit® P	Ylopan Folien GmbH 34537 Bad Wildungen, DE	4212	411
TransVelbex	British Industrial Plastics Film Div. Turner & Newall, Manchester, GB	131	43, 54
Travertine	Marley Floors Ltd., Lenham, Maidstone, Kent, ME17 2DE, GB	132	444

8.3 Trade Names

Name	Company (see 8.1, page 479)	Resin (see 8.2.1, page 481)	Forms of delivery (s. 8.2.2, p. 484)
Traytuf Ultra-Clear	Goodyear Tire & Rubber Co. Akron, OH 44 316, US	2412	1, 2
Treafilm	Trea Ind. Inc. Kingstown, RI 02 852, US	1127 med	
Trefsin	Advanced Elastomer Systems L.P. St. Louis, MO 63 167, US	112/332	1127 med
Tregalon	Bayer AG, 51368 Leverkusen, DE	237	72
Trespaphan®	Hoechst AG 65929 Frankfurt, DE	112	411, 413
Trevira®	–,,–	2412	72
Triafol®	Bayer AG, 51368 Leverkusen, DE	4223, 4224	411, 413
Triax®	Monsanto Chemical Co., Plastics Div. St. Louis, MO 63 167, US	1221+261	21, 224, 226
Tribit	Samyang, Seoul, Korea	2412	2
Triform	Goodyear Tire & Rubber Co. Akron, OH 44 316, US	131	41, 5
Trilene	Uniroyal, Mishawaka, IN 46 544, US	3411 s	11
Trirex	Samyang, Seoul, Korea	2411	2
Tritherm 981	P. D. George Co. St. Louis, MO 63 147, US	272	11 el
Trivoltherm N	August Krempel Soehne GmbH & Co. 71655 Vaihingen, DE	2412+261	412, 413, 437
Trixene	Baxenden Chemical Co. Ltd. Accrington, Lancs. BB5 2SL, GB	16/263, 2631	114
Trocal®	HT Troplast AG 53840 Troisdorf, DE	131	3411, 433, 434, 435
Trocellen®	–,,–	111, 1116	71
Trolit®	Hüls AG, 45764 Marl, DE	5	2
Trolon	–,,–	212	113
Trosifol®	HT Troplast AG 53840 Troisdorf, DE	155	43
Trosiplast®	Hüls AG, 45764 Marl, DE	1311, 1313, 1314, 132	21
Trovicel®	–,,–	131	5/71
Trovidur®	HT Troplast AG 53840 Troisdorf, DE	131	51, 52, 53

8.3 Trade Names

Name	Company (see 8.1, page 479)	Resin (see 8.2.1, page 481)	Forms of delivery (s. 8.2.2, p. 484)
Trusurf®	Owens Corning Fiberglas Corp. Toledo, OH 43 659, US	242	113
Tubolit	Armstrong World Ind. Inc. Lancaster, PA 17 604, US	111	71
Tufcote	Speciality Composites Corp. Newark, DE 19 713, US	263	414
Tuff-a-tex®	Clopay Corp. Plastics Products Div. Cincinnati, OH 45 202, US	111, 112	4
Tuffak®	Atohaas, 92062 Paris, La Défense 10, FR	163	11, 2
Tufnol®	Tufnol Ltd. Birmingham B42 2TB, GB	21, 223, 233, 27, 28	61
Tufpet	Toyobo Co. Ltd., Osaka 530, JP	2412	2
Tufrex®	Mitsubishi Monsanto Chemical Co. Tokyo, JP	1221	2
Tufset	Tufnol Ltd. Birmingham B42 2TB, GB	263	33, 51
Tufsyn	Goodyear Tire & Rubber Co. Akron, OH 44 316, US	321	11
Tuftane®	London Artid Plastics, London, GB	264	4, 5
TufX®	Hoechst AG 65929 Frankfurt, DE	231	224
Tungophen®	Bayer AG, 51368 Leverkusen, DE	214	114
Tuplin	Union Carbide Corp. Danbury, CT 06 817, US	1111 L	2
Twaron®	Nippon Aramid Yugen Kaisha, Tokyo, JP	261 s	226
Tybon	Georgia-Pacific Atlanta, GA 30 348, US	21	11, 224
Tygaflor	Cyanamid Aerospace Products Ltd. Wrexham, Clwyd LL13 9UF, GB	1351	436/614
Tygan®	–„–	133	43
Tylose®	Hoechst AG 65929 Frankfurt, DE	423	111, 116, 13
Tynex®	DuPont Co. Inc. Wilmington, DE 19 898, US	261	35, 36
Typar®	–„–	112	721

8.3 Trade Names

Name	Company (see 8.1, page 479)	Resin (see 8.2.1, page 481)	Forms of delivery (s. 8.2.2, p. 484)
Tyril®	The Dow Chemical Company Midland, MI 48641-2166, US	1222	2
Tyvek®	DuPont Co. Inc. Wilmington, DE 19 898, US	111	721
U			
U Sheet	Tahei Chem. Prod. Co., Tokyo, JP	2414	4
U-pica	Japan U-Pica Co. Ltd., Osaka, JP	242	123
U-polymer	Unitika Ltd., Osaka, JP	2414	11
Ubec®	Ube Industries Ltd. (America) Inc. Ann Arbor, MI 48 108, US	1111, 1111 L	2, 22 el
Ubepol	–„–	3211	11
Ubetex®	–„–	27	3, 5
UCAR	Union Carbide Corp. Danbury, CT 06 817, US	1111	145, 2 el
Ucarsil FR	–„–	11	22 se
Ucrete®	ICI PLC Welwyn Garden City, Herts. AL7 1 HH, GB	263	146
Udel®	Amoco Performance Products Atlanta, GA 30 350, US	2361	2 tra
Uformite®	Reichhold Chemicals Inc. Jacksonville, FL 32 245, US	22, 223	114, 115
Ugiflex®	Arco Chemical Co. Newton Square, PA 19 073, US	2631	117
Ugipol®	–„–	2633	111
Ulon	Unitex Ltd., Knaresborough, North Yorks., HG5 OPP, GB	263	–
Ulon®	British Vita Co. Ltd. Manchester M24 2D3, GB	263	5
Ultem®	GE Plastics Pittsfield, MA 01 201, US	274	2
Ultimet	Hercules Powder Company Inc. Wilmington, DE 19 894, US	112	41 s
Ultra Rib®	Uponor Innovation AB 51300 Fristad, SE	131	31 s
Ultra Wear®	Polypenco GmbH 51437 Bergisch Gladbach, DE	1113	31, 33, 342, 51

8.3 Trade Names

Name	Company (see 8.1, page 479)	Resin (see 8.2.1, page 481)	Forms of delivery (s. 8.2.2, p. 484)
Ultrablend®S	BASF Aktiengesellschaft 67063 Ludwigshafen, DE	2411/2412	113, 124, 224
Ultrac®	Allied Signal Corp. Morristown, NJ 07 962, US	1113	2
Ultracast®	Baxenden Chemical Co. Ltd. Accrington, Lancs. BB5 2SL, GB	263	112
Ultracel	Union Carbide Corp. Danbury, CT 06 817, US	263	117
Ultradur®	BASF Aktiengesellschaft 67063 Ludwigshafen, DE	2412	113, 21, 224
Ultraform®	–„–	2312	21, 224, 113
Ultralen®	Lonza-Folien GmbH 79576 Weil am Rhein, DE	112	41
Ultramid®	BASF Aktiengesellschaft 67063 Ludwigshafen, DE	261	113, 21, 2212, 224
Ultramid® RC	BASF Aktiengesellschaft 67063 Ludwigshafen, DE	261	224r
Ultramoll®	Bayer AG, 51368 Leverkusen, DE	–	91
Ultrapek®	BASF Aktiengesellschaft 67063 Ludwigshafen, DE	2351	113, 21, 224, 226
Ultraphan®	Lonza-Folien GmbH 79576 Weil am Rhein, DE	422	41, 43 s
Ultrason® E	BASF Aktiengesellschaft 67063 Ludwigshafen, DE	2361	113, 21, 224
Ultrason® S	–„–	236	113, 21, 224
Ultrastyr®	ECP EniChem Polimeri srl. 20 124 Mailand, IT	11, 122	2
Ultrathene®	Quantum Chemical Corp. Cincinnati, OH 42 249, US	1115	2
Ultrex®	Spiratex Comp. Romulus, MI 48 174, US	1113	2
Ultzex®	Mitsui Petrochemical Ind. Ltd. Tokyo, JP	1111 L	2
Uni-Flow	Fortin-Industries Inc. Sylmar, CA 91 342, US	233	2251
Unican	British Vita Co. Ltd. Manchester M24 2D3, GB	263	146

8.3 Trade Names

Name	Company (see 8.1, page 479)	Resin (see 8.2.1, page 481)	Forms of delivery (s. 8.2.2, p. 484)
Unichem®	Colorite Plastics Co. Ridgefield, NJ 07 657, US	131, 131/1314, 132	3
Uniclene	Nippon Unicar Co. Ltd., Tokyo, JP	111	11
Unidene®	International Synthetic Rubber Co. Ltd., Southampton SO9 3AT, GB	322	123
Unifoam	British Vita Co. Ltd. Manchester M24 2D3, GB	263	71
Unileaf®	−,,−	263	4
Unilok®	−,,−	163	113, 115, 73
Unimoll®	Bayer AG, 51368 Leverkusen, DE	−	91
Uniseal	British Vita Co. Ltd. Manchester M24 2D3, GB	263	145
Uniset	Nippon Unicar Co. Ltd., Tokyo, JP	111	11, 2
Unithane®	Cray Valley Products Int. Farnborough, Kent, BR6 7EA, GB	263	11
Unival®	Union Carbide Corp. Danbury, CT 06 817, US	1113 s	11
Upilex®	ICI PLC, Welwyn Garden City, Herts. AL7 1 HH, GB	27	41, 412
Upirex	Ube Industries Ltd. (America) Inc. Ann Arbor, MI 48 108, US	27	431
Upodur	Uponor, Stourton, Leeds, LS10 1UJ, GB	1311	31
Upolar	−,,−	1113	411
Uponal	−,,−	1311	31
Uponyl®	−,,−	131	31
Upotel	−,,−	131	32 el
Upoten	−,,−	1111	411
Urac®	American Cyanamid Corp. Wayne, NJ 07 470, US	221	113, 115
Uracron	DSM Resins BV, 8000 AP Zwolle, NL	16	114
Uradur	−,,−	263	111
Urafil®	Akzo Engineering Plastics Inc. Evansville, IN 47 732, US	264	224
Uraflex	DSM Resins BV, 8000 AP Zwolle, NL	37	11
Uralac	−,,−	244	114

8.3 Trade Names

Name	Company (see 8.1, page 479)	Resin (see 8.2.1, page 481)	Forms of delivery (s. 8.2.2, p. 484)
Uralane	Ciba Geigy AG, 4002 Basel, CH	263	114, 115
Uramex	DSM Resins BV, 8000 AP Zwolle, NL	22	114
Uramul	–,,–	15	114
Uranox	–,,–	233	114
Uravar	–,,–	21	114
Urebade	Newage Industries Willow Grove, PA 19 090, US	263	321
Urecoll®	BASF Aktiengesellschaft 67063 Ludwigshafen, DE	221	113, 115, 123
Ureol®	Ciba Geigy AG, 4002 Basel, CH	264	11
Urepan®	Bayer AG, 51368 Leverkusen, DE	2632	11
Urochem	Chemiplastics S.p.A. 20 151 Mailand, IT	221	2
Uroform	Chromos – Ro Polimeri 4100 Zagreb, Croatia	221	2
Uromix	Ubbink Nederland BV 4700 BG Roosendaal, NL	242	224
Uroplas®	AMC-Sprea S.p.A., 20 101 Milano, IT	222	2
Uropreg	Ubbink Nederland BV 4700 BG Roosendaal, NL	242	225
Urotuf	Reichhold Chemicals Inc. Jacksonville, FL 32 245, US	263	11
Ursus	J.H.R. Vielmetter GmbH & Co. KG 12277 Berlin, DE	132	32
Urutuf	Reichhold Chemicals Inc. Jacksonville, FL 32 245, US	263	11
Uthane	Urethanes India Ltd. (Chemicals and Plastics India), IN	264	11
Uvex®	Eastman Chemical Intern. Kingsport, TN 37 662, US	4224	51, 53
V			
Vac pac	Richmond Technology, Inc. Redlands, CA 92373, US	1353	411
Vacuflex®	Techno-Chemie Kessler & Co., GmbH 64546 Mörfelden-Walldorf, DE	112, 1127, 1315, 132, 264, 343	32, 321

Name	Company (see 8.1, page 479)	Resin (see 8.2.1, page 481)	Forms of delivery (s. 8.2.2, p. 484)
Valeron®	Van Leer Plastics Inc. Houston, TX 77 240, US	1113	411, 412
Valite	Lockport Thermosets Inc. Lockport, LA 70 374, US	211	113, 2
Valox®	GE Plastics Pittsfield, MA 01 201, US	2412	2, 22, 224, 4
Valtec®	Montell Polyolefins Wilmington, DE 19 894, US	112 s	1, 2
Valvac®	Van Leer Plastics Inc. Houston, TX 77 240, US	412	
Vamac®	DuPont Co. Inc. Wilmington, DE 19 898, US	345	11
Vandar®	Hoechst AG 65929 Frankfurt, DE	2412 s	1, 2, 224
Vantel	Porvair P.L.C. King's Lynn, Norfolk PE30 2 HS, GB	262	424, 712
Varcum	Reichhold Chemicals Inc. Jacksonville, FL 32 245, US	21	114
Variopox EP-C	BAVG GmbH, 51381 Leverkusen, DE	233	112/146
Varlan®	DSM, Polymers & Hydrocarbons 6130 AA Sittard, NL	1311, 1314	2
Vaycron	Hydro Polymers Ltd., Newton Aycliffe Co. Durham DL5 6EA, GB	1315	110 TE, 2
Vector	Dexco Polymers Houston, TX 77 079, US	1253	21
Vectra®	Hoechst AG 65929 Frankfurt, DE	2412	119
Vegetalite	Sintesi s.r.l., 21050 Borsano, IT	212	231
Vegon	Hüls AG, 45764 Marl, DE	112	72
Vekaplan®	Veka AG 48324 Sendenhorst, DE	131	71, 761
Vekton®	Norton Performance Plastics Wayne, NJ 07 470, US	261	112, 3, 51
Velbex	Wradle Storeys PLC Manningtree, Essex, CO11 1NJ, GB	1311	41, 42, 423, 424, 51

Name	Company (see 8.1, page 479)	Resin (see 8.2.1, page 481)	Forms of delivery (s. 8.2.2, p. 484)
Velite	Atohaas, 92062 Paris, La Défense 10, FR	163	11, 2
Velkor®	Alkor GmbH, 81479 München, DE	121, 131, 132	421, 423, 4241, 452, 52, 53
Veloflex®	Veloflex Carsten Thormählen 25337 Köln, DE	111, 131	423
Velva-flex	Clopay Corp. Plastics Products Div. Cincinnati, OH 45 202, US	111	42
Venilia®	Solvay S.A., 1050 Bruxelles, BE	132	422, 451
Venipak®	Alkor GmbH, 81479 München, DE	1311	411, 412
Verdur	August Krempel Soehne GmbH & Co. 71655 Vaihingen, DE	233	413, 612, 614
Veriskin®	Veritex B.V., 7300 AA Appeldoorn, NL	132	4241
VersAcryl	Gen. Corp. Polymer Products Newcomerstown, OH 43 832, US	131/16	53
Versalon®	Henkel Corp. Ambler, PA 19 002, US	261	115
Versamid®	–,,–	2611	112, 113, 114, 115
Verton®	ICI PLC, Welwyn Garden City, Herts. AL7 1 HH, GB	237, 26	224 s
Vespel®	DuPont Co. Inc. Wilmington, DE 19 898, US	27	3, 5
Vestagon	Hüls AG, 45764 Marl, DE	263	141
Vestamelt®	–,,–	2413	2/73
Vestamid®	–,,–	261	11, 120, 141, 224, 73
Vestanat	–,,–	2631	114
Vestenamer®	–,,–	341	11
Vesticoat	–,,–	263	11
Vestiform®	–,,–	163	9
Vestinol®	–,,–	–	91
Vestoblend®	–,,–	235+261	2
Vestodur®	–,,–	2412	2, 9
Vestogrip	–,,–	331	1

Name	Company (see 8.1, page 479)	Resin (see 8.2.1, page 481)	Forms of delivery (s. 8.2.2, p. 484)
Vestolen A®	Hüls AG, 45764 Marl, DE	1113	11, 2, 224
Vestolen EM®	–„–	1127	2
Vestolen P®	–„–	112	2, 224
Vestolit®	–„–	131	11, 117, 2
Vestopal®	–„–	242	11
Vestoplast®	–„–	11 s	14
Vestopren	–„–	3411	110 TE
Vestoran®	–„–	235	2
Vestosint®	–„–	261	141
Vestoson®	–„–	2612	2
Vestowax®	–„–	111	94
Vestypor®	–„–	121	117
Vestyron®	–„–	121, 1222, 1223	11, 2
Vetrelam®	Micafil AG, 8048 Zürich, CH	233	611, 614
Vetresit®	Micafil AG, 8048 Zürich, CH	242, 233	614, 6141
Vetronit	Von Roll Isola 4226 Breitenbach, CH	21, 233, 28	614, 6141
Vibrathane®	Uniroyal, Mishawaka, IN 46 544, US	2632	112, 117
Viclan®	ICI PLC, Welwyn Garden City, Herts. AL7 1 HH, GB	133	14
Viclon	Kureha Chemical Industry Co. Ltd. Tokyo, JP	131	72
Vicora S®	J. H. Benecke GmbH 30007 Hannover, DE	132	436
Vicotex®	Brochier S. A. 69 152, Decines Charpieù, FR	2	2251, 2261
Victrex®	Victrex Sales Ltd., Thornton Cleveleys, Lancs FY5 4QD, GB	2351, 2361, 2414	11, 119
Vidar®	Solvay S.A., 1050 Bruxelles, BE	1354	146
Videne	Goodyear Tire & Rubber Co. Akron, OH 44 316, US	2412	437
Vigopas®	Raschig AG, 67061 Ludwigshafen, DE	242	51
Vilit	Hüls AG, 45764 Marl, DE	1313, 133	11, 121, 14
Vinacel	Goodyear Tire & Rubber Co. Akron, OH 44 316, US	131	71

8.3 Trade Names

Name	Company (see 8.1, page 479)	Resin (see 8.2.1, page 481)	Forms of delivery (s. 8.2.2, p. 484)
Vinagel	Vinatex Ltd. (Norsk Hydro a.s.) Havant, Hampsh. PO9 2NQ, GB	132	122
Vinakon®	−,,−	1253	2
Vinalit	Emil Keller AG 9220 Bischofszell/TG, CH	1311	55 se
Vinalit®	Buna AG, 06258 Schkopau, DE	151	121
Vinamold	Vinatex Ltd. (Norsk Hydro a.s.) Havant, Hampsh. PO9 2NQ, GB	132	2
Vinarol®	Hoechst AG 65929 Frankfurt, DE	152	116, 123
Vinatex	Hydro Polymers Ltd., Newton Aycliffe Co., Durham DL5 6EA, GB	131	122
Vinavil	Synthesis S.p.A. EniChem 20 138 Mailand, IT	14	121
Vinelle	Goodyear Tire & Rubber Co. Akron, OH 44 316, US	132	424
Vinex®	Air Products & Chemicals Inc. Allentown, PA 18 105, US	152	11 a
Vinidur®	BASF Aktiengesellschaft 67063 Ludwigshafen, DE	1314	113, 21
Vinika®	A. Schulman Inc. Akron, OH 44 309, US	1314	2
Vinitex	Buna AG, 06258 Schkopau, DE	133	121
Vinitex®	Werkstofftechnik Dr. Ing. H. Teichmann Nachf. GmbH, 82538 Geretsried, DE	1111, 1115, 1311, 132, 261, 264	32, 321, 34, 342
Vinloc	Shrink Tubes & Plastics Ltd. Redhill, Surrey, RH1 2 LH, GB	1314	32 el
Vinnapas®	Wacker-Chemie GmbH 81737 München, DE	1115, 151	11, 121, 123
Vinnolit®	Vinnolit Kunststoff GmbH 85737 Ismaning, DE	131, 1311, 1313, 1314	11
Vinnylan®	Werkstofftechnik Dr. Ing. H. Teichmann Nachf. GmbH, 82538 Geretsried, DE	1111, 1115, 1311, 132, 261, 264	32
Vinofan®	BASF Aktiengesellschaft 67063 Ludwigshafen, DE	151	121
Vinoflex®	−,,−	1311	11

8.3 Trade Names

Name	Company (see 8.1, page 479)	Resin (see 8.2.1, page 481)	Forms of delivery (s. 8.2.2, p. 484)
Vinophane®	BCL, Bridgewater, Somerset, TA6 4PA, GB	132	411
Vinora	Vinora AG, 8645 Jona, CH	1111, 1111 L, 1112, 1113	411, 431, 432
Vintex®	Werkstofftechnik Dr. Ing. H. Teichmann Nachf. GmbH, 82538 Geretsried, DE	1111, 1115, 1311, 132, 261, 264	32, 321, 34, 342
Vinuran®	BASF Aktiengesellschaft 67063 Ludwigshafen, DE	11	
Vinychlon	Mitsui Toatsu Chemicals Inc. Tokyo, JP	131, 1314	11, 2
Vinychlore®	Saplast S.A. 67 100 Straßburg-Neuhof, FR	132	2
Vinyclair®	Rhône-Poulenc Films 69 398 Lyon, Cedex 3, FR	131	4, 53
Vinyfoil	Mitsubishi Gas Industries Ltd. Tokyo, JP	1311	411, 53
Vinylaire	Marley Floors Ltd., Lenham, Maidstone, Kent, ME17 2DE, GB	132	444
Vinylec-F	Chisso Corp., Tokyo, JP	153	11
Vinylite®	Union Carbide Corp. Danbury, CT 06 817, US	151	11
Vipac®	Rhône-Poulenc Films 69 398 Lyon, Cedex 3, FR	131	4, 53
Vipathene®	–,,–	1113+1311	412
Vipla®	ECP EniChem Polimeri srl. 20 124 Mailand, IT	1311	2
Viplast®	–,,–	132	11
Viplavil®	–,,–	1313	11
Viplavilol®	–,,–	152	11
Vipolit®	Wacker-Chemie GmbH 81737 München, DE	151	121
Vipophan®	Lonza-Folien GmbH 79576 Weil am Rhein, DE	131	41
Viscacelle	BCL, Bridgewater, Somerset, TA6 4PA, GB	4222	411
Visico	Neste Oy Chemicals, 02 151 Espoo, FI	1116/28	21 el

8.3 Trade Names 627

Name	Company (see 8.1, page 479)	Resin (see 8.2.1, page 481)	Forms of delivery (s. 8.2.2, p. 484)
Visqueen®	ICI PLC, Welwyn Garden City, Herts. AL7 1 HH, GB	1111, 1111 L, 1115	411, 43
Vistaflex®	Advanced Elastomer Systems L.P. St. Louis, MO 63 167, US	1117, 11	2, 120
Vistalon®	–„–	1115/112, 1115/112/341	1
Vistanex®	Advanced Elastomer Systems L.P. St. Louis, MO 63 167, US	114	11
Vitabond	British Vita Co. Ltd. Manchester M24 2D3, GB	131, 161, 261	721
Vitacel	–„–	131	71
Vitacom® TPE, TPO	–„–	1117, 1253	2
Vitafilm	Goodyear Tire & Rubber Co. Akron, OH 44 316, US	132	411
Vitafoam	British Vita Co. Ltd. Manchester M24 2D3, GB	263	712
Vitapol®	British Vita Co. Ltd. Manchester M24 2D3, GB	131	21
Vitawrap	–„–	2633	71
Vitel	Goodyear Tire & Rubber Co. Akron, OH 44 316, US	241	11
Viton®	DuPont Co. Inc. Wilmington, DE 19 898, US	348	11
Vitradur®	Stanley Smith & Co. Plastics Ltd. Isleworth, Middx. TW7 7AU, GB	1113 s	5
Vitralene®	–„–	112	5
Vitralex	–„–	1221, 1354	51
Vitrapad®	–„–	111, 112, 131	5
Vitrathene®	–„–	111	3, 5
Vitron®	ECP EniChem Polimeri srl. 20 124 Mailand, IT	111	21
Vitrone®	Stanley Smith & Co. Plastics Ltd. Isleworth, Middx. TW7 7AU, GB	131, 132	41, 43, 5
Vitrosil®	Sirlite Srl., 20 161 Milano, IT	21	11, 113
Vivak®	Axxis N.V., 8700 Tielt, BE	2412	52

8.3 Trade Names

Name	Company (see 8.1, page 479)	Resin (see 8.2.1, page 481)	Forms of delivery (s. 8.2.2, p. 484)
Vivak®-UV	Axxis N.V., 8700 Tielt, BE	2412	52
Vivyfilm	ECP EniChem Polimeri srl. 20 124 Mailand, IT	2412	41
Vivyform C®	–,,–	2412	4
Vivypak	–,,–	2412	4
Vixir®	Sirlite Srl., 20 161 Milano, IT	131	121
Vliespreg	Isovolta AG 2355 Wiener Neudorf, AT	233	2512
Volara®	Sekisui Chemical Co. Ltd. Osaka 530, JP	111	414, 71
Vole®	Tufnol Ltd. Birmingham B42 2TB, GB	21	612
Volex	Comalloy Intern. Corp. Nashville, TN 37 027, US	2412	2
Volex®	Sekisui Chemical Co. Ltd. Osaka 530, JP	111	414, 120
Volloy 100	Comalloy Intern. Corp. Nashville, TN 37 027, US	1121	11 se
Volony®	A. Schulman Inc. Akron, OH 44 309, US	112, 1221, 237, 2361, 2411, 2411, 2412, 261	
Voltaflex	Isovolta AG 2355 Wiener Neudorf, AT	151	614
Voltalef®	Elf Atochem 92 091 Paris, La Défense 10, FR	1352	2
Voltis	Isovolta AG 2355 Wiener Neudorf, AT	212	611, 612
Votafix	–,,–	233	614
Votastat	–,,–	233	614
Votastop	–,,–	233	616
Vova Tec	Polialden Petroquimica Camaćari 42810-BA, BR	1113	11
Vulcapas	Vulcascot, London, GB	1223	611
Vulkaprene®	ICI PLC, Welwyn Garden City, Herts. AL7 1 HH, GB	37	112
Vulkaresen®	Hoechst AG 65929 Frankfurt, DE	214	113

8.3 Trade Names

Name	Company (see 8.1, page 479)	Resin (see 8.2.1, page 481)	Forms of delivery (s. 8.2.2, p. 484)
Vulkide®	ICI PLC, Welwyn Garden City, Herts. AL7 1 HH, GB	132	424, 53
Vulkollan®	Bayer AG, 51368 Leverkusen, DE	37	112
Vybak	Wardle Storeys PLC Manningtree, Essex CO11 1NJ, GB	1311	51, 52
Vycell	Goodyear Tire & Rubber Co. Akron, OH 44 316, US	131	117/713
Vyflex	Plascoat-Systems Ltd. Farnham, Surrey GU9 9NY, GB	131	141
Vygen	Continental General Tire Inc. Akron, OH 44 329, US	131	2
Vyloglass®	Toyobo Co. Ltd., Osaka 530, JP	242	224
Vylon®	–„–	2413	115, 142, 73
Vylopet	–„–	2412	224
Vynalast®	ICI PLC, Welwyn Garden City, Herts. AL7 1 HH, GB	1314	51
Vynaloy®	B. F. Goodrich Chemical Group Cleveland, OH 44 131, US	131	53
Vynathene®	Quantum Chemical Corp. Cincinnati, OH 42 249, US	140M1, 2	
Vynide®	ICI PLC, Welwyn Garden City, Herts. AL7 1 HH, GB	132	424
Vynoid	Plastic Coatings Ltd. Melbourne, AU	1311, 132	41, 42
Vyon®	Porvair P.L.C., King's Lynn, Norfolk PE30 2 HS, GB	1113, 112	5 porous
Vyram®	Advanced Elastomer Systems L.P. St. Louis, MO 63 167, US	322, 3411	110 TE

W

Name	Company	Resin	Forms of delivery
Wacker Si-Dehäsive®	Wacker-Chemie GmbH 81737 München, DE	28	146
Wacker Sil Gel®	–„–	28	112
Wacker Silicone	–„–	38	11, 112, 146
Wacosit	August Krempel Soehne GmbH & Co. 71655 Vaihingen, DE	233, 242	342, 614, 6141
Walkiflex	UPM Walki Pack 37 601 Valkeakoski, FI	111/261/111	412

8.3 Trade Names

Name	Company (see 8.1, page 479)	Resin (see 8.2.1, page 481)	Forms of delivery (s. 8.2.2, p. 484)
Walkivac	UPM Walki Pack 37 601 Valkeakoski, FI	111/261	412
Walocel® C	Wolff Walsrode AG 29655 Walsrode, DE	4231, 4233	115, 116
Walocel® M	–„–	4234	115, 116
Walomer®	–„–	165	123
Walomid-Combi®	–„–	111, 112, 152, 24, 26	411, 412
Waloplast®	–„–	26	41, 411
Walopur®	–„–	111	411, 412, 437
Waloran®	–„–	4221	114
Walothen®	–„–	112	411, 412, 437
Walsroder NC	–„–	4221	114
Warcétal	Isobelec S.A., Sclessin/Liège, BE	231	31, 33, 43, 5
Warcide	–„–	21	615
Warlène	–„–	111	31, 33, 43, 5
Warlon	–„–	261	33, 43, 5
Warolite	Isobelec S.A., Sclessin/Liége, BE	21	611
Wartex	–„–	21	612
Wavelene	Flexible Reinforcements Ltd. Clitheroe, Lancaster, GB	1111, 261	247/436
Wavitube	Wavin-Repox GmbH, 49763 Twist, DE	242	6141
Wefapress®	Wefapress-Werkstoffe GmbH 48691 Vreden, DE	1113	33, 34, 342, 51
Weholite Spiro	KWH Pipe GmbH, 47877 Willich, DE	1113	312
Wellamid	CP-Polymer Technik GmbH 27721 Ritterhude, DE	261	2
Wellamid®	Wellmann Inc. Johnsonville, SC 29 555, US	261	22, 224
Welvic®	ICI PLC, Welwyn Garden City, Herts. AL7 1 HH, GB	1311, 132	2
Werkstoff „S"	Murtfeldt GmbH & Co. KG 44309 Dortmund, DE	1113 s	33, 51, 342
Werzalit®	Buna AG, 06258 Schkopau, DE	21, 22	5, 622

8.3 Trade Names

Name	Company (see 8.1, page 479)	Resin (see 8.2.1, page 481)	Forms of delivery (s. 8.2.2, p. 484)
Westoplan®	Westag & Getalit AG 33378 Rheda-Wiedenbrück, DE	–	761
Whale®	Tufnol Ltd. Birmingham B42 2TB, GB	21	612
Wicothane	Witco Corp. Organics Div. New York, NY 10 017, US	263	11
Wilflex	Flexible Products Co. Marietta, GA 30 061, US	131	121
Wilkoplast	Wilke-Säurebau 30179 Hannover, DE	132	433, 434
Winlon	Winzen International Inc. Minneapolis, MN 55 420, US	1113	42
WIPAK ...	Wihuri Oy Wipak, 15561 Nastola, FI	11, 2412, 261	412
Wirutex	Wirus-Werke Pfleiderer Industrie GmbH, 33332 Gütersloh, DE	223	761
Witamol®	Hüls AG, 45764 Marl, DE	–	91
Wolfin®	Chemische Werke Grünau, DE 89251 Illertissen, DE	132	43
Wolpryla	Märkische Faser AG 14723 Premnitz, DE	161	72
Wood-Stock®	Solvay S.A., 1050 Bruxelles, BE	112	618
Woodlite	Sekisui Plastics Co. Ltd., Tokyo, JP	121	713
Woodstock®	GOR App. Speciali SpA, Buriaso, IT	112	76
Wopadur®	Gurit-Worbla AG 3063 Ittigen-Bern, CH	13	52, 53
Wopal®	–,,–	13	41, 42, 43, 45, 52, 53, 54
Wopavin®	–,,–	13	52, 53, 54
Worblex-Electra®	–,,–	11, 12	44, 52, 53 el
Worblex®	–,,–	11, 12, 13	41, 43, 52, 53, 54, 55
Worpack®	–,,–	13	41, 52, 53, 54

X

X-sheet	Idemitsu Petrochemical Co. Ltd. Tokyo, JP	112 s	224, 614

8.3 Trade Names

Name	Company (see 8.1, page 479)	Resin (see 8.2.1, page 481)	Forms of delivery (s. 8.2.2, p. 484)
X-tal	Custom Resins, Div. of Bemics Co. Henderson, KY 42 420, US	261 s	11
X-TPL®	DuPont Co. Inc. Wilmington, DE 19 898, US	–	–
Xantar®	DSM, Polymers & Hydrocarbons 6130 AA Sittard, NL	2411	2
Xap® haroco	Guttacoll Klebstoffe GmbH & Co. 21614 Buxtehude, DE	11	73 s
Xenalak	Baxenden Chemical Co. Ltd. Accrington, Lancs. BB5 2SL, GB	16	11
Xenoy	GE Plastics Pittsfield, MA 01 201, US	2411+2412, 2415	2, 224
Xirocoll® puro	Guttacoll Klebstoffe GmbH & Co. 21614 Buxtehude, DE	11	73 s
Xironet®	Sarnatech-Xiro AG 3185 Schmitten, CH	11	73 s
Xiropac®	–,,–	11	411r
Xtrabond	Allied Signal Corp. Morristown, NJ 07 962, US		
Xycon®	Amoco Performance Products Richfield, CT 06 877, US	242/263	11 s
Xydar®	Amoco Performance Products Richfield, CT 06 877, US	2414 s	119, 2, 2213
Xylan®	Whitford Corp. West Chester, PA 19 380, US	1351, 1354	14
Xylon®	Akzo Engineering Plastics Inc. Evansville, IN 47 732, US	261	224 se
Xyron®	Asahi Chemical Ind. Co. Ltd. Tokyo, JP	235	2
Xytrabond	Allied Signal Corp. Morristown, NJ 07 962, US	261 s	11
Y			
Yery-or	Mayer Enterprises Ltd., Coating Dept., Tel Aviv, IL	132	421
YF-Serie	LNP Plastics Nederland BV 4940 Raamsdonksveer, NL	271	2

8.3 Trade Names 633

Name	Company (see 8.1, page 479)	Resin (see 8.2.1, page 481)	Forms of delivery (s. 8.2.2, p. 484)
Ylopan	Ylopan Folien GmbH 34537 Bad Wildungen, DE	1111, 112	411, 431, 432, 437
Z			
Zellamid®	Zell-Metall GmbH, 5710 Kaprun, AT	231, 2311, 2412, 261	31, 32, 33, 34, 433, 51
Zellidur	–,,–	1311	55
Zelux®	Westlake Plastics Co. Lenni, PA 19 052, US	2411	33, 54
Zemid 600®	DuPont Co. Inc. Wilmington, DE 19 898, US	111	22
Zetpol	Nippon Zeon Co. Ltd., Tokyo, JP	323	11
Zimek®	DuPont Co. Inc. Wilmington, DE 19 898, US	1115	2
Zitex®	Norton Performance Plastics Wayne, NJ 07 470, US	135	4, 5 porous
Zyex®	ICI PLC Welwyn Garden City, Herts. AL7 1 HH, GB	2351	72
Zylar®	Novacor Chemicals Inc., Plastics Div. Leominster, MA 01 453, US	1641	21
Zypet	American Filtrona Corp. Richmond, VA 23 233, US	2412	5
Zytel®	DuPont Co. Inc. Wilmington, DE 19 898, US	261/341, 264	2, 224
Zytex®	Norton Performance Plastics Wayne, NJ 07 470, US	135	4, 5 porous

Subject Index

A

ABS copolymers 205
ABS pipe 305
ABS sheet 305
accumulator head 108
acetal homopolymers and copolymers 247, 321
acetone cyanohydrin 16
acetylene 16, 17
acrylates 16, 171, 240
acrylic ester rubbers 413
acrylic foil 316
acrylic sheet, applications 316
acrylic sheet, grades 319
acrylonitrile 16
acrylonitrile cop. 16, 245
adaptive clamping systems 85
addition polymerization 42
addition reaction 353
addition reactions of isocyanates 360
additives 6, 22, 24, 47
additives for polyurethane 359
adipic acid 17
aeronautical engineering 25
Aerosil 51
aerospace 22
AH salt 17
AKW process 292
alcoholisis 293
aliphatic diglycidyl ether 353
alkyd resins 17
alloys 5, 44
allyl ester 356
allylalcohol 17
alternating molds 108
alternating copolymer 4
ε-aminocaprolactam 17
aminoplastic molding compound 374
11-aminoundecanoic acid 17
ammonia 16, 17
amorphous 6
amorphous polyamides 259
amorphous polyarylates 277
analytical chemical methods 433
Andritz process 292
aniline resins 17
annealing 146
antistatic agents 48
aramide fibers 53
armatic-heterocyclic polymer systems, structures 401
aromatic polyesters 265

aromatic polymers 406
artificial horn 333
artificial leather 310
Astralon process 121
atactic arrangement 43
atactic polypropylene 168
autoclave process 68
automated injection molding plants 84
automated presses 87
automated SMC processing 89
automobile 22, 24
auxiliaries 47

B

back pressure 78, 115
benzene 17
benzidine 405
benzimidazol 405
benzylalcohol 16
benzyl cellulose 174
bi-functional polymer structure 4
biaxial orientation 139
biaxially stretched, OPP film 303
bifunctional 3
binders 341
biodegradable plastics 295
bismaleimide 405
bismaleimide-triazine resins 404
bisphenol A 353, 404
bisphenol A based resins 351
bisphenol-glycidyl-ether type 353
bitumen blends 169
blends 5
blister packaging 150
block-copolymer 4
block copolymerization 42
blow molding 72, 107
blowing agents 57, 396
blown films 135
blown film, biaxially oriented 308
blown film plant 136
BMC bulk-molding compound 93, 377
bonding of plastics 155
books 32
branched chains 4, 42
bristles 134
British Standards Institution 420
Brownian motion 9
Burger-Kelvin model 8
butanediol 17
butyl rubber 411
n-butylene 16

C

CAB sheet 326
cable covering 135
calender lines 142
calendering 141
Campus 45, 432
caprolactam 166
carbodiimid-diisocyanat 358
carbon ceramic 7
carbon dioxide 17
carbon fibers 14, 409
carbon monoxide 17
carbonization 409
carbonylchloride (phosgene) 17
carboxymethyl cellulose 175
cascade extruder 129
casein 16
casein plastics 333
cast PMMA 317
cast resin products 381
casting films 122
castor oil 17
catalysts 47
cellophane 331
cellular plastics 55, 400
celluloid 16, 326
Celluloid process 121
cellulose 16, 173
cellulose acetate 173, 287
cellulose acetobutyrate 16, 173, 287
cellulose esters 287, 326
cellulose-glycol ether 175
cellulose nitrate 16, 173
cellulose propionate 173, 287
cellulose triacetate 16
cellulosics, properties 289
cement concrete 344
cenospheres 52
centrifugal casting 63, 65
chain transmission 42
chalk 51
chemical resistance 451
chill-roll-process 138, 303
chinoline 405
chinoxaline 405
chlorinated polyethylene 167, 414
chlorinated rubber 172
chlorofluorinated alkanes 16
chloroprene polymers 413
chlorosilanes 17
chlorosulfonated polyethylene 414
chromosulfuric acid 189
CIM 82
Cinpress process 105

cis-polybutadiene 411
cis-polyisoprene 411
clamping force 77
clamping system 84
coal 16, 17
coal tar 17
coating 143, 330, 390
coating machine 143
coextrusion 137
cold-forming 146
cold compression molding 68
colorants 48
composite components 68
composites 6
compounding 54
compression molding 68, 71, 86
compression molds 95
computer aided data bases 431
computer aided design (CAD) 476
condensation polymerization 17, 42, 44
conducting additives 48
coordination catalysts 43
copolymerization 5, 50
copolymers 43
copolymers with vinyl alcohol 192
corrosion resistance 14
cotton linters 16
coupling agents 50
CP sheet 326
CPL sheets 390
cracked gas 16
creep curve 463
creep extension stress 466
creep modulus 467
creep rupture strength 465, 466
creep rupture test 462
cresol 333
cresol resins 17
cross-linked network 3
cross-linked polyethylene 190
cross-linking 4, 44, 405
cross-linking reaction resins 117
crystalline 6
curing agents 349
cutting 161
cyanamide 17
cycloaliphatic diglycidyl ester 354

D

data banks of material properties 45
data blocks (ISO) 423
decorating 162
decorative laminated materials 151
decorative laminates 387

Subject Index 637

decorative profiles 121
deep drawing, mechanical 147
deflashing 87
degassing 78
degradable plastics 295
degradation 1
design of plastics products 14
designation 423
designing plastics parts 72
Deutsches Institut für Normung 420
diallyl phthalate molding compounds 380
dian resins 17
diaphragm gate 94
dicyandiamide resins 17
dielectric curing 350
dielectric welding 154
diffusion bonding 156
diglycidyl amine derivatives 353
dioxidiphenyl propane 17
dip blow molding 110
diphenylmethane-diisocyanate 357
dispersion adhesives 158
dispersions 174
double-bond polymerization 42
double toggle 80
double-walled sheet 316
doubling 142
downstroke hydraulic presses 85
dry compounds 379

E

E/VA-copolymers 192
ejection of moldings 84
elastic modulus 13
elastomeric hot-casting resins 365
elastomeric materials 12
elastomers 2, 11, 411
electrical conductivity 24
electrical insulation 16
electrical insulation films 328
electroplating 164
electrostatic coating 61
embossing 146
emulsifying agents 47
emulsion PVC 211
encapsulating resins 365
engineering plastics 22, 25, 280, 321
enhancers 50, 51
entropy-elastic 12
epichlorohydrin 17, 353
epichlorohydrin rubber 414
epoxide chemistry 352
epoxy resins 17, 352
epoxy resin compound 379

epoxy(cresol) novolac type 353
EPS 209, 304
EPS boards and sheeting 305
ethyl benzene 16
ethyl cellulose 174, 288
ethylalcohol 16
ethylene 16, 17, 167
ethylene acrylic acid copolymer 194
ethylene chloride 16
ethylene-chlorotrifluoroethylene
 copolymer 235
ethylene copolymers 167, 191
ethylene cyanohydrin 16
ethylene ethyl acrylate copolymer 194
ethylene glycol 17
ethylene oxide 16
ethylene propylene polymer 183, 184
ethylene-propylene rubber 412
ethylene propylene vinyl ether 234
ethylene-tetrafluoroethylene copolymer 235
ethylene-vinylacetate copolymers 413
Euromap 77, 82, 107
expandable polystyrene see EPS
expanded plastics 55
extenders 50
extruder 122
extruder head with pipe die 133
extruder screws 127
extrusion 122, 139
extrusion plants 132
extrusion blow molding 107
extrusion of pipes 132
extrusion of profiles 133
extrusion of semifinished
 thermoplastics 122
extrusion of sheets 137
extrusion of webs 141

F

faults in injection moldings 98
feed hopper 123
fiber-reinforced composites 382
fiber-reinforced plastic sheets 383
fibers 140
fiber-spray process 67
fibrillated filaments 140
fibrous reinforcements 52
filaments 16, 51, 134, 140
filament textile glass products 53
filament winding 69
filled compounds 372
fillers 50
film gate 94
films 137, 141, 301, 306, 329

Subject Index

films made from solutions 141
fire performance 444
fish-tail die 139
flame retardants 47, 49, 362
flame-retarded building materials 305
flat filaments 303
flexibilizers 48
flexible decorative laminates 117
flexible foams 368
flexural strengh 348, 460
flock finishing 164
flotation process 292
flow behavior 47, 434
flow-casting process 79
flow index 47
flow rate 46
flow rate test, conditions 436
flow restriction bushes 124
fluorinated rubbers 414
fluoropolymers 234, 311
foam principles 396
foamed PE 190
foamed profiles 134
foams (summary) 395
formaldehyde 16, 17, 334
friction 453
friction welding 154
furane resins 341
furane resins, properties 342
furniture foils 308

G

gas-phase polymerization 42
gas transmission 457
gating systems 90
gel coat 67, 348
geopolymer composites 7
glass-cloth prepregs 378
glass fibers 50
glass fiber reinforced plastics 290, 381
glass mat reinforced thermoplastics (GMT) 290
glass transition 6
glass transition temperature 9, 11
glycerine 17
grafting 5
granulating 54
GRP-laminates 347
GRP sheets and sandwiches 393
GRP tanks 70

H

hand-lay-up process 67
hardening 115

hardness test 473
haze 437
headlamp glasses 25
heat distortion temperature 442
heated tool welding 152
heterocyclic aromatic polymers 406
hexamethylene diamine 17
hexamethylene diisocyanate 17
hexane-1,6-diisocyanate 358
high-frequency preheating 85
high-frequency welding 154
high-impact PVC 215
high-pressure PE (low density) 176
high-pressure molding techniques 71
high-pressure reaction-casting machine 63
high-temperature plastics 397, 407
high-voltage discharge plasma 189
holding pressure 78
hollow articles 60, 63
homopolymer 4
Hooke's law 7
hot gas welding 151
hot runner 90
HPL sheets 389
hybrid resins 351
hydantoin 405
hydraulic press 86
hydrogen bonding 252
hydrogenation 294
hydrolysis 293
hydroxy-ethyl cellulose 175
hydroxypropyl cellulose 175

I

IMD decoration 93
impact modifiers 50
impact resistance 470
incineration 294
induction welding 155
industrial laminates 151, 385
injection blow molding 110
injection molding 71
injection molding machines 76, 77
injection molding machines, fully-electric 83
injection-molding process 80
injection molds 90
injection stretch blow molding 111
injection unit 76
integral foam RIM-systems 363
interface layer binding 50
International Standardization Organization ISO 418
interpenetrating networks 6

Subject Index 639

intra-molecular combined polymers 5
ionomers 6, 194
irradiation 191
isobutyl alcohol 16
isobutylene 16
isochore 82
isochronous stress-strain curves 463
isocyanates 361
isocyanurate triisocyanate 358
isophorene-diisocyanate 358
isophthalic acid 17
isotactic arrangement 43
Izod impact test 470

J
joining 144
journals 28

L
lacquer resins 16
lacquering 162
ladder polymers 397
laminated sheets 393
laminates 330, 384
laminating 117, 143, 150
lauroyl lactam 166
LCP 279, 402
light curing resins 350, 378
light resistance 449
linear chains 4
linear macromolecules 42
linear polyarylene ethers 402
linear polyarylenes 403
linear polyesters 17
literature on plastics technology 27
living polymers 42
locking unit 76
long-term behavior 466
low-alkali glass 52
low-pressure PE (high density) 176
low-pressure molding 60
low-pressure processes 65
lubricants 47, 48

M
machining 161
Maillefer screw 129
maleic acid anhydride 17
manufacturing costs 75
Martens test 443
mass polymerization 42, 211
mass PVC 211
masterbatches 49
mastication 411

MDI 357
mechanical loss factor 11
mechanical properties 458
mechanical separation 292
mechanical testing standards 459
medical use 26
medium-pressure PE (high density) 176
melamine 17
melamine-formaldehyde resins 339
melamine-phenolic resin molding
   compounds 375
melamine-polyester molding
   compounds 375
melamine resin foam 341
melamine resin molding compounds 374
melamine resins, application 339
melt filtration 293
melt flow index 8, 46
melt volume index 46
melting temperatures 6
metal inserts 73
metallizing 163
methacrylates 165
methacrylic esters 16
methanol 16, 17
methyl-2-cyano acrylate 167
methyl cellulose 175
methylmethacrylate 240, 314
4-methylpentene-1 16
mica flakes 50
microelectronic components 24
microspheres 52
migration 456
mineral fillers 51
mixing 54, 128
MMA molding compounds 245
modified polyphenylene oxide (PPO) 322
molding compounds 54, 69
molding techniques for thermoplastic
   materials 95
molding techniques for thermosets and
   elastomers 114
molding tools 89
molecular weight 3, 213
molecular-weight distribution 45
monomers 1
mounting platen 79
moving platen 79
multi-channel die 140
multi-component casting 62
multi-component injection molding 104
multi-component spraying 63
multifunctional 15
multifunctional resins 355

Subject Index

multi-layer 139
multi-layered webs 142
multi-layer film 329
multi-screw extruders 129
multi-section screws 127

N

naphthylene-diisocyanate 357
national standards institutes 418
natural gas 16
natural products 16, 172
natural rubber 410
Neopolen process 190
nets 134
nitrile rubber 413
nitrobenzene 17
nitrocellulose 16
non-return valve 79
no-pressure molding 60
Novolak 334
nucleating agents 58
number average molecular weight 45
nylon see polyamides
Nyrim process 166

O

oil 16
olefin polymers 167
olefinic thermoplastic elastomers 196
one-stage blow molding 107
open casting 61
open injection nozzles 79
optical properties 437
orientation 12
oxidative coupling 44
oxidizing flame 189

P

PA see polyamides
packaging films 327
packaging materials 329
packaging sheeting 307
paraban acid 405
partially crystalline polymers 12
particle foam 210
particle structure of PVC powder 212
parting agents 48
PC + ABS blends 269
PC + PBT blends 269
PE see polyethylene
pendulum hammer equipment 472
perfluoro-alkoxy copolymer 234
permeability coefficient 456
permeation 456

PET see polyethylene terephthalate
phenol 333
phenol-aralykyl resins 335
phenolic molding compounds 370
phenolic resin laminated paper 384
phenolic resins for corrosion protection 374
phenolic resins, applications 337
photodegradable plastics 295
photographic film 16
phthalic acid anhydride 17
physical properties 438
pin gate 93
pipe joints 298
piping material 296
planetary gear extruder 129
plasticized PVC 308
plasticizers 50
plasticizing 122
plasticsols 232
PMC powder mold coating 93
PMMA + ABS blends 245
PMMI 247
pneumatic forming 148
polishing 162
polyacetylene 24
polyacrylate resins 16, 171
polyacrylonitrile 402
polyaddition 15, 44
polyalkylene terephthalates 272
polyamide copolymers 172, 260
polyamides (PA, nylons) 17, 251, 321
polyamide 6 252
polyamide 66 252
polyamide 11, 12 260
polyamides, amorphous 259
polyamide elastomers 260
polyamine products 172
polyaniline 24
polyaryl-ether-sulfide 17
polyarylates (aromatic polyesters, APE) 321
polyarylates (LCP), self-reinforced 278
polyarylates 402
polyarylene amides 403
polyarylene ethers 278, 281, 282, 403
polyarylene ether ketones 278
polyarylene ketones 281
polyarylene polyether ketones 282
polyarylene sulfides 278, 281, 402
polyarylene sulfones 278, 281, 402
polyblends 44
polybutene-1 16, 201
polybutylene terephthalate 272
polycaprolactones 359

polycarbonate 17, 266, 271, 323
polycarbonate blends 269
polychlorotrifluoroethylene 235
polycondensation 15
polyester imide 405
polyester polyols 359
polyester resin compounds 375
polyester resin mats 378
polyester resins 17
polyether-block amides 260, 277
polyether polyols 359
polyethylene 176
polyethylene, chlorinated 191
polyethylene, crosslinked 190
polyethylene, foam 396
polyethylene, high density 176
polyethylene, linear low density 177
polyethylene, low density 176
polyethylene terephthalate 273, 324
polyethylene oxide 175
polyfluorocarbons and copolymers 311
polyimides 284, 404
polyisobutylene 16, 168
polyisocyanates 357
polyisocyanurates 362
polymer concrete 344
polymer flow 7
polymer syntheses 42
polymerization 42, 45
polymethacrylates 16, 240
polymethacrylate molding compound 242
polymethacrylate semi-finished products 315
polymethacrylate sheet 314
polymethacrylimide compounds 246
polymethacrylimide foam 319
polymethylmethacrylate 240
polymethylmethacrylate semi-finished products 315
polymethylmethacrylate sheet 314
polymethylpentene 16, 202
polyolefins 176
polyoxymethylene 16, 247
polyphenylene sulfide 282, 322
polypropylene 16, 196, 303
polypyrrole 24
polystyrene 16, 205, 304
polysulfide rubber 413
polysulfones 322
polytetrafluoroethylene 16, 234, 312
polyurethanes 17, 356
polyurethane chemistry 356
polyurethane flexible foam 368
polyurethane rigid foam 367

polyurethane RIM systems 363
polyurethane structural foam 364
polyurethane thermoplastic elastomers 262
polyvinyl acetals 171
polyvinyl acetate 16, 170
polyvinyl alcohol 16, 170, 175
polyvinyl copolymers 170
polyvinyl ethers 16, 171
polyvinyl methyl ether 175
polyvinyl pyrrolidone 175
polyvinylchloride see PVC
polyvinylfluoride 235
polyvinylidene chloride 16, 215
polyvinylidene fluoride 16, 235
pore size regulators 58
post-shrinkage 75
powder coating 54, 61
pre-pelleting 85
precipitation polymerization 42
precision instruments 25
preform thickness 110
prepolymer binder resins 367
prepregs 69, 380
presses 85
printing 162
processing data 435
processing temperature ranges 97
propeller shafts 25
propyl vinyl ether copolymer 234
propylene 16, 17
prosthetics 26
PU-PIR foams 367
pultruded tubes and profiles 382
pultrusion 121
punching 161
PVC blends 215
PVC coated metal sheets 311
PVC compounds 214
PVC copolymers 215
PVC flooring, wall and table coverings 310
PVC foams 311
PVC homopolymers 215
PVC-P compounds 226
PVC pipes 299, 306
PVC pastes 232
PVC plasticizers 226
PVC profiles 309
PVC stabilizer systems and fillers 218
PVC tubes 309
PVC-U compounds 219
pvt diagram 82
pyrolysis 293
pyromellitic acid anhydride 405

Q

quartz sand 17
quick-change systems 84

R

radiation resistance 453
ram extrusion 123
random-copolymer 4
reactive-resin castings 62, 345
reactive-resin moldings, indicatives values 377
reactive resins for concrete 343
reactive resins 341
rearrangement polymerization 44
reciprocating screw 77
reclaiming 291
recycling 291
redispersible powders 175
regenerated cellulose 16, 331
reinforced prepregs 372
reinforcements 51
reprocessing 291
resin transfer molding process 68
resistance to chemicals 450
resistance to environmental effects 448
resistance to fungi, bacteria and animals 449
resistance to high-energy radiation 453, 455
resistance to mechanical wear, friction and sliding 453
resit 336
resitol 336
resols 335
rigid foams 398
RIM moldings 363
RIM reaction injection molding 62
Rockwell hardness 473
Rohacell 319
roll mill pressing 146
rotational molding 63
rotor blades 25
rovings 69
RRIM reinforced reaction injection molding 62
rubber-like state 12
rubber hydrochloride 173
rubber latex 16
rubbers 16, 411

S

sandwich injection molding 104
Schramm–Zebrowski incandescent rod resistance method 446
screw cylinders 78
screw joints 159
screw types 128
sebacid acid 17
secondary plasticizers 226
self-degating injection systems 84
self-reinforced 189
self tapping screws 159
semi-finished acrylics, application 320
semi-flexible foams 363
semi-flexible integral foams 365
semi-rigid foams 399
semi finished thermoplastics 296
semicrystalline thermoplasts 11
semifinished thermosets 117, 381
separating 161
shear modulus and mechanical loss factor of thermoplastics 441
shear modulus temperature curves 185
sheet extrusion 137
sheet molding compounds 369
Shore A and D hardness tests 473
short-term compression test 461
short-term tests 460
shot speed 78
shrinkage 74, 146, 351
side-feed extruder head 125
silanes 415
silicone oils 416
silicone pastes 417
silicone molding compounds 381
silicone rubbers 414
silicones 17, 415
single-screw extruder 125
single-stage-stretching 114
single-value tests 443
sintering 61
skin packaging 150
slabstock-foam 368
slot die extrusion 138
snap-in joints 159
solution polymerization 42
specific heat 440
specific viscosity 7, 46
spheres 51
spiral flow test 47
spraying machines 62
spreading 143
sprue gate 93
sprueless hot runners 84
stabilizers 47, 48
standardized test methods 422
stick-slip effect 453
strainer plates 124
stress-strain properties 460
stress-time curves 469

stress cracking 315
stretch blow molding 113
stretch films 139
stretching 146
structural foam 105
styrene-based terpolymers and alloys 206, 271
styrene-butadiene rubber 411
styrene copolymers 169
styrene polymers 202, 304
styrene 16, 165
Styropor 210
surface protection resins 355
surface treatment 162
suspension polymerization 43, 211
suspension PVC 211
swelling 75
syndiotactic arrangement 43
synthetic fibers 16, 52
synthetic rubber 410

T

tacticity 43
talc 50, 51
tar 17
TDI 357
teleblock copolymers 203
temperature-resistant thermoplastics 58, 392
temperature-time limits 400, 443
temperature indices 445
tensile strength 13, 462
terephthalic acid 17
tests and their significance 430
tests for properties 437
tetrafluoroethylene-perfluoropropylene copolymer 234
thermal analysis 443
thermal conductivity 14
thermal expansion 14
thermal resitance 22
thermal stability 24
thermodiffusion printing 163
thermoelasticity 146
thermofixing 149
thermoforming 145
thermogram 401
thermogravimetric analysis 397
thermoplastic composites 290
thermoplastic compounds 54
thermoplastic elastomers 2, 6, 17, 196, 260, 262
thermoplastic materials, designatory properties 427

thermoplastic piping systems 297
thermoplastic poly(ether) ester elastomers 275
thermoplastic polyesters 272, 323
thermoplastic polyimide compounds 286
thermoplastic polyurethane elastomers 262
thermoplastic polyurethanes 251, 321
thermoplastic semi-polyimides 402
thermoplastics 1, 9, 165, 296
thermosets 1, 9, 55
thermosetting molding compounds 55, 369
thermosetting plastics, semi-finished products 333
three-layer coextrusion 109
three-plate mold 95
three-stage screws 78
TMC 377
tolerances 74
tolylene-diisocyanate 357
tool resins 355
tools 65
torsional pendulum test 6, 11, 439
trade names index 479
transfer-injection stretch molding 106
transfer molding 71, 86
transfer molding tools 93
trifunctional linkages 3
triallyl cyanurate 167
trichloroethane 16
trimellitic acid anhydride 405
trioxane 16
triple-walled sheet 316
tunnel gate 94
twin screw extruder 131
two-dimensional stretching of films 145
two-stage stretch molding 113
two-stage stretching 140

U

ultra high molecular weight polyethylenes 189
ultrasonic welding 153
umbrella gate 94
Underwriters Laboratory safety standard 446
unsaturated polyesters 17, 348
upstroke presses 85
urazol 405
urea foam 338
urea resins 17, 338
urea resin molding compounds 374
urethane rubbers 413

V

vacuum forming 148
vacuum injection molding process 68
vented extruder 128
vibration welding 154
Vicat softening temperature (VST) 443
vinyl acetate copolymers 170
vinyl acetate 16
vinyl carbazole 165
vinyl chloride copolymers 170, 215
vinyl chloride polymers 211, 306
vinyl chloride 16
vinyl ethers 16
vinylester resins 351
vinylidene chloride 16
vinylidene fluoride terpolymers 234
vinylidene fluoride 16
viscoelasticity 8
viscosity number 46, 47
visible light curing of GF-UP 350
vulcanization 115, 411
vulcanized fiber 151, 331
Vulkollan 365

W

wall tapering 72
wall thickness regulation 108, 109
water-soluble polymers 175
water absorption 448
water vapor transmission 458
wear 453
webs 309
weight average molecular weight M_w 45
welding 151
whiskers 50
winding machines 69
wire covering 135
wollastonite 50
wood 16
wood based material sheet 390
wood-chip molding materials 372
wood flour 51
wound fiber reinforced profiles 384

X

xylene 16
xylenol resins 333

Z

Ziegler-Natta catalysts 43
zinc alloys 89

Buyer's Guide

Delivery and service offers for all plastics and plastics-related branches and groups. The companies are liable for the costs of their listing. This register does not claim to be complete.

Group 501: Raw Material for Plastics/Rubbers 1–3
Group 502: Chemical agents, auxiliaries and additives 4–14
Group 504: Plastics processing Services – Job processing 15
Group 505: Semifinished plastics products 16–18
Group 506: Finished plastics products 19–22
Group 507: Machinery, appliances, moulds and tools for plastic processing 23–33
Group 508: Auxiliary agents for plastics processing 34
Group 509: Analysis, measurement and testing equipment, incl. acce 35–36

The editor cannot be held accountable. Further details from
Hanser Publishers, PO Box 860420, 81631 Munich, Germany

Group 501: Raw Material for Plastics/Rubbers

Aminoplast moulding compounds

Perstorp Chemitec
S-28480 Perstorp, Sweden
Tel. +46/435/38000, Fax /38805

Cresylic resins

Raschig AG
D-67061 Ludwigshafen
Tel.+49(621)56180 / Fax.582885

Furane resins

Raschig AG
D-67061 Ludwigshafen
Tel.+49(621)56180 / Fax.582885

Glass-reinforced thermoplastics

Elastogran GmbH
GB Halbzeug, Bauteile (H)
PF 1140, 49440 Lemförde
Tel. 05443/12-0, Fax/12-2370

Granulates

Elastogran GmbH
GB Elastomere (E)
PF 1140, 49440 Lemförde
Tel. 05443/12-0, Fax/12-2555

Isocyanates

Elastogran GmbH
GB Grundprodukte (G)
BASF Polyurethane
67056 Ludwigshafen
Tel. 0621/60-0, Fax/60-43960

Melamine resin moulding compounds

Perstorp Chemitec
S-28480 Perstorp, Sweden
Tel. +46/435/38000, Fax /38805

Raschig AG
D-67061 Ludwigshafen
Tel.+49(621)56180 / Fax.582885

Phenolic resin moulding compounds

Perstorp Chemitec
S-28480 Perstorp, Sweden
Tel. +46/435/38000, Fax /38805

Raschig AG
D-67061 Ludwigshafen
Tel.+49(621)56180 / Fax.582885

Phenolic resins

Raschig AG
D-67061 Ludwigshafen
Tel.+49(621)56180 / Fax.582885

Polphthalamides

Nyltech Deutschland GmbH
Postfach 5786
79025 Freiburg
Tel. 0761/511-0 Fax. /3677

Polyacetals (POM)

Nyltech Deutschland GmbH
Postfach 5786
79025 Freiburg
Tel. 0761/511-0 Fax. /3677

Polyamide copolymers

Nyltech Deutschland GmbH
Postfach 5786
79025 Freiburg
Tel. 0761/511-0 Fax. /3677

Polyamides

Nyltech Deutschland GmbH
Postfach 5786
79025 Freiburg
Tel. 0761/511-0 Fax. /3677

POLYMER UND FILAMENT GmbH
Breitscheidstraße 103
D-07407 Rudolstadt
Tel:03672/52461 Fax/52282

Polyester (unsaturated)

Raschig AG
D-67061 Ludwigshafen
Tel.+49(621)56180 / Fax.582885

Polyester polyols

Elastogran GmbH
GB Grundprodukte (G)
BASF Polyurethane
67056 Ludwigshafen
Tel. 0621/60-0, Fax/60-43960

Polyester resin moulding compounds

Raschig AG
D-67061 Ludwigshafen
Tel.+49(621)56180 / Fax.582885

Polyether polyols

Elastogran GmbH
GB Grundprodukte (G)
BASF Polyurethane
67056 Ludwigshafen
Tel. 0621/60-0, Fax/60-43960

Polyethylene terphthalate

POLYMER UND FILAMENT GmbH
Breitscheidstraße 103
D-07407 Rudolstadt
Tel:03672/52126 Fax/52279

Polyurethan systems

Elastogran GmbH
GB Spezialsysteme (P)
Industriestr. 82140 Olching
Tel. 08142/178-0, Fax/178-213

Elastogran GmbH
GB Hartschaumsysteme (R)
PF 1140, 49440 Lemförde
Tel. 05443/12-0, Fax/12-2474

PUR elastomers

Elastogran GmbH
GB Elastomere (E)
PF 1140, 49440 Lemförde
Tel. 05443/12-0, Fax/12-2555

PUR, thermoplastic

Elastogran GmbH
GB Elastomere (E)
PF 1140, 49440 Lemförde
Tel. 05443/12-0, Fax/12-2555

PUR starting material

Elastogran GmbH
GB Grundprodukte (G)
BASF Polyurethane
67056 Ludwigshafen
Tel. 0621/60-0, Fax/60-43960

Group 502: Chemical agents, auxiliaries and additives

Additive concentrates

Hoechst Aktiengesellschaft
Business Unit Additive
Postfach 10 15 67
D-86005 Augsburg
Tel.: 0821/479-0
Fax : 0821/496639

Additives

Bärlocher GmbH
Riesstr. 16, 80992 München
Tel. 089/14373-0 FS 5215773
Fax: 089/14373-312

Byk-Chemie GmbH
Abelstr. 14, D-46483 Wesel
Tel. 0281/670-0 Fax: /65735

ciba
ciba Additive GmbH
Chemiestr., 68623 Lampertheim
Tel.: 06206/502-0, Fax /502-368

Hoechst Aktiengesellschaft
Business Unit Additive
Postfach 10 15 67
D-86005 Augsburg
Tel.: 0821/479-0
Fax : 0821/496639

Nemitz, D-48341 Altenberge
Tel. 02505/674, Fax /3042

E. & P. Würtz GmbH & Co.
55411 Bingen-Sponsheim
Tel. 06721/96900-Fax 41589

Adhesion promoters

Hoechst Aktiengesellschaft
Business Unit Additive
Postfach 10 15 67
D-86005 Augsburg
Tel.: 0821/479-0
Fax : 0821/496639

Aluminium hydroxide

Martinswerk GmbH
Postf. 1209, D-50102 Bergheim
Tel. 02271/9020, Fax /902555

Antiblocking agent

MECO GmbH, Postfach 224
D-78473 Allensbach/Germany
Tel. 07533/1611, Fax./4435

E. & P. Würtz GmbH & Co.
55411 Bingen-Sponsheim
Tel. 06721/96900-Fax 41589

Antioxidants

ciba
ciba Additive GmbH
Chemiestr., 68623 Lampertheim
Tel.: 06206/502-0, Fax /502-368

HASCO® Mouldmaking-Technology for the Plastics Processing Industry

HASCO-Normalien · D-58505 Lüdenscheid
Tel. (02351) 9570 · Fax (02351) 957237

IHNE&TESCH: Plastic Industry's supplier for engineered Electric Heaters

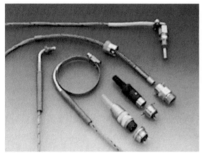
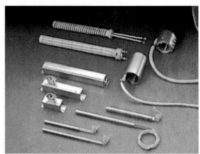

Our products of high quality are designed to meet even tomorrow's demands of Manufacturers of Original Equipment and their Customers.

Your choice of our Products:

- BAND HEATERS
- NOZZLE HEATERS
- FRAME HEATERS
- PLATEN HEATERS
- CATRIDGE HEATERS
- CAST-IN HEATERS
- TUBULAR HEATERS
- PREHEATERS AND DRYERS
- ELETRICAL CONTROL GEAR
- TEMPERATURE SENSORS AND ACCESSORY
- TERMINAL BOXES AND CONNECTORS
- HEAT PROTECTION COVERS

ELEKTRO - WÄRMETECHNIK

Ihne & Tesch GmbH
Postfach 1863
Am Drostenstück 18
Tel. 0 23 51 / 6 66-0

D-58468 Lüdenscheid
D-58507 Lüdenscheid
Fax 0 23 51 / 6 66-24

Hoechst Aktiengesellschaft
Business Unit Additive
Postfach 10 15 67
D-86005 Augsburg
Tel.: 0821/479-0
Fax : 0821/496639

Nemitz, D-48341 Altenberge
Tel. 02505/674, Fax /3042

Raschig AG
D-67061 Ludwigshafen
Tel.+49(621)56180 / Fax.582885

Antistats

Dr. Th. Böhme KG, Chem. Fabrik
Isardamm 79, 82538 Geretsried
Tel.: 08171/628-0, Tlx. 526312
Telefax: 08171/628-388

ciba
ciba Additive GmbH
Chemiestr., 68623 Lampertheim
Tel.: 06206/502-0, Fax /502-368

Hoechst Aktiengesellschaft
Business Unit Additive
Postfach 10 15 67
D-86005 Augsburg
Tel.: 0821/479-0
Fax : 0821/496639

MECO GmbH, Postfach 224
D-78473 Allensbach/Germany
Tel. 07533/1611, Fax./4435

Nemitz, D-48341 Altenberge
Tel. 02505/674, Fax /3042

E. & P. Würtz GmbH & Co.
55411 Bingen-Sponsheim
Tel. 06721/96900-Fax 41589

Barytes, very finely ground

Seitz + Kerler, 97816 Lohr

Calcium carbonates

Plüss-Staufer AG
CH-4665 Oftringen
Telefon ++41 99 11 11
Fax ++41 99 20 77
neu ab 5.11.95:
Telefon ++41 789 29 29
Fax ++41 789 20 77

Catalysts

Elastogran GmbH
GB Grundprodukte (G)
BASF Polyurethane
67056 Ludwigshafen
Tel. 0621/60-0, Fax/60-43960

Chemical products for the paint industry

ciba
ciba Additive GmbH
Chemiestr., 68623 Lampertheim
Tel.: 06206/502-0, Fax /502-368

Chemical products for the plastic industry

ciba
ciba Additive GmbH
Chemiestr., 68623 Lampertheim
Tel.: 06206/502-0, Fax /502-368

Chlorinated paraffins

Hoechst Aktiengesellschaft
Business Unit Additive
Postfach 10 15 67
D-86005 Augsburg
Tel.: 0821/479-0
Fax : 0821/496639

Coating resins

Wacker-Chemie GmbH
Hanns-Seidel-Platz 4
81737 München

Colour concentrates

Georg Deifel KG, Farbenfabrik
POB 4066, 97408 Schweinfurt

KARL FINKE GmbH & Co. KG
Hatzfelder Str. 174
42281 Wuppertal
Tel. (0202) 70906-0
Telefax (0202) 703929

Colour pastes

KARL FINKE GmbH & Co. KG
Hatzfelder Str. 174
42281 Wuppertal
Tel. (0202) 70906-0
Telefax (0202) 703929

Colourants and pigments

Georg Deifel KG, Farbenfabrik
POB 4066, 97408 Schweinfurt

Conductivity improvers

Potters-Ballotini GmbH
Morschheimerstr. 11, Pf. 1226
D-67292 Kirchheimbolanden
T.06352/8484 Fax 06352/1853

Deactivators for metal

Hoechst Aktiengesellschaft
Business Unit Additive
Postfach 10 15 67
D-86005 Augsburg
Tel.: 0821/479-0
Fax : 0821/496639

Deaerating agents

Byk-Chemie GmbH
Abelstr. 14, D-46483 Wesel
Tel. 0281/670-0 Fax: /65735

Dicyandiamide

SKW TROSTBERG AG
P.O.Box 1262
D-83303 Trostberg
Phone: (0)8621/86-0

Dispersants

Dr. Th. Böhme KG, Chem. Fabrik
Isardamm 79, 82538 Geretsried
Tel.: 08171/628-0, Tlx. 526312
Telefax: 08171/628-388

Byk-Chemie GmbH
Abelstr. 14, D-46483 Wesel
Tel. 0281/670-0 Fax: /65735

Dispersion

Wacker-Chemie GmbH
Hanns-Seidel-Platz 4
81737 München

Elastomers, thermoplastic

Gummiwerk Kraiburg GmbH & Co.
PF. 1160, D-84464 Waldkraiburg
Tel. (08638)610, Fax /61-310

Deutsche Shell Chemie GmbH,
65760 Eschborn,Tel. 06196/474-0

Epoxidized oils

ciba
ciba Additive GmbH
Chemiestr., 68623 Lampertheim
Tel.: 06206/502-0, Fax /502-368

Extenders

Plüss-Staufer AG
CH-4665 Oftringen
Telefon ++41 99 11 11
Fax ++41 99 20 77
neu ab 5.11.95:
Telefon ++41 789 29 29
Fax ++41 789 20 77

Fillers

C.F.F. D-41065 Mönchengladbach
Fax 02161-656.200 Tx 852102
TECHNOCEL Cellulose Füllst.

Chemie-Mineralien AG & Co. KG
Pf. 106523, 28065 Bremen

Gebr. Dorfner GmbH & Co. KG
Scharhof, D-92242 Hirschau
Tel. ++49/(0)9622/82-0
Telefax ++49/(0)9622/82-69

Martinswerk GmbH
Postf. 1209, D-50102 Bergheim
Tel. 02271/9020, Fax /902555

Naintsch Mineralwerke GmbH
A-8045 Graz, Postfach 35
Statteggerstr. 60
Tel. 0316/693650, Tlx. 311223
Fax. 0316/693655

Plüss-Staufer AG
CH-4665 Oftringen
Telefon ++41 99 11 11
Fax ++41 99 20 77
neu ab 5.11.95:
Telefon ++41 789 29 29
Fax ++41 789 20 77

Fillers, aluminium hydroxide as

Martinswerk GmbH
Postf. 1209, D-50102 Bergheim
Tel. 02271/9020, Fax /902555

Fillers, fumed silica as

Wacker-Chemie GmbH
Hanns-Seidel-Platz 4
81737 München

Fillers, Kaolin as

Chemie-Mineralien AG & Co. KG
Pf. 106523, 28065 Bremen

Fillers, lightweight

Expancel, Fax +46 60 56 95 18
Box 13000 Sundsvall Sweden

Fillers, marbles as

Potters-Ballotini GmbH
Morschheimerstr. 11, Pf. 1226
D-67292 Kirchheimbolanden
T.06352/8484 Fax 06352/1853

Fillers, organic

Jelu-Werk, D-73494 Rosenberg
Tel. 07967/411, Fax /8525
Füllst. Holz-u. Cellulosebasis

Fillers, silicates as

Hoffmann Mineral
Franz Hoffmann & Söhne KG
Münchener Str. 75
86633 Neuburg (Donau)
Telefon (08431) 53-0
Telefax (08431) 53-330
Sillitin,Sillikollid,Silfin,
Aktisil

Flame retardants, general

ALCOA Moerdijk B.V.
Vlasweg 19, P.O. Box 16
NL-4780 AA Moerdijk
Tel. +31(0)1680-24865
Fax +31(0)1680-24565
Aluminium Hydroxide

Dr. Th. Böhme KG, Chem. Fabrik
Isardamm 79, 82538 Geretsried
Tel.: 08171/628-0, Tlx. 526312
Telefax: 08171/628-388

C.F.F. D-41065 Mönchengladbach
Fax 02161-656.200 Tx 852102
TECHNOCEL Cellulose Füllst.

Hoechst Aktiengesellschaft
Business Unit Additive
Postfach 10 15 67
D-86005 Augsburg
Tel.: 0821/479-0
Fax : 0821/496639

Martinswerk GmbH
Postf. 1209, D-50102 Bergheim
Tel. 02271/9020, Fax /902555

Glass fibre products

Schuller GmbH
Faserweg 1
97877 Wertheim
Tel. 09342/801-0, Fax /801-140

Hardeners for epoxy resins

Deutsche Shell Chemie GmbH,
65760 Eschborn,Tel. 06196/474-0

SKW TROSTBERG AG
P.O.Box 1262
D-83303 Trostberg
Phone: (0)8621/86-0

Heat stabilizers

Rohm and Haas Deutschland GmbH
In der Kron 4
60489 Frankfurt
Tel. 069/78996-0 Fax 7895356

Impact modifiers for PVC

Rohm and Haas Deutschland GmbH
In der Kron 4
60489 Frankfurt
Tel. 069/78996-0 Fax 7895356

Kaolin

Chemie-Mineralien AG & Co. KG
Pf. 106523, 28065 Bremen

Light spar, very finely ground

Seitz + Kerler, 97816 Lohr

Light stabilizers

Hoechst Aktiengesellschaft
Business Unit Additive
Postfach 10 15 67
D-86005 Augsburg
Tel.: 0821/479-0
Fax : 0821/496639

Lubricants

Bärlocher GmbH
Riesstr. 16, 80992 München
Tel. 089/14373-0 FS 5215773
Fax: 089/14373-312

Dr. Th. Böhme KG, Chem. Fabrik
Isardamm 79, 82538 Geretsried
Tel.: 08171/628-0, Tlx. 526312
Telefax: 08171/628-388

Hoechst Aktiengesellschaft
Business Unit Additive
Postfach 10 15 67
D-86005 Augsburg
Tel.: 0821/479-0
Fax : 0821/496639

Nemitz, D-48341 Altenberge
Tel. 02505/674, Fax /3042

E. & P. Würtz GmbH & Co.
55411 Bingen-Sponsheim
Tel. 06721/96900-Fax 41589

Marbles

Potters-Ballotini GmbH
Morschheimerstr. 11, Pf. 1226
D-67292 Kirchheimbolanden
T.06352/8484 Fax 06352/1853

Master batches

Color Service GmbH
63512 Hainburg, Telex 4184569
Offenbacher Landstr. 107-109
Tel. 06182-4034-37, Fax -66886
Spez. Universal Masterbatch

KARL FINKE GmbH & Co. KG
Hatzfelder Str. 174
42281 Wuppertal
Tel. (0202) 70906-0
Telefax (0202) 703929

Nemitz, D-48341 Altenberge
Tel. 02505/674, Fax /3042

Plüss-Staufer AG
CH-4665 Oftringen
Telefon ++41 99 11 11
Fax ++41 99 20 77
neu ab 5.11.95:
Telefon ++41 789 29 29
Fax ++41 789 20 77

Matting agents

Rohm and Haas Deutschland GmbH
In der Kron 4
60489 Frankfurt
Tel. 069/78996-0 Fax 7895356

Mica

Chemie-Mineralien AG & Co. KG
Pf. 106523, 28065 Bremen

Montanic acid ester waxes

Hoechst Aktiengesellschaft
Business Unit Additive
Postfach 10 15 67
D-86005 Augsburg
Tel.: 0821/479-0
Fax : 0821/496639

Nucleation agents

Hoechst Aktiengesellschaft
Business Unit Additive
Postfach 10 15 67
D-86005 Augsburg
Tel.: 0821/479-0
Fax : 0821/496639

Naintsch Mineralwerke GmbH
A-8045 Graz, Postfach 35
Statteggerstr. 60
Tel. 0316/693650, Tlx. 311223
Fax. 0316/693655

Organic phosphites

Hoechst Aktiengesellschaft
Business Unit Additive
Postfach 10 15 67
D-86005 Augsburg
Tel.: 0821/479-0
Fax : 0821/496639

Pastes, pigments

ARICHEMIE GMBH
Valterweg 21/22
D-65817 Eppstein
Tel.: 06198-59120
Fax : 06198-32527
Heliocolor Feinpulver
Vocaplast Feinteige
Vulcogran Gummifarben

Pigment pastes

ARICHEMIE GMBH
Valterweg 21/22
D-65817 Eppstein
Tel.: 06198-59120
Fax : 06198-32527
Heliocolor Feinpulver
Vocaplast Feinteige
Vulcogran Gummifarben

Pigments

KARL FINKE GmbH & Co. KG
Hatzfelder Str. 174
42281 Wuppertal
Tel. (0202) 70906-0
Telefax (0202) 703929

Pigments, pearly lustre

E. Merck, Postfach 4119
D-64271 Darmstadt
Tel. 06151/72-0, Tx 419328-0
Fax: 06151/727684
IRIODIN

Pigments, titanium dioxide as

Kronos International, Inc.
PF. 100720, 51307 Leverkusen
Tel. 0214/356-0, Tx 8510823
Telefax 0214/42150

Plasticizers

Dr. Th. Böhme KG, Chem. Fabrik
Isardamm 79, 82538 Geretsried
Tel.: 08171/628-0, Tlx. 526312
Telefax: 08171/628-388

Plastics auxiliaries

ciba
ciba Additive GmbH
Chemiestr., 68623 Lampertheim
Tel.: 06206/502-0, Fax /502-368

Polyacrylates and Copolymers

Dr. Th. Böhme KG, Chem. Fabrik
Isardamm 79, 82538 Geretsried
Tel.: 08171/628-0, Tlx. 526312
Telefax: 08171/628-388

Polybutene

Deutsche Shell Chemie GmbH,
65760 Eschborn, Tel. 06196/474-0

Polyisocyanates

Deutsche Shell Chemie GmbH,
65760 Eschborn, Tel. 06196/474-0

Polyols

Deutsche Shell Chemie GmbH,
65760 Eschborn, Tel. 06196/474-0

Polystyrene, expandable

Deutsche Shell Chemie GmbH,
65760 Eschborn, Tel. 06196/474-0

Polyurethane systems

Deutsche Shell Chemie GmbH,
65760 Eschborn, Tel. 06196/474-0

Polyvinylacetate

Wacker-Chemie GmbH
Hanns-Seidel-Platz 4
81737 München

Polyvinylbutyral

Wacker-Chemie GmbH
Hanns-Seidel-Platz 4
81737 München

Processing agents

Henkel KGaA
COK Plastics Technology
D-40191 Düsseldorf
Tel.:0211/7977414, Fax:/7989638

Rohm and Haas Deutschland GmbH
In der Kron 4
60489 Frankfurt
Tel. 069/78996-0 Fax 7895356

Propellants/blowing agents

Expancel, Fax +46 60 56 95 18
Box 13000 Sundsvall Sweden

Hoechst Aktiengesellschaft
Business Unit Additive
Postfach 10 15 67
D-86005 Augsburg
Tel.: 0821/479-0
Fax : 0821/496639

PVC additives

Dr. Th. Böhme KG, Chem. Fabrik
Isardamm 79, 82538 Geretsried
Tel.: 08171/628-0, Tlx. 526312
Telefax: 08171/628-388

Henkel KGaA
COK Plastics Technology
D-40191 Düsseldorf
Tel.:0211/7977414, Fax:/7989638

Hoechst Aktiengesellschaft
Business Unit Additive
Postfach 10 15 67
D-86005 Augsburg
Tel.: 0821/479-0
Fax : 0821/496639

PVC copolymers

Wacker-Chemie GmbH
Hanns-Seidel-Platz 4
81737 München

PVC, unplasticized

Deutsche Shell Chemie GmbH,
65760 Eschborn,Tel. 06196/474-0

Reactive thinner

Raschig AG
D-67061 Ludwigshafen
Tel.+49(621)56180 / Fax.582885

Reinforcing agents

C.F.F. D-41065 Mönchengladbach
Fax 02161-656.200 Tx 852102
TECHNOCEL Cellulose Füllst.

Resins, epoxy

Deutsche Shell Chemie GmbH,
65760 Eschborn,Tel. 06196/474-0

Resins, silicone

Wacker-Chemie GmbH
Hanns-Seidel-Platz 4
81737 München

Rubber, silicone

Wacker-Chemie GmbH
Hanns-Seidel-Platz 4
81737 München

Rubber, styrene-butadiene (SBR)

Deutsche Shell Chemie GmbH,
65760 Eschborn,Tel. 06196/474-0

Silanes

Wacker-Chemie GmbH
Hanns-Seidel-Platz 4
81737 München

Silicone fluids

Wacker-Chemie GmbH
Hanns-Seidel-Platz 4
81737 München

Silicones

Dr. Th. Böhme KG, Chem. Fabrik
Isardamm 79, 82538 Geretsried
Tel.: 08171/628-0, Tlx. 526312
Telefax: 08171/628-388

Wacker-Chemie GmbH
Hanns-Seidel-Platz 4
81737 München

Solvents

Deutsche Shell Chemie GmbH,
65760 Eschborn,Tel. 06196/474-0

Stabilizers

Bärlocher GmbH
Riesstr. 16, 80992 München
Tel. 089/14373-0 FS 5215773
Fax: 089/14373-312

ciba
ciba Additive GmbH
Chemiestr., 68623 Lampertheim
Tel.: 06206/502-0, Fax /502-368

Stabilizers for PVC

ciba
ciba Additive GmbH
Chemiestr., 68623 Lampertheim
Tel.: 06206/502-0, Fax /502-368

Elastogran GmbH
GB Grundprodukte (G)
BASF Polyurethane
67056 Ludwigshafen
Tel. 0621/60-0, Fax/60-43960

Hoechst Aktiengesellschaft
Business Unit Additive
Postfach 10 15 67
D-86005 Augsburg
Tel.: 0821/479-0
Fax : 0821/496639

Stabilizers, UV-

Hoechst Aktiengesellschaft
Business Unit Additive
Postfach 10 15 67
D-86005 Augsburg
Tel.: 0821/479-0
Fax : 0821/496639

Surfactants

Raschig AG
D-67061 Ludwigshafen
Tel.+49(621)56180 / Fax.582885

Deutsche Shell Chemie GmbH,
65760 Eschborn,Tel. 06196/474-0

Talcum

Gebr. Dorfner GmbH & Co. KG
Scharhof, D-92242 Hirschau
Tel. ++49/(0)9622/82-0
Telefax ++49/(0)9622/82-69

Naintsch Mineralwerke GmbH
A-8045 Graz, Postfach 35
Statteggerstr. 60
Tel. 0316/693650, Tlx. 311223
Fax. 0316/693655

UV absorber

Hoechst Aktiengesellschaft
Business Unit Additive
Postfach 10 15 67
D-86005 Augsburg
Tel.: 0821/479-0
Fax : 0821/496639

UV stabilizers

Nemitz, D-48341 Altenberge
Tel. 02505/674, Fax /3042

Viscosity depressants

Dr. Th. Böhme KG, Chem. Fabrik
Isardamm 79, 82538 Geretsried
Tel.: 08171/628-0, Tlx. 526312
Telefax: 08171/628-388

Byk-Chemie GmbH
Abelstr. 14, D-46483 Wesel
Tel. 0281/670-0 Fax: /65735

Waxes

Hoechst Aktiengesellschaft
Business Unit Additive
Postfach 10 15 67
D-86005 Augsburg
Tel.: 0821/479-0
Fax : 0821/496639

Waxes, lubricants

Hoechst Aktiengesellschaft
Business Unit Additive
Postfach 10 15 67
D-86005 Augsburg
Tel.: 0821/479-0
Fax : 0821/496639

Waxes, polyethylene

Hoechst Aktiengesellschaft
Business Unit Additive
Postfach 10 15 67
D-86005 Augsburg
Tel.: 0821/479-0
Fax : 0821/496639

Wood flour

C.F.F. D-41065 Mönchengladbach
Fax 02161-656.200 Tx 852102
TECHNOCEL Cellulose Füllst.

Zinc stearate

Seitz + Kerler, 97816 Lohr

Group 504: Plastics processing Services - Job processing

Extrusion

Matthias Oechsler & Sohn
D-91781 Weißenburg
Tel. 09141/990-0

Injection moulding

formplast Lechler GmbH
Kunststofftechnik
PF 450153, 90212 Nürnberg
Tel. 0911/99455-0, Fax -50

Tank construction

Eichholz, 48480 Schapen

Group 505: Semifinished plastic products

Composites

AUGUST KREMPEL SOEHNE
POB 1240, D-71655 Vaihingen
Tel. 07042/915-0, Fax /15985

Copper clad laminates for printed circuits

Isola Werke AG
52348 Düren
Tel. 02421/8080, Fax / 808-389

Epoxy resin prepregs

Isola Werke AG
52348 Düren
Tel. 02421/8080, Fax / 808-389

Fluoroplastic semi-finished products

Symalit AG
CH-5600 Lenzburg/Schweiz
Tel. 064/508150, Fax 064/519104

FRP laminates

AUGUST KREMPEL SOEHNE
POB 1240, D-71655 Vaihingen
Tel. 07042/915-0, Fax /15985

GMT-Glass mat reinforced thermoplastic sheet

Symalit AG
CH-5600 Lenzburg/Schweiz
Tel. 064/508150, Fax 064/519104

GRP fibre glass profiles

FIBROLUX GMBH
65719 Hofheim
Tel. 06122/91000, Fax./15001

GRP hardboards

Isola Werke AG
52348 Düren
Tel. 02421/8080, Fax / 808-389

Insulating materials

AUGUST KREMPEL SOEHNE
POB 1240, D-71655 Vaihingen
Tel. 07042/915-0, Fax /15985

Laminated fabrics

Isola Werke AG
52348 Düren
Tel. 02421/8080, Fax / 808-389

Laminates, technical

Isola Werke AG
52348 Düren
Tel. 02421/8080, Fax / 808-389

Lining laminates

Symalit AG
CH-5600 Lenzburg/Schweiz
Tel. 064/508150, Fax 064/519104

Polyamide (semi-finished-products)

Licharz GmbH, Fax:02241/336581
D-53757 St. Augustin

Polycarbonate plates

Axxis NV, Wakkensesteenweg 47
B-8700 Tielt, Belgien
Fax (32) 51.40.48.18

Polyester (semi-finished products)

Isola Werke AG
52348 Düren
Tel. 02421/8080, Fax / 808-389

Polyester sheet

Axxis NV, Wakkensesteenweg 47
B-8700 Tielt, Belgien
Fax (32) 51.40.48.18

Polyethylen semifinished products

TERBRACK KUNSTSTOFF
GmbH & Co.
KG, D-48686 Vreden
Tel. 02564/3930 Fax /39360

Polyethylene pipe

Symalit AG
CH-5600 Lenzburg/Schweiz
Tel. 064/508150, Fax 064/519104

Polyethylene sheets (PE-HMW und PE-UHMW)

Wefapress-Werkstoffe
D-48691 Vreden,Tel. 02564/93290
Fax 02564/6372

Polypropylene (semi-finished products)

TERBRACK KUNSTSTOFF
GmbH & Co.
KG, D-48686 Vreden
Tel. 02564/3930 Fax /39360

Polytetrafluoroethylene (PTFE semi-finished product)

Fietz GmbH, PTFE Produkte
51388 Burscheid
Tel. 02174/674-0, Fax /674-222

Prepregs

Isola Werke AG
52348 Düren
Tel. 02421/8080, Fax / 808-389

AUGUST KREMPEL SOEHNE
POB 1240, D-71655 Vaihingen
Tel. 07042/915-0, Fax /15985

Profiles

G. Binder GmbH u. Co.
D-71084 Holzgerlingen, PF 1161
Tel. 07031/683-0, Fax /683-179

AUGUST KREMPEL SOEHNE
POB 1240, D-71655 Vaihingen
Tel. 07042/915-0, Fax /15985

PVC profiles

G. Binder GmbH u. Co.
D-71084 Holzgerlingen, PF 1161
Tel. 07031/683-0, Fax /683-179

Rods

FIBROLUX GMBH
65719 Hofheim
Tel. 06122/91000, Fax./15001

Semi-finished products, extruded

Metzeler Plastics GmbH
Postfach 17 60
52407 Jülich
Tel. 02461/64-0
Fax 02461/64-210
Marken:
METZOPLAST+TROVIDUR PO

Symalit AG
CH-5600 Lenzburg/Schweiz
Tel. 064/508150, Fax 064/519104

Semi-finished products of PUR elastomers

Elastogran GmbH
GB Halbzeug,Bauteile (H)
PF 1140, 49440 Lemförde
Tel. 05443/12-0, Fax/12-2370

Sheetings

Axxis NV, Wakkensesteenweg 47
B-8700 Tielt, Belgien
Fax (32) 51.40.48.18

Technical profiles

Matthias Oechsler & Sohn
D-91781 Weißenburg
Tel. 09141/990-0

Thermoforming sheet and film general

Axxis NV, Wakkensesteenweg 47
B-8700 Tielt, Belgien
Fax (32) 51.40.48.18

Tubes

AUGUST KREMPEL SOEHNE
POB 1240, D-71655 Vaihingen
Tel. 07042/915-0, Fax /15985

Group 506: Finished plastic products

Cable conduit

Symalit AG
CH-5600 Lenzburg/Schweiz
Tel. 064/508150, Fax 064/519104

Chain guides

Wefapress-Werkstoffe
D-48691 Vreden,Tel. 02564/93290
Fax 02564/6372

Chain tensioner

Murtfeldt Kunststoffe GmbH
Heßlingsweg 14,44309 Dortmund
Tel. 0231/20609-0, Fax 251021

Closures

G. Binder GmbH u. Co.
D-71084 Holzgerlingen, PF 1161
Tel. 07031/683-0, Fax /683-179

Digit drums

Matthias Oechsler & Sohn
D-91522 Ansbach,Tel. 0981/18070

Edge protection

ROGA KG, 50380 Wesseling
Tel. 02236/47011, Fax -/47013

Fibre-composite mouldings

ACT Hochleistungs-
kunststofftechnik G.m.b.H.
Hauptstr. 2
A-2630 Ternitz/Österreich
Tel. 02630/35161631, Fax/35156

Finished articles

Murtfeldt Kunststoffe GmbH
Heßlingsweg 14,44309 Dortmund
Tel. 0231/20609-0, Fax 251021

Flexible insulating tubing

Elfoflex Isolierschlauchfabrik
Dipl.Ing. Helmut EBERS
Huttenstr. 41/44, PF 210467
10504 Berlin
Tel. 030/3444024
Telefax 030/3441659

Gaskets/Seals

Fietz GmbH, PTFE Produkte
51388 Burscheid
Tel. 02174/674-0, Fax /674-222

Gearwheels

Matthias Oechsler & Sohn
D-91522 Ansbach,Tel. 0981/18070

Glassfibre reinforced plastic products

Isola Werke AG
52348 Düren
Tel. 02421/8080, Fax / 808-389

GRP covers

Isola Werke AG
52348 Düren
Tel. 02421/8080, Fax / 808-389

GRP mouldings

Isola Werke AG
52348 Düren
Tel. 02421/8080, Fax / 808-389

Hoses

ROGA KG, 50380 Wesseling
Tel. 02236/47011, Fax -/47013

Injection moulded parts

Rauschert GmbH & Co KG
54578 Oberbettingen
T. 06593/1031, Fax 9001

Reiher GmbH
D-38116 Braunschweig
Tel. 0531/52081-82

Injection mouldings, thermoplastic

Matthias Oechsler & Sohn
D-91522 Ansbach, Tel. 0981/18070

Insulating boards

Isola Werke AG
52348 Düren
Tel. 02421/8080, Fax / 808-389

Linings

TERBRACK KUNSTSTOFF
GmbH & Co.
KG, D-48686 Vreden
Tel. 02564/3930 Fax /39360

Machine elements

Isola Werke AG
52348 Düren
Tel. 02421/8080, Fax / 808-389

Moulded parts

Isola Werke AG
52348 Düren
Tel. 02421/8080, Fax / 808-389

Mouldings in accordance with sample, drawing or cust. tool

G. Binder GmbH u. Co.
D-71084 Holzgerlingen, PF 1161
Tel. 07031/683-0, Fax /683-179

Elastogran GmbH
GB Halbzeug, Bauteile (H)
PF 1140, 49440 Lemförde
Tel. 05443/12-0, Fax/12-2370

Presswerk Köngen GmbH
Postfach 1165, 73253 Köngen
Tel. 07024/808-0, Tlx 7267214
Telefax 07024/808-111

®Hoechst High Chem — Additives for Plastics

Additives for Plastics from Hoechst for better products and higher performance

®Hostanox
Antioxidants

®Hostastat
Antistatic Agents

Hoechst Waxes and ®Hostalub
Internal and External Lubricants

®Hostavin
Light Stabilizers

®Hostaflam
Flame Retardents

®Hostastab
Heat Stabilizers

®Hostatron
Foaming Systems

Additive Systems Hoechst

Hoechst Aktiengesellschaft
Business Unit Additive
Postfach 10 15 67
D-86005 Augsburg
Tel.: 08 21 479-0
Fax: 08 21 49 66 39

PIGMENTS
MASTERBATCH
ADDITIVES

Georg Deifel KG

Buntfarbenfabrik
Alte Bahnhofstraße 11
D-97422 Schweinfurt
Tel. (0 97 21) 17 74
Fax (0 97 21) 18 50 99

YOUR PARTNER FOR COLORATIONS

IN
SERVICE
FOR
A
WORLD
OF
COLOUR
FOR
MORE
THAN
70 YEARS

THE JOURNAL FOR THE WORLD'S POLYMER PROCESSING LEADERSHIP

International Polymer Processing represents both the first truly international scientific journal in polymer processing and the official journal of the Polymer Processing Society. It publishes original innovative technical articles from engineers and scientists throughout the world. In addition, it presents rapid communications of recent advances and occasional timely review papers.

International Polymer Processing also publishes important news on technological innovations and industrial ventures as well as news of the Polymer Processing Society and its meetings.

The editors of **International Polymer Processing** are based around the world in three major polymer application centers – Far East, Europe and North America. The international spirit of the journal may be seen in that the contributions come from each of these centers providing a unique distribution of perspectives in its research papers.

See for yourself! We will gladly send you a free sample copy.

Carl Hanser Verlag
P. O. Box 86 04 20
81631 Munich, Germany
Fax: 089/98 48 09

AUGUST KREMPEL SOEHNE
POB 1240, D-71655 Vaihingen
Tel. 07042/915-0, Fax /15985

Matthias Oechsler & Sohn
D-91522 Ansbach,Tel. 0981/18070

Rauschert GmbH & Co KG
54578 Oberbettingen
T. 06593/1031, Fax 9001

Polyethylene finished articles

Wefapress-Werkstoffe
D-48691 Vreden,Tel. 02564/93290
Fax 02564/6372
machined

Profiles

G. Binder GmbH u. Co.
D-71084 Holzgerlingen, PF 1161
Tel. 07031/683-0, Fax /683-179

Matthias Oechsler & Sohn
D-91522 Ansbach,Tel. 0981/18070

ROGA KG, 50380 Wesseling
Tel. 02236/47011, Fax -/47013

PTEF products

Fietz GmbH, PTFE Produkte
51388 Burscheid
Tel. 02174/674-0, Fax /674-222

Refrigerator parts

ROGA KG, 50380 Wesseling
Tel. 02236/47011, Fax -/47013

Resin-bonded paper sleeves

Isola Werke AG
52348 Düren
Tel. 02421/8080, Fax / 808-389

Rotary parts

Fietz GmbH, PTFE Produkte
51388 Burscheid
Tel. 02174/674-0, Fax /674-222

Sliding elements

Wefapress-Werkstoffe
D-48691 Vreden,Tel. 02564/93290
Fax 02564/6372

Technical injection mouldings

Matthias Oechsler & Sohn
D-91522 Ansbach,Tel. 0981/18070

Reiher GmbH
D-38116 Braunschweig
Tel. 0531/52081-82

Technical mouldings of GRP

Isola Werke AG
52348 Düren
Tel. 02421/8080, Fax / 808-389

Technical parts

Isola Werke AG
52348 Düren
Tel. 02421/8080, Fax / 808-389

Technical parts according to sample, drawing or cust.tool

Elastogran GmbH
GB Halbzeug,Bauteile (H)
PF 1140, 49440 Lemförde
Tel. 05443/12-0, Fax/12-2370

Presswerk Köngen GmbH
Postfach 1165, 73253 Köngen
Tel. 07024/808-0, Tlx 7267214
Telefax 07024/808-111

Rauschert GmbH & Co KG
54578 Oberbettingen
T. 06593/1031, Fax 9001

Tube, general

Isola Werke AG
52348 Düren
Tel. 02421/8080, Fax / 808-389

Vehicle fittings

Presswerk Köngen GmbH
Postfach 1165, 73253 Köngen
Tel. 07024/808-0, Tlx 7267214
Telefax 07024/808-111

Group 507: Machinery, appliances, moulds & tools f. plast. processing

Annealing lines

Herbert Olbrich GmbH & Co. KG
46395 Bocholt, T. 02871/283-0
Fax 02871/283-189, Tx 813 807

Automatic blow moulding eyuipment

R. Stahl Blasformtechnik GmbH
70771 Le-Echterdingen
Tel. 0711/9472-0
Fax 0711/9472-222

Automatic tube manufacturing equipment

Ossberger-Turbinenfabrik GmbH
& Co Abt. Kunststoffmaschinen
Pf. 425, 91773 Weissenburg
Tel. 09141/977-0, Tx 624672
Fax 09141/977-20

Calenders embossing

Herbert Olbrich GmbH & Co. KG
46395 Bocholt, T. 02871/283-0
Fax 02871/283-189, Tx 813 807

Calenders, polishing

Herbert Olbrich GmbH & Co. KG
46395 Bocholt, T. 02871/283-0
Fax 02871/283-189, Tx 813 807

Cleaning lines, exhaust-air

Herbert Olbrich GmbH & Co. KG
46395 Bocholt, T. 02871/283-0
Fax 02871/283-189, Tx 813 807

Cleaning systems and baths

Telsonic AG, Industriestraße
CH-9552 Bronschhofen
Tel. 073/225353 Fax 073/225357

Coating plants

Herbert Olbrich GmbH & Co. KG
46395 Bocholt, T. 02871/283-0
Fax 02871/283-189, Tx 813 807

Coiling machines

Graewe GmbH, 79395 Neuenburg
Tel. 07631/7944-0, Fax /7944-22

Comp. such as rotary feeders conveying-tube gates, etc.

Bühler GmbH, 38023 Braunschweig
Postf. 3369, Tel. 0531/5940

Compounding & Blending Plants

Erema GmbH
Unterfeldst.3, A-4052 Ansfelden
Tel: 0732/311761-0
Tx: 222 300 erema a
Telefax: 0732/311764

Thyssen Henschel GmbH
D-34112 Kassel, Pf. 102969
Tel. 0561/801-01; Fax/801-6943

Control equipment for temperature

EOC Normalien GmbH + Co.KG
58463 Lüdenscheid/PF. 1380
Tel. 02351/437-0/Fx.-356

HOTSET Heizpatronen und
Zubehör GmbH, PF 1860,
D-58468 Lüdenscheid,
Tel.+49/2351/4302-0

Conveying and metering equipment

Elastogran GmbH
GB Spezialsysteme (P)
Industriestr., 82140 Olching
Tel. 08142/178-0, Fax/178-213

Conveying plants, mechanical

Eichholz, 48480 Schapen

Conveying plants, pneumatic

Bühler Ltd.
CH-9240 Uzwil/Switzerland
Tel. +41 (0) 73 50 11 11
Fax. +41 (0) 73 50 33 79

Bühler GmbH, 38023 Braunschweig
Postf. 3369, Tel. 0531/5940

Motan GmbH, P.O. Box 1363
D-88307 Isny/Allgäu
Phone 07562/76-0
Fax 07562/76-111

Cooling mixers

Papenmeier GmbH Mischtechnik
D-32832 Augustdorf
Vertrieb: Elsener Str. 7-9
D-33102 Paderborn
Phone: +49/(0)5251/309-112
Fax: +49/(0)5251/309-123

Thyssen Henschel GmbH
D-34112 Kassel, Pf. 102969
Tel. 0561/801-01; Fax/801-6943

Discontinous mixers

Papenmeier GmbH Mischtechnik
D-32832 Augustdorf
Vertrieb: Elsener Str. 7-9
D-33102 Paderborn
Phone: +49/(0)5251/309-112
Fax: +49/(0)5251/309-123

Dryers, vacuum

Eichholz, 48480 Schapen

Drying plants for granulates

Bühler Ltd.
CH-9240 Uzwil/Switzerland
Tel. +41 (0) 73 50 11 11
Fax. +41 (0) 73 50 33 79

Bühler GmbH,38023 Braunschweig
Postf. 3369, Tel. 0531/5940

Motan GmbH, P.O. Box 1363
D-88307 Isny/Allgäu
Phone 07562/76-0
Fax 07562/76-111

C.F.SCHEER & CIE GmbH
Postfach 301020,70450Stuttgart
Tel. 0711/8781-0, Fax./8781-295

Dust collectors

Bühler GmbH,38023 Braunschweig
Postf. 3369, Tel. 0531/5940

Ejector pins

DREI-S-Werk, 91124 Schwabbach
Tel. 09122/1505-50, Fax 1505-54

EOC Normalien GmbH + Co.KG
58463 Lüdenscheid/PF. 1380
Tel. 02351/437-0/Fx.-356

Electric heating elements

HOTSET Heizpatronen und
Zubehör GmbH, PF 1860,
D-58468 Lüdenscheid,
Tel.+49/2351/4302-0

Embossing presses

FRICLA-Fritz Claussner
D-90257 Nürnberg,P.O.B. 840106
Tel. 0911/313293, Fax./311502

Extruders

Hans Weber Maschinenfabrik GmbH
Postfach 1862, 96308 Kronach
Bamberger Straße 19-21
Telefon: 09261/4090, Telefax: 09261/409199, Telex: 642636

Extruders& extrusions equipment,Single&Twin screw extr.

Paul Kiefel
Extrusionstechnik GmbH
Cornelius-Heyl-Straße 49
D-67547 Worm
Tel.: 0 62 41 / 9 02-0
Fax : 0 62 41 / 9 02-100

Reifenhäuser GmbH & Co
D-53839 Troisdorf
Tel. 02241/481-0, Fax./408778

Extrusion lines, general

Erema GmbH
Unterfeldst.3,A-4052 Ansfelden
Tel: 0732/311761-0
Tx: 222 300 erema a
Telefax: 0732/311764

KLÖCKNER ER-WE-PA GMBH
Mettmanner Strasse 51
D-40699 Erkrath/Düsseldorf
Telefon 0211/24040
Telefax 0211/2404281

Extrusions plants for profiles

Hans Weber Maschinenfabrik GmbH
Postfach 1862, 96308 Kronach
Bamberger Straße 19-21
Telefon: 09261/4090, Telefax: 09261/409199, Telex: 642636

Filament winding machines

Josef Baer Mafa GmbH
Große Halde 6
88279 Amtzell-Korb
Tel. 07520/953-0
Fax 07520/953-250

Film welding machines

Widmann Schweißmaschinen GmbH
Siemensstraße 19
73278 Schlierbach
Tel:(07021)9733-0, Fax:/973380

Fluidised-bed driers

Bühler Ltd.
CH-9240 Uzwil/Switzerland
Tel. +41 (0) 73 50 11 11
Fax. +41 (0) 73 50 33 79

Follow-up machines for extruders

Graewe GmbH, 79395 Neuenburg
Tel. 07631/7944-0, Fax /7944-22

Reifenhäuser GmbH & Co
D-53839 Troisdorf
Tel. 02241/481-0, Fax./408778

C.F.SCHEER & CIE GmbH
Postfach 301020,70450Stuttgart
Tel. 0711/8781-0, Fax./8781-295

Granulating plants

Erema GmbH
Unterfeldst.3,A-4052 Ansfelden
Tel: 0732/311761-0
Tx: 222 300 erema a
Telefax: 0732/311764

C.F.SCHEER & CIE GmbH
Postfach 301020,70450Stuttgart
Tel. 0711/8781-0, Fax./8781-295

Granulators

Papenmeier GmbH Mischtechnik
D-32832 Augustdorf
Vertrieb: Elsener Str. 7-9
D-33102 Paderborn
Phone: +49/(0)5251/309-112
Fax: +49/(0)5251/309-123

Hans Weber Maschinenfabrik GmbH
Postfach 1862, 96308 Kronach
Bamberger Straße 19-21
Telefon: 09261/4090, Telefax: 09261/409199, Telex: 642636

Heat-contact welding machines and equipment

Widmann Schweißmaschinen GmbH
Siemensstraße 19
73278 Schlierbach
Tel:(07021)9733-0, Fax:/973380

Heating cartridges

HOTSET Heizpatronen und
Zubehör GmbH, PF 1860,
D-58468 Lüdenscheid,
Tel.+49/2351/4302-0

Ihne & Tesch GmbH
POB 1863, D-58468 Lüdenscheid
Tel. 02351/666-0 Fax 666-24

Heating equipment

Ihne & Tesch GmbH
POB 1863, D-58468 Lüdenscheid
Tel. 02351/666-0 Fax 666-24

Heating mixer

MIXACO
Dr. Herfeld GmbH & Co. KG
Ein Unternehmen der
Kraftanlagen Heidelberg AG
D-58803 Neuenrade, POB 1147
Tel. (+49)2392/6347-9, Fax/62013

Papenmeier GmbH Mischtechnik
D-32832 Augustdorf
Vertrieb: Elsener Str. 7-9
D-33102 Paderborn
Phone: +49/(0)5251/309-112
Fax: +49/(0)5251/309-123

Thyssen Henschel GmbH
D-34112 Kassel, Pf. 102969
Tel. 0561/801-01; Fax/801-6943

High-Speed Mixer

Papenmeier GmbH Mischtechnik
D-32832 Augustdorf
Vertrieb: Elsener Str. 7-9
D-33102 Paderborn
Phone: +49/(0)5251/309-112
Fax: +49/(0)5251/309-123

Hot channel systems for injection moulds

EOC Normalien GmbH + Co.KG
58463 Lüdenscheid/PF. 1380
Tel. 02351/437-0/Fx.-356

Impregnating machines

Herbert Olbrich GmbH & Co. KG
46395 Bocholt, T. 02871/283-0
Fax 02871/283-189, Tx 813 807

Infrared radiators

Krelus AG
CH-5042 Hirschthal, Tx 982204
Fax 064/813256, Tel. /812777

Injection blow moulding machines

Ossberger-Turbinenfabrik GmbH
& Co Abt. Kunststoffmaschinen
Pf. 425, 91773 Weissenburg
Tel. 09141/977-0, Tx 624672
Fax 09141/977-20

Injection moulding machines

Engel Vertriebsges. m.b.H.
A-4311 Schwertberg/Österreich
Tel. 07262/620-0
Fax 07262/620-308

Injection moulds

BRAUN & KELLER GMBH
Präzisionsformenbau
Unter Gereuth 14
D-79353 Bahlingen
Tel. 07663/9320-0 Fax. 3727

Engel Vertriebsges. m.b.H.
A-4311 Schwertberg/Österreich
Tel. 07262/620-0
Fax 07262/620-308

formplast Lechler GmbH
Kunststofftechnik
PF 450153, 90212 Nürnberg
Tel. 0911/99455-0, Fax -50

Kneader mixers

Papenmeier GmbH Mischtechnik
D-32832 Augustdorf
Vertrieb: Elsener Str. 7-9
D-33102 Paderborn
Phone: +49/(0)5251/309-112
Fax: +49/(0)5251/309-123

Laminating plant

Herbert Olbrich GmbH & Co. KG
46395 Bocholt, T. 02871/283-0
Fax 02871/283-189, Tx 813 807

Magnets

S+S Metallsuchgeräte
und Recyclingtechnik GmbH
94509 Schönberg
Tel. 08554/308-0 Fax /2606

Metal detectors

S+S Metallsuchgeräte
und Recyclingtechnik GmbH
94509 Schönberg
Tel. 08554/308-0 Fax /2606

Metal separator

S+S Metallsuchgeräte
und Recyclingtechnik GmbH
94509 Schönberg
Tel. 08554/308-0 Fax /2606

Metering eyuipment, multi-component

LEWA Herbert Ott GmbH + Co
Pf 1563, D-71226 Leonberg
Tel. 07152/14-0, Fax 14-303

Metering plant

Motan GmbH, P.O. Box 1363
D-88307 Isny/Allgäu
Phone 07562/76-0
Fax 07562/76-111

Mixers, high-speed

MIXACO
Dr. Herfeld GmbH & Co. KG
Ein Unternehmen der
Kraftanlagen Heidelberg AG
D-58803 Neuenrade, POB 1147
Tel. (+49)2392/6347-9, Fax/62013

Mixers, static

Sulzer Chemtech AG
CH-8404 Winterthur, Schweiz
Tel. 052-262 67 20

Mixing, ensiling, conveying and metering equipment

Bühler GmbH, 38023 Braunschweig
Postf. 3369, Tel. 0531/5940

Eichholz, 48480 Schapen

Motan GmbH, P.O. Box 1363
D-88307 Isny/Allgäu
Phone 07562/76-0
Fax 07562/76-111

Mixing plant

MIXACO
Dr. Herfeld GmbH & Co. KG
Ein Unternehmen der
Kraftanlagen Heidelberg AG
D-58803 Neuenrade, POB 1147
Tel. (+49)2392/6347-9, Fax/62013

Mould building

Matthias Oechsler & Sohn
D-90613 Großhabersdorf
Tel. 09105/302

Mould manufacture

BRAUN & KELLER GMBH
Präzisionsformenbau
Unter Gereuth 14
D-79353 Bahlingen
Tel. 07663/9320-0 Fax. 3727

Nozzles

HOTSET Heizpatronen und
Zubehör GmbH, PF 1860,
D-58468 Lüdenscheid,
Tel.+49/2351/4302-0

Ovens for coating equipment

Herbert Olbrich GmbH & Co. KG
46395 Bocholt, T. 02871/283-0
Fax 02871/283-189, Tx 813 807

PAD Printing Machines

TAMPONCOLOR
TC-Druckmaschinen GmbH
Dornhofstr. 14
D-63263 Neu-Isenburg

Pipe extrusion lines

Hans Weber Maschinenfabrik
GmbH
Postfach 1862, 96308 Kronach
Bamberger Straße 19-21
Telefon: 09261/4090, Telefax:
09261/409199, Telex: 642636

Pipe extrusion plants

Fränkische D-97486 Königsberg
Tel. (09525)88-0, Fax /88-411

Polishing equipment and appliances

EOC Normalien GmbH + Co.KG
58463 Lüdenscheid/PF. 1380
Tel. 02351/437-0/Fx.-356

Prepeg plants

Herbert Olbrich GmbH & Co. KG
46395 Bocholt, T. 02871/283-0
Fax 02871/283-189, Tx 813 807

Presses

Maschinenfabrik
J.Dieffenbacher GmbH & Co.
Postfach 162
75020 Eppingen

Presses for polyester

Maschinenfabrik
J.Dieffenbacher GmbH & Co.
Postfach 162
75020 Eppingen

Presses for reinforced plastic parts

Maschinenfabrik
J.Dieffenbacher GmbH & Co.
Postfach 162
75020 Eppingen

Pressure machines

Herbert Olbrich GmbH & Co. KG
46395 Bocholt, T. 02871/283-0
Fax 02871/283-189, Tx 813 807

Pressure transducer

Kistler Instrumente AG Winter.
CH-8408 Winterthur/Schweiz
Tel. +41-52-2241111 Fax/2241414

Pumps, metering

LEWA Herbert Ott GmbH + Co
Pf 1563, D-71226 Leonberg
Tel. 07152/14-0, Fax 14-303

Pxx-compounder

Papenmeier GmbH Mischtechnik
D-32832 Augustdorf
Vertrieb: Elsener Str. 7-9
D-33102 Paderborn
Phone: +49/(0)5251/309-112
Fax: +49/(0)5251/309-123

Reclaiming plants

Erema GmbH
Unterfeldst.3,A-4052 Ansfelden
Tel: 0732/311761-0
Tx: 222 300 erema a
Telefax: 0732/311764

Recycling plant

Erema GmbH
Unterfeldst.3,A-4052 Ansfelden
Tel: 0732/311761-0
Tx: 222 300 erema a
Telefax: 0732/311764

SIKOPLAST Maschinenbau
Heinrich Koch GmbH
Jakobstr. 84-88,53721 Siegburg
Tel. 02241/65011-13, Fax/52454

Zimmer AG
Borsigallee 1, 60388 Frankfurt
Tel. 069/4007-01, Tlx 417172
Fax 069/4007-546

Removal eyuipment for injection moulding machines

Engel Vertriebsges. m.b.H.
A-4311 Schwertberg/Österreich
Tel. 07262/620-0
Fax 07262/620-308

Re-rolling machines

Herbert Olbrich GmbH & Co. KG
46395 Bocholt, T. 02871/283-0
Fax 02871/283-189, Tx 813 807

Rheological Testing Instruments

COESFELD D-44149 Dortmund
Materialtest, Fax 0231/179885

Göttfert Werkstoff-Prüf-
maschinen GmbH, Postfach
12 61, D-74711 Buchen
Tel. 06281-4080, Fax-40818

Rheometer

Göttfert Werkstoff-Prüf-
maschinen GmbH, Postfach
12 61, D-74711 Buchen
Tel. 06281-4080, Fax-40818

Roll-coating plant

Herbert Olbrich GmbH & Co. KG
46395 Bocholt, T. 02871/283-0
Fax 02871/283-189, Tx 813 807

Rolls

Leonhard Breitenbach GmbH
Postf. 100349, 57003 Siegen
Tel. 0271/37580 Fax/3758290

Rubber injection moulding plants

Maschinenfabrik
J.Dieffenbacher GmbH & Co.
Postfach 162
75020 Eppingen

Screen-changing devices

Erema GmbH
Unterfeldst.3,A-4052 Ansfelden
Tel: 0732/311761-0
Tx: 222 300 erema a
Telefax: 0732/311764

Secondhand equipment

Engelking, 32602 Vlotho

PPM Vertriebs GmbH
Freischützstr.75,81927 München
Tel. 089/9570073, Fax/9577376
Extruder- u. Mischanlagen

Sheet stretching equipment

Brückner-Maschinenbau
Gernot Brückner GmbH & Co. KG
POB 1161, D-83309 Siegsdorf
Tel. 08662/63-0
Fax: 08662/63-220

Sheet stretching machines

Lindauer Dornier GmbH
88129 Lindau/Bodensee
Telefax (08382) 703-378

Silo

Eichholz, 48480 Schapen

Slit dies

VERBRUGGEN N.V.
Jan de Malschelaan 2
B-9140 Temse
Tel. 03/711 19 97
Fax 03/711 54 90

Solid-State Polycondensation (SSP) for PET and PA

Bühler Ltd.
CH-9240 Uzwil/Switzerland
Tel. +41 (0) 73 50 11 11
Fax. +41 (0) 73 50 33 79

Spread coating machines

Herbert Olbrich GmbH & Co. KG
46395 Bocholt, T. 02871/283-0
Fax 02871/283-189, Tx 813 807

Stamping and punching machines

Hans Naef AG, CH-8045 Zürich
Talwiesenstr. 17
Tel. 01/4510801 Fax 4621919

Standard parts for injection moulding

EOC Normalien GmbH + Co.KG
58463 Lüdenscheid/PF. 1380
Tel. 02351/437-0/Fx.-356

Standards for tool manufacture

HASCO-Normalien
Hasenclever GmbH + Co
Im Wiesental 77
D-58513 Lüdenscheid
Tel. 02351/9570, Fax 957237

Strand granulators

C.F.SCHEER & CIE GmbH
Postfach 301020, 70450 Stuttgart
Tel. 0711/8781-0, Fax./8781-295

Synthetic fibre lines

Zimmer AG
Borsigallee 1, 60388 Frankfurt
Tel. 069/4007-01, Tlx 417172
Fax 069/4007-546

Testing machines and equipments

COESFELD D-44149 Dortmund
Materialtest, Fax 0231/179885

Thickness controller

Elektro-Physik Köln
Pasteurstr. 15
50735 Köln
Tel. 0221/752040 Fax /75204-68

Tool standards

HASCO-Normalien
Hasenclever GmbH + Co
Im Wiesental 77
D-58513 Lüdenscheid
Tel. 02351/9570, Fax 957237

Ultrasonic welding devices

Telsonic AG, Industriestraße
CH-9552 Bronschhofen
Tel. 073/225353 Fax 073/225357

Ultrasonics Steckmann GmbH
Hauptstrasse 24
D-61279 Grävenwiesbach
Tel. 06086/1818 Fax 06086/3166

Ultrasonic welding machines

Telsonic AG, Industriestraße
CH-9552 Bronschhofen
Tel. 073/225353 Fax 073/225357

Ultrasonics Steckmann GmbH
Hauptstrasse 24
D-61279 Grävenwiesbach
Tel. 06086/1818 Fax 06086/3166

Varnishing plants

Herbert Olbrich GmbH & Co. KG
46395 Bocholt, T. 02871/283-0
Fax 02871/283-189, Tx 813 807

Vulcanisations presses

Maschinenfabrik
J.Dieffenbacher GmbH & Co.
Postfach 162
75020 Eppingen

Welding lines

Ultrasonics Steckmann GmbH
Hauptstrasse 24
D-61279 Grävenwiesbach
Tel. 06086/1818 Fax 06086/3166

Wet mixer

Papenmeier GmbH Mischtechnik
D-32832 Augustdorf
Vertrieb: Elsener Str. 7-9
D-33102 Paderborn
Phone: +49/(0)5251/309-112
Fax: +49/(0)5251/309-123

Group 508: Auxiliary agents for plastics processing

Adhesives for plastics

Wacker-Chemie GmbH
Hanns-Seidel-Platz 4
81737 München

Antifoam agents

Wacker-Chemie GmbH
Hanns-Seidel-Platz 4
81737 München

Impregnating agents, silicone

Dr. Th. Böhme KG, Chem. Fabrik
Isardamm 79, 82538 Geretsried
Tel.: 08171/628-0, Tlx. 526312
Telefax: 08171/628-388

Wacker-Chemie GmbH
Hanns-Seidel-Platz 4
81737 München

Lubricants, silicone

Wacker-Chemie GmbH
Hanns-Seidel-Platz 4
81737 München

Paint auxiliaries

Wacker-Chemie GmbH
Hanns-Seidel-Platz 4
81737 München

Paints for plastics surface treatment

E. Peter & Sohn GmbH
Postfach 25 51, 32015 Herford
Tel. 05221/9625-0 / Fax -44

Release agents, general

E. & P. Würtz GmbH & Co.
55411 Bingen-Sponsheim
Tel. 06721/96900-Fax 41589

Silicone mould release agents

Wacker-Chemie GmbH
Hanns-Seidel-Platz 4
81737 München

Group 509: Analysis, measurement and testing equipment, incl. acce.

Instrumentation for melt-flow indes

BRABENDER OHG, Kulturstr. 51
D-47055 Duisburg, Germany
Tel. -49-203-7788-0
Fax -49-203-7788-100

Laboratory calenders

Dr. Collin GmbH, Sportpark-
str. 2, D-85560 Ebersberg
Tel. 08092/2096-0, Fax /20862

Laboratory extruders

Dr. Collin GmbH, Sportpark-
str. 2, D-85560 Ebersberg
Tel. 08092/2096-0, Fax /20862

Laboratory kneaders

Dr. Collin GmbH, Sportpark-
str. 2, D-85560 Ebersberg
Tel. 08092/2096-0, Fax /20862

Laboratory mixers

MIXACO
Dr. Herfeld GmbH & Co. KG
Ein Unternehmen der
Kraftanlagen Heidelberg AG
D-58803 Neuenrade, POB 1147
Tel. (+49)2392/6347-9, Fax/62013

Laboratory presses

Dr. Collin GmbH, Sportpark-
str. 2, D-85560 Ebersberg
Tel. 08092/2096-0, Fax /20862

Laboratory punches

Hans Naef AG, CH-8045 Zürich
Talwiesenstr. 17
Tel. 01/4510801 Fax 4621919

Laboratory roll-mills

Dr. Collin GmbH, Sportpark-
str. 2, D-85560 Ebersberg
Tel. 08092/2096-0, Fax /20862

Testing equipment

Kistler Instrumente AG Winter.
CH-8408 Winterthur/Schweiz
Tel. +41-52-2241111 Fax/2241414

Testing equipment and instrumentation

Kistler Instrumente AG Winter.
CH-8408 Winterthur/Schweiz
Tel. +41-52-2241111 Fax/2241414

Testing equipment for rheological properties

BRABENDER OHG, Kulturstr. 51
D-47055 Duisburg, Germany
Tel. -49-203-7788-0
Fax -49-203-7788-100

Testing equipment for rubbers and plastics

BRABENDER OHG, Kulturstr. 51
D-47055 Duisburg, Germany
Tel. -49-203-7788-0
Fax -49-203-7788-100

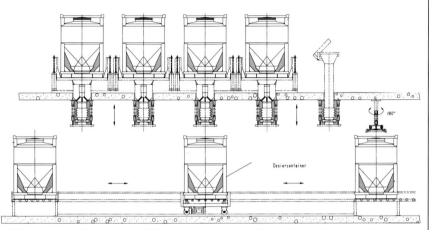

Container Systems Suitable for Automatic Processing of Bulk Materials

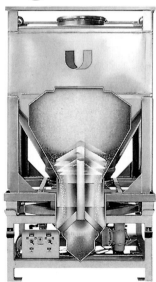

Industrial processing of bulk materials often involves precise composition and blending of different substances, before these can be packed or further developed.

The centre piece of the system is the special discharge facility, which is docked dust-free to the alternating container. The difference of flow speed and weight of the material determines the opening of the container outlet. The desired final weight is achieved by the strong stream activating the medium stream. The fine stream is activated last by the vibrating containers' cone valve which is controlled by signals from the weighing system.

Ask for further informations
Fax +49 7831/77-209

Umformtechnik Hausach GmbH, Postfach 1180, D-77750 Hausach
Telefon (+49) 78 31/77-0, Telefax (+49) 78 31/77-209

Attention to detail makes all the difference

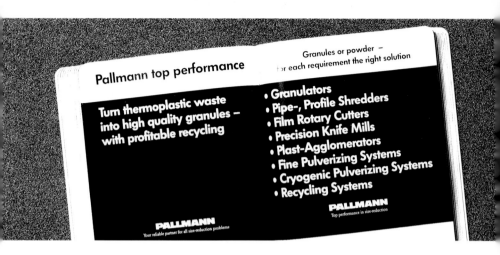

No. 1

Pallmann manufactures size-reduction machines and installations for clean-cut, homogenous granules, for free-flowing powders and systems for agglomerating and compounding. Tell us, which thermoplastic material you want to utilize to 100% – most economic way. With service. And with utmost essential in this domain. simply call us and you formation for your spe-

The complete size-reduction technology:
- Know-how
- Machinery program
- Service

Pallmann Maschinenfabrik GmbH & Co. KG
P.O. Box 1652
D-66466 Zweibrücken
Phone (0 63 32) 8 02-0
Fax (0 63 32) 80 21 06

and we show you the know-how, machines and care for detail which is Send us a letter or fax or will soon receive all incial size-reduction task.

PALLMANN
Top performance in size-reduction

For large and small feeding tasks in the plastics processing industry, you will find quick and economical solutions with K-TRON SODER modular feeders and K-TRON's well-developed network of services.

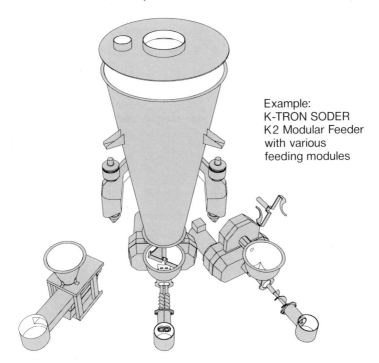

Example:
K-TRON SODER
K2 Modular Feeder
with various
feeding modules

Delivery Program:
A complete feeder program for the volumetric and gravimetric feeding in your industry process.

Sales Offices:
More than 80 K-TRON Reps worldwide for your support.

K-TRON SODER products are ISO 9001 certified worldwide.

K-TRON Deutschland GmbH
Geschäftsbereich Soder
Postfach 14
D-61371 Friedrichsdorf 2

K-TRON (Schweiz) AG
Division K-Tron Soder
Industrie Lenzhard
CH-5702 Niederlenz

K-TRON Great Britain Ltd.
4, South Link Business Park
GB-Oldham, OL4 1DE

K-TRON Asia Pacific Pte. Ltd.
438, Alexandra Road
#03-02 Alexandra Point
Singapore 0511

Index of Advertisers

Brabender OHG, Duisburg/Germany	A 7
Böhler Edelstahl GmbH, Kapfenberg/Austria	opposite title page
Deifel KG, Schweinfurt/Germany	between BG 20/BG 21
Engel, Schwertberg/Austria	opposite Foreword
Göttfert GmbH, Buchen/Germany	Front endpaper
Hanser Verlag, München/Germany	between BG 20/BG 21, Back inside cover
Hasco, Lüdenscheid/Germany	between BG 4/BG 5
Hoechst AG, Augsburg/Germany	between BG 20/BG 21
Hoffmann Mineral, Neuburg/Germany	Bookmark
IBS Brocke GmbH, Morsbach/Germany	opposite copyright page
Ihne & Tesch GmbH, Lüdenscheid/Germany	between BG 4/BG 5
K-Tron Soder AG, Niederlenz/Switzerland	A 17
Maplan, Ternitz/Austria	Front inside cover
Pallmann GmbH, Zweibrücken/Germany	A 16
Peroxid-Chemie GmbH, Pullach/Germany	A 5
Umformtechnik GmbH, Hausach/Germany	A 15
Weingärtner Ges.m.b.H., Kirchham/Austria	A 6
Werner & Pfleiderer GmbH, Stuttgart/Germany	Front endpaper